MW00386557

ORIGINS

First published in the United Kingdom in 2000 by Cassell & Co

A CIP catalogue record for this book is available from the British Library

ISBN 0 304 35403 1

Designed by Harry Green
Printed and bound in Italy

Cassell & Co
Wellington House
125 Strand
London WC2R 0BB

ORIGINS

THE EVOLUTION OF CONTINENTS, OCEANS AND LIFE

WRITTEN AND PHOTOGRAPHED BY RON REDFERN

CASSELL&CO

Contents

I FIRST LIGHT

Just two unifying theories have radically transformed the quest for knowledge about the behavior of the Earth. The first such revolution was initiated in the 16th century by Nicolaus Copernicus with his hypothesis that the Earth is in daily motion about its axis and in annual rotation about the Sun. It became effective in the 17th century when Isaac Newton's law of gravitation explained the observed behavior of the planets in the solar system. The second revolution was sparked by an early 20th-century theory that the continents are adrift. *First Light* shows how this concept proved to be fact and thus became the starting point of an ever-widening reappraisal of natural science.

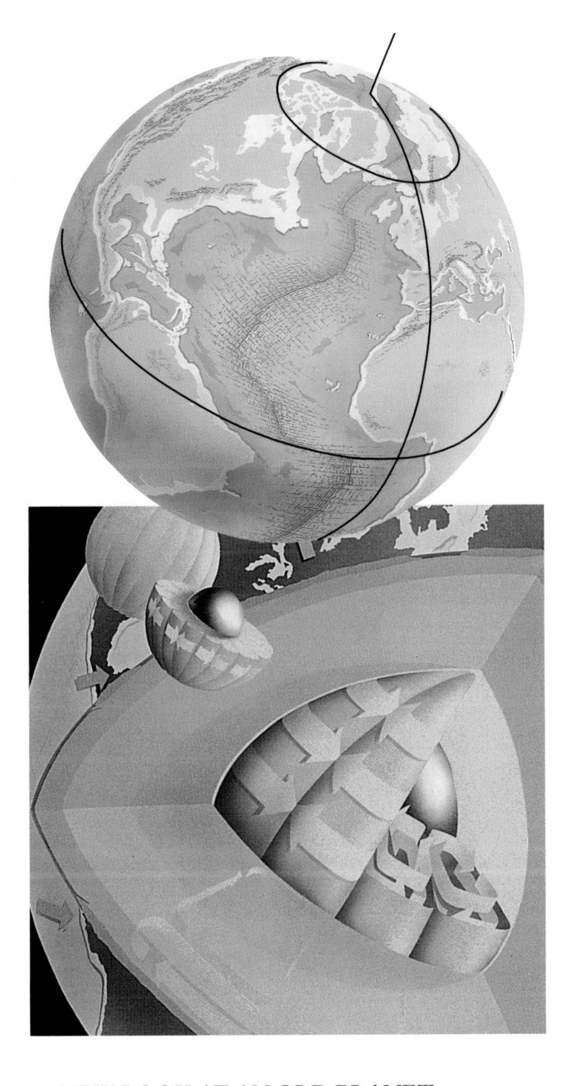

II NEW LOOK AT AN OLD PLANET

Tiny annual increments of crustal movement on the Earth's surface build mountains, create and destroy oceans, cause continents to flood, and influence global climatic change. The Earth has a magnetic field that intermittently reverses itself. The tilt and wobble of the planet and the irregularities in its orbit around the Sun determine the timing and intensity of seasonal change in opposing hemispheres. In combination and in permutation, all these events have affected the pace of evolution over billions of years and have contributed to the present diversity of life. Indeed, it seems that nothing about the Earth's modus operandi is irrelevant to the future of life on Earth. How then does the Earth work? What drives the movement on its surface? How is its magnetic field generated? Why does it tilt and wobble—and why should it matter that it does? How have living organisms responded to repeated crises?

III IAPETUS AND AVALONIA

This book's review of "the evolution of continents, oceans and life" begins 650 million years ago (MYA). At this time two physical events dominated the Earth's environment: the continuing breakup of a supercontinental clustering called "Rodinia" that had existed 1,100 MYA, and an ice age so severe and extensive that the planet at that time is now described as the Snowball Earth. The earliest-known animals began to appear in the oceans 650 MYA, but evolution did not escalate until after the ice age around 550 MYA.

Rodinia's continental elements rotated and resolved into several megacontinents: Laurentia (primarily North Amercia); Baltica (primarily Scandinavia and Euro-Russia); and Gondwana (primarily Africa), all of which ultimately re-assembled to form Wegener's Pangea. The evolution and diversity of life progressed exponentially during this time, suffering several severe mass extinctions and resurgence in the process. At 250 MYA, there came the most devastating mass extinction of all time.

One important fragment of Gondwana is called "Avalonia" and one of the oceans that opened between Laurentia, Baltica and Gondwana during Rodinia's breakup is called the "Iapetus Ocean." These are the "Iapetus and Avalonia" of this chapter's title.

IV TORNQUIST'S SEA

Around 480 MYA a narrowing sea separated proto-Scandinavia and other parts of Baltic Europe from the northern margin of Avalonia. Today, certain geological features mark the boundary between Scandinavia on the one hand, northern Germany and the Low Countries on the other. These are thought to mark the line of suture formed in the collision that followed the sea's closure. Both the suture and the sea are named after the man who first described the line of the demarcation, a Polish geologist named Tornquist, who lived and worked in Denmark in the early 19th century.

V THE THIRD AGE

The titles of this and the following chapter have been adopted from J.R.R. Tolkien's novel *The Lord of the Rings*. The reason for this is that in his story of the Hobbits Tolkien's setting for his "Third Age of Middle-earth" bears an uncanny resemblance to the mountainous central region of the supercontinent called Pangea (meaning "all-earth").

Tolkien's kingdoms of Middle-earth are separated by mountain ranges stretching from Northern Waste to Far Harad. The mountains of Central Pangea stretched on either side of the Equator from the Pacific Coast of Mexico via North America to Western Europe and beyond. All the continental elements that make up the modern political Western World around the North Atlantic Ocean were then in close proximity. In the Third Age the final assembly of Pangea was near, a time of destruction of ocean floors and deformation of continental margins.

VI MIDDLE EARTH

Wegener's vision was of an Earth with one continent surrounded by one ocean. In reality Pangea was much more complex: it was an ever-shifting congregation of jostling parts that approached their closest fit about 250 MYA. During the assembly of Central Pangea, the *Middle Earth* of the chapter title, the region may have been Mediterranean-like in character. By 250 MYA it was a contiguous and restless amalgam of Western World elements. Throughout its existence Central Pangea straddled the Equator and was broadly divided northeast to southwest by almost unbroken chains of spectacular coastal and continental mountain ranges—certainly Alpine, and possibly Himalayan in stature.

VII THE NEW WORLD

The first signs of breakup had appeared in Central Pangea before 208 MYA (the commencement of the Jurassic Period). Nevertheless, the supercontinent was not physically separated into post-Pangean Laurasia and post-Pangean Gondwana until after 165 MYA. But the very existence of a fracturing Pangean supercontinent and its reshaping into the post-Pangean world had profound implications for the evolution and diversity of life on Earth during this period.

The therapsids (mammal-like reptiles) were the dominant terrestrial animals through much of Pangean time. With the breakup of Pangea they were superseded by dinosaurs. The dinosaurs won the battle for command of the post-Pangean world but lost the war of ultimate possession. Dinosaurs became extinct, but the mammals that populate the world today are descended from Pangean therapsids.

VIII ATLANTIC REALM

The Atlantic system of oceans formed in three stages and several substages, each with its own specific effect on global circulation, climate, and the distribution of marine and terrestrial species. The first stage was completed by the formation of the Central Atlantic Ocean and the establishment of an equatorial pattern of circulation. *Atlantic Realm* describes the continued development of the Atlantic from that point, and explores the marine world in Cretaceous time.

During the course of these events, astronomical numbers of minute marine organisms formed the enormously thick and extensive chalk deposits of the Anglo-Paris Basin—best known today for its white cliffs in the English Channel and its Champagne wines. Other organisms built the Florida-Bahama limestone structures that are now three miles high, and yet others formed the substance of about 70 percent of the world's oil.

1 **Central Atlantic**
2 **Anglo-Paris Basin**
3 **Fringing Reefs**
4 **Interior Platform**
5 **Caribbean Sea**
6 **South Atlantic**
7 **Bay of Biscay**
8 **Labrador Sea**

IX MARITIME WEST

The events of the previous chapter took place when low-lying continental regions were flooded by shallow seas. They covered the equivalent of up to 60 percent of the Earth's present terrestrial surface. But why was this so? And why did the North American West (the *Maritime West*) steadily drain while much of Europe became ever more deeply inundated?

During this period the region north of Mexico to Alaska held the most diverse population of dinosaurs known to have existed. But how did they live in such numbers? Did they advance the evolution of flowering plants? And why did the dinosaurs become extinct?

1 **Great Inundation**
2 **Western Uplift**
3 **Angiosperm Advance**
4 **Dinosaur Regimes**
5 **Mongolian Connection**
6 **Cooling Climate**
7 **The End of an Era**

X MIDLAND AND NORDIC SEAS

The Midland Sea of this chapter is a remnant of an ocean reduced to a seaway. The term "Nordic Sea" refers to the extreme reaches of the North Atlantic beyond Iceland. This is the region where North America detached from Europe about 36 MYA and thus severed all landbridges between North America and Europe. The evolutionary picture in the Northern Hemisphere changed dramatically from this point in time. The closure of the Midland Sea to form the almost totally enclosed Mediterranean, and the opening of the Nordic Sea connection to the Arctic Ocean, led to the third and final stage in the formation of the Atlantic system of oceans. It also led to the onset of the present ice age.

1 **Midland Seaway**
2 **Northern Connections**
3 **Greenland Hot Spot**
4 **Nordic Seaway**
5 **Grasses and Savannahs**
6 **Family of Man**

XI FOUNTAINS OF YOUTH

On Easter Sunday, March 27, 1513, Juan Ponce de León sailed from the Bahama island of San Salvador with three ships. He set out to discover a spring said to restore "age'd men to youths" that was to be found on the island of Bimini. He did discover a "spring" of sorts—a swift-flowing ocean river called the Florida Current, part of the Gulf Stream system of currents, which swept his ships far off course.

On most maps the Florida Current, the Gulf Stream, and the North Atlantic Current are shown as a continuous current. They act in unison but are in fact individual currents with different characteristics. Additionally there are many other surface, intermediate, and bottom currents in the North and South Atlantic Oceans. They are all part of a three-dimensional system of interpolar currents that lace the Atlantic Ocean system. They are indeed "Fountains of Youth", in the sense that collectively they are a crucial part of an elaborate interactive system that drives the global climate.

XII CHILDREN OF THE APPLE TREE

The closure of the Isthmus of Panama around 3.5 MYA was a relatively modest tectonic event. Nevertheless, it not only had a profound and lasting effect upon the climate of the Northern Hemisphere, but also upon the interchange of Southern and Northern Hemisphere animals.

Another thread in this story is that our own ecological and social evolution took place during the northern ice age that was triggered by the closure of the isthmus. This final chapter retraces the ascent of hominids in the last 3.5 million years. It does so in the context of the last glacial age when what we now call the "Western World" was first populated by the *Children of the Apple Tree* (T.S. Eliot: *Little Gidding*).

BETWEEN TWO WAVES OF THE SEA

We have followed the events of 700 million years of physical and biological change. The time scale involved is virtually unimaginable. If one inch represents a million years, the scale would require sixty feet. The last chapter focused down to human technological adaptation to the conditions of an ice age, a span that barely stretches a tenth of an inch on this scale. But the 500 years that have seen the development of the present world from the time of Columbus would occupy just one two-thousandths of an inch on the scale! This epilogue is a resume of events that led to that development "between two waves of the sea" in an ocean of time.

About the photography

The photographs in this book are landscapes. They give due aesthetic consideration to color, light, balance, and mood. But their primary purpose is interpretive. The collective result is that you will see "the evolution of continents, oceans and life" not through an archival anthology, but directly through the eyes of your author. The moment the shutter was triggered, the film exposed, and the image transfixed was a particular expression of a scene that we will share on the printed pages that follow this table of contents.

This photography had to be planned and researched just as rigorously as the data that are the basis for the text and illustrations. Preparatory work began with an elaborate outline of the book and months of library research; it resulted in a formidable list of potential subjects ranging in location from the Arctic to the Equator, and from Alaska to West Africa.

The list was refined into specific localities and further categorized into a series of expeditions carried out over three years. The work was planned with the seasons, traveling from farthest north to farthest south each year. For instance, in one year individual sorties included the Canadian High Arctic in summer; Saharan Morocco, the Azores, Madeira, and Canary Islands in autumn; Trinidad and the Lesser Antilles in early winter. The following year it was Spitsbergen in early summer, then northern Norway, Iceland, and the North American coast from Newfoundland to the Gulf of Mexico by November—and so on. The circles on the map define the specific regions in which the photographs were taken.

All these places, and every one of the photographic locations circled, contributed in some way to the ninety or so photo-essays that follow. The main picture in an essay is usually a panorama and the inset picture a medium-format shot. My objectives were both to capture the technical significance of a scene and to share the often overwhelming magnificence of the place. If you are curious to know where a picture was taken, a mark on the miniature globe overlapping each main picture will show you. If your curiosity is further roused to wonder where a particular place was in terms of the geography of the past, when this is possible an arrow on a paleomap will point it out.

Local scientists and others at national institutions and universities gave generously of their time to discuss their science in the context of my personal objectives for this book. They pinpointed the best locations for photography on large-scale maps. Their tutorials were never less than a few hours in duration. In special circumstances, such as visits to Canadian research institutions in Calgary and Ottawa, discussion was spread over days and often led to an invitation to join or to visit a particular expedition in the field. Pertinent scientific papers, books, and introductions to other specialists were offered freely. These many contributors are acknowledged in another section of the book, but all deserve much more than the warmest thanks and appreciation that I can express. This book and its photography are tributes to their generosity of mind.

Ron Redfern *April, 2000*

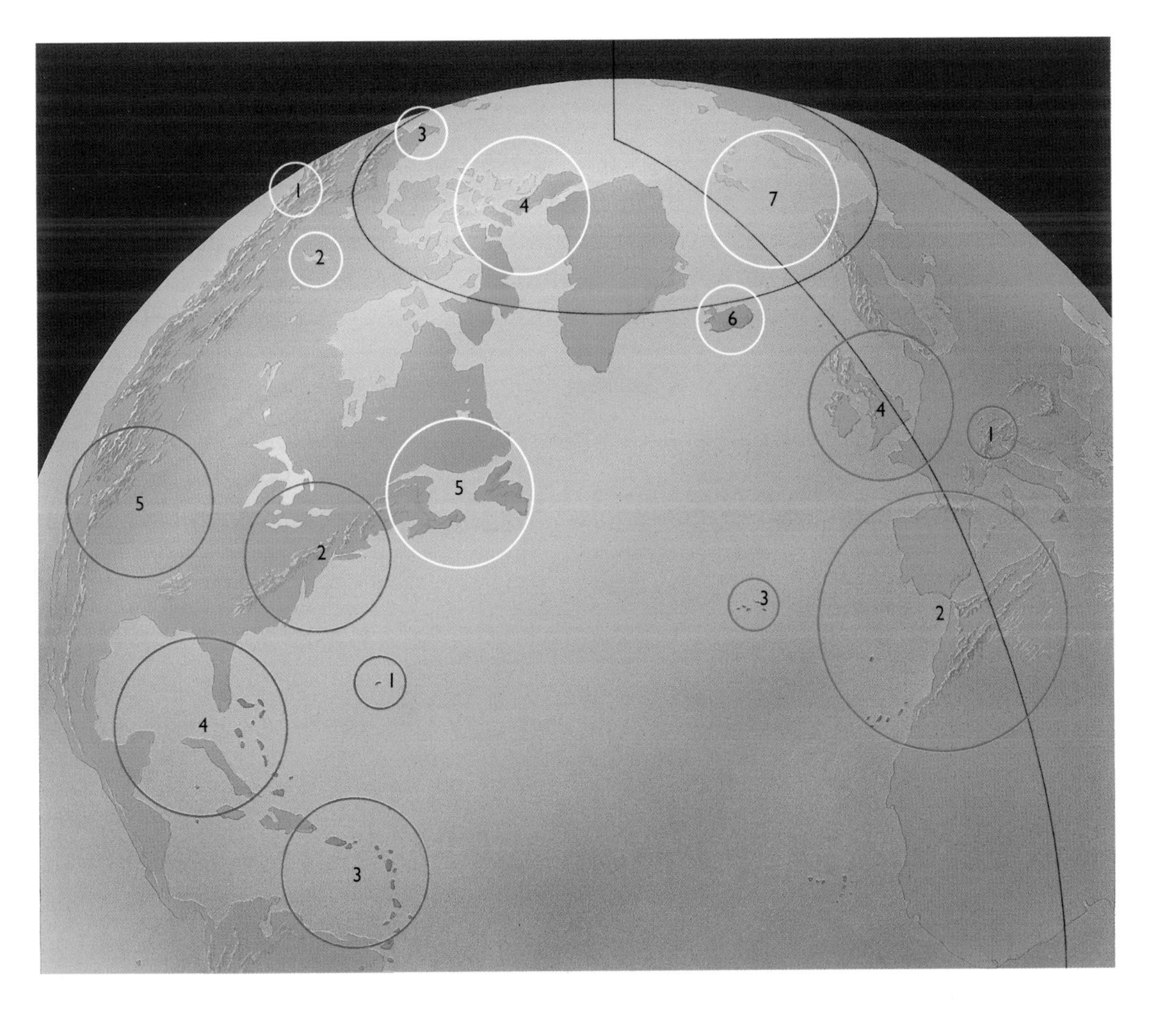

NORTHERN LOCALITIES

1 Pacific Northwest
2 Northwest Territories
3 Alaska
4 Canadian High Arctic
5 Newfoundland, Nova Scotia, New Brunswick
6 Iceland
7 Spitsbergen (Svalbard), Northern Norway

EASTERN LOCALITIES

1 Switzerland
2 Pyrenees, Iberia, Morocco, Madeira, Canary Islands
3 Azores
4 British Isles, Northern France

WESTERN LOCALITIES

1 Bermuda
2 Ontario, North Carolina, New York, Maine
3 Lesser Antilles
4 Louisiana, Florida, Yucatán, Bahamas
5 Colorado Plateau

Introduction

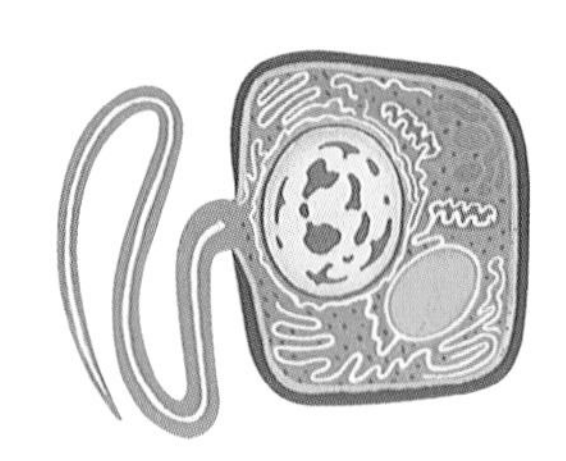

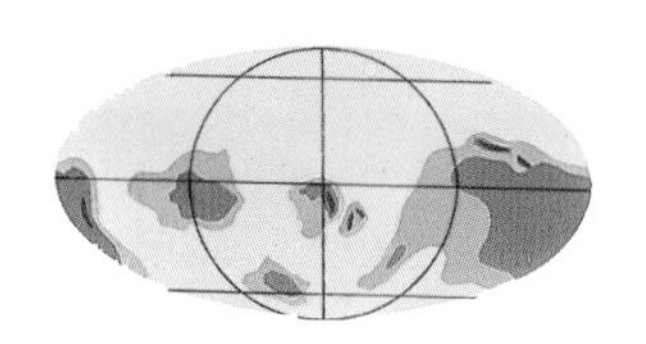

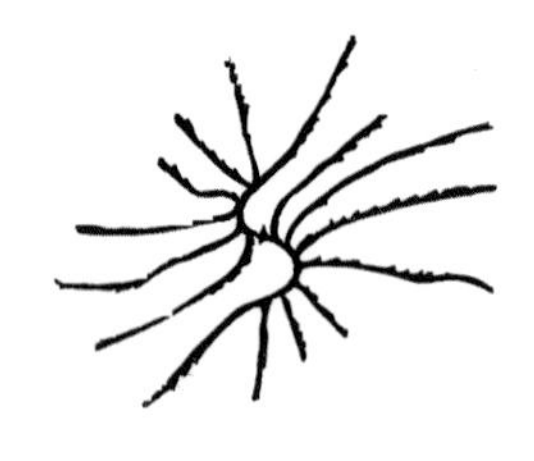

From the beginning global change has been the key factor that sets the conditions for life on Earth. All living things have always had to conform to such change or suffer extinction. Humankind is no exception: it too will have to adapt or perish as, in future centuries, we face coincident superhuman problems. Many of these are already grave concerns. For example, we now have a global population that doubles every fifty years and simultaneously a progressive mass extinction of other species. We face the probability of a significant rise both in average global temperature and average sea level this century. In addition to the risk of impact from a wayward comet or asteroid, there is little doubt that we will plunge into the next glacial advance late in the next millennium or shortly thereafter. Is all this the fault of humankind, or is this the way that evolution works and a natural state of affairs?

This book presents and explains, so far as it is possible to do so, the principal known causes and consequences of such change. It does so by reviewing what is known of 700 million years of inexorable continental drift and a rapidly evolving biosphere. It does this in the context of a scientific revolution, the tectonic revolution, the discovery that we are indeed living on continents adrift in ephemeral oceans. This apparently academic facet of the Earth's system has implications of Copernican magnitude. It has lifted the lid off a Pandora's Box of never-ending multi-disciplinary discoveries about how the physical Earth works, and how life itself survives dynamic change. According to Greek myth, the last thing to emerge from Pandora's Box, after a string of evils that afflicted the world, was "Hope!" . . . and perhaps, human ingenuity and adaptability are indeed our only hope?

Although worldwide in general scope and implication, the book's ideas, and the theories that it presents, are expressed in terms of the modern Western World—that is, North America and Europe plus adjacent oceans and seas: the North Atlantic, Arctic, Mediterranean, and the Gulf of Mexico. The object in so doing is to present for a non-specialist audience, a well-researched and structured "big picture" synthesis of a new, still-growing understanding of the Earth's interactive physical and biological systems.

In developing the book, two primary methods have been used to achieve this objective. The book's photography and explanatory figures are presented in the form of clearly individual photo-essay spreads, consisting of six or more per chapter. In contrast, the book's running text columns are illustrated by elegantly painted and numbered icons—exemplified by the illustrations shown on these pages. In addition to cross referencing text subjects with photo essays, the icon numbers are repeated in bold type at an exact point of reference in the text. The result of this approach is, in effect, two books with parallel stories presented in one volume—for reading on one level and browsing on the other. And because the book is chronological in its approach and structure, its chapters can be read in any order of interest without loss of context.

Each chapter is preceded by a photographic title spread with an overview of the content. This is followed by a chapter introduction that describes the paleogeography of the time frame covered in the chapter. Access to the book as a whole is through an illustrated table of contents supported by a comprehensive index, a glossary of terms, and a detailed bibliography.

In combination the text and essays demonstrate that the old idea of a solid Earth, on which life evolved incidentally and independently, has largely been invalidated. We live on a dynamic and interactive planet. One on which life's evolution for at least the last 700 million years of multicellular development has largely been determined by a range of factors: the disposition of continents; the formation and destruction of oceans; radical changes in global climate from ice age to greenhouse age; and extraordinary fluctuations in global sea level, mostly stemming from the side effects of "continents adrift in ephemeral oceans." And most probably by the occasional visitor from outer space—the impact of a ten-mile-wide asteroid or a comet the size of a city.

A crucial step in writing and photographing the book was to have particular subjects reviewed and criticized by specialists in different disciplines at several stages during progress of work. The objective was, wherever possible, to catch errors of presentation, to clear ambiguities, to update the science where ideas had been superseded, and to seek advice in steering a fair course in controversial waters. My purpose here is to acknowledge those who so generously contributed their help and personal time to this end, during the initial and the final stages of production. Additional acknowledgement of the many scientists who contributed their "local" expertise in particular regions is expressed from page 339 onwards. Although from this it can be seen that a strenuous effort has been made to ensure an authoritative statement herein, responsibility for its accuracy as it goes to print is mine and mine alone.

First and foremost, I am indebted to Dr. Kevin C.A. Burke of Houston University, Houston, Texas, for his general reading of all text and his review of all illustrations relating to the geological and planetary sciences. In similar vein, I owe a debt to Dr. William B.F. Ryan of the Lamont-Doherty Geological Observatory, Columbia University, New York, for his reading of the history of tectonic science in Chapter I and the essays in Chapter II concerning the early Earth and its evolution. In Europe I was advised by Dr. Clive Bishop OBE and by Frederick W. Dunning OBE of the Natural History Museum in London, England. To both I owe a great deal for their unflagging interest and guidance over many years.

The intricate world of cellular biology is to many the most inaccessible of all the natural sciences—as one should expect of a science that is deciphering the origins of cellular structures and taxonomy. I am indebted to Dr. Lynn Margulis of the University of Massachusetts at Amherst who read and fine-tuned both the written and graphic interpretation of her work that is reproduced in the two "sections" that conclude Chapter II. Particular thanks too, to Jenny Stricker MA, of Colorado Springs, one of Dr.Margulis' graduate students, who further read, checked, and discussed ways of clarifying key points.

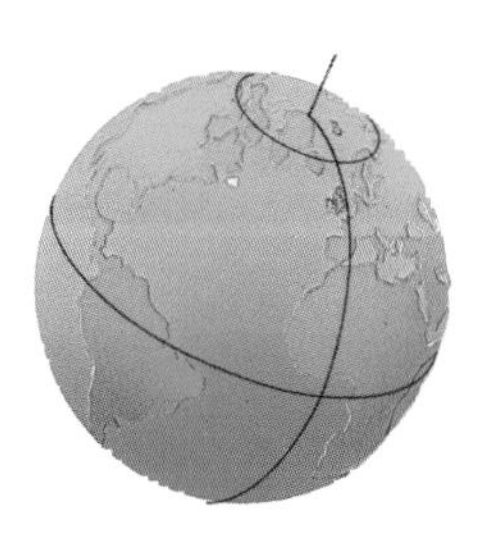

Paleobotanists study the origin of plants, particularly the evolution of the flowering plants on which life on Earth largely depends. All the book's paleobotany was reviewed by Dr. Leo J. Hickey of the Peabody Museum of Paleontology, Yale University. In the last fifteen years or so, considerable advances have been made in the determination of the origin of angiosperms; some of these discoveries have not yet achieved general circulation. Dr. Hickey, who contributed to this research, guided me through the new theories incorporated in both text and essays.

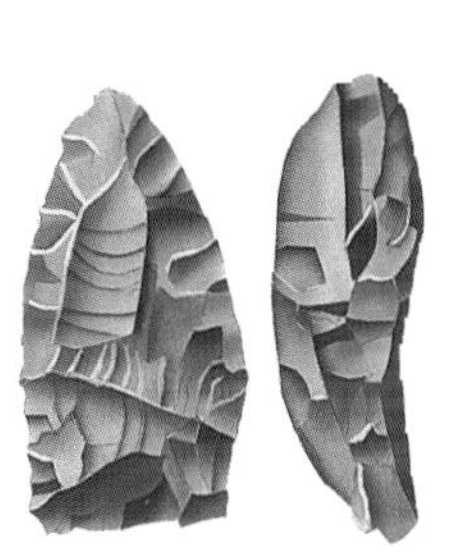

If anyone deserves the accolade of "father figure" of vertebrate paleontology today, it is Dr. Edwin H. Colbert, of the Museum of Northern Arizona. It has been my good fortune to have known Dr. Colbert for many years, both personally and through his books on dinosaurs and other reptiles stranded on continents adrift. Fortunately Dr. Colbert was available to read and to criticize my rendering of vertebrate paleontology in all its aspects. I am deeply grateful for his help and advice. Similarly Professor John E. Warme, of the Colorado School of Mines, read the invertebrate paleontology, particularly that related to the Cretaceous Period, and had several constructive suggestions to make. Prof. Warme also advised on a new interpretation of the primary cause of the Great Unconformity of Grand Canyon that appears on pages 138–139, for which I am very grateful.

A summary of the origin and distribution of human populations from the Old to the New World is presented in the last chapter and Epilogue. The text and photo essays on ice age paleoanthropology were reviewed by Dr. James Adovasio, Director of the Mercyhurst Archeological Institute, Erie, Pennsylvania, who spent a great deal of his time in so doing. I am of course greatly indebted to Dr. Adovasio for making this very special effort on my behalf.

The Epilogue, which tells of the discovery by Europeans of an unexpected New World, was read by Dr. Carol Urness, who specializes in 15th- and 16th-century history at the University of Minnesota. As Assistant Curator of the James Ford Bell Library, Dr. Urness made it possible to reconstruct the Martin Waldseemüller globe of 1507 on page 323, from prints of the original woodcut gores owned by the library. So far as is known, this is the first reconstruction of its kind.

I felt that it would be appropriate to end the book with an illustration of this 16th-century globe—the first to use the name "America." This because all the other globes and maps in the book are redrawn from late-20th-century computer-generated models of the Earth's paleogeography. Since these projections span 600 million years, they are a far cry from the first projection of an Earth with two strange and unknown continents separating Europe from Asia. The Waldseemüller globe and the paleoglobes together provide a succinct measure of human progress in the understanding of the Earth's geography.

Indeed, the paleogeography of the book is its mainstay. Although the work of a number of specialists in particular areas of the Western World has been used for reference, there are just a half-dozen or so scientists worldwide who specialize in global reconstructions covering the timespan of this book. Because of the differences in interpretation of available data, particularly Precambrian reconstructions, it was necessary to focus on the work of one such specialist. Thus the paleogeography portrayed here is based entirely on the published work of Dr. Christopher R. Scotese of the University of Texas, Arlington.

Dr. Scotese most generously acted as my adviser on a chapter-by-chapter basis as work progressed and made a final review before publication. A more detailed account of how this work was accomplished is in the acknowledgments section, but no account can fully repay his extraordinary generosity of mind. Or the generosity of mind, the freedom with data, the engaging enthusiasm, and the encouragement of the scientific fraternity for over a decade.

RON REDFERN
Ermington, South Devon, England
April, 2000

I FIRST LIGHT

Pond Inlet near Bylot Island, north Baffin Island, Canadian High Arctic

During the last few decades there has been a major revolution in science, but not many among us have had an opportunity to participate in this truly momentous advance in human knowledge. This book reveals the complex, dynamic, and biologically interactive planet on which we live—a long-secret world known until now to only a growing number of specialists. This picture of the Arctic Sun at a few minutes past midnight perhaps encapsulates the notion that we are indeed at the dawn of a new understanding of the Earth and its biosphere.

FIRST LIGHT

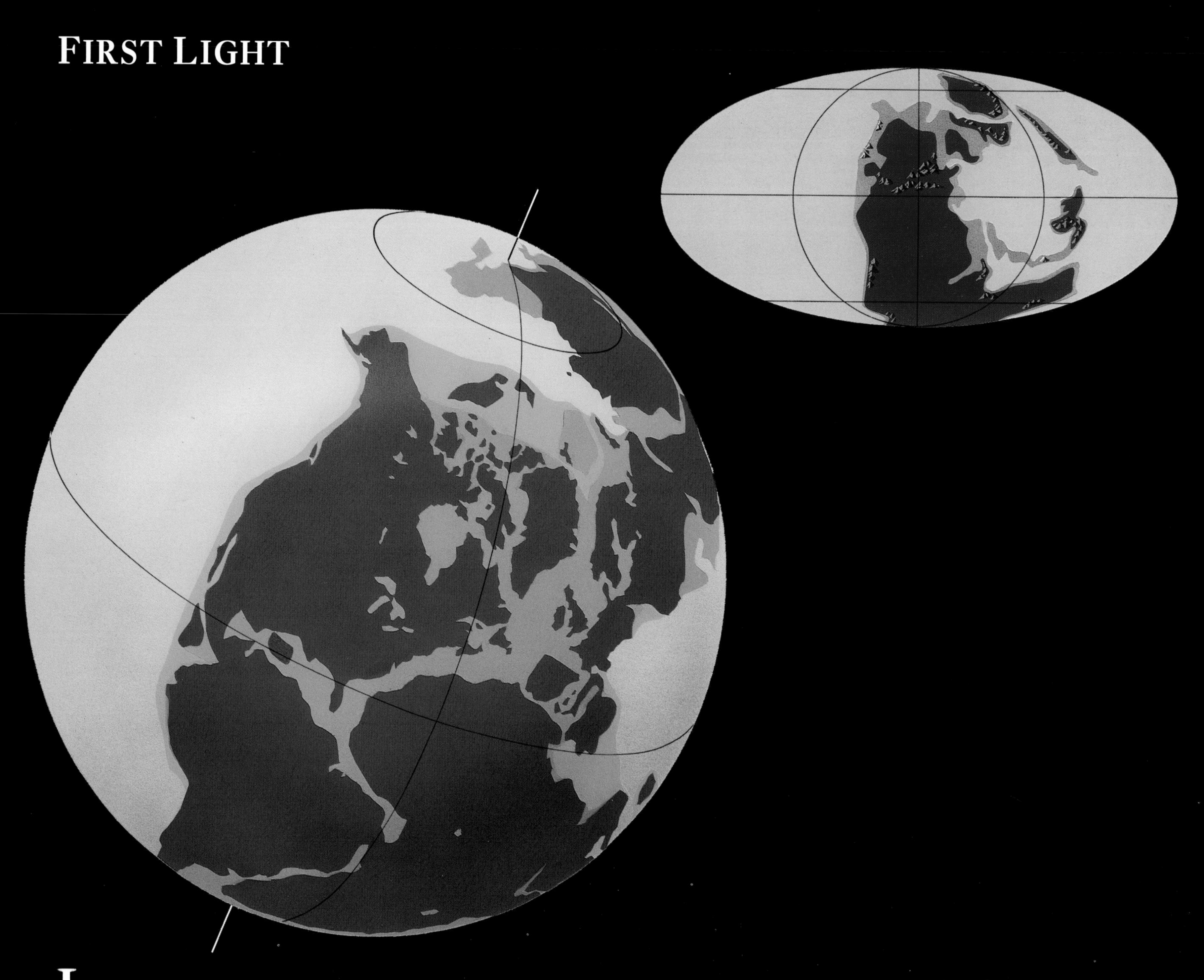

In November 1911, a German meteorologist named Alfred Lothar Wegener had a novel idea that seemed to explain a curious contradiction: why there is evidence of past glaciation in Africa near the Equator, and why there are fossils of tropical plants in Spitsbergen near the North Pole. A few months later, Wegener proposed that all of today's continents had once been joined in a single supercontinent that had moved from south to north before dividing into the present continents.

Wegener called his hypothetical supercontinent "Pangea" (from the Greek *pan gaia*, all-earth), and the single ocean that had surrounded it, Panthalassa (Gr. *pan thalassa*, all-ocean). *First Light* tells the story of the hornets' nest this theory stirred up in the geoscience community. It shows how in the 20th century that controversy resolved into a fundamentally new understanding of the mechanisms of the planet and its biosphere.

The globe and the oval projection reproduced here show a modern reconstruction of the geography of Pangea at about 250 million years ago (henceforward abbreviated to "MYA," with "MY" for a million years). The text to the right explains the use of similar global projections that precede each chapter.

The large globe is a "paleoglobe" (Gr. *palaios*, ancient). It shows a western aspect of the Earth at about 250 MYA when Wegener's Pangea was at its zenith. But the projection is not a true representation in the sense that the shapes of the modern continents have been retained so that particular localities of today can easily be identified. The familiar dark green outlines of modern continents formed in geologically recent times did not exist in this form 250 MYA. The low-lying and generally sea-covered margins of paleocontinents, continental shelves, are colored in light green. Deep seas and oceans are colored blue.

The globe is turned at an angle that emphasizes the Northern Hemisphere relative to the North Pole, the Arctic Circle, the Equator, and the Greenwich Meridian. These standard points of reference, together with the Antarctic Circle where visible, will appear on all reconstructions. This includes the miniature paleoglobes and maps adjacent to the text. Thus the reader can always equate modern geography with the paleogeography—even when a southern aspect of the Earth is shown. Whenever possible the paleolocation of the main feature in a photograph will be marked with an arrow on the associated paleomaps. The present-day locality of the feature is indicated on a small modern globe set at the edge of the picture.

The oval projection is more accurate in the sense that it shows the whole surface of the Earth as it is believed to have looked 250 MYA. The central circle presents one hemisphere as if we were directly over the intersection of the Equator and the Greenwich Meridian. The two crescent-shaped regions on either side of the circle show adjacent regions of the opposing hemisphere, areas that are out of sight on the reverse side of the globes. The oval projection also shows, here and elsewhere, the continents (dark terracotta) and mountain chains of the age, and the extent of flooded parts of those continents (light terra-cotta). Sea-filled basins and deep oceans are colored one shade of blue. These features may not appear always to correspond to those of companion globes. The former are all flat projections from a fixed point above the Earth. The latter are spherical: the viewpoint varies, and the curvature of the globe distorts the geography.

Other than near-equatorial Central Pangea—the northeast-southwest width of the supercontinent Pangea was not quite the tightly knit amalgamation that Wegener visualized. The rest of the landmass consisted of a clustering rather than an assembly of continents, with some continents indeed sutured, while others were gathered in closely associated groups or strings. But the extraordinary fact is that the geopolitical world that borders the North Atlantic Ocean today consists of parts of Central Pangea. Thus, for easy and convenient reference we will use the terms "East Central Pangea" for Europe and adjacent parts of North America, "Mid Central Pangea" for the North American East Coast locality, and "West Central Pangea" for the Florida-Bahamas and Gulf of Mexico region.

His skis stood stark and firmly upright in the gathering darkness of an arctic night. His tent, pitched high on the Greenland ice sheet, was already half-buried by early winter drift. He lay peacefully on a folded sleeping bag. His eyes were wide open and he had a wistful, contemplative smile on his face. His nose was touched by frostbite, but he looked younger than he had in life. The rest of his body was enveloped in a makeshift shroud made from two sleeping bag covers joined by tight, neat, intricate stitches, sewn in the Eskimo tradition. It was November, in the year 1930.

Early in May the following year a search party, from a German expedition base at Umanak Bay in Baffin Bay, spotted his skis silhouetted in low-angled light reflected from the now sunlit flanks of the ice sheet. The dead man was their leader. His name was Alfred Lothar Wegener. He was born on November 1, 1880, in Berlin, and at the time of his death he was a well-known and distinguished meteorologist and explorer.

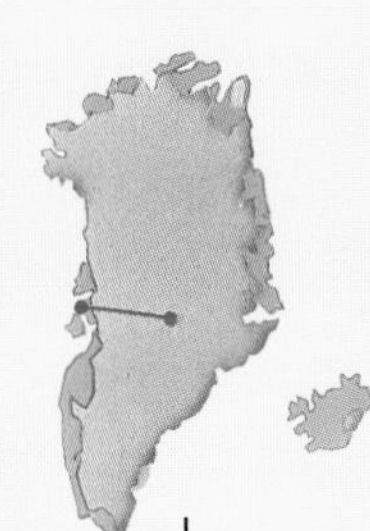

It is believed that Wegener died of a heart attack brought on by overexertion in acutely arduous weather. He and his companion, a young Greenlander named Rasmus Villumsen, had been trying to reach the expedition's coastal base from a meteorological observatory set up at "Mid-Ice," 3,000 m (10,000 ft) above sea level at the very center of the ice sheet (**1**). Because of delays in establishing Mid-Ice, and limited space and food for a winter's stay with the other members of the expedition, Wegener and Villumsen had made the attempt to return to the coast—but far too late in the season. It was Villumsen who had erected the tent, sewn the shroud, and planted Wegener's skis after his death. He had then continued his way down off the ice sheet towards the expedition base 380 km (240 miles) away. But the onset of near-perpetual darkness and the violence of the first blizzards of an arctic winter were too much for any man to bear alone: Villumsen was never seen again. Wegener's remains were buried where they were found, his grave marked with a cross of iron. Grave and cross and Wegener's body too have long since been enveloped by another shroud—the 3 km (2 mile) sheet of blue-green ice that covers much of Greenland.

WEGENER'S NAME IS INDELIBLY linked to an extraordinary hypothesis, the theory of "continental drift," now the science of plate tectonics. The word "tectonics" is from the Greek *tektonikos*, building, and in geoscience it refers to the structures that result from crustal plate movements. At first glance the concept of drifting continents seems to have little to do with the science of climate. But it was Wegener's interest in ancient climates that led him by accident into fields of discovery and science not of his province: his discovery was entirely serendipitous. Little did he know that his cross-disciplinary approach would one day become the very substance of a new science—Earth System science. This is the multidisciplinary science that, in the course of the 21st century, should show us how to maintain conditions for life on a dynamic planet in the vacuum of space.

As a professor of meteorology and navigational astronomy at Marburg University (located near Frankfurt am Main, Germany, 1908–18), Wegener had access to geological papers dating back to the early 19th century. Many of these papers reported irrefutable data that could only be explained by a radical reinterpretation of ancient climate patterns. There were, for example, reports of score-marks made by glacial ice sheets on ancient rocks in the equatorial Sahara and India. In the Southern Hemisphere, "paleoglacial" debris of a later age had been found in regions of South Africa and South America. There were also reports of paleoclimate anomalies in the Northern Hemisphere. Conditions there in the geological past seemed to have been the opposites of those in the southern continents. For example, Spitsbergen, now near the North Pole, has coal seams—the products of ancient temperate or tropical forests. [*Paleo* is a combining form, from the Greek *palaios*, meaning ancient, early, or prehistoric; thus one gets paleogeography, paleobotany, and so on.]

[*continued on page 20*]

1 The Gondwana Equation

Some 19th-century geologists believed that in the distant past continents in the Southern Hemisphere had been linked by land-bridges. They termed this interconnected group "Gondwana" (land of the Gonds) because the group included India, and *wana* means "land" in the Indian Gondi language. The presence of land-bridges that had subsequently submerged seemed the only explanation for the presence of *Glossopteris* (a primitive seed-bearing plant), *Mesosaurus* (a primitive aquatic reptile), and other animals that originally lived in temperate climates and whose fossils are now found in widely separated tropical or polar regions of the southern continents. The presence of *Lystrosaurus*, illustrated here, in Antarctica was unknown in Wegener's time.

As a meteorologist, Alfred Wegener was intrigued by the 19th-century approach. He suggested that the apparent climatic

anomaly and matching fossils would be better explained if, instead of landbridges, all the continents around the present North and South Atlantic Oceans had been joined. If the supercontinent (Pangea) had been farther south at the time it existed, it would explain the presence of anomalous fossilized tropical plants on Spitsbergen, now an arctic island in the Svalbard archipelago. Wegener's argument is exemplified here by *Araucarioxylon*, the fossilized tree trunk in Arizona's Petrified Forest.

Araucarioxylon (a genus of conifer wood) did not have annual growth rings; it was therefore a *tropical* plant. It grew in a swampy area of equatorial Pangea. But Arizona is a long way north of the Equator today. In contrast, the *Glossopteris* tree was deciduous and the woody stem of its trunk was ringed. It therefore grew in a region with a *seasonal* climate, yet its fossils are often found in tropical localities where leaves do not fall all at once and annual rings do not form. Thus both Arizona and the tropical regions where *Glossopteris* fossils are found have moved northwards since the living plants existed.

Wegener also used the jigsaw fit of the continents to support his hypothesis but could not satisfactorily explain how they moved. The two globes show that when today's southern continents are fitted together they do indeed match. There is also a perfect match between the broad localities in which *Glossopteris*, *Mesosaurus*, and *Lystrosaurus* are found.

2 Mountain Building

Other 19th-century discoveries also tantalized Wegener. Different "styles" of mountains had been recognized around the North Atlantic and in related regions. Some mountains were composed of volcanic rock, like many in North Wales—the birthplace of modern geology. Most were made up of highly contorted sedimentary rocks, of which the Swiss Alps were an outstanding example. Later it was recognized that the "alpine style" seen in this panorama also characterized the Appalachian Mountains of North America, as well as mountain belts in the British Isles, Norway, and many other localities around the North Atlantic. Extraordinarily they all appeared to have been folded, overthrust, and folded back on themselves into what are termed "nappes."

In the key diagram at right, the black lines outline wedges of rock that have been overthrust; arrows indicate the direction of thrusting. Other formations have been completely overturned and the extent of "napping" can be judged by the mauve-colored wraps around orange-colored strata. Yet all these rocks were originally formed by eroded sediments laid down and compressed in a horizontal plane. We now know that the architect of such "alpine" mountain building is the inconceivable force of continental collision. The Alps are caused by the collision of Africa with Eurasia, just as most of the mountain belts around the North Atlantic were formed as opposing continents collided during the assembly of Pangea.

Nappe des Diablerets above Taveyanne, near Montreux, Switzerland

[*continued from page 15*]

THE POTENTIAL FIT of continents on either side of the Atlantic Ocean had been noticed and commented upon since its geography had first been mapped with reasonable accuracy. Wegener knew this, of course, though at first he dismissed the idea. Then, according to his own account,

In the autumn of 1911, I became acquainted (through a collection of references, which came into my hands by accident) with the paleontological [fossil] evidence of the former land connection between Brazil and Africa, of which I had not previously known.

Wegener was referring to the work of some 19th-century geologists, Eduard Suess and others. Between 1885 and 1909 they had hypothesized the prior existence of a linked assembly of Southern Hemisphere continents—Africa, South America, India, and Australia (**2**). They used the term "Gondwanaland" instead of "Gondwana" to describe this figurative assembly of continents (see explanation below). Suess and his associates based their conclusions on the fact that independent regions of these southern continents, now separated by oceans, possessed similar fossils of plants and animals. When the continents were assembled in accordance with matching-fossil localities, they fitted together like pieces in a jigsaw puzzle. However, as a meteorologist, Wegener was more impressed by the fact that the paleoglacial regions he had read about also matched up in the Gondwana assembly. ["Gondwana" means land of the Gonds. The word was derived from the Gondi language of India by H.B. Medlicott in 1872, and first published by O. Feistmantel in 1876.]

It was then that Wegener made his great intuitive leap: if *all* the continents around the North and South Atlantic were reassembled into one huge jigsaw, and if the whole assembly was then moved down, southwards across the equator, then every one of the paleoclimate anomalies that he and others had puzzled over could be explained. The ice sheets that had simultaneously covered adjacent regions of South Africa and Brazil would be comprehensible if these regions had been located nearer to the South Pole. And the tropical coal forests of Spitsbergen would also be explained because that region would then lie nearer to the Equator (**3**). Furthermore, the much older glacial remnants found in the Sahara region of Africa would make meteorological sense if there had been a considerable northward movement of the supercontinent over an extended period of time.

ONE OF WEGENER'S SENIOR COLLEAGUES, Emanuel Kayser, was head of the geology department at Marburg. Kayser was also the chairman of a breakaway geological society dedicated to broader views of the science than those of the very conservative and long-established Geological Society of Germany. He invited Wegener, then aged 31, to address the second annual general meeting of the fledgling group (the Geological Association of Frankfurt am Main) on January 6, 1912. Kayser knew what Wegener would propose and had high expectations for a stimulating lecture and discussion. In fact he was about to set in motion the development of one of the most significant theories about the Earth since Nicolaus Copernicus (1473–1543). Copernicus had proposed that the Sun, not the Earth, was at the center of the solar system, that the Earth rotated about its axis each day and orbited the Sun annually.

In his Frankfurt address Wegener presented his hypothesis that there had once been a single primordial continent, not just a "linked assemblage" of southern continents. He suggested that all the present continents had been as one, forming a supercontinent he called "Pangea" (Gr. *pan gaia*, all-earth). Wegener further reasoned that if there had been a Pangea, then at that time there had also been only one ocean, which he called "Panthalassa" (Gr. *pan thalassa*, all-ocean: **4**). Subsequently, Wegener theorized, Pangea had drifted northwards and had fragmented into the present continents, which had then separated and drifted into their present positions, forming the North and South Atlantic Oceans in the process.

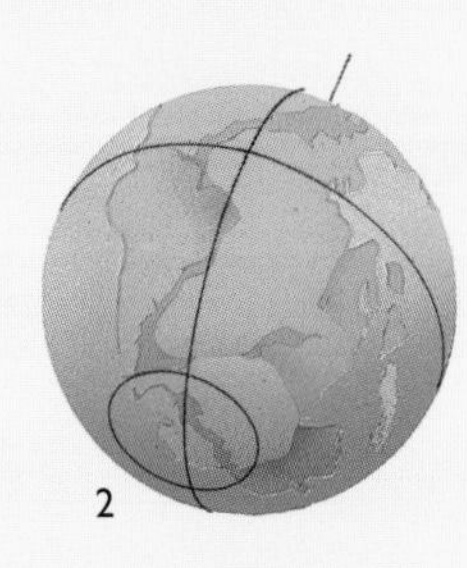
2

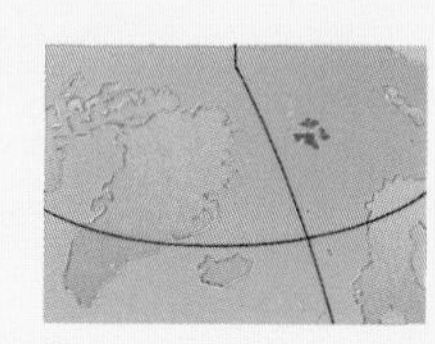
3

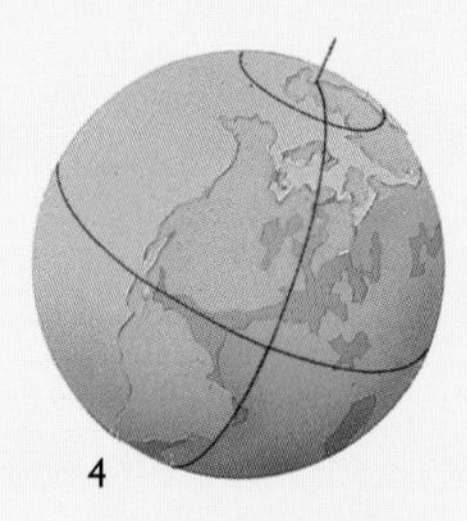
4

One crucial explanation was missing from this "hurried" thesis (Wegener's own description of the very few months it entailed). He could not provide an acceptable model of how the modern continents might have "drifted" (Wegener himself used the term "continental displacement"). His idea was that the ocean floors had "stretched like rubber" as the continents responded to the centrifugal force from the Earth's spin. Another suggestion was that the continents gradually migrated from the poles towards the equator as a consequence of the Moon's gravitational pull (F.B. Taylor, 1908). Still others suggested that we live on a cooling, and therefore contracting, Earth, while a few scientists still suggest the exact opposite—that the Earth is in fact expanding.

In order to explain the existence of a Gondwana, Eduard Suess in 1885 had embraced the notion that as the Earth cools its surface shrinks and wrinkles like the withering skin of an aging apple. We live on the "high plateaus of the apple" (the continents) while sea-water fills the wrinkles between the plateaus (the ocean basins). Many geologists were willing to consider this concept seriously because in essence it conformed to their idea of an Earth with a rigid surface that could be moved vertically but not horizontally. It was evident to them that some continental regions had been uplifted and deformed into mountain chains, and that other parts had sunk, and had formed sea-filled basins. A contracting Earth also permitted the possibility of the previous existence of "landbridges" between the "plateaus." This would explain how animals had migrated from one continent to another before the landbridges had "withered" and sunk, leaving shallow seas or open ocean between.

Wegener's Pangea theory provided the basis for at least as good an explanation of intercontinental animal and plant dispersion as any other. But the basic premise—that the continents were formed from the fragments of a parent body that had subsequently drifted apart, was quite unacceptably outrageous to most of his peers. Geologists had been taught that the Earth's upper crust is rigid, and much of their science was based on this view. The fact that there was no scientifically provable explanation for lateral (as distinct from vertical) movement of the continents became the focal point of a bitter controversy. The dispute split the scientific community for more than half a century.

THE FIRST ENGLISH EDITION of Wegener's *The Origin of Continents and Oceans* was published in 1924, nine years after its first German publication. The claims Wegener made in this book encapsulated the points that separated the "drifters" from the "antidrifters." Although Wegener made many shrewd and telling observations that deserved serious attention, he also made many statements that earned him ridicule and reduced his overall credibility.

For instance, he estimated the rates of separation between many continental locations around the world and focused particularly on the rate of separation of Europe from North America. He quoted from the results of latitudinal (north/south) and longitudinal (east/west) geodetic observations (measurements of the shape of the Earth) made 85 years apart by Edward Sabine in 1823, and by the Danish Greenland Expedition of 1906–08. From the differences in these observations Wegener calculated that North America was separating from Europe at a phenomenal rate.

The year before his death he published a re-calculation. This was based on the results of a shortwave radio method of calculating longitude between fixed points on either side of the Atlantic. He now claimed that "Scotland and Greenland are separating at the rate of 18 to 36 meters per year." (60 to 120 ft per year.) This would require continents to drift halfway around the Earth in a million years at the lower rate, or all the way around at the higher rate! The true rate of separation is about 40 km (23 miles) per million years. This is the rate at which fingernails grow—about 4 cm (1.5 in) per year.

At the time of Wegener's death the "antidrifters" far outnumbered the "drifters." Yet in the face of vociferous and often contemptuous criticism, Wegener remained unflinching and absolute in his belief in the supercontinent, Pangea. While publishing numerous papers and books on meteorology and geophysics, he continued to try to unravel the tangled skein of clues that could prove his drifting-continent theory correct. His book, *The Origin of Continents and Oceans*, was revised and expanded several times. The fourth and last edition was published in 1929.

That we now understand as much as we do about the dynamism of the Earth is primarily due to the two hallmarks of a great scientist that Alfred Lothar Wegener undoubtedly possessed: intuition and tenacity. His intuitive thought, together with the dedication of hundreds of others in the half-century following his death in 1930, led to a revolutionary change in our perception of the mechanisms that drive the Earth. They also led to a better understanding of the distribution of species and the evolution of life on Earth.

As soon as Wegener's hypothesis had been proved viable, the science of plate tectonics was born. As it became clear that the dynamism of the planet, as well as its relationship with the Sun, set the conditions for life on Earth, a second science was founded: interdisciplinary Earth System science. Before these new sciences were established, the proof of continental drift had to be forthcoming. Such proof was indeed achieved in the 1950s and 1960s. But as extraordinary as these achievements were, to view them in isolation is misleading. The surprising fact is that literally hundreds of scientific papers were written from the 18th century onwards that from today's viewpoint offered very real clues in support of Wegener's ideas.

Perhaps the earliest and one of the most telling clues came from the Swiss mathematician Leonhard Euler, who is considered by many mathematicians today to have been, along with Euclid, their most distinguished predecessor. Euler is known for his work on the calculus, logarithmic and trigonometric functions, and other school-day agonies. In 1775 he evolved a theorem that proved that a section of the rigid surface of a sphere can only move by rotating that section about an independent axis. When translated into terms of the Earth's spherical surface, this simply means that as continents "drift" and separate they can only swing away from each other at an ever-increasing angle. In other words, on the surface of a sphere, elements cannot separate and still maintain a parallel relationship.

There is no better example of this than the present shape of the North Atlantic Ocean. As a result of several stages of separation of the continents on either side of it (**5**), the ocean is now triangular: it has an apex towards the North Pole and a broad base near the Equator. Euler's theorem is fundamental to the mathematics of tectonic science—and would, of course, have been available to anyone who happened to realize its significance in Wegener's time.

Many other facts that established the science of plate tectonics were also known considerably before Wegener's time. For example, it was discovered that the Hawaiian chain of volcanic islands are lined up in strict order of age (**6**). The youngest and most active are in the lead and the oldest in the rear—about to be eroded away to the point of submersion (J.D. Dana, 1849). Modern scientists have shown that the Hawaiian Islands are located over a deep-seated "mantle plume" beneath the ocean crust. The volcanoes result from a series of "holes" being "burned" through the ocean floor as the Pacific plate moves over the stationary plume, like a sheet of metal moving over a fixed acetylene torch.

It had also been noted that the structure of the Appalachian Mountains near Tennessee consists of overthrust blocks like a number of "thick slates or tiles on a roof" (J.M. Safford, 1856). Similar discoveries were made in many other localities around the North Atlantic and particularly in the Swiss Alps. Here the "tiles" were not only overthrust but folded back on themselves, termed "nappes" (H. Schardt, 1893). Today these nappes are known to result from "shortening" of the Earth's crust—like the folds of a closing concertina. They formed when Africa collided with North America long ago, and more recently in the Alps and elsewhere, when Africa collided with Eurasia. During these events the continental crust in the regions of collision is shortened by hundreds of kilometers [Essay 2].

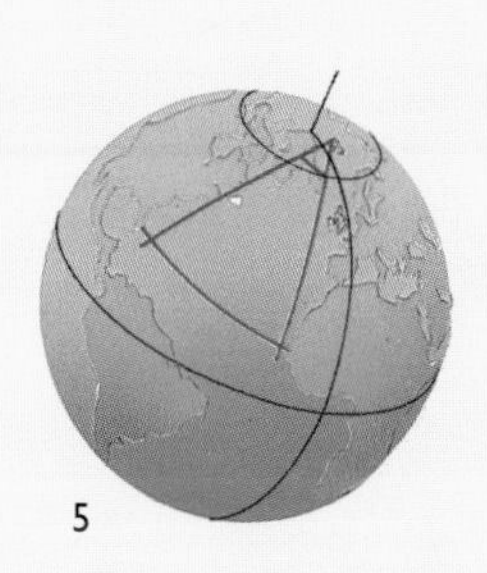

5

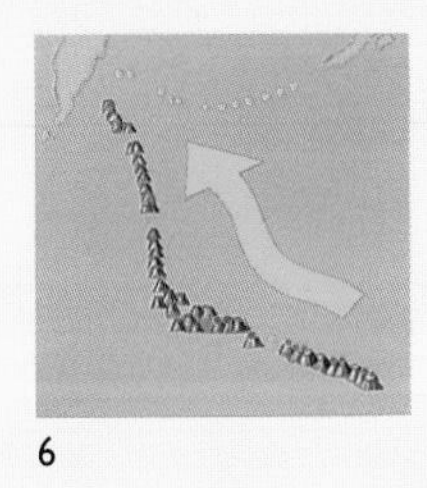

6

7

8

Another vital clue that, as we have seen, Wegener learned in the autumn of 1911, took the form of overwhelming fossil evidence. This included the presence of *Glossopteris* (a primitive seed-bearing plant: **7**), *Mesosaurus* (a primitive aquatic reptile), *Lystrosaurus* (an amphibious mammal-like reptile: **8**), and other terrestrial fossils found abundantly in once-interconnected regions of the southern continents [Essay 1]. This had already persuaded a large number of geologists that there had at least been landbridges connecting Africa, Madagascar, India, South America, and Australia. As we have seen, this group of continents was named "Gondwanaland" (but to avoid confusion we will continue to use "Gondwana" the H.B. Medlicott 1876 version). From 1904 to 1924 Eduard Suess produced a five-volume treatise in which he summed up all the known facts about and arguments for the megacontinent.

The Antarctic continent became a potential member of Gondwana with the discovery of *Glossopteris* in the rocks of Mount Buckley (E.A. Wilson, 1912) during Robert Falcon Scott's ill-fated expedition to the South Pole. But the presence of *Glossopteris* alone was not considered sufficient to justify certain inclusion of Antarctica in Gondwana. Antarctica was not confirmed until more than 60 years later, when fossils of *Lystrosaurus* were found in the rocks of Coalsack Bluff (E.H. Colbert, 1967). *Lystrosaurus* was an amphibious animal that could hardly have swum across the intervening ocean between Africa and Antarctica.

Magnetism in rocks is one of the most important components of tectonic science today. Almost 150 years ago it was discovered by experiment that when rocks containing iron compounds cool down from a high temperature they assume a magnetic field aligned to the present location of the north magnetic pole (M. Melloni, 1853). Subsequently it became apparent that the magnetic field of certain rock formations in widely separated places did not line up with the present field. It was also known that the location of the north magnetic pole had changed substantially in the 73 years between its discovery in 1831 (J.C. Ross) and its redefinition (R. Amundsen, 1904). It was then concluded from this observation, and from the evidence of varying magnetic direction in rocks on different continents, that the magnetic pole wanders about the Earth's surface [Essay 6].

It also appeared that some rock formations had somehow reversed their magnetic fields altogether: extraordinarily, their fields pointed 180 degrees in the opposite direction to neighboring rock formations with normal polarity. That the magnetic pole reversed its polarity from time to time was demonstrated by the discovery of baked clays immediately beneath recently solidified lava flows (B. Brunhes and P. David, 1904–06). Both the baked clay and the lava from the recent flow were shown to have normal (north) polarity. Nearby clay that had not been baked by the red-hot lava had a reversed (south) polarity. The direction of the magnetic pole had therefore reversed from south to north after the formation of the underlying rocks.

Important clues about the magnetism of rocks also came from a different quarter. The fact that tiny flakes of lodestone embedded in a sliver of wood point north if floated on a liquid had been known since classical times ("lodestone" is a form of magnetic iron ore). The fact that heat can destroy the magnetism of bricks when they are baked in a kiln, and that these magnetic properties return to the bricks as they cool, was known in the 19th century. Indeed, the French physicist Pierre Curie earned his doctorate in 1895 for his thesis describing the laws that relate some magnetic properties to changes in temperature. Bricks, and many rocks, contain iron minerals. These minerals lose their magnetism when heated as high as

[*continued on page 26*]

Part of the dormant Reykjanes Ridge near Langisjór, southern Iceland

3 Interactive Boundaries

The upper layer of the solid Earth consists of the "tectosphere," which has two dynamic elements: a passive plastic region, called the "asthenosphere," and a rigid surface, the "lithosphere," which is made up of seven huge "plates" (and a number of lesser plates: see page 46). Their principal boundaries are shown here. The continents are embedded in these plates "like icebergs in a frozen sea" (Wegener's expression) and, with very few exceptions, plate boundaries are located on the ocean floors. The boundaries interact with each other as the rigid plates are moved around by the slow circulation of material in the aesthenosphere. The plate motions that result from this activity, and their effects on the Earth's crust, comprise the subject of the science of "plate tectonics." At some boundaries, plates are actively separating, while at others they plunge beneath or scrape past each other, interacting like floes in an ice field. The overall movement is continuous but spasmodic, so one part of the same boundary can be inactive while a neighboring part several hundred miles away is active.

Where boundaries separate, usually beneath the oceans, they form "spreading-centers". Volcanic, seismically hyperactive, "mid-ocean spreading-centers" run for some 58,000 km (36,000 miles) in total—although we cannot see them, they are the most prominent physical feature on the Earth's surface. Their spasmodic submarine eruptions are the marine equivalent of terrestrial volcanoes, pumping material from the aesthenosphere into the depths of the global ocean. The effect of this activity on the deposition of minerals on the ocean floor, and particulary upon the chemistry of the ocean itself, is of great significance to the setting of conditions for marine and terrestrial life.

Unusually, the spreading-center between the Eurasian and North American plates surfaces from the ocean to cross Iceland from south to north. The main picture shows a presently volcanically inactive part of this ridge. The boundary region here has cooled, shrunk, and shriveled, temporarily forming a sunken valley (a "graben") riven with gorges. The inset picture is of a presently active part of the same boundary. The North American plate is to the top left of these pictures and the Eurasian plate is to the right.

The active Reykjanes Ridge at Krafla, northern Iceland

Grímsvötn caldera, Vatnajökull, Iceland

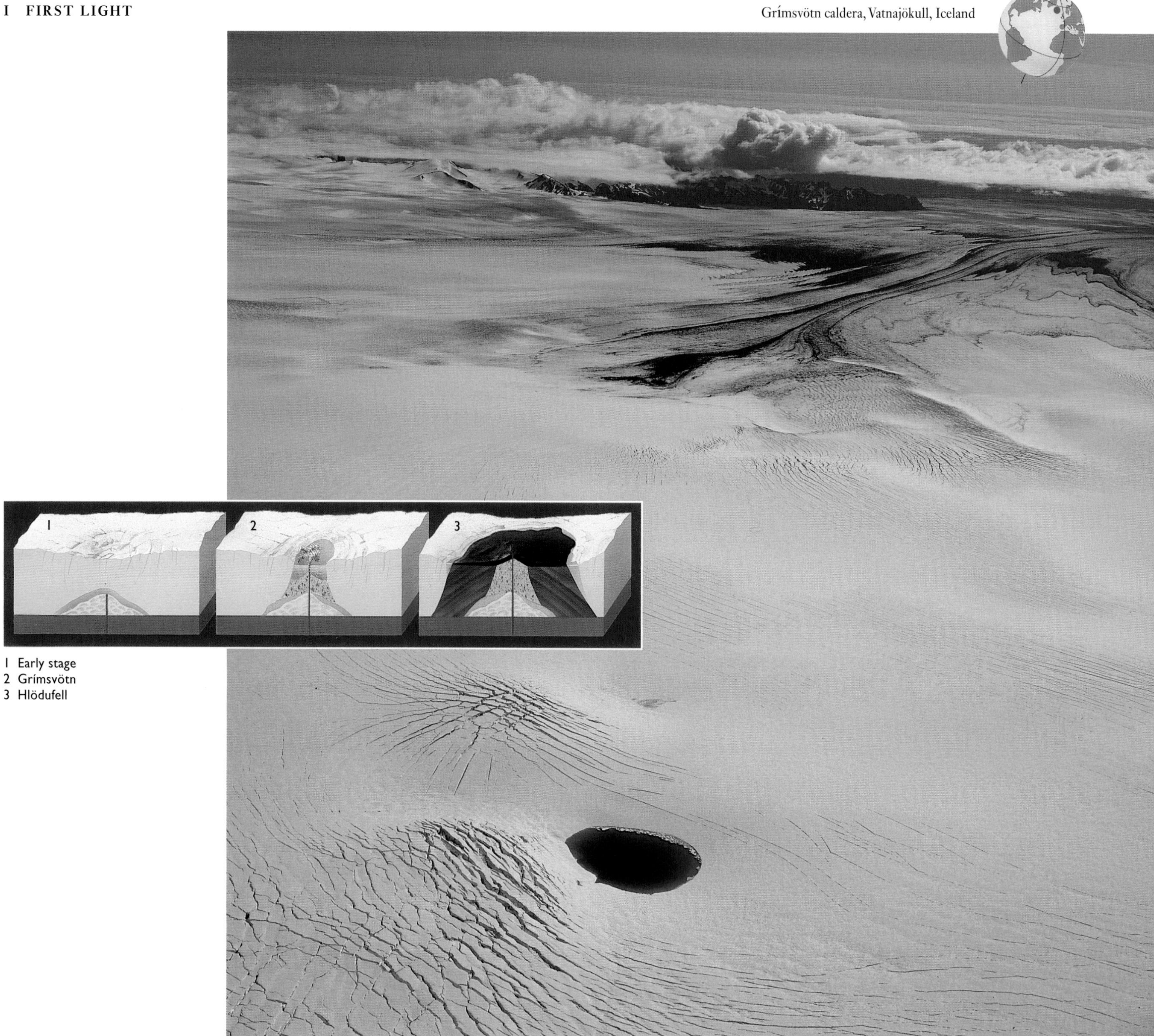

1 Early stage
2 Grímsvötn
3 Hlödufell

4 Subglacial Volcanoes

The largest and thickest ice cap remaining in Iceland (Vatnajökull) is astride an active spreading-center. The surface of the ice cap we see here is tending to fracture, sink, and accumulate meltwater in a depression—the relatively small ice-blue lake of Grímsvötn—above a vast, partially water-filled ice chamber. This chamber was formed within the ice cap as it was melted from inside by volcanic activity almost 1.6 km (1 mile) beneath the lake. Grímsvötn no longer exists—a rare event called a jökulhlaup (ice burst) ocurred in November 1996 after this photograph was taken. The contents of the interior lake discharged catastrophically along a 50-km (30-mile) front at the base of the ice cap, pouring across the alluvial floodplain to the sea. However, before this event, the depth of accumulated water in the chamber had exerted enough pressure to prevent gases trapped in the erupting volcanic lava from exploding. As a consequence, the lava behaved like toothpaste squirting from a tube. The lava "paste" broke off from an "extrusion" to form thin-skinned "pillow lavas," which accumulated on the floor of the chamber and built up to form the base of a mountain.

If we could see it, the mountain would appear as if cast from a mold and would look like Hlödufell, the mountain pictured on the opposite page. Most of Iceland's central mountains were formed in this way—a style of mountain building that is analogous to the formation of seamounts on the seafloor.

Hlödufell, central Iceland

Surtsey, Vestmannaeyjar, Iceland

Terrestrial Seamounts

At active spreading-centers on the seafloor, pillow lavas accumulate from 0.4 to 0.8 km (0.25–0.5 mile) in thickness as they are conveyed slowly away from either side of the spreading-centers. Subsequently the seafloor fractures into mountains that form parallel ridges on either side of the spreading-center. The ridges are simply sections of the seafloor that have been faulted and tilted during plate separation.

The seafloor is a kilometers-thick sandwich of pillow lavas at the surface, vertical volcanic dikes in the middle, and gabbro beneath [Essay 5]. But pillow lavas can also accumulate to form "seamounts" (also called "guyots"), which sometimes peak above the surface of the sea. For example, when volcanic activity occurs in shallow seawater off the coast of Iceland, the water pressure on the lava is much lower than normal. At a certain depth, the pressure drops to a point where the trapped gas inside the lava can boil out, and the lavas explode into glassy nodules. These nodules can rapidly accumulate to form the base of a new volcanic island emerging from the sea—such as Surtsey in the Vestmann Islands (inset).

The Azores islands in the central North Atlantic are good examples of deep-ocean seamounts. More typically, seamounts are eroded down to sea level by wave action as they cool, contract, and subside beneath the waves. In general appearance seamounts resemble Hlödufell, the tabletop Icelandic mountain pictured here.

[*continued from page 21*]

750°C (1,382°F), and regain it at the temperature at which they lost it when they cool. This temperature, the Curie point, varies according to the composition of the minerals.

The only explanation for the apparent reversal of the magnetized minerals in rocks had to be that the Earth's magnetism changed episodically from north to south, then, after an unpredictable period of time, changed back to north again. It was recognition of this very odd phenomenon that later proved to be vital in the final unraveling of Wegener's tangled skein.

Meanwhile, in 1872 the British research vessel HMS *Challenger*, with Captain George Strong Nares in command, made a monumental four-year voyage of discovery in the world's oceans. The voyage was made under the scientific eye of Charles W. Thomson, a professor of natural philosophy from Edinburgh. While making a series of soundings for the future laying of telegraph cables they hauled up a great number and variety of hitherto unknown specimens of marine life. The scientists on board also discovered an extraordinary range of flat-topped mountains beneath the sea surface [Essay 4]. These formed part of a chain on either side of a ridge that bisected the Atlantic Ocean from north to south. Thomson and his associates took soundings with weights attached to vast lengths of hemp rope. They discovered that the plateaus on the mountaintops had an average height of about 1,830m (6,000 ft) above the ocean floor. The system ran from about 50°N between the British Isles and Newfoundland to about 40°S off the Falkland Islands in the South Atlantic. This became known as the "Mid-Atlantic Ridge" (MAR). Later still, the MAR (**9**) was discovered to be seismically active—constantly subject to earthquakes (E. Rudolph, 1887).

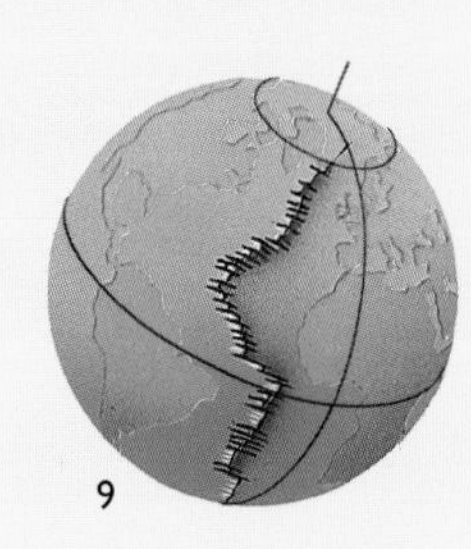
9

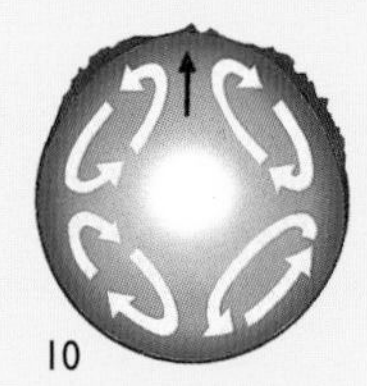
10

POST-WEGENER BREAKTHROUGHS in continental drift theory were initiated by A.L. Du Toit, a South African geologist and one of the leading geoscientists of his time. He had written a number of important papers from 1921 onwards providing additional evidence for linking South Africa with South America. His papers presented data on the similarity in rock formations on either side of the South Atlantic, the continuity of structural trends, the similarity of fossil flora and fauna, the synchronicity of paleoglaciation, and more. In 1937 Du Toit went much further in his support of Wegener. He published a book, *Our Wandering Continents*, in which he summarized all the evidence then available in favor of Wegener's general hypothesis.

The next important step came in 1944 when the British geologist Arthur Holmes proposed a convective mechanism for continental drift in a book called *The Principles of Physical Geology*. Holmes's initial work in this field was published in 1928. His thesis had in fact been anticipated even earlier. In 1914, a geophysicist named J. Barrell proposed, in a little-known paper in the *Journal of Ecology,* that the rigid crust of the Earth floats on a plastic layer in the upper mantle. This proposal, together with Holmes's ideas, ultimately became the basis of the present-day view. Had Holmes known about Barrell's ideas, the modern theory of plate tectonics might have been advanced by a decade.

Nevertheless, Holmes's contribution was invaluable: he described the concept of rising convection currents (like hot water in a heating system), which he had concluded were located beneath ocean ridges (**10**). He proposed that new seafloor might be formed as the continents move away from each other, and went on to suggest that if this was so, ultimately the continents would collide, with the result that mountain ranges would form at their margins. A workable theory based on Wegener's hypothesis was now in place. But proof of the theory was elusive.

Du Toit's and Holmes's publications, the first immediately before, and the second towards the end of World War II, were followed by a peace that made naval ships and equipment readily available. These ships were used for global research in the unknown ocean basins of the world: three-fifths of the Earth's surface were still unexplored. From the 1940s on, the rate of discoveries supporting Wegener's propositions accelerated. Nevertheless, many of the postwar discoveries were not entirely new. As we have seen, a number of key factors had already been identified but their significance had not been recognized. These factors were now rediscovered and, although puzzling, they were thought to be potentially important. In due course they were fitted in to the jigsaw.

We will follow a necessarily abbreviated chronology of the "plate tectonic" revolution here, beginning the story in Iceland—which Wegener had visited several times on his way to Greenland but found "uninteresting." According to his biographer (M. Schwarzbach), Wegener dismissed the island as a "waste product" lacking in tectonic interest. Ironically, at one time Wegener was at the port of Akureyri just 50 km (30 miles) west of Krafla. This is a currently active section of the Reykjanes spreading-center and part of the MAR [Essay 3].

BECAUSE OF ITS VOLCANISM Iceland was considered to be an ideal place to investigate the phenomenon of magnetic reversals in volcanic rocks. In 1951 J. Hospers found a series of four lava flows and sediments that were magnetized either to the south (like that baked clay discovered by Brunhes in 1904) or to the north (like Brunhes's unbaked clay): the magnetic polarity had apparently flipped three times during the formation of the series. Later, when dependable dating techniques had become available, it was shown that magnetic reversal in rocks is a worldwide phenomenon: rocks of the same age are magnetized in accordance with the direction of the magnetic pole of their time.

Following Hospers's discovery, the paleomagnetic orientation of large numbers of specimens of sedimentary and volcanic rocks of various ages in the British Isles was noted. These dates were plotted on a map according to their relative ages (K.M. Greer, E. Irving, S.K. Runcorn, 1954). The plot produced traces, one for each age, and each trace described the wandering path of an ancient magnetic pole. From the varying paths of the traces it looked as if the magnetic pole had moved arbitrarily through time. But when in 1956 Irving plotted similar traces from data obtained from several localities on separate continents, the polar-wander curves did not correspond. The notion of an arbitrarily wandering magnetic pole capable of tracing different paths in different places at the same time simply did not make sense. However, Irving realized that if the continents moved independently and the magnetic pole remained fixed, the apparent anomalies of polar wander could be explained.

Runcorn then showed (also in 1956) that the polar-wander curves could indeed be made to correspond. First, one had to assume that North America had moved 24°W relative to Europe in the last 250 million years (MY). Then one had to restore the continents to their relative positions at that time. Runcorn also noted that the polar-wander curves could be matched whether the magnetic pole was normally magnetized as now (a "positive anomaly") or reversed.

THE FIRST OF MANY DISCOVERIES in the ocean was made in 1952. This was during the mapping (by echo-sounder) of the North Atlantic seafloor on a line between Dakar in Senegal and Barbados in the Lesser Antilles (B.C. Heezen, M. Ewing, et al., 1953). An extraordinary pattern of alternating peaks and valleys was revealed, a pattern that bore no known relationship to any terrestrial landscapes. Over the next several years it was discovered that the deepest valleys and the highest peaks in the system coincided with the Mid-Atlantic Ridge, and also that the central "valley" of the ridge was both a seismically active rift and part of an earthquake epicenter belt (M. Tharp and B.C. Heezen, 1954, and independently, M.N. Hill, 1954). From other evidence, Heezen and Ewing implied (in 1956) that the mid-ocean ridge system is a world-encircling feature. Ewing later confirmed this, further observing that the mid-ocean ridge usually bisects the ocean floor. Also that it probably has earthquake epicenters concentrated in the central rift valley throughout its length.

He concluded that a "positive magnetic anomaly" may be characteristic of all mid-ocean ridges at the present time (M. Ewing, B.C. Heezen, and J.R. Hirschman, 1957).

A second series of key discoveries in the ocean came as a consequence of the adaptation of a sensitive instrument called a magnetometer. This had been used in wartime to measure the intensity of magnetic fields from the air and therefore to detect submarines beneath the sea. It was housed in a pod suspended above the sea at the end of a cable so as to be unaffected by the aircraft's magnetic field. A similar technique was now used over land to distinguish regional "hard" rock from "soft" rock. Hard volcanic rocks have a stronger magnetic field than soft sedimentary rocks (which might contain oil). Why not use a similar means of mapping the disposition of hard and soft rocks on the ocean floor?

An intensive deep-water mapping survey was carried out in 1955–56 by the United States Coast and Geodetic Survey off the Pacific West Coast. Visiting scientists were permitted to tow a magnetometer behind the research vessel *Pioneer*, provided it did not interfere with the Survey's mapping work. Fortunately this work required the ship to steam back and forth without deviation across an area about 240 km by 2,000 km (150 miles by 1,250 miles). The magnetometer could therefore produce its own map of magnetic intensities for this huge area of seafloor. The result was astonishing: the map was striped like a zebra (**11**)—black stripes indicated positive anomalies, while white stripes showed negative ones. No one could explain how stripes of differently magnetized rocks could have been produced on such a scale (A.D. Raff and R.G. Mason, 1961).

AT THIS POINT during such a rapid series of extraordinary discoveries (we have selected only a few of the many), a unifying scientific synthesis was becoming essential. It would be an antidote to the now-angry controversy over Wegener's ideas. The discoveries badly needed to be put into a general perspective.

Arthur Holmes's seafloor-spreading hypothesis of 1944 was resurrected and enlarged upon in the light of the most recent discoveries by H.H. Hess (1960). Hess had directed echo-sounding surveys of the Pacific Ocean floor during World War II and had discovered and explained the formation of flat-topped seamounts on the seafloor [Essay 4]. But he had also recognized the significance of "island arcs" in the Pacific—literally arcs of volcanic islands surmounting a very deep trench and subject to frequent earthquakes. Hess had pondered the relative youth of the ocean floor compared to the age of the continents and the extraordinary length of the mid-ocean rift system. He now proposed that the seafloor is indeed created at the mid-ocean ridges, and that after formation, the rigid seafloor spreads towards the trenches, there to descend beneath island arcs into the Earth's mantle. Thus, the continents did not drift independently of each other: they were in effect embedded in the seafloors and were carried away from each other as the seafloor spreads apart. And finally Hess reintroduced the idea that the overall system is driven by convection cells in the Earth's mantle that act upon the underside of both the seafloors and the continents.

Some margins, Hess reasoned, appeared to be "passive," like those on either side of the Atlantic Ocean, and some were obviously volcanically "active," like those surrounding the Pacific Ocean (**12**). If the Atlantic was opening while the Pacific Ocean was closing, the Pacific Ocean crust must be being destroyed beneath the continents converging upon it. This would explain the island arcs surrounding the Pacific, the absence of volcanism and earthquakes at the Atlantic continental margins, and the relative youth of all the world's ocean crust. The ocean crust must be continually destroyed at its edges between converging continents and be rejuvenated at the spreading centers between diverging continents. (The exception in the North Atlantic region is the active margin manifested by the Lesser Antilles island arc: here the North Atlantic plate is descending beneath the Caribbean plate.)

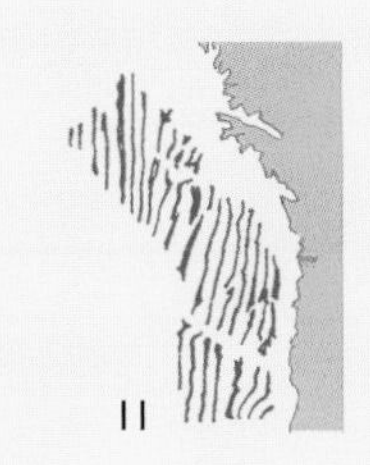

11

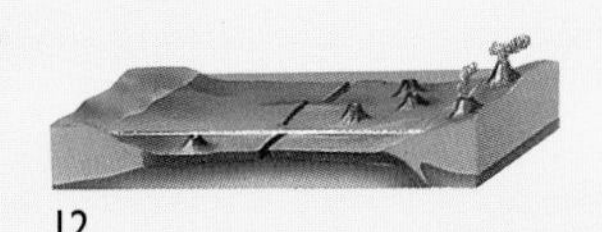

12

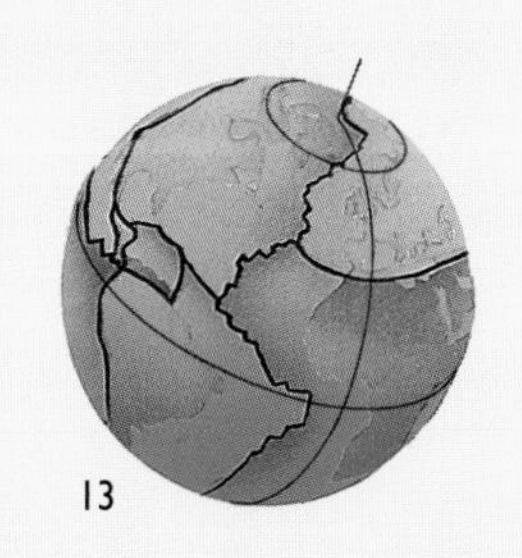

13

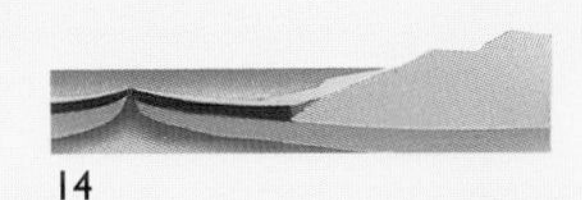

14

THIS "GEOPOETRY" (Hess's expression) inspired a research student at Cambridge University (F. Vine) and, quite independently, a Canadian Geological Survey paleomagnetist (L.W. Morley), to look more closely at the paleomagnetism of mid-ocean ridges and the magnetic anomalies commented upon by Bruce Heezen, Maurice Ewing, and others. Vine and Morley also considered the significance of those extraordinary geomagnetic zebra stripes on the Pacific Ocean floor that, it now seemed, were universally present on all ocean floors. By this time ocean research vessels habitually towed magnetometers wherever they sailed.

In essence, both Vine and Morley proposed that the symmetry of the stripes, on either side of and parallel to the mid-ocean ridges, could be explained only if one assumed that the seafloor was separating along the whole length of the ridges. Thus the crust increases in age with distance from the mid-ocean ridges, out to the continental shelves where the seafloor would be oldest. Furthermore, they proposed that as liquid magma (heavy, iron-rich basaltic minerals from beneath the ocean crust) welled up to fill the gap at the spreading center, the newly solidified basalt would assume the magnetic polarity of the day. As Curie had shown, bricks that contain iron minerals lose their magnetism when heated and regain it when they cool. It had been clearly established that the Earth's magnetic pole periodically switches from north to south. One could therefore expect that in addition to the parallel symmetry of successive ridges, successive series of ridges should be either positively or negatively magnetized according to their age. Thus the chronology for the periods when the Earth's magnetism had been reversed was magnetically imprinted on the seafloor.

The seafloor was drilled at points increasingly far removed from the spreading-center. Cores were taken from regions where magnetism is normal and from others where it is reversed. The cores were dated, and it was found that no ocean floor at its most extreme distance from a spreading-center was older than about 170 million years.

Vine was fortunate enough to have his paper published in *Nature*—a leading scientific journal (F. Vine and D. Matthews, 1963). Vine was filled with trepidation before publication: he was asked if *Nature* had their articles reviewed "or do they publish almost anything?" Vine replied, "Well, we're about to find out. I've just put my paper in, and if they publish that they'll publish anything." Morley was unlucky—his paper was rejected by *Nature* and dismissed by a reviewer of another journal as being "the kind of thing you talk about at a cocktail party!" An excerpt of Morley's paper appeared in 1967, but it was not published in full until 1981.

Today it is accepted that the Earth's rigid outer part, its lithosphere, is fractured into enormous plates [Essay 3]. The plates are resting on a semiplastic region, called the asthenosphere, in the upper part of the Earth's mantle. There are seven such plates covering most of the Earth's surface (**13**). In addition there are a variety of medium-sized plates and a large number of very small fragments called "microplates." Individual plates are moved in two main ways (or so it is generally theorized). They are pulled by their own weight as they sink into the asthenosphere beneath island arcs, and they are moved from below by the stirrings of the semiplastic material beneath the solid crust—rather like pack ice planted with icebergs being moved by sea currents.

BUT WHY ARE SEAFLOORS lower than the continents? Compounds of silica are the most commonplace materials that make up the bulk of both the basaltic rocks that form the seafloors and the granitic rocks that form the continents (**14**). But basalt is about 2 percent heavier than granitic rock, primarily because the former is richer in iron and the latter is richer in aluminum. This small difference in density ensures that the comparatively young and thin basaltic seafloors are low-lying and are covered by the sea [Essay 5].

The relationship between the seafloor and the continents bears an uncanny resemblance to deep-keeled icebergs in a field of pack ice.

[*continued on page 32*]

Svartáfoss, Ishólsvatn, northern Iceland

5 Continents and Seafloors

Continental basement rocks are "granitic"; they contain a high proportion of silica and aluminum. The seafloor crust is "basaltic"; it contains less silica and a high proportion of iron. Continents are therefore lighter than the ocean floors—like icebergs trapped in sea ice. This accounts for the average 5 km (3 miles) by which the continents are elevated above the seafloors. Since water always gravitates to the lowest point on any surface, it also accounts for the location of the oceans.

As plate boundaries move apart at a spreading-center, iron-rich basaltic lava wells up to fill the gap, cooling and solidifying to form new seafloor. The picture above shows an example of such basalt that has erupted onto the surface in Iceland in the region of the active spreading-center. This basalt has cooled at different rates, forming bands of three very different-looking rocks that are identical in composition. The bands are analogous to the structure of the seafloor, which also is composed of three different forms of basalt layered on top of each other: pillow lavas on top, vertical dikes between, and "gabbro" (intruded basaltic material) at the bottom. Discounting the sediment layers that cover the deep ocean floor, their overall thickness also averages 5 km (3 miles).

The aerial picture at right shows a region of Iceland where melted granitic rock has been ejected onto the surface from volcanoes as "rhyolite." When melted rock of the same composition as rhyolite does not reach the surface but is injected between other rocks, it is called "granite". Rhyolite is extruded onto the Earth's surface and granite is intruded beneath the surface. But how can granitic rhyolite, a continental material, form in the center of a seafloor structure—Iceland? The rhyolite pictured here is a rare mixture of minerals that were produced from basaltic magma after prolonged retention in a magma chamber. This witches' brew then spewed upon the Earth's surface here in Iceland. Similar formations can be seen in a few other places such as Ascension Island near the Equator in the center of the South Atlantic Ocean and in Yellowstone National Park, Wyoming (back cover).

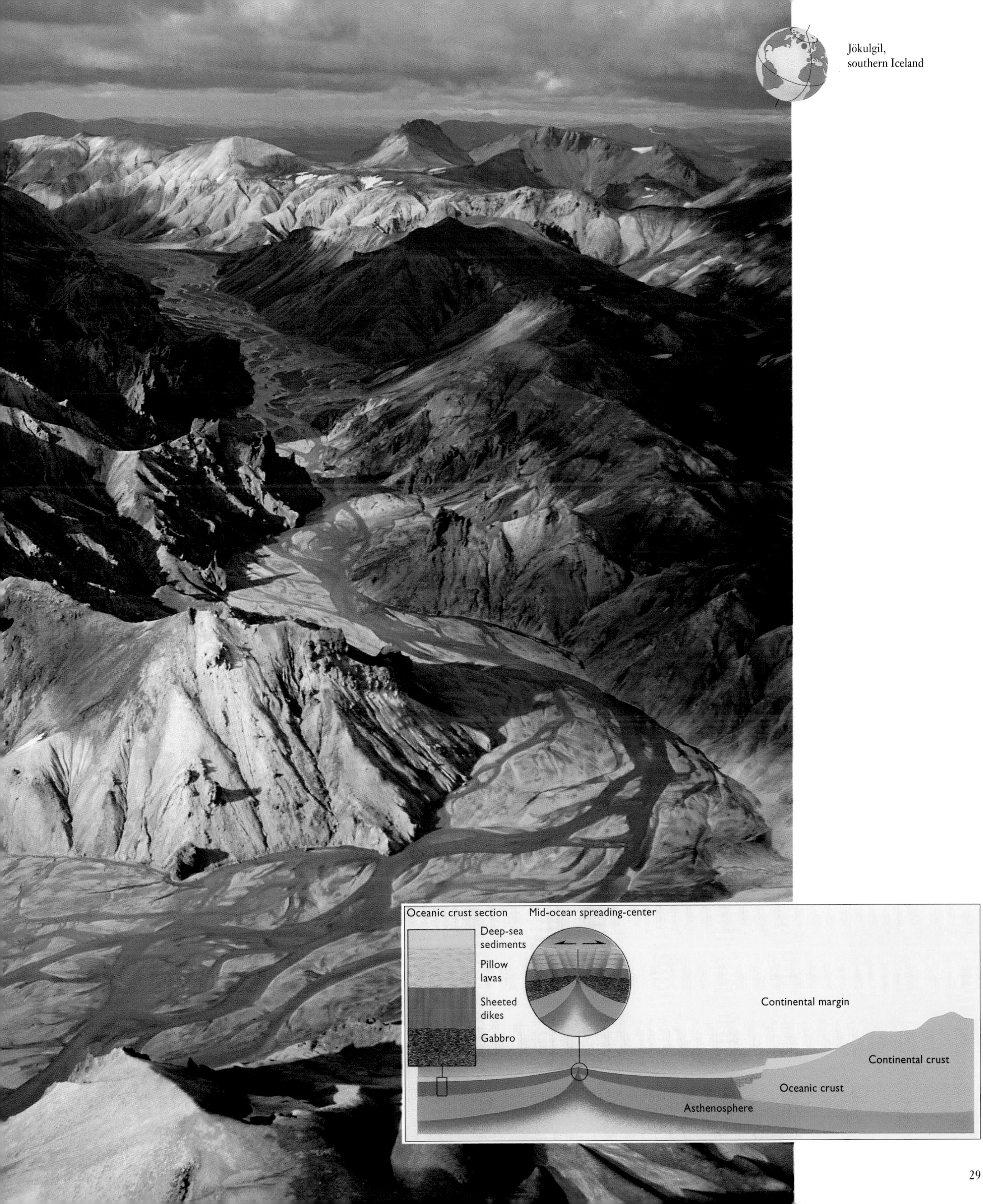

Jökulgil,
southern Iceland

Bathurst Island, Canadian High Arctic

6 Wandering Poles

The Earth's north *magnetic* pole was discovered in 1831. It is now known that the magnetic pole "pulses" like a dynamo and is difficult to locate. It moves on an hour-to-hour and day-to-day basis but is confined within an oval-shaped region. This aerial picture was taken within that area, a region in which all magnetic compasses are useless. This is a desolate, featureless, even dangerous place, but a crucial reference point on the Earth's surface.

After its first discovery, the magnetic pole was rediscovered in 1903. It had moved a considerable distance. Today's magnetic polar region is nearly 800 km (500 miles) NNW of its 1831 position. Such movement, during a relatively short period of time, may be a prelude to a complete reversal of the Earth's magnetic field. The field has switched irregularly from north to south and back again innumerable times in the geological past in response to changing dynamo currents in the magnetic molten iron of the Earth's core. The last reversal (from south to north) occurred about 700,000 years ago. The erratic path of a previous reversal at 15 MYA is traced on the globe below [see Essay, page 182].

As plate boundaries separate beneath the ocean, iron compounds in the newly formed seafloor adopt a magnetism corresponding to the direction of the existing magnetic pole as the new material cools. When the magnetic pole switches to a south polarity, new seafloor also adopts south polarity. If the North Atlantic Ocean floor could be striped according to these periods, black for normal and white for reversed polarity, one would see matching pairs of parallel stripes at ever-increasing distances from either side of the Mid-Atlantic Ridge. The discovery of such zebra-like magnetic striping on the seafloor and its correct interpretation in 1963 was ultimately accepted as the final proof of seafloor spreading and therefore of continental drift.

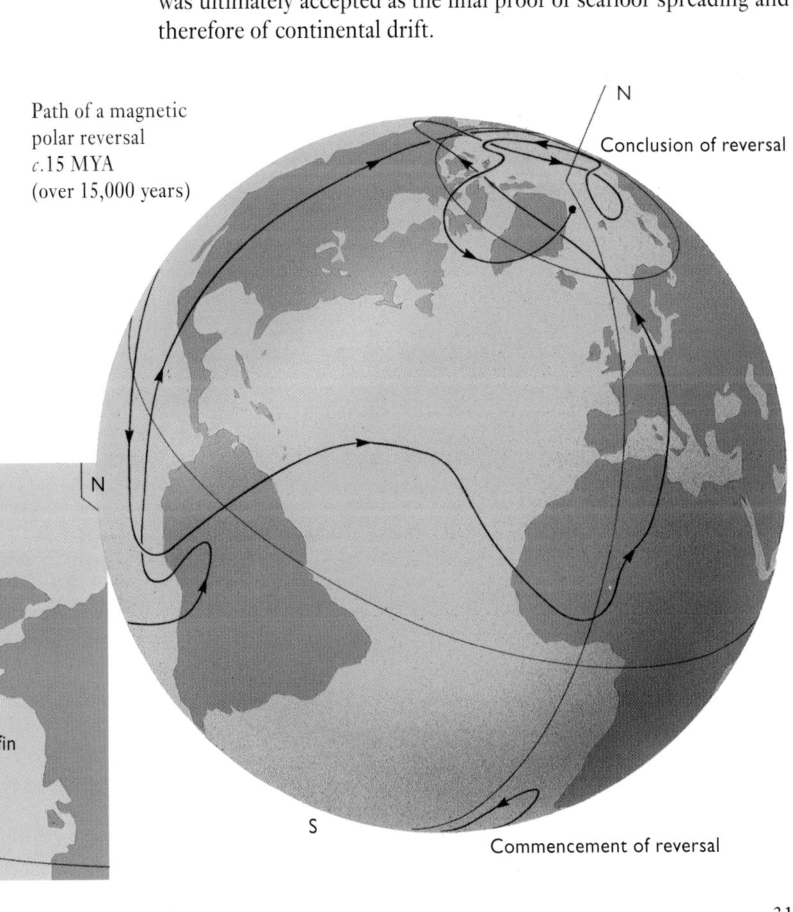

Path of a magnetic polar reversal *c*.15 MYA (over 15,000 years)

[*continued from page 27*]

Pack ice is frozen seawater, which is very slightly heavier, volume for volume, than that which forms icebergs. This is because icebergs are calved off glaciers and are therefore frozen fresh water. Just as a field of pack ice has a low profile relative to the icebergs it has ensnared, so the sheets of basalt formed at the edges of separating plates have a low profile relative to the high-standing and deep-rooted continents.

As new basaltic ocean crust is formed at the plate boundaries on the seafloor, the distant plate edges interact and jostle each other. The low-lying basaltic elements of these faraway edges are destroyed either by being forced one beneath another, or beneath the edges of approaching continents. When all the low-lying basaltic material has been consumed the continents themselves collide and form mountain ranges (**15**).

As a consequence, over the several thousands of millions of years during which these apparently linked events have taken place, there have been times when microcontinents collided to form continents. There have been times when continents collided to form megacontinents and others when megacontinents combined to form a supercontinent, a Pangea. These huge masses of continental material were like incredibly large icebergs, with their keels rooted in the "tectosphere"—a word used to describe the Earth's interactive outer casing (the lithosphere and the asthenosphere). In this perpetually turbulent ocean of semi-plastic rock, supercontinents and megacontinents were always vulnerable to fragmentation. In the course of time they fractured into continents—like those of today.

So it is not so much that continents "drift" but that the Earth's internal dynamism causes its crustal plates to interact. The interaction constantly causes the plates to change size and shape. The continents play a passive role. Wegener's basic premise, that the modern continents were derived from a supercontinent, has been vindicated. But it *is* necessary to qualify the word "supercontinent." From modern reconstructions it appears that Wegener's Pangea was not the tightly knit amalgamation that he visualized. There was a point in time 250 MYA when the continents that surround the Atlantic Ocean today were indeed "close-knit." Even so, parts of Asia and the Middle East were strung out in a great arc stretching from one hemisphere to the other. So the ultimate "Pangea" was a *clustering* of continents rather than an assembly of continents (**16**), a clustering in which some continents were indeed sutured, while others gathered in associated groups or strings. In this sense it is believed by a growing number of leading scientists that Pangea had a predecessor termed "Rodinia" and that the makeup and breakup of such a clustering might possibly be a repeated feature of the Earth's incredibly long history [discussed in Chapter III: *Iapetus and Avalonia*].

But how do we know that this continent and that continent were once joined? Or that this or that continental fragment, now an island archipelago, was once attached to a now distant continent? Or indeed, how is it known today that this or that major element of the North Atlantic realm was once resident in the Eastern Hemisphere beyond the South Pole? One way of answering these rhetorical questions is to follow in Wegener's own footsteps, for he was the first scientist to act on his belief in continental drift.

Of the many arguments for drift that Wegener used, the most compelling were those he took from Eduard Suess regarding the southern group of Pangean continents—Gondwana. But as we have seen, Wegener went much further than Suess: figuratively, instead of depending on landbridge connections between the southern continents, he had sutured them together coastline to coastline and had then added the northern continents to the supercontinental assembly, incorporating the entire surface of the Earth into his concept of Pangea surrounded by Panthalassa. The evidence for both the Gondwanan and Pangean conjectures lay in matching rocks, related mountain belts, glaciated regions, and similar past and present fauna and flora on opposing sides of both the North and South Atlantic Oceans. Wegener put it this way:

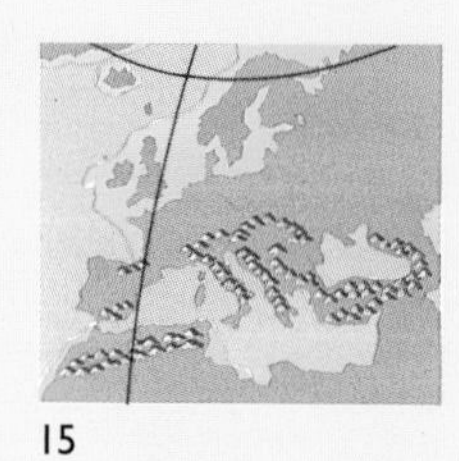

15

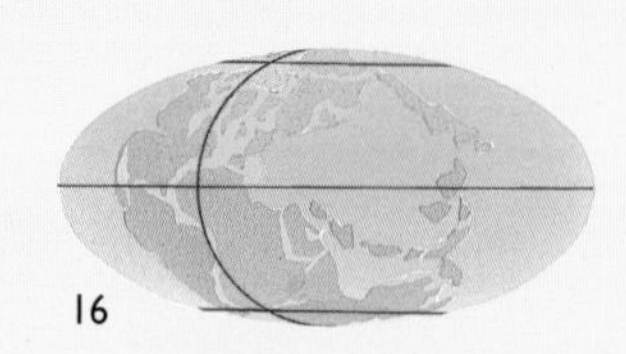

16

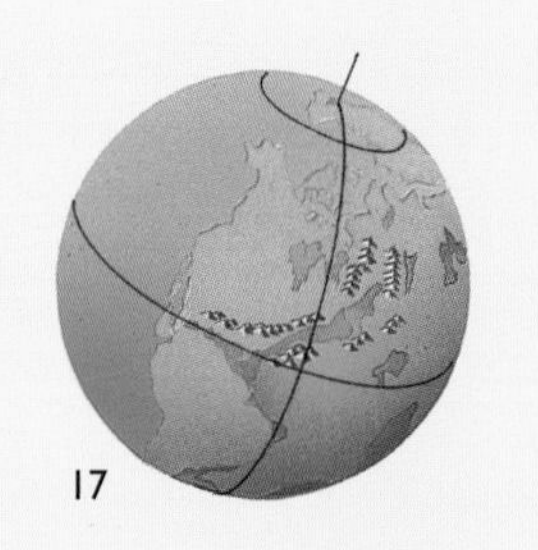

17

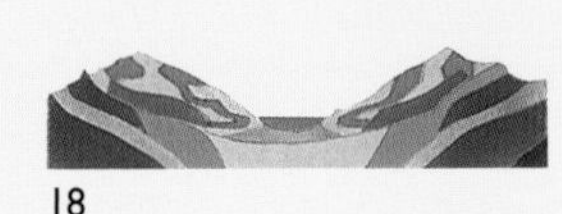

18

It is just as if we were to refit the torn pieces of a newspaper by matching their edges and then check whether the lines of print run smoothly across. If they do, there is nothing left but to conclude that the pieces were in fact joined this way. If only one line was available for the test, we would still have found a high probability for the accuracy of fit, but if we have n *lines, this probability is raised to the* n*th power.*

Today's geoscientists study Wegener's "torn pieces of a newspaper" by reconstructing the general paleogeography of the Earth at different times, and particularly the fit of continental shapes into the geography of Pangea. They do this by looking at the detailed relationships between now widely separated continental margins or bits and pieces of continents that have become separated from their parent bodies. Individual results might suggest the possibility of a link, but it is the collective results from a number of different kinds of data that establish a scientific consensus [Essay 7].

As we have seen, when the Earth's tectonic plates interact at their boundaries the result can be a collision of continents, such as the present and ongoing collision of Africa with Eurasia. This has left the Mediterranean Sea a minor remnant of a major ocean that has almost completely closed—and is still closing. When continents collide, mountains are thrown up in belts along the respective margins. In the case of the Mediterranean, collision has created the Alpine belt that stretches from Spain and Morocco on either side of the Straits of Gibraltar to the Balkans. The same scenario applies to the North Atlantic Ocean (**17**), which has four vast and closely related mountain belts around its perimeter, formed during stages in the assembly of Pangea. When Pangea separated, these mountain belts were also split apart and are now widely separated on opposing shores.

Such mountain belts have their own tectonic "styles," characteristic trends, folds, overthrusts, and nappes. When Pangea broke apart, rifting frequently started in the very regions where the pre-Pangean continents had collided. It is these once-shared structural features that relate the now-separate and often very distant regions (**18**). In addition, many rocks in now widely separated mountain ranges retain some degree of magnetic imprint, "footprints" that orient them to their original position at the time of deformation into mountains. Moreover, such deformation itself sometimes causes minerals in mountain rocks to be heated above their Curie point, and thus resets their magnetic fields as they cool—establishing their distinct paleomagnetic identity. The decay of radioactive trace elements (isotopes) found in these rocks is a form of radioactive "clock" from which the geologic age of the rock formation or its deformation can be read. If the structural likenesses of now-separate continental margins seem to correspond, if their "footprints" correlate and their "compass" and "clock" settings also tally, there is a strong likelihood that the continental margins were once contiguous.

Other clues can be collected from the unique chemical composition of rocks, which provides the equivalent of a "fingerprint." These prints must also correspond for earlier contiguity to be claimed. As mountains wear down, the products of their erosion accumulate as sediments (desert sands, lake sediments, and seashore sand and silt), and compress to form layered rock; these sedimentary sequences characterize the environments in which they were formed. Such sequences can be compared, and sometimes matched, in widely separated regions.

In most sedimentary rocks minute particles of magnetically attracted grains have aligned themselves with the Earth's magnetic poles at the time of their deposition. This direction, the rock formation's paleomagnetism, can be measured, plotted, and related to the record of past polarities. Limestone can also preserve paleomagnetism through the presence of certain types of bacteria that left aligned magnetic signatures in ancient coral reefs.

Sedimentary rocks may contain fossils of plants and animals, or their traces and tracks, laid down in a succession that directly corresponds to their order of evolution. These fossils can also offer clues to the environment in which they lived in the distant past. When fossilized pollen or spores are found in sedimentary rocks, identification of the pollen or spore helps to determine the age and environment in which the sediments were deposited. All such evidence of life's presence on distant and opposing continental margins has to match to prove previous conjunction of the margins.

THE "LINES OF PRINT" from Wegener's "torn newspaper" appear in corresponding margins both above and beneath the sea. Above the sea they appear as fossil reefs, rift valleys, and mountain ranges. Where there are no outcrops to reveal a cross section of the story, the layers and textures of the rock sequence beneath the two continental margins can be charted by exploding underground charges, or other seismic vibration techniques. The travel times of the seismic waves that are generated and then reflected by the interfaces of different layers of rock are registered on a seismic graph. The changing velocities of the sound waves as they pass through the different kinds of rock are also recorded. The seismic graph creates an image of the stratigraphy under the surface. A similar technique, with shipboard airguns instead of explosives, is used to obtain a cross section of sediments on the seafloor and the structure of its basement rock.

These different data sets are used either separately or in various combinations to demonstrate the original connection between two distant continental shores. Nevertheless, scientific proof of juxtaposition, as distinct from strong likelihood of it, has not been established for every mile of opposing shoreline around the North Atlantic. But, as Wegener suggested, one perfect match along an extensive shoreline is sufficient to establish the probability of a relationship; several matches increases that probability exponentially.

THERE ARE FEW LANDMASSES around the North Atlantic Ocean that afford a better opportunity to demonstrate the effectiveness of these techniques than the huge, subcontinental island of Greenland. It is related to Europe by an extensive mountain belt called the Caledonides (from Caledonia, the Roman name for Scotland, where the distinct features of this Pangean mountain range were first recognized). The Caledonides were the first and largest mountain belt formed during the assembly of Pangea. Some of the principal correlating features between Greenland and Europe are therefore structural in character—with identical types of folding, overthrusting, and angles of faulting and inclination (**19**). They also contain some matching rocks that were "cooked" to the same degree before they were separated—rather like the interfaces of a medium-rare steak cut from end to end and separated. The separate cuts reveal matched gradients of cooked meat from pink to well done! The basement rocks of this long-eroded mountain chain, the once deeply buried foundations on which the northern section of the Caledonide mountains stood, can be seen today above the surface of the sea in northwest Scotland and in southeast Greenland. These rocks are perfectly matched in many of the ways described above, even though now separated by 1,700 km (1,100 miles) of North Atlantic Ocean.

THE NARROW STRAIT separating Greenland from North America is Nares Strait—named after Captain George Nares of HMS *Challenger*, the ship that made that first historic research voyage. In 1875–76 Nares conducted the first successful navigation of the Strait from the Kane Basin at the head of Baffin Bay to the Polar Sea, a discovery that resulted in the subsequent race to be the first to reach the North Pole (achieved by R.E. Peary and M. Henson in 1909). But as we shall see later [Chapter XII Essay 12] the Vikings had established a base at the entrance to the Strait over 800 years before the Nares expedition. Significant human events—including the discovery of Greenland by the Inuit—had taken place many centuries before the Vikings had arrived in this remote corner of the Western World.

Geologically this strait is a sea-filled fault. It was named the "Wegener Fault" by J. Tuzo Wilson, a Canadian "antidrifter" who

19

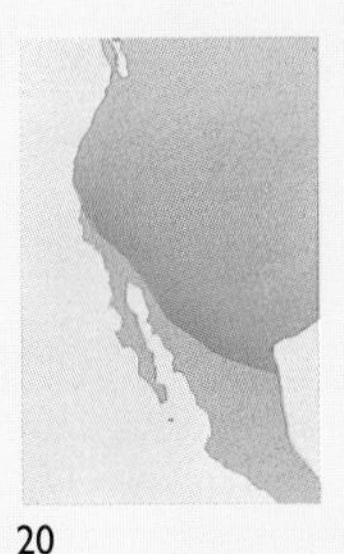

20

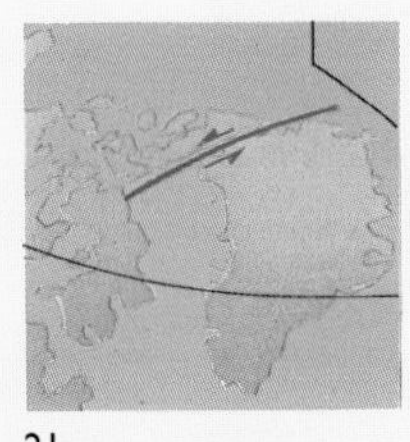

21

became a convert to Wegener's concept and later a father-figure of tectonic science. The Wegener fault is the result of one opposing shoreline moving relative to the other parallel to the direction (the "strike") of the fault—a phenomenon called a "strike-slip fault." The best contemporary example of an active strike-slip fault is in California. Here, a large piece of North America about 1,600 km (1,000 miles) long, stretching from Baja California in the south to Cape Arena in northern California (**20**), is being moved up the West Coast. It is conveyed along the San Andreas Fault by the movement of the Pacific Ocean floor. This huge continental sliver has moved over 480 km (300 miles) from its original position (Baja California was attached to the coast of Mexico) in a period of about 30 million years. Unless there are major changes in the relative movement of the Earth's tectonic plates, this sliver will eventually "dock" with Alaska—as many other continental and oceanic bits and pieces have done in the past.

THE EXTENT OF A SIMILAR strike-slip movement between Greenland (a micro-continent) and Ellesmere Island (part of the North American continent) along the opposing shores of Nares Strait (**21**), is believed by some geoscientists today also to have been considerable (as much as 400 km or 250 miles). By others it is thought to have been negligible (50 km, 30 miles or less relative to Ellesmere Island). Physical evidence in support of these opposing points of view is in apparent contradiction, and argument between experts is therefore vigorous. Perhaps the most likely theory is that as Greenland moved northwards relative to North America, parts of the Canadian High Arctic archipelago were also rifting apart. Perhaps they stretched like the bellows of a concertina relative to the passing Greenland block. [Essay 8].

However, in recognizing the character of the fault between Greenland and Ellesmere Island, Wegener was again well ahead of his time. In 1924 he wrote:

The displacement [Nares Strait] does not exist in the form of a dragging away from each other of the margins of the rift . . . in the Northwest of Greenland, but in a lateral displacement of great dimension, a so-called tear fault [strike-slip fault]. Grinnell Land [now Ellesmere Island] has slid along Greenland producing the striking straightline boundaries of both the blocks.

The characteristics of strike-slip faulting—for these are what Wegener was describing—are now known to be crucial to the understanding of the movement of what are called "suspect" or "displaced terranes." The Baja California block is already on the way to becoming one. [The noun "terrane" spelt in this way has been adopted by geoscientists to mean "an anomalous body of rocks bounded by faults, and with a completely different geological history to adjacent regions."]

SO IT WAS THAT WEGENER'S serendipitous hypothesis made in 1912 sparked the dawn of a new understanding of how the Earth works. His strong intuitive sense caused him to explore his ideas tenaciously and with passionate persistence, often in the face of professional ridicule—for some of his ideas were wrong-headed and their expression ill-advised.

In the years that followed Wegener's death in Greenland in 1930, many startling and far-reaching discoveries were made by those who decided to test his ideas rather than to reject them. As a consequence, it was proved later in the 20th century that the principal premise of Wegener's hypothesis is correct: the present continents were indeed once part of a supercontinent, and did rift and separate into their present-day configurations. Even so, just as the profound implications of the 16th-century Copernican revolution took centuries to be fully understood, it will take much of the 21st century for the full implications of the Wegenarian revolution to become fully clear to us.

7 Displaced Terranes

The very fact that Spitsbergen appeared to have been subtropical while parts of Gondwana had been glaciated at different times in the past was one of Wegener's main arguments for Pangea. The part of Spitsbergen shown here is now about 1,200 km (750 miles) from the North Pole. Even so, there are rich coal deposits in the mountains to the right of the picture, and coal is the product of at least a temperate climate. But the mountains to the far left have no coal and are of a completely different geological structure. At the left center there is a glacier-filled valley that is the line of suture joining these two otherwise unrelated regions. These, because of their geological differences, are termed "displaced" terranes. The word "terrane", as distinct from "terrain", is used by tectonic scientists to mean a fault-bounded body of rock, blocks, and fragments that has a different history from adjacent regions.

Today's Spitsbergen originated off the tip of the north coast of Greenland and remained attached to the Eurasian plate when Greenland separated from Europe about 37 MYA. The diagram at the left shows how the three displaced terranes that make up the island as a whole may originally have been related to each other (and to Greenland) about 400 MYA. The "sliver" with the coal deposits is thought to have originated off Ellesmere Island in the Canadian High Arctic, 1,600 km (1,000 miles) due west of its present position.

Svalbard: present
Kongs-
fjorden
Spitsbergen
0
100 miles
0
100 km
400 MYA
Present

8 The Wegener Fault

Ellesmere Island in the Canadian High Arctic is separated from Greenland (far horizon) by Nares Strait. The narrow strait is over 480 km (300 miles) long and at its narrowest point only 25 km (15 miles) wide. This is the site of an ancient suture between Greenland and North America. It is thought that about 80 MYA Greenland detached, began to slide along the edge of Ellesmere Island and has subsequently moved 320 km (200 miles) or more. This is similar to the way in which western California has been sliding up the coast along the San Andreas Fault. Some scientists are convinced that the present sedimentary rocks on opposite sides of the strait match too closely for there to have been such appreciable movement. Another theory suggests that as Greenland slipped northwards relative to North America, parts of the Canadian High Arctic archipelago were rifting and extending. The theory is that the edge of North America expanded like the bellows of a concertina as Greenland moved northwards. This would account for the apparently small relative movement between Greenland and neighboring Ellesmere Island and the claimed distance of travel.

The map shows the position of the fault today. This "strike-slip" fault relieves the continental stress caused by the lateral interaction of tectonic plates. A similar lateral fracturing occurs in the seafloors as a result of unequal rates of spreading. Stress is relieved by the "transform" faulting of the great number of fracture zones that cut across the zigzagging mid-ocean spreading-centers. These can be seen clearly on the seafloor in the illustration of the Earth that appears on page 38. The Nares Strait boundary is called the Wegener Fault—a fitting memorial to a man with ideas that once bitterly divided the world of geoscience.

Nares Strait from
Ellesmere Island,
Canadian High Arctic

II New Look at an Old Planet

This view of the Earth encapsulates all the geographical features discussed through the book, except that here we see them without sea, snow, or ice. Later, in other maps and globes, we will see them as they are thought to have appeared during innumerable stages of their assembly into today's world. The globe above identifies the main terrestrial features, the ocean-basins, and their "working parts." The globe on the opposite page identifies the mountainous regions of the Western World (using the term for the mountain-building period in which particular ranges formed).

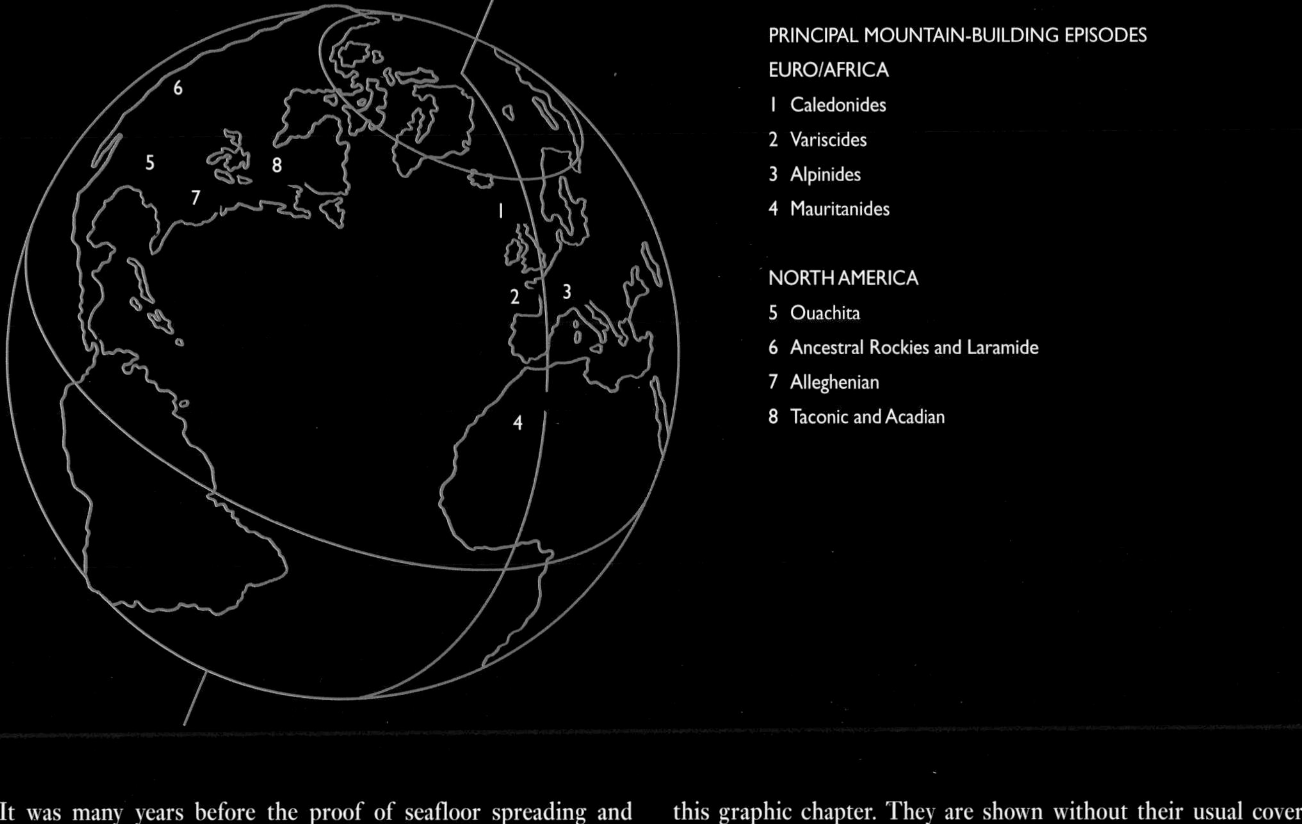

It was many years before the proof of seafloor spreading and relative continental movement was broadly accepted by the geoscience community; the proof was in place by 1965, but the theory of plate tectonics that evolved from these discoveries was not embraced by the majority until the early 1970s. Since that time, the concept of a dynamic Earth implied by the theory has intrigued many researchers, both in related and unrelated disciplines. The principle of a dynamic Earth promised solutions to a multitude of age-old and seemingly unrelated geological problems. It also transformed research in fields as far removed from each other as biogeography (the distribution of life on Earth) and planetary geology (the geology of terrestrial planets).

In the space of just a few decades at the end of the 20th century, the elaboration of this dynamic model culminated in a drive for a new perception of natural science. The groundswell of research resulted in many breakthroughs in the concept of how this marvelous Earth of ours works. The tectonic theory has not provided answers to all the questions asked of it in its own specialized realm. But it has provided new routes for problem solving in the biological and physical sciences of the biosphere, the atmosphere, the cryosphere, and the hydrosphere.

So many new ideas about the Earth and its dynamic processes have emerged, and new scientific disciplines to match them, that one is reminded of Henrik Ibsen's hero Peer Gynt. He tried to discover what lay at the center of an onion, and as he removed one skin after another, he only found yet another skin beneath. So it is with these extraordinary and far-reaching end-of-century discoveries. They make it necessary to summarize the perception of the natural science of the Earth at the start of the third millennium, before we launch into the main storyline of the book.

Although a global view is taken in the account that unfolds in the following pages, its focal point is the North Atlantic and its surrounding continents. The storyline is conjointly that of this region's geological evolution from elemental and widely distributed parts, and biological evolution from primitive organisms. The globe opposite defines the continents and continental margins we will discuss in the text and illustrate in the photo-essays that follow this graphic chapter. They are shown without their usual cover of sea and ice but with related Arctic, Mediterranean, and Caribbean basins.

This chapter opens with a review of the formation of the primordial Earth, followed by illustrations of the anatomy of the Earth and of the dynamic features of its interior and surface. It concludes with a two-phase reconstruction of the development of life on Earth.

The first-phase reconstruction of life's micro-evolution illustrates how life began and how, several billion years later, its course was profoundly affected by a global environmental crisis. This was caused by the overpopulation of the Earth with cyanobacteria. This primitive life form initiated an irreversible change in the composition of the Earth's atmosphere by excessive production of oxygen as a waste product. Oxygen is a gas that was poisonous to most forms of bacteria that had evolved up to that time, including cyanobacteria itself.

The second phase illustrates how the changed composition of the atmosphere led to macro-evolution, the evolution of a completely new form of life. This form depended upon endosymbiosis (Gr. *endo*, within; *sym biosis*, together manner-of-life); symbiosis between organisms within a single cell. Endosymbiosis was the key step that led to the appearance of life forms that had nucleated cells. These include all micro-marine creatures, plants, animals, and fungi. In fact all the living things that we can actually see around us on Earth today evolved during this second phase (macro-evolution). All the living things that we cannot see evolved in the first phase (micro-evolution). Even so, it seems that all organisms that exist today, at all levels and at all times, are interdependent. We ourselves are dependent upon bacteria and they upon us. We live in a perpetual state of symbiosis with all living things and with the physical world about us.

As we move through the chapter we will draw closer to the surface of the Earth until we see the planet in the final illustrations as if from a south polar aspect at about 700 MYA. This is the point in time and the place where this book's preliminaries end and discussion of the evolution of continents, oceans, and life can begin.

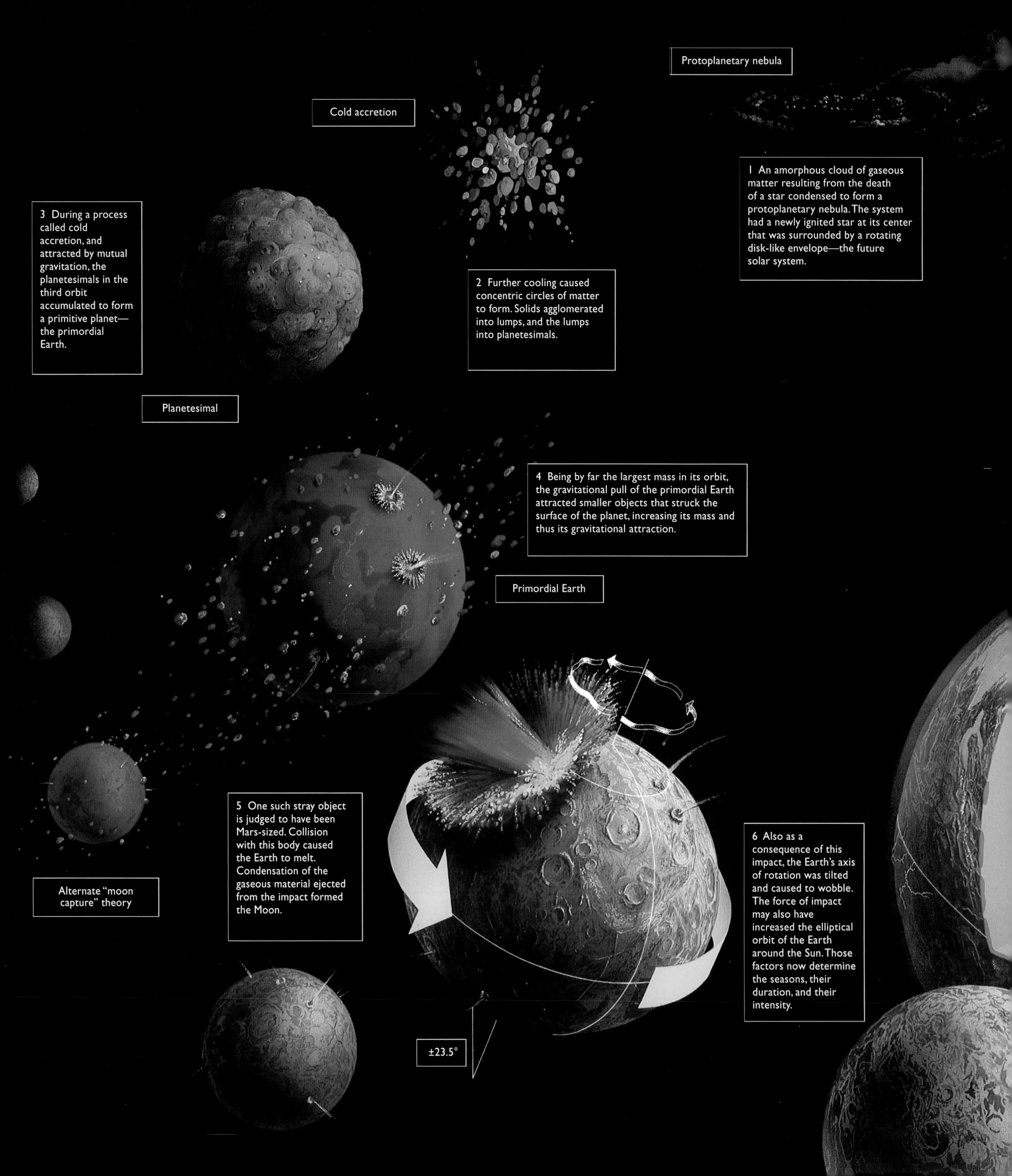

Protoplanetary nebula

Cold accretion

1 An amorphous cloud of gaseous matter resulting from the death of a star condensed to form a protoplanetary nebula. The system had a newly ignited star at its center that was surrounded by a rotating disk-like envelope—the future solar system.

2 Further cooling caused concentric circles of matter to form. Solids agglomerated into lumps, and the lumps into planetesimals.

3 During a process called cold accretion, and attracted by mutual gravitation, the planetesimals in the third orbit accumulated to form a primitive planet—the primordial Earth.

Planetesimal

4 Being by far the largest mass in its orbit, the gravitational pull of the primordial Earth attracted smaller objects that struck the surface of the planet, increasing its mass and thus its gravitational attraction.

Primordial Earth

5 One such stray object is judged to have been Mars-sized. Collision with this body caused the Earth to melt. Condensation of the gaseous material ejected from the impact formed the Moon.

Alternate "moon capture" theory

6 Also as a consequence of this impact, the Earth's axis of rotation was tilted and caused to wobble. The force of impact may also have increased the elliptical orbit of the Earth around the Sun. Those factors now determine the seasons, their duration, and their intensity.

±23.5°

9 As the surface cooled the Earth's lithosphere began to form and the first continents appeared. Water vapor condensed to form seas and oceans, while fractionation of the interior led to the formation of a solid inner core.

8 As the Earth cooled from its liquid state its elemental parts began to separate by fractionation—the heaviest tended to gravitate to the core and the lighter elements to remain nearer to the surface.

7 By 3.8 BYA major bombardment of the Earth and its moon had ended and the surface had begun to cool

The solar system formed from a nebula, the remnants of a dyn star that collapsed and exploded over five billion years ago. H gases from the blast formed into rotating disk-like structures, eac a potential new solar system. One of these "protoplaneta nebulas" had a newly ignited star at its center—the Sun. As parti ulates condensed, they agglomerated to form unstructured chun of matter orbiting the Sun. Collisions between these chun tended to force them into circular rings. Some of the bigger, mo gravitationally attractive objects in the third ring from the Su eventually amalgamated into "planetesimals." By continued ama gamation, one particularly large planetesimal (the forming Eart gained sufficient mass to attract another planet-sized but mu smaller planetesimal. It is thought that the two bodies collided wi such force that the collision had many extraordinary results, wi major consequences for setting the conditions for life on Earth.

The collision hypothesis explains the speeds of rotation and re ative movements of Earth and Moon. Another crucial result was 10 percent increase in the size of the early Earth, and this was fu ther increased by vast numbers of lesser impacts by other objec The sum total of all these collisions largely determined the Earth present mass, and thus its gravity and ability to retain an atm sphere of sufficient density and pressure for water to condense ar for terrestrial life to evolve.

Another result was the ejection of a filament of gaseous mater from the Earth's surface. Some gaseous material was lost to spa but the remainder formed an orbital ring around the Earth. As th condensed, the condensates formed fragments that amalgamat into "moonlets" that combined to form the Moon. The impact also believed to have altered the angle of the Earth's axis from ve tical relative to the Sun, to a varying obliquity on a wobbling ax The sum total of these effects, together with the Earth's eccentr orbit around the Sun, varying from elliptical to near-circular a back to elliptical every 90,000 years or so, determined both the se sonal nature of the Earth's climate and long-term climatic cyc (called Milankovitch cycles).

Within moments of impact both colliding bodies were heated melting point by the energy released in their collision. The Eart sphere "fractionated" as it cooled from a semi-fluid to a solid sta lighter semi-fluid materials rose to the sphere's surface and heav materials gravitated towards the center, forming the Earth's prim tive core. At the same time, the spin of the now semi-fluid Ear caused the planet to become oblate—flatter at the Poles and d tended at the Equator. The Earth was also devolatilized by t impact; its content of water vapour, nitrogen, carbon dioxide, a other elementary volatiles was depleted. These were later restor by a bombardment of meteorites and water-ice comets pepper with cosmic dust. Indeed, swarms of comets from the theoreti Oort cloud, that is believed to envelop the solar system at a d tance of about 95 *billion* miles, produced an "ice flux" that was s ficient to provide the Earth's primitive hydrosphere, literally forms of water on the primordial Earth—marine, terrestrial, a atmospheric. The Moon suffered a similar fate but its mass w insufficient for it to retain volatiles in the form of an atmosphere a hydrosphere—though it is thought that some water may ha been retained at the Poles.

The Earth and the Moon, now much the largest objects in third orbit from the Sun, had more gravitational pull than indiv ual planetesimals, meteors, and meteorites in adjacent orbits. I sweeping-up operation together they attracted "debris" wh bombarded both until about 3.8 billion years ago—the time t major cratering ended and life on Earth began.

Structure

ne Earth is indeed structured like Peer Gynt's onion, for it is
ade up of a series of spheres one inside the other. Here we deal
th the onion's form and dimensions. How the spheres are
ought to interact with each other is discussed in following essays.
The lightest elements form the Earth's very thin outer covering
its atmosphere and hydrosphere. Air and water can only retain
atively minute quantities of heat, and for this reason they form
e coldest as well as the most tenuous of the Earth's concentric
heres. By contrast, the innermost sphere of the Earth, its core, is
e densest region of the Earth and is a heat source. It generates
d releases enormous amounts of energy to the mantle.
The mobile upper region of the Earth, the tectosphere, is about
0 km (250 miles) thick, and is the Earth's interactive outer
sing. It consists of a rigid outer shell, the lithosphere (Gr. *lithos*,
k), which rests on a hot plastic region of the Earth's upper
antle called the asthenosphere (Gr. *asthenes*, weak).
The oceanic and continental lithospheres both vary in thickness.
ne oceanic lithosphere is thinnest at spreading-centers where the
afloor is created. Generally the ocean crust, a thin layer of the
hosphere, averages between 5 and 10 km (3–6 miles) in thickness.
en at its most distant from a spreading-center the global ocean
ust is never more than 100 km (60 miles) thick, nor more than
out 170 MY old.

Continents are believed to have formed like thin sheets of slag
the surface of the Earth before 4,000 MYA. For billions of years
ese sheets collided and amalgamated with each other, separated,
d rejoined. During this process accumulated volcanic rocks were
ded. The products of erosion formed sediments which became
imentary rock. Meanwhile, congealing ocean floors were con-
ntly subducted (L. *sub ducere*, to lead under) and destroyed
neath continents or other ocean floors.
Because of their age and structural evolution the continents vary
thickness according to the topography of their mountains,
ins, and basins, and have correspondingly deep or shallow pro-
s beneath the surface. The continental lithosphere therefore
ges between 40 and 200 km (25–125 miles) in thickness, but its
face at its highest point (Mount Everest: 8,848 m; 29,028 feet) is
y 19.88 km (12.35 miles) above the lowest point on the ocean
or (Mariana Trench: -11,034 m; -36,201 feet).
t is a popular misconception that the Earth's surface "floats" on
ed-hot liquid: this is not so. Liquefaction (and sometimes
canic eruption) only happens when discrete regions in an
remely hot *semiplastic* region of the upper mantle are decom-
ssed. The malleability of this upper mantle, the "astheno-
ere," accommodates the varying thicknesses of overlying
anic and continental lithosphere. The lower mantle is solid and
ut 2,400 km (1,500 miles) thick. Its upper 650 km (400 miles) is
sidered to be a transition region between "solid" mantle rocks
the "plastic" asthenosphere.

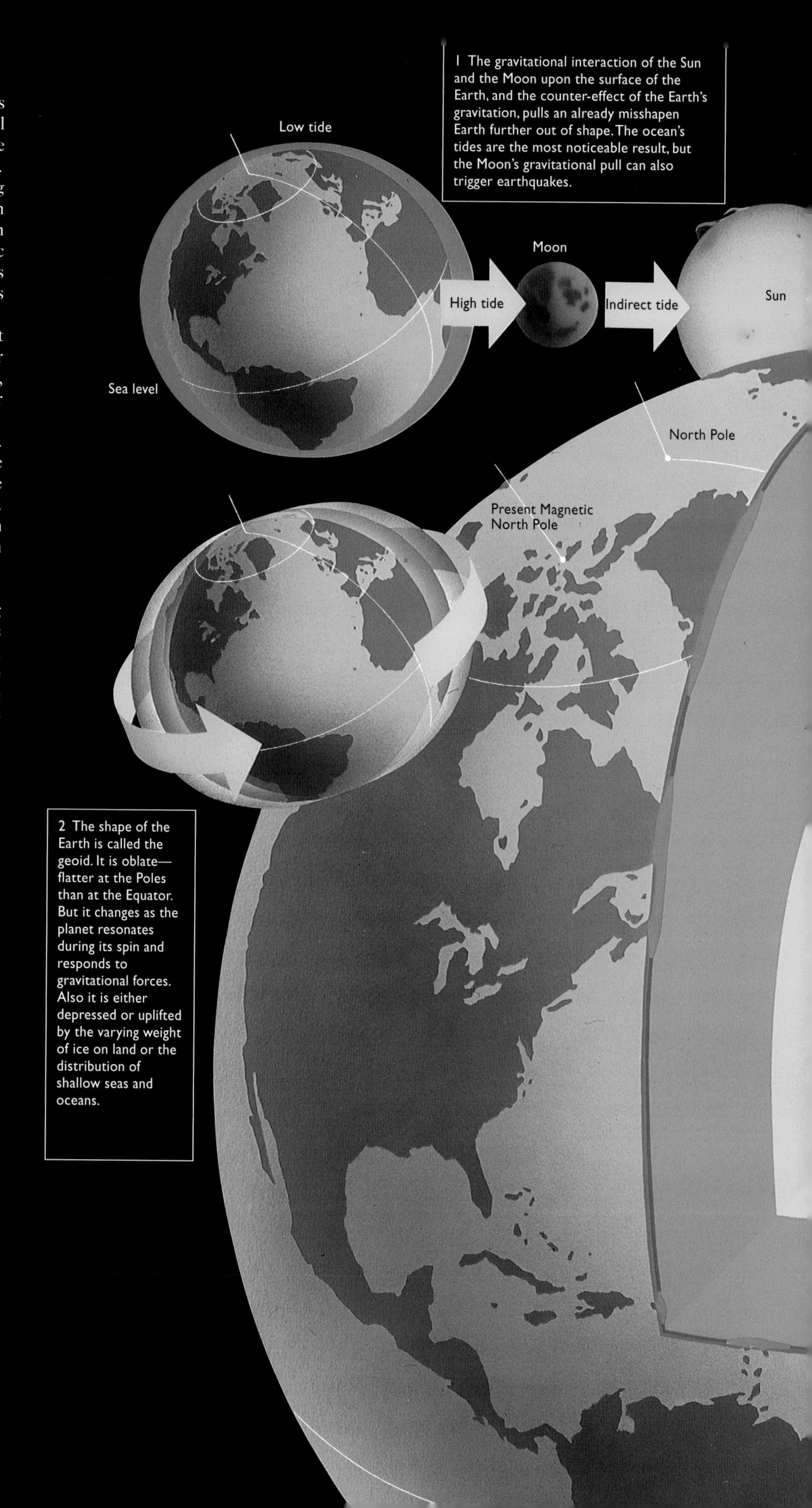

1 The gravitational interaction of the Sun and the Moon upon the surface of the Earth, and the counter-effect of the Earth's gravitation, pulls an already misshapen Earth further out of shape. The ocean's tides are the most noticeable result, but the Moon's gravitational pull can also trigger earthquakes.

2 The shape of the Earth is called the geoid. It is oblate—flatter at the Poles than at the Equator. But it changes as the planet resonates during its spin and responds to gravitational forces. Also it is either depressed or uplifted by the varying weight of ice on land or the distribution of shallow seas and oceans.

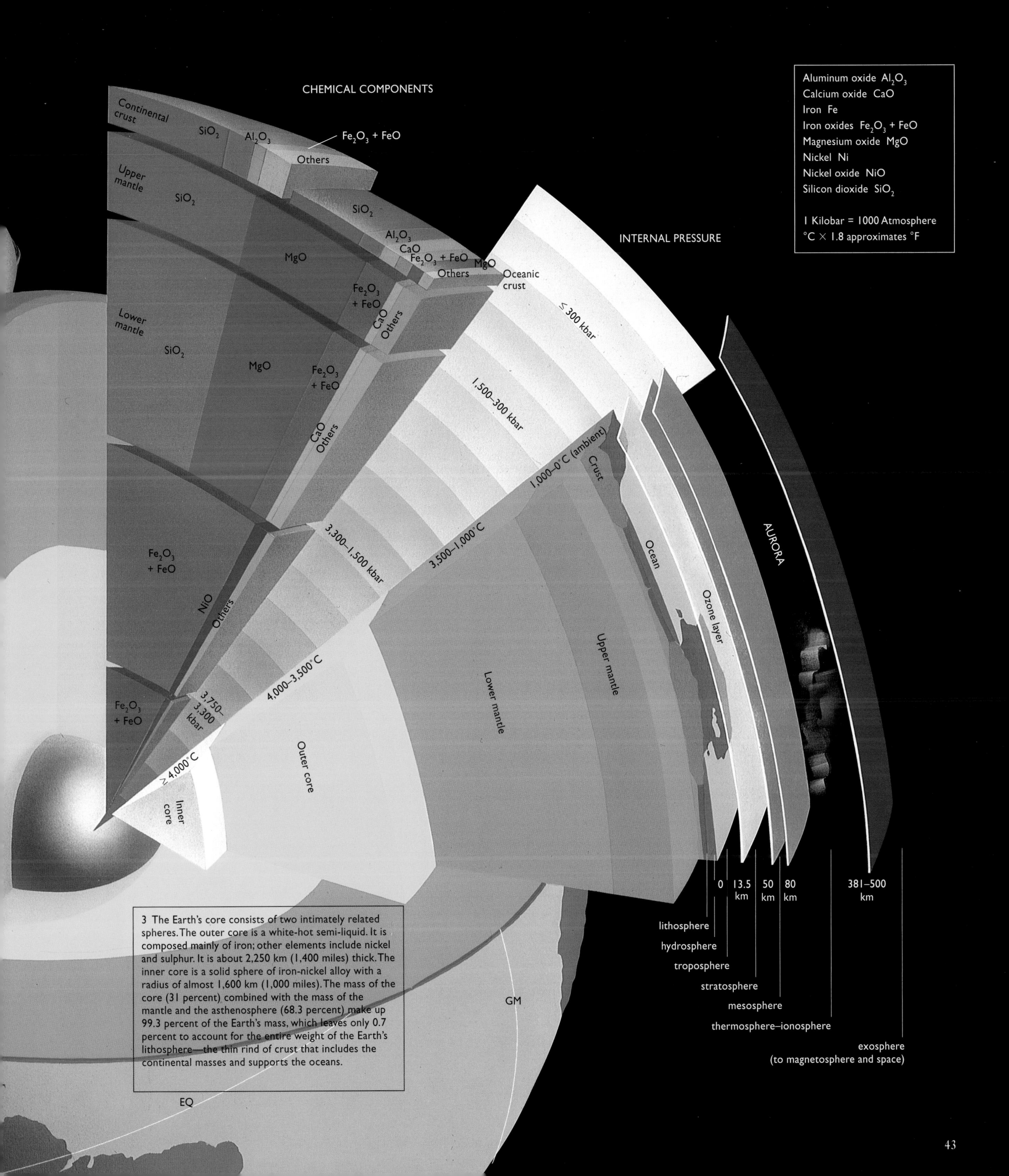

3 The Earth's core consists of two intimately related spheres. The outer core is a white-hot semi-liquid. It is composed mainly of iron; other elements include nickel and sulphur. It is about 2,250 km (1,400 miles) thick. The inner core is a solid sphere of iron-nickel alloy with a radius of almost 1,600 km (1,000 miles). The mass of the core (31 percent) combined with the mass of the mantle and the asthenosphere (68.3 percent) make up 99.3 percent of the Earth's mass, which leaves only 0.7 percent to account for the entire weight of the Earth's lithosphere—the thin rind of crust that includes the continental masses and supports the oceans.

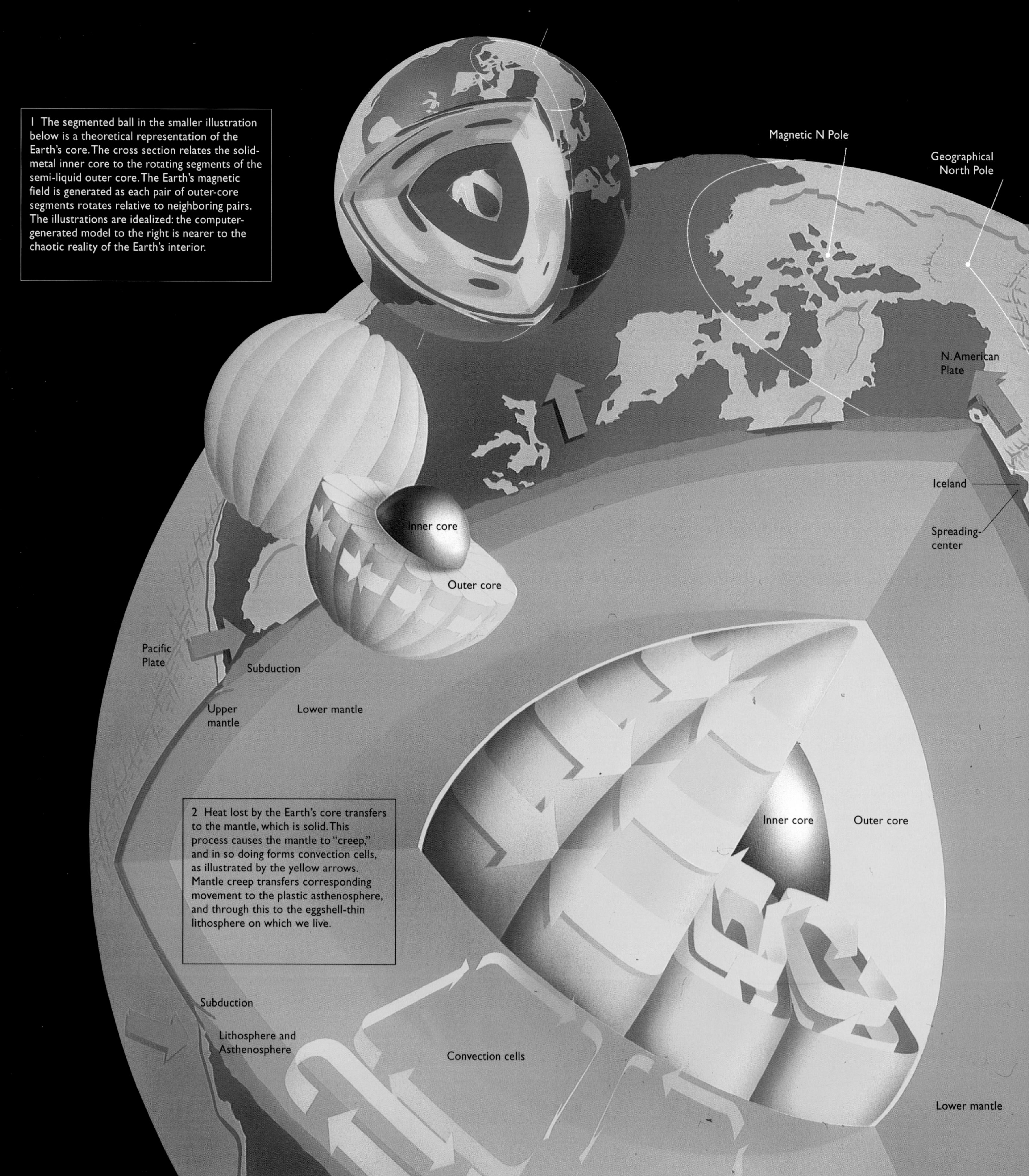

1 The segmented ball in the smaller illustration below is a theoretical representation of the Earth's core. The cross section relates the solid-metal inner core to the rotating segments of the semi-liquid outer core. The Earth's magnetic field is generated as each pair of outer-core segments rotates relative to neighboring pairs. The illustrations are idealized: the computer-generated model to the right is nearer to the chaotic reality of the Earth's interior.
Magnetic N Pole
Geographical North Pole
N. American Plate
Iceland
Spreading-center
Inner core
Outer core
Pacific Plate
Subduction
Upper mantle
Lower mantle
2 Heat lost by the Earth's core transfers to the mantle, which is solid. This process causes the mantle to "creep," and in so doing forms convection cells, as illustrated by the yellow arrows. Mantle creep transfers corresponding movement to the plastic asthenosphere, and through this to the eggshell-thin lithosphere on which we live.
Inner core
Outer core
Subduction
Lithosphere and Asthenosphere
Convection cells
Lower mantle

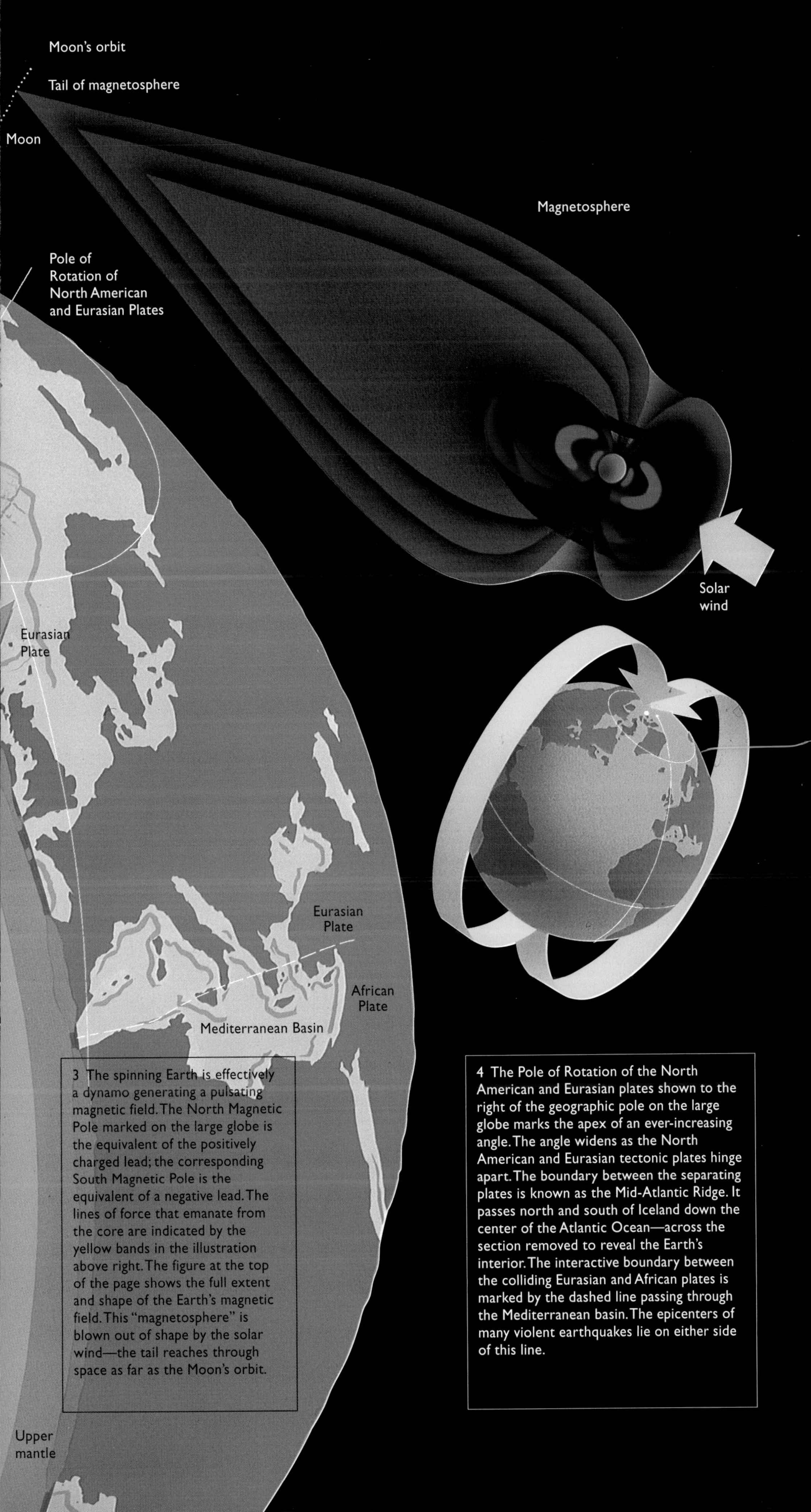

3 Interior

If one physical property above all others dictates how the Earth works, it is that gases, liquids, and solids contract as they cool and expand when they are heated. The notable exception to this edict is freezing water, which expands as it freezes. Otherwise, from the Earth's atmosphere to the surface of its solid inner core, materials change in relative density according to their temperature. As a consequence, the coolest regions of a liquid or plastic material tend to sink, and warmer regions of the same mass tend to rise. This physical response to heat lost or gained causes movement throughout the mass, the motion of convection.

It is difficult to visualize this Hadean world in terms of the pressures and temperatures at the Earth's surface. For instance, if one could suddenly remove a very large area of the Earth's lithosphere several miles in diameter, there would be a monumental explosion of the plastic asthenosphere beneath as it decompressed. But the pressure in the Earth's inner core is estimated to be twelve times that of the asthenosphere, and about three million times that of the atmosphere at the surface. Penetration to the core might result in the destruction of the planet—like pricking a balloon with a pin.

According to seismic reflection images, the Earth's inner core is a solid metal sphere. This is enveloped by the 2,250 km (1,400 miles) of turbulently convecting outer core—a mix of molten iron, nickel, sulphur, and other elements. Although the outer core is thought to be kept at white heat by the energy generated by gravitation, this is insufficient to prevent overall cooling. Consequently, some of the outer core's metallic content (iron and nickel) forms crystals that gravitate and settle as a mush on the surface of the solid inner core.

The "ceiling" above the liquid outer core is formed by the underside of the solid enveloping mantle. This has inverted mountain ranges etched on its surface—possibly scoured by the force of the convective motion of the liquid material in the outer core.

The pattern of the convection in the outer core has long been the subject of scientific conjecture. An experiment was conducted aboard a spacecraft in orbit, which showed that when a rotating sphere of hot, thick, liquid material, analogous to the outer core, was allowed to cool in weightless conditions, it formed counter-rotating pairs of banana-shaped cells. These cells clustered around the sphere's axis of rotation.

This experiment added support to theoretical models of the outer core: some scientists are modeling the outer core as a structure of at least nine such pairs of cells surrounding the inner core, as illustrated here. Since the cells contain a high proportion of iron it is also believed that interaction between the pairs of "banana" cells produce the Earth's magnetic field. The cells are estimated to be rotating at a speed of four to five inches per day.

The rotating convection cells in the outer core transfer heat to the underside of the surrounding rocky mantle. This causes a special form of convection within the 1,850 km (1,150 miles) of solid rock between the lower mantle and the underside of the upper mantle. This is known as "solid-state creep"—a movement of a few inches per *year*. The mantle's convection cells are broadly polygonal in cross section, and are thought to be structured like gigantic honeycombs. But it is most important to emphasize, particularly in view of the necessarily specific cutaway illustration, that terms like "banana-shaped," "honeycomb," and "convection cells" are of course idealized expressions only; the reality of the Earth's interior will at best be a grossly deformed version of these models—as shown in the separate computer-model image.

Moon's orbit

Tail of magnetosphere

Moon

Magnetosphere

Pole of Rotation of North American and Eurasian Plates

Solar wind

Eurasian Plate

Eurasian Plate

African Plate

Mediterranean Basin

Upper mantle

3 The spinning Earth is effectively a dynamo generating a pulsating magnetic field. The North Magnetic Pole marked on the large globe is the equivalent of the positively charged lead; the corresponding South Magnetic Pole is the equivalent of a negative lead. The lines of force that emanate from the core are indicated by the yellow bands in the illustration above right. The figure at the top of the page shows the full extent and shape of the Earth's magnetic field. This "magnetosphere" is blown out of shape by the solar wind—the tail reaches through space as far as the Moon's orbit.

4 The Pole of Rotation of the North American and Eurasian plates shown to the right of the geographic pole on the large globe marks the apex of an ever-increasing angle. The angle widens as the North American and Eurasian tectonic plates hinge apart. The boundary between the separating plates is known as the Mid-Atlantic Ridge. It passes north and south of Iceland down the center of the Atlantic Ocean—across the section removed to reveal the Earth's interior. The interactive boundary between the colliding Eurasian and African plates is marked by the dashed line passing through the Mediterranean basin. The epicenters of many violent earthquakes lie on either side of this line.

4 Tectosphere

As the immense interior pressures within the Earth decrease near the Earth's surface, the upper regions of the mantle, the transition zone, and the overlying upper crust become more plastic and resilient. They are therefore more responsive to the driving force of convection. This upper region is called the "tectosphere." It has two dynamic components: a plastic "asthenosphere," and a rigid surface, the "lithosphere," which is made up of seven vast tectonic "plates" and a large number of lesser plates—as illustrated by the small globe at bottom right.

The asthenosphere is thought to be not less than 100 km (60 miles) thick, and acts as a cushion between the mantle's transition zone and the lithosphere—the ocean floors and continents. Thus the energy of the slow but inexorable "solid-state creep" in the mantle is transferred via convection cells through the cushion of plastic rock to the Earth's rigid surface.

Although movement of the tectonic plates is now proven, the mechanism that drives them is not yet clearly known—but there are several theories. One is that the weight of subducted ocean crust descending into the asthenosphere beneath continents, or beneath the seafloor at opposing plate boundaries, pulls upon itself. The subducting seafloor acts like a tablecloth slipping off a polished table—as the cloth begins to slip it rapidly gathers momentum from its own weight and falls to the floor. Another theory suggests a role for the general down-hill thickening of ocean lithosphere as it cools and ages away from spreading-centers. This is thought to generate a downslope force like that of snow sliding down a hillside.

Whether or not these or other models are correct, it is generally believed that the lithosphere does participate in the convection process, and that it also responds to convection in the asthenosphere below. The result of all these forces acting in concert is that the rigid lithosphere fractures into numerous crustal plates that are endlessly forced to interact with each other.

However, only a limited range of interactions are possible between moving plates on a sphere—five of which are illustrated at far right. They can separate, or collide, or shear by "transform" faulting (slide past each other), or subduct (slide beneath each other). But the *consequences* of the limited range of plate interactions are infinite. For example, continent–ocean boundaries on separating plates have passive margins free from volcanic activity (like those of the North Atlantic). Such boundaries on colliding plates have active margins with volcanoes and violent earthquakes (like those in the Mediterranean and Aegean Seas). Plate movement can lead to the assembly of megacontinents (as Africa and Eurasia have almost become), and to the breakup of supercontinents (the earlier separation of Eurasia and Africa from North America and South America). Interactive plate movements have also had a profound effect upon the Earth's climate and its biogeography, and consequently also upon life's diversity and evolution.

The four globes illustrated here demonstrate some of the first principles of plate tectonics. The top pair show that because the plates are parts of a spherical surface, they can only move within the constraints that are inherent to the surface of a sphere. They must necessarily rotate away from each other at an ever-increasing angle from a common pole of rotation. Thus continents cannot remain parallel as they separate along their pivotal axis of rotation.

The bottom pair of globes show that the lithosphere has to compensate for the stress of such movement by faulting at right angles to the axis of the movement—called transform faulting. The bottom-left globe also illustrates the symmetrical magnetic striping of the seafloor as it moves away from either side of a spreading-center. Individual segments within a spreading-center appear to be out of line, but projections from each segment would meet at a common pole of rotation.

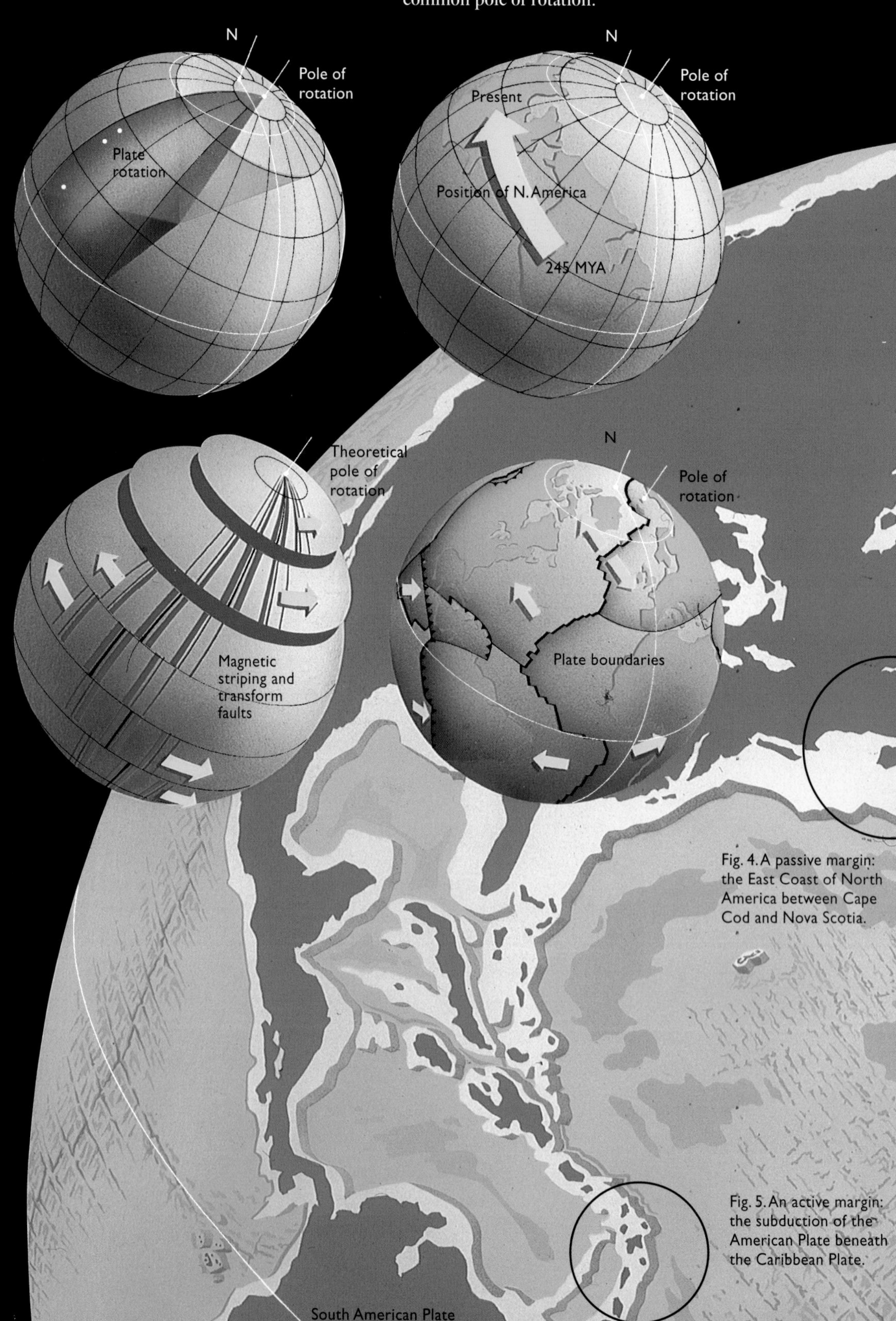

Fig. 4. A passive margin: the East Coast of North America between Cape Cod and Nova Scotia.

Fig. 5. An active margin: the subduction of the American Plate beneath the Caribbean Plate.

The five diagrams to the right represent hypothetical "cores" removed from the lithosphere at the points circled on the large globe. They illustrate a few of the consequences of plate movements in the Atlantic realm, but they exemplify the type of interaction that takes place over the whole surface of the Earth. Thus some oceans are widening, in this case the North Atlantic (Figure 1: a detail of the Mid-Atlantic Ridge south of Iceland). Some plates are shearing past one another, like the North American, Eurasian, and African plates at their junction in the region of the Azores (Figure 2). Some oceans are closing, causing continents to collide—as in the Mediterranean (Figure 3), where Africa meets Eurasia . Some ocean margins are inactive—passive margins, illustrated here by part of the East Coast of North America (Figure 4). And some plates are interacting violently as they descend beneath other plates, in this case the North American plate under the Caribbean plate, resulting in the Lesser Antilles island-arc—an active margin (Figure 5).

Thus some ocean floors are destroyed at plate boundaries when they are subducted beneath converging continents (as around the Mediterranean). They can also be subducted beneath other seafloors (like the North Atlantic seafloor beneath the Caribbean seafloor). Meanwhile, new ocean floor is being created at mid-ocean spreading-centers, such as the Mid-Atlantic Ridge. The system is in perpetual motion. It destroys and replaces itself continuously.

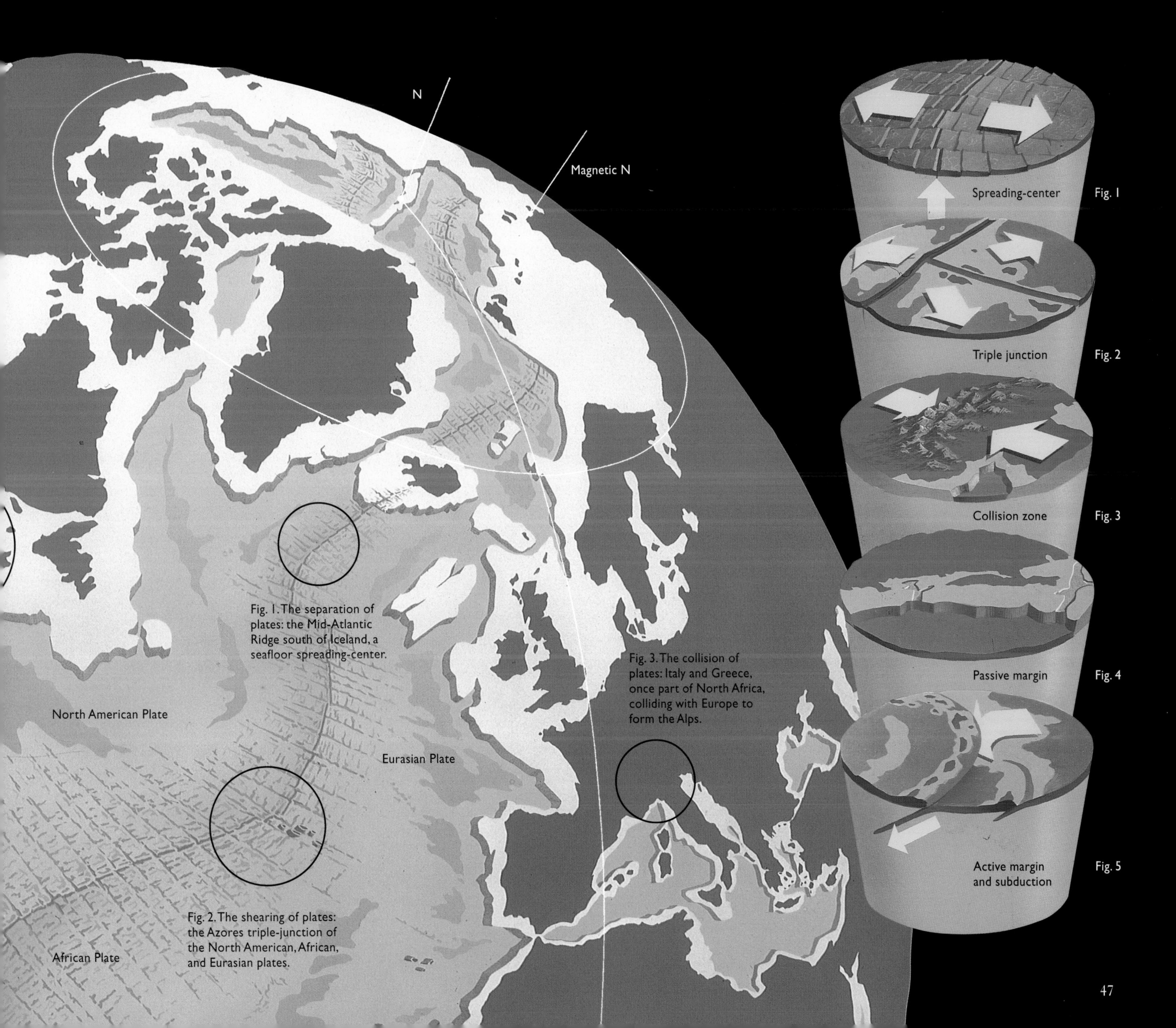

Fig. 1. The separation of plates: the Mid-Atlantic Ridge south of Iceland, a seafloor spreading-center.

Fig. 2. The shearing of plates: the Azores triple-junction of the North American, African, and Eurasian plates.

Fig. 3. The collision of plates: Italy and Greece, once part of North Africa, colliding with Europe to form the Alps.

5 Life Begins

The hypothesis presented in the essay "Formation" [pages 40–41] describes the impact of a Mars-sized body with the Earth. In this event, the primitive Earth was deprived of essential volatiles like H_2O, N_2, and CO_2. The impact hypothesists suggest that these, and other elements and compounds lost by vaporization, were later reintroduced to the Earth's lithosphere. One theory is that this happened through a phenomenal bombardment of icy comets originating from the Oort cloud (a huge reservoir of comets believed to envelope the solar system at a considerable distance).

The Earth's semi-molten surface continued naturally to cool (from left to right of the diagram depicting the early Earth's surface). Meanwhile, icy comets and asteroids were pounding the Earth. However, water vapor produced from comet-ice caused a cooling effect. The vapor condensed at high altitudes as it rained down through the atmosphere. Gradually the altitude at which the water re-vaporized fell as the Earth's surface cooled. When the surface had finally cooled sufficiently to permit condensation at ground level, the Earth's "hydrosphere" began to form. This process took the form of a prolonged and uninterrupted downpour of near-boiling acid rain, a downpour that may have lasted millions of years. The water-cooling towers of power stations operate in an analogous way today.

The acid rain eroded and interacted with the Earth's volcanic rocks, resulting in the precipitation of clay-like sediments and metallic salts. The downpour was accompanied by violent thunder and lightning. The energy released by electrical discharge, together with intense ultraviolet radiation from the still-youthful Sun, transformed inorganic compounds into organic ones. Organic compounds are those with carbon, hydrogen, and oxygen atoms as the key to their molecular structure. These compounds are thought by some scientists to have been supplemented by organic compounds transported from outer space by impacting bolides.

Pools of warm water grew to form lakes, then seas, and ultimately oceans. By four billion years ago oceans lapped the shores of primitive continents and flooded their tidal estuaries. As clay and other minerals were stripped from the exposed parts of the Earth's surface they formed sediments in lakes, in estuaries, and on ocean basin floors. These sediments were rich in "polymers" (Gr. *poly meros*, many parts)—large, simple, but often highly reactive carbon-based molecules that spontaneously paired and bonded to form a variety of complex compounds—nucleic acids, proteins, and other building blocks of life (upper band: group 1).

The oldest known terrestrial fossils that are accepted to be those of living organisms are microscopic. They are found in rocks in Swaziland, South Africa and at Warrawooma, Western Australia. They vary in age from 3.6 to 3.3 billion years. Very early organisms included types of fermenting bacteria and autotrophic bacteria capable of obtaining energy for metabolism from existing inorganic compounds such as methane, and compounds of sulphur and nitrogen (upper band: group 2). These single-celled "non-nucleated" organisms are termed "prokaryotes," (Gr. *pro*, before; *karyon*, kernel, i.e., nucleus), and their cells reproduced by budding or by simply dividing in two. The prokaryote cells illustrated here represent modern bacteria thought to be typical of their ancestors—they include prokaryotic organisms called "cyanobacteria."

By 3.0 BYA cyanobacteria colonies—microbial mats called "stromatolites" (L. *stroma*, mattress)—were abundant on tidal flats near seashores. They converted carbon dioxide into carbohydrates using sunlight—the process of photosynthesis. Free oxygen was produced as a waste product.

Oxygen (O_2) is a hyperactive gas that readily combines with other elements or compounds to form oxides. Ultimately, all elemental substances that were exposed on the Earth's primordial surface were oxidized (iron, for example) by the "waste" oxygen produced by photosynthesis. When all exposed materials that were vulnerable to oxidation had indeed been oxidized, excess oxygen, called free oxygen, began to accumulate in volume. Because oxygen is much lighter than carbon dioxide it tended to rise to the top of the atmospheric mix. In the stratosphere 15–30 km (9–18 miles) above the surface, some free oxygen combined to form ozone (O_3), a "super-oxygen," by the action of sunlight. An ozone layer now began to accumulate in the stratosphere that partially reflected the Sun's powerful ultraviolet radiation back into space.

As protection from ultraviolet light increased, stromatolite populations increased their rate of reproduction, and by 2.5 BYA stromatolite colonies dominated the "biomass." (The biomass is the total weight of living things that the Earth's resources can support at any one time.) As indicated by the changing color of the atmosphere in the diagram, by 2.0 BYA the Earth's nitrogen-based atmosphere contained a higher proportion of oxygen than it did of carbon dioxide.

Because oxygen was poisonous to most prokaryotes, both the rate and the direction of life's evolution now began an extraordinary change. Prokaryotes called "aerobes" (oxygen-dependent) evolved—but many bacteria did not make this adaptation and remained "anaerobic" (upper band: group 3). By 1.5 BYA some of the latter had responded to the problem of survival in an increasingly hostile oxygen-based environment by entering into symbiotic relationships at cellular level with aerobic bacteria (cross-link I). Such relationships are known as "endosymbiotic" (Gr. *endo*, within; *sym biosis*, together manner-of-life). These cellular communities could combine attributes contributed by individual prokaryotes—for example, the ability to propel themselves (motility), to utilize oxygen, or to photosynthesize. This cellular community of individuals contributed to the makeup of a

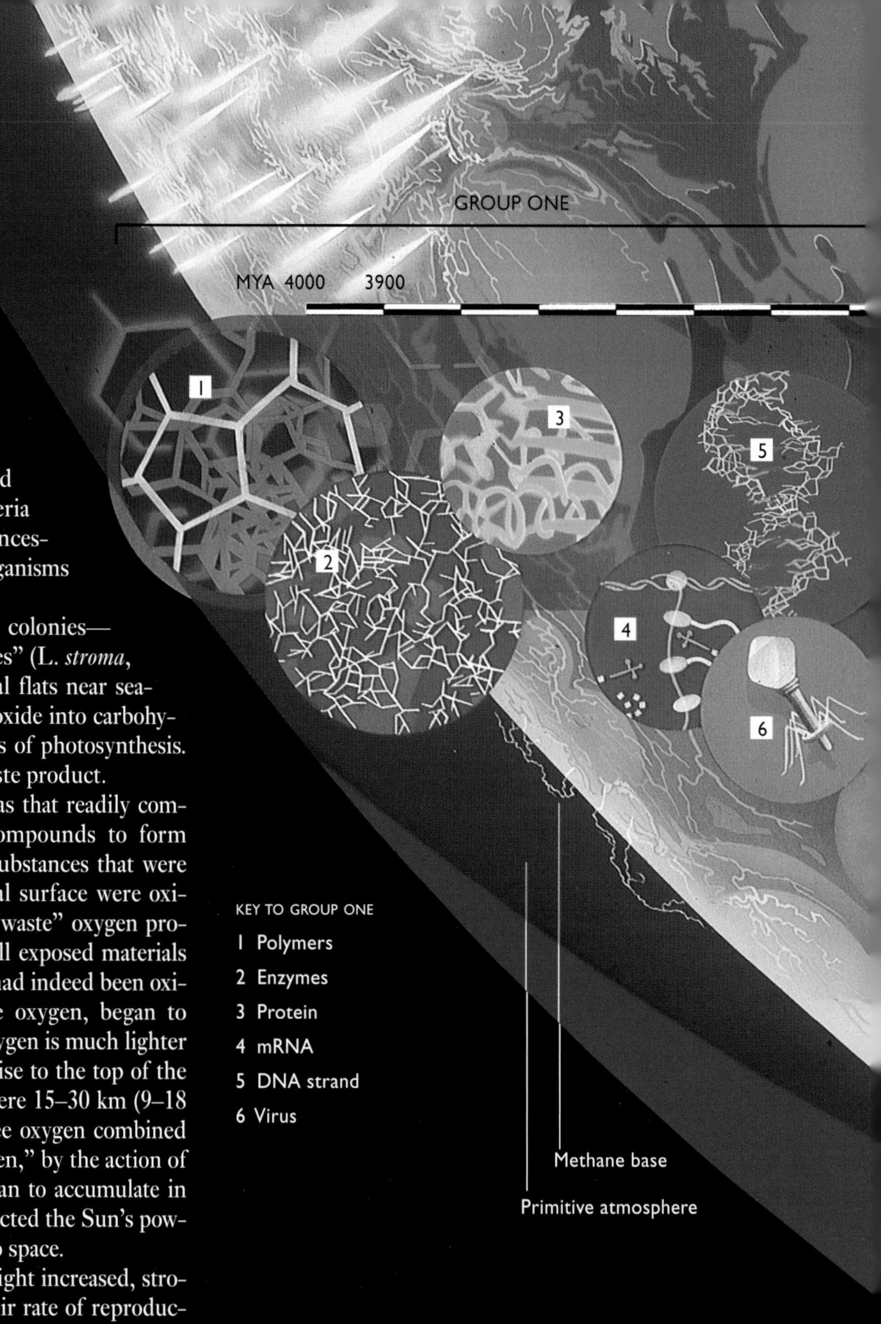

GROUP ONE

MYA 4000 3900

KEY TO GROUP ONE

1 Polymers
2 Enzymes
3 Protein
4 mRNA
5 DNA strand
6 Virus

Methane base

Primitive atmosphere

nucleus, a genetic center. Such nucleated cells proved to be the keys to the evolution of more complex organisms called "eukaryotes" (Gr. *eu karyon*, true kernel).

But the question cellular biologists are asking today is how the individual components came together. Take for instance the molecular "zip-fasteners" that transmit genetic information. These evolved from polymers and enzymes about 3.8 billion years ago (upper band: group 1). But which came first: the genetic "zip" of DNA (deoxyribonucleic acid), or the extraordinary "fastener," messenger RNA (ribonucleic acid), which ensures that the "teeth" of DNA match and interlock? Or did the zip and its fastener develop together as an integrated whole? And how did cell membranes develop from the microspheres that amino acids form when wet? And possibly the most important question of all—how did the whole genetic apparatus get into this prebiotic package to become the first *living* cell?

The morphologically simple but chemically complex prokaryotes in the upper band of the diagram on this and the next pages had no nucleus. They reproduced when two intertwined strands of DNA opened and duplicated themselves—a process called binary fission. Prokaryotes could build colonies such as the

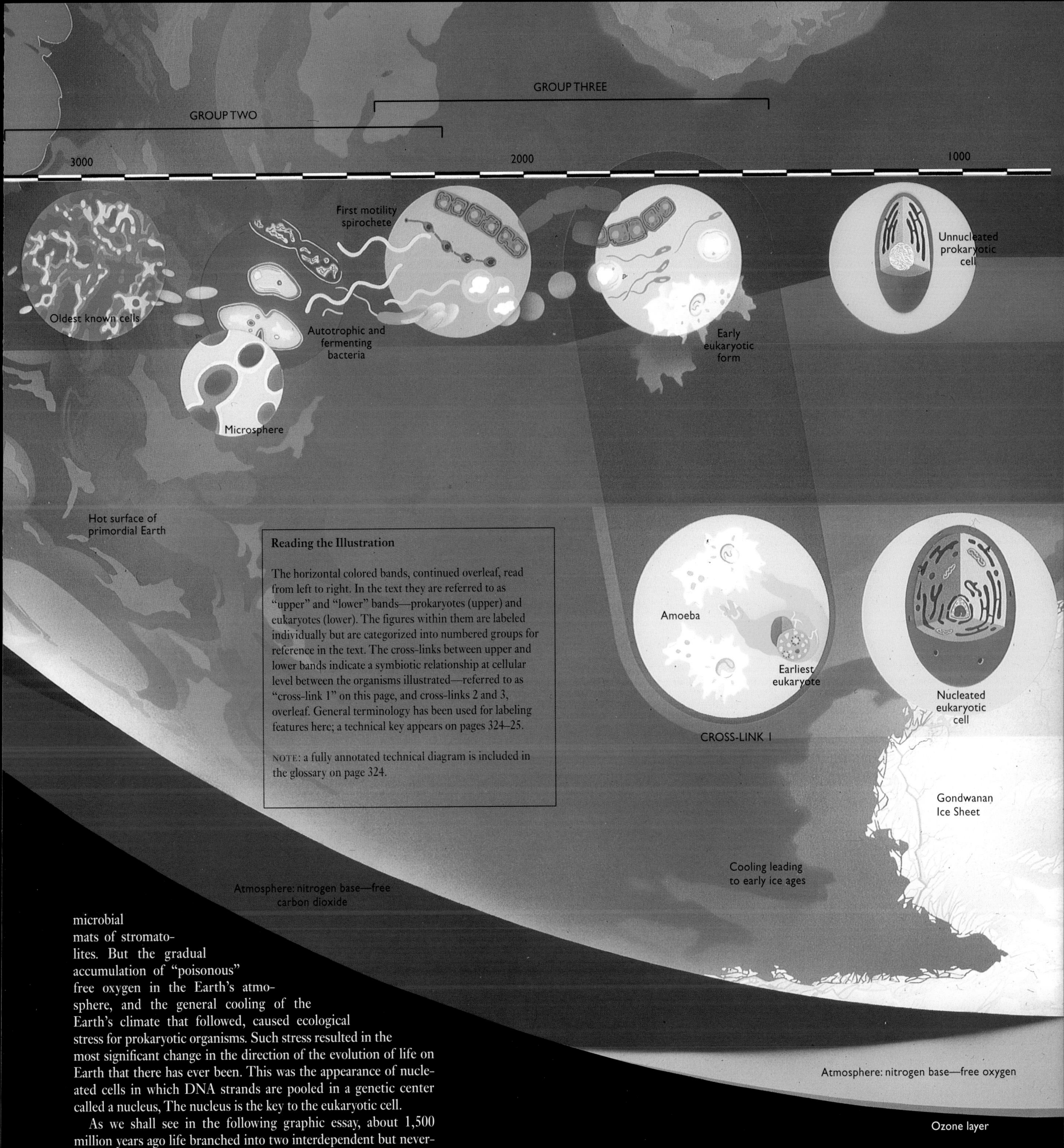

GROUP THREE

GROUP TWO

3000 2000 1000

Oldest known cells

Microsphere

First motility spirochete

Autotrophic and fermenting bacteria

Early eukaryotic form

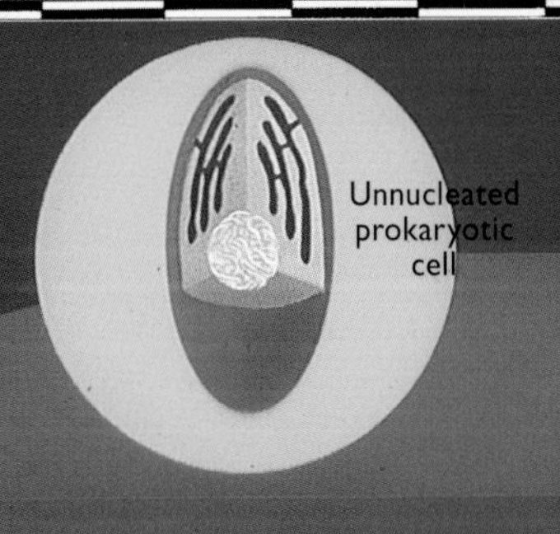

Hot surface of primordial Earth

Reading the Illustration

The horizontal colored bands, continued overleaf, read from left to right. In the text they are referred to as "upper" and "lower" bands—prokaryotes (upper) and eukaryotes (lower). The figures within them are labeled individually but are categorized into numbered groups for reference in the text. The cross-links between upper and lower bands indicate a symbiotic relationship at cellular level between the organisms illustrated—referred to as "cross-link 1" on this page, and cross-links 2 and 3, overleaf. General terminology has been used for labeling features here; a technical key appears on pages 324–25.

NOTE: a fully annotated technical diagram is included in the glossary on page 324.

Amoeba

Earliest eukaryote

CROSS-LINK 1

Nucleated eukaryotic cell

Gondwanan Ice Sheet

Cooling leading to early ice ages

Atmosphere: nitrogen base—free carbon dioxide

microbial mats of stromatolites. But the gradual accumulation of "poisonous" free oxygen in the Earth's atmosphere, and the general cooling of the Earth's climate that followed, caused ecological stress for prokaryotic organisms. Such stress resulted in the most significant change in the direction of the evolution of life on Earth that there has ever been. This was the appearance of nucleated cells in which DNA strands are pooled in a genetic center called a nucleus, The nucleus is the key to the eukaryotic cell.

As we shall see in the following graphic essay, about 1,500 million years ago life branched into two interdependent but nevertheless distinct pathways that transformed the evolution of life on Earth.

Atmosphere: nitrogen base—free oxygen

Ozone layer

NOTE: the diagram continued over pages 48–49 and 50–51 is a south polar view, transitional from the earliest Earth to recent geological time

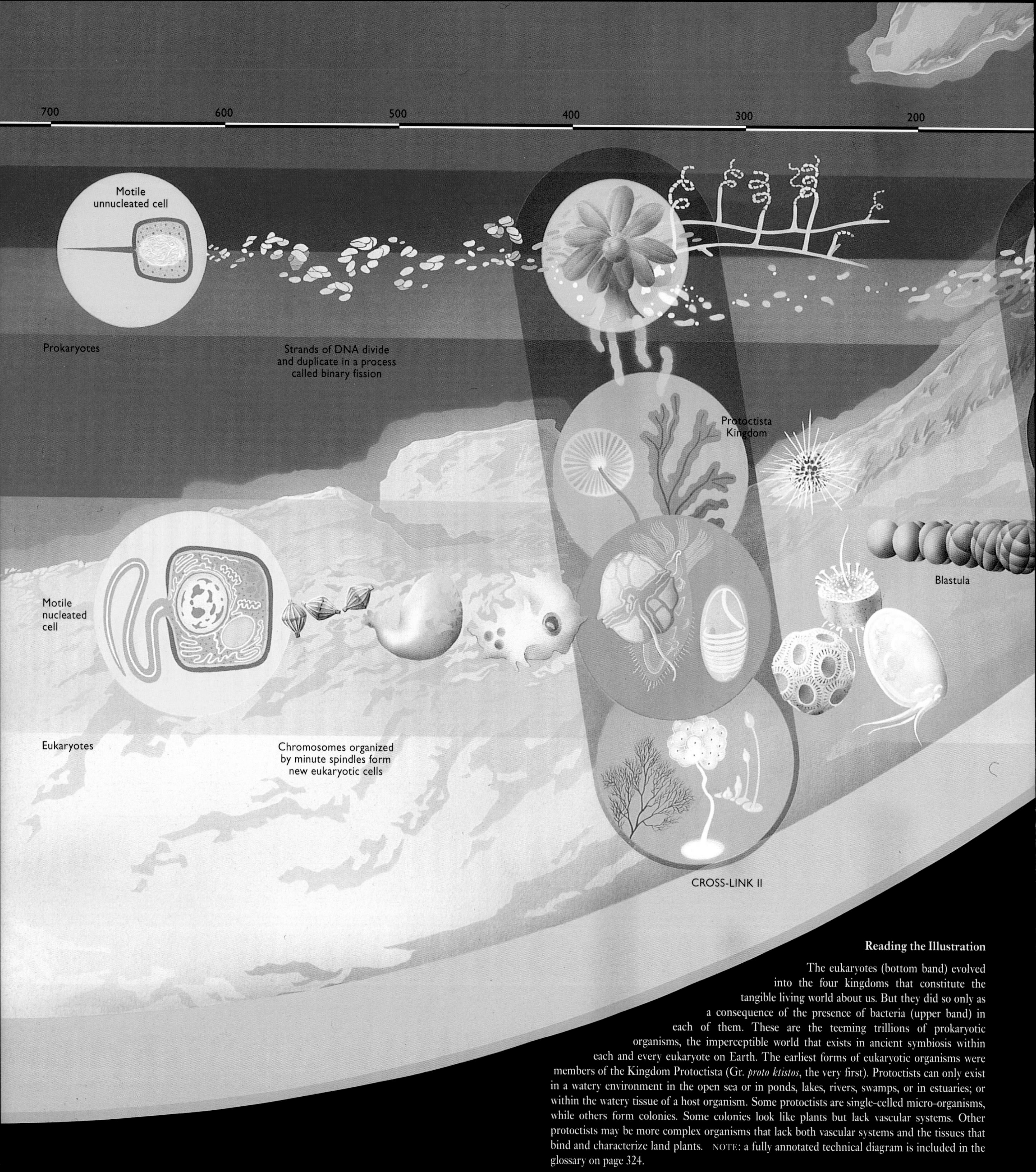

Reading the Illustration

The eukaryotes (bottom band) evolved into the four kingdoms that constitute the tangible living world about us. But they did so only as a consequence of the presence of bacteria (upper band) in each of them. These are the teeming trillions of prokaryotic organisms, the imperceptible world that exists in ancient symbiosis within each and every eukaryote on Earth. The earliest forms of eukaryotic organisms were members of the Kingdom Protoctista (Gr. *proto ktistos*, the very first). Protoctists can only exist in a watery environment in the open sea or in ponds, lakes, rivers, swamps, or in estuaries; or within the watery tissue of a host organism. Some protoctists are single-celled micro-organisms, while others form colonies. Some colonies look like plants but lack vascular systems. Other protoctists may be more complex organisms that lack both vascular systems and the tissues that bind and characterize land plants. NOTE: a fully annotated technical diagram is included in the glossary on page 324.

6 Crisis and Diversity

The final illustrations of the previous page suggest how bacteria such as the anaerobic prokaryotes, to which oxygen is poisonous, may have responded to the challenge of the oxygen crisis, precipitated by changes in the composition of the Earth's atmosphere. Aerobic prokaryotes, for example, contributed their ability to survive in an oxygen-based atmosphere to the growing eukaryotic cellular community. In the process of such endosymbiosis, it is thought that anaerobes simply invaded aerobic prokaryotes, and that the attributes of invader and invaded became parts of a growing community within a single nucleated cell. Generally, characteristics were perpetuated by a genetic center in the cell—the nucleus of the "eukaryote" (Gr. *eu karyon*, true kernel). Meanwhile, separate lineages of prokaryotes have continued to play a crucial role in the evolution of life.

As we saw in the preceding essay, prokaryotic bacteria reproduce replicas of themselves by binary fission. Consequently, prokaryotes are prolific in numbers but limited in morphological forms. There are far more individual prokaryotes in the gut of a single animal than there are people on Earth. Prokaryotes can survive incredibly diverse and hostile environments, including temperatures well above the boiling point of water, or near absolute zero. It is also likely, at least to some degree, that prokaryotes play a part in the metabolism of all living organisms. They inhabit every nook and cranny of life's existence. This crucial role is implied by linking illustrations in the top band of the diagram with cross-links to the horizontal green band illustrating the evolution of eukaryotic organisms.

The eukaryotic cell comprises a genetic nucleus surrounded by a profusion of once-independent units that vary in kind and number with the type of cell. For example, the denizens of some cells are responsible for the cell's respiration, others for its photosynthesis, and yet others for its movement. In all eukaryotic cells the nucleus contains the chromosomes (the "printed circuits") and the mechanisms to ensure the reproduction of the cell as a whole. Among other functions, the nucleus ensures that opposing teeth in newly synthesized DNA strands are interlocked in the correct order.

Eukaryotes reproduce either sexually or asexually. The asexual process, called mitosis, involves replicating DNA strands exactly and then dividing them equally, to produce identical offspring or clones. The alternative is sexual reproduction, in which replicated DNA strands are segregated through meiosis into gametes such as sperm or ova, each containing half the information required to make a complete individual. The DNA strands of different parents are combined togethe through the process of fertilization.

Complex cellular division is a process that ofte involves sexual interchange of genes, and this metho occasionally results in accidental imperfection calle mutation that may lead to new species. Mutations ar also possible during the binary fission process i prokaryotes, but it is the eukaryotic sexual process tha has resulted in the extraordinary diversity of plant and animals. The exchange of genetic informatio from different parents permits eukaryotes to adopt, t adapt, to improve survival rates, or to discard unuse characteristics. All this depends upon circumstance that are dictated by the physical condition of th hydrosphere, the atmosphere, and the lithosphere ove aeons of time.

The prokaryotes in the top band (all bacteria) have kingdom of their own—Monera, and play a crucial rol in the chemical transformations of life. But the viruses thought to be rogue pieces of RNA or DNA, are no generally classified with living organisms. They are no "microbes," rather they are complex pieces of chem istry that assume particular shapes. They need t invade a living cell before they replicate.

Relative to prokaryotes, the eukaryotes in the lowe band are few in number, but they have demonstrate an amazing morphological diversity. They are divide into four kingdoms, all illustrated here: Protoctista i cross-link II; Plants, Animals, and Fungi in cross-lin III. Since they first appeared in the microfossil recor around 1.5 BYA, eukaryotes have proliferated into ten of millions of species— some scientists suggest hun dreds of millions.

Members of the Protoctista Kingdom are noncon formists that are classified by exclusion—see below. A members of the Plant Kingdom produce an embry within maternal tissue at some stage in their reproduc tion. All members of the Animal Kingdom, in whic we all have a particular interest, produce embryos fro a hollow ball of cells called a blastula. Members of th Fungi Kingdom produce spores that do not develo embryos at any stage in their reproduction. And herei lies the key to "definition by exclusion": if a newly dis covered eukaryote does not reproduce by using one o these three methods of reproduction, it is classifie into the Protoctista.

The six graphic presentations in this chapter sho the Earth to be a complex and dynamic planet that ha profoundly influenced the way in which life ha evolved upon it. The last two graphic spreads hav demonstrated that apart from the origin of life itsel the development of the nucleated cell was the mos innovative, the most significant, and the most explosiv step in life's evolution. It is for these two reasons tha the storyline of this book begins with the first appear ance in the fossil record of animal-like forms of nucle ated organisms, about 650 MYA.

Coincidentally, this date is about as far back in tim as it is possible to go to portray the evolution of conti nents and oceans with reliability. Fortunately, from thi very distant point in time, it is also possible for the sto ryline to follow the concurrent evolution of life i ephemeral oceans and on restless continents up to th appearance of the most advanced form of nuclear organism—humankind.

III IAPETUS AND AVALONIA

Pointe de Penhir and Pointe du Toulinquet (horizon), Crozon Peninsula, Brittany, France

The rocks of Brittany's Crozon peninsula pictured here were laid down as sediments in ancestral West Africa between 1,000 and 650 MYA. Towards the end of this period West Africa was just north of the Antarctic Circle and was separated from other parts of modern Africa to the north of it by a Pan African Ocean. West Africa was one of several elements that were jostling and coalescing into the megacontinent of Gondwana as this evolving assembly straddled

IAPETUS AND AVALONIA 650–490 MYA

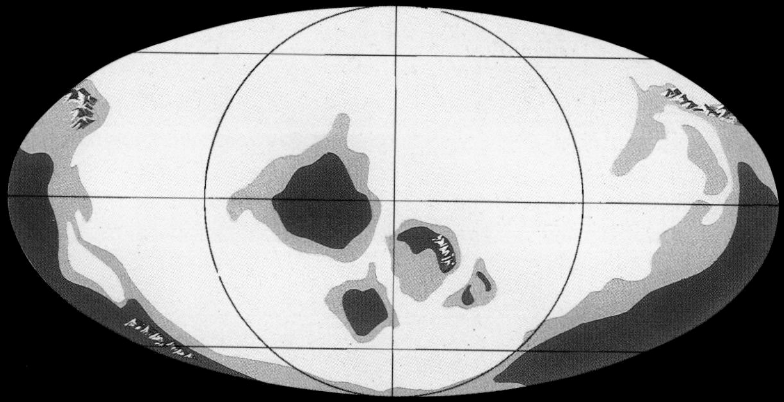

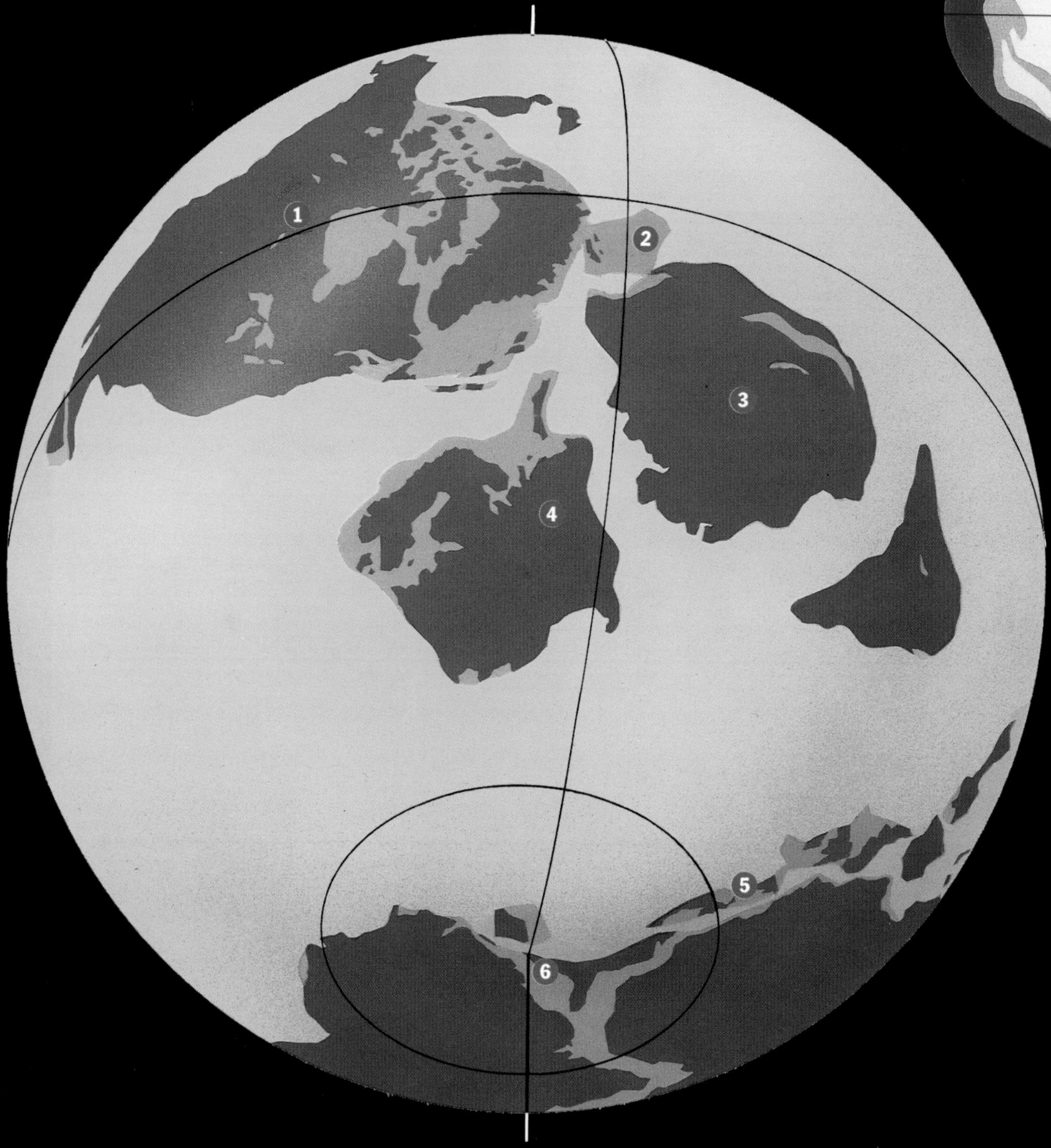

1. LAURENTIA
   North America, North Slope Alaska, Canadian Archipelago, Greenland, Northern Newfoundland, Scotland and Northern Ireland, Mexico

2. BARENTSIA
   Svalbard (Spitsbergen)
   Barents Shelf

3. SIBERIA

4. BALTICA
   Northern Europe
   Norway, Sweden, Denmark (part), European Russia, Northern Germany (part)

5. WEST AFRICAN GONDWANA
   AVALONIA
   England (excluding southwest), Wales, Southern Ireland, The Low Countries, Southern Newfoundland
   Displaced Terrane
   Southern Germany (part), Eastern New Brunswick, Southeastern Nova Scotia, New England
   ARMORICA
   Brittany
   IBERIA
   Spain, Portugal
   NORTH AFRICAN GONDWANAN MARGIN
   Greece/Yugoslavia, Italy, Sardinia

6. SOUTH AMERICAN GONDWANA
   Piedmont Basement
   Carolinas, Georgia, Florida, Cuba, Yucatán

The paleoglobe is a South Polar projection at 543 MYA. We are looking at the fragments of a postulated pre-Pangean continent called "Rodinia" (from the Russian *Rodina*, motherland). These continental elements eventually recombined to form Pangea, which ultimately fragmented into the modern continents.

While the longitudinal relationships (north-south) of the continents shown on this and other diagrams in the book are reliable, the latitudinal positions (east-west) relative to the Greenwich Meridian are approximate. They are included here and on all paleoglobes in the book to provide the reader with perspective relative to the continents' present geographical positions.

The corresponding oval projection shows the whole surface of the globe at 543 MYA in two dimensions and consequently may not always appear to correspond to the detail on the three-dimensional globe. Also, the oval diagram shows the estimated shape and relative positions of the then existing continents without superimposed modern outlines but with reconstructed continental shelves and mountain ranges. The circle within the oval is marked with a vertical line representing the Greenwich Meridian; the horizontal lines represent the Arctic and Antarctic Circles and the Equator, and the segments on either side of the circle show adjacent areas hidden from view on a globe.

The North Atlantic realm at 543 MYA

One can hardly use the description "North America" or "Europe" to describe the embryonic modern continents that did not exist in their present form. The outlines of well-known features like Hudson Bay will orient the reader, but of course these features did not exist at the time. Therefore, tectonic scientists have developed a special nomenclature to deal with the problem of identification. For instance, the isolated equatorial North American continent shown here is called "Laurentia" while Scandinavia and European Russia to its south is termed "Baltica." When both paleocontinents and Siberia are combined that combination is termed "Laurussia."

We can see the edge of Gondwana at the South Pole and can judge its enormous extent by looking at the oval projection. On the paleoglobe we can also see that the Florida-Piedmont province was once part of Gondwana, but it is important not to be misled by the modern outline. Neither Florida nor the Piedmont existed in this form 543 MYA—but the present basement rock of these localities is Gondwanan in origin. Similarly as we move along the Gondwanan coast away from the South Pole we can see Avalonia, Iberia and Armorica. These regions of Gondwana are now key elements of the present North Atlantic realm and are the prime subjects of this chapter, which opens at 650 MYA.

By this time, the oldest known soft-bodied animals, called the Ediacara fauna, had appeared in the seas. As the story progressed towards the end of Cambrian time (490 MYA), the continents visible on the globe opposite—excepting Siberia—began to reassemble into what would ultimately become Central Pangea.

[Readers interested in specifics should note that the dates for the Cambrian Period, always a controversial subject, have been revised to 543–490 MYA (GSA revision of December 1999) from the original, longstanding dates of 570–505 MYA.]

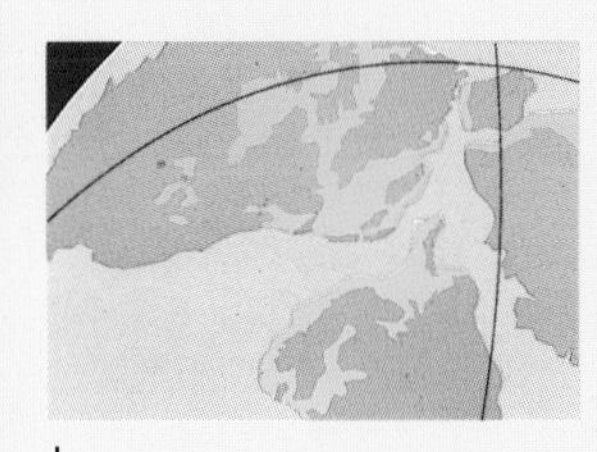

1

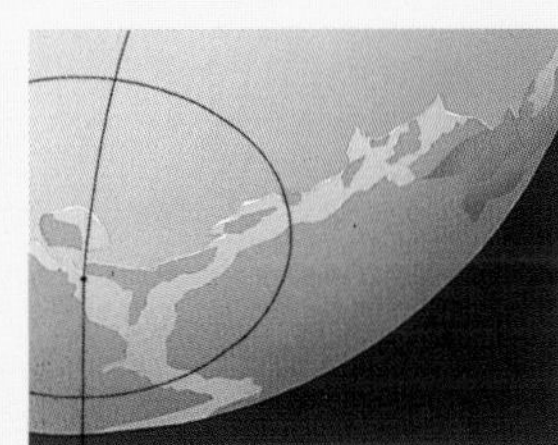

2

In the 19th century, the traditional use of Greek or Latin words to mint scientific names was an orderly, flexible shorthand that effectively described and categorized a host of new discoveries. In the 20th century, tectonic scientists were faced with the need to name oceans that no longer exist and ancestral continents that were not only differently configured but also far removed from their descendants' present location. They also had to identify and relate the priority blocks and fragments—the bits and pieces of modern continents that originated elsewhere. Paleomaps were annotated with names drawn from Greek, Roman, and Celtic mythology, and neo-classical names were based on familiar place names.

The Atlantic Ocean, for example, had been named after the Greek god Atlas, who was condemned by Zeus to carry the sky on his shoulders. Atlas was the son of Iapetus, the supposed son of Heaven and Earth. When the notion of an ancestral North Atlantic Ocean was first advanced it was called the "Iapetus Ocean," to imply a filial relationship. Such imaginary relationships arising from Greek mythology became an established tradition for naming paleo-oceans, while paleocontinents and displaced continental blocks tended to be named after actual places. For instance, the oldest Precambrian supercontinental clustering that is generally agreed to have existed before Pangea is called "Rodinia"—from *Rodina*, the Russian word for motherland. It is thought to have existed 1,100 MYA.

ALTHOUGH RODINIA'S one-time existence is controversial, the majority of geoscientists take the view that there was indeed such a supercontinent. Controversy arises from the fact that paleomagnetism in Rodinian and other ancient rocks can indicate longitude but not latitude, and that the older the rocks are, the more unreliable this particular clue to global orientation becomes. Since the matching of fossils in Rodinian rocks has to be discounted, it is understandable that ancestral supercontinental reconstructions are controversial. But there is one outstanding common factor on which all agree. The global extent of glacial deposits in such rocks is extraordinary. Huge deposits of "tillites" (see essay on page 65) and other clues to heavy glaciation, are found in Rodinian rocks dated between 750 and 500 MYA in formations all over the Earth.

These universal deposits suggest that during Rodinia's breakup the Earth as a whole was subjected to a series of severe global ice ages from tropical latitudes to the Poles—a period termed the "Snowball Earth" [see Essay 8 pages 138–139]. Such episodes account for fluctuations in sea level, and in part for the changes in the chemistry of the oceans, that set the scene for the explosion of multicellular life that followed. After 550 MYA climate changed dramatically. Glaciers disappeared, and continents were flooded by the rising seas. It was springtime for multicellular life on Earth.

THE IAPETUS OCEAN began to form in a region of the disintegrating Rodinia that lay between ancestral North America (**1**), called "Laurentia," and the Baltic countries of Northern Europe. Laurentia is named after the Laurentian mountains of Quebec Province in Canada. The Baltic countries were part of a subcontinent called "Baltica." As the Iapetus opened it developed into a major ocean. Although most of its seafloor was destroyed during its closure in the assembly of Pangea, remnant parts of the Iapetus Ocean floor still exist. They can be found in New Brunswick [see pages 56-57], New-Foundland (see pages 58-59), Ireland, and the Lowlands of Scotland.

By 543 MYA, Laurentia (including Greenland) straddled the Equator from what is now Arctic Canada to Mexico—a 90° clockwise shift from today's orientation. Baltica lay to the south of Laurentia with the Iapetus Ocean widening between them. Parts of the modern North Atlantic realm, the basement rocks of present Iberia and Brittany for example, were north of the Antarctic Circle in the *Eastern* Hemisphere (**2**). Fragments now parts of Sardinia, Italy, and Greece in the Mediterranean were east of Iberia and Brittany on the African Gondwanan coast. The basement rocks of Florida and Yucatán were

[*continued on page 60*]

Nantcol, Harlech Dome, North Wales

1 Iapetus and Avalonia

The maps at right are based on those originally drawn in 1966 by the Canadian scientist J. Tuzo Wilson to show that there had been an "ancestral" North Atlantic Ocean—now called the Iapetus Ocean. The paleoglobe is a modern reconstruction of the geography then. The main photograph is of part of the Iapetus Ocean floor that now forms the walls and bed of the St. John River in New Brunswick.

The first map shows the Atlantic, with the mixed colored regions on opposite margins showing the location of incompatible fossils over 500 MY in age, including two kinds of trilobites:

Olenellus, in theory found only in North America, and *Paradoxides*, which should be found exclusively in northern Europe. The inset picture shows part of the Harlech Dome in North Wales, where *Paradoxides* can be found.

The second map shows that the colored regions match when the margins are joined into a Pangean arrangement. When this conjunction is separated according to color (the third map), the previously mixed assemblage becomes two distinct fossil realms. From this Wilson concluded that an ocean had closed during the formation of Pangea, and that during the subsequent breakup of the supercontinent the modern North Atlantic had opened along a different line. This process had left complete regions of "displaced terranes" on the "wrong" side of the new ocean.

The rocks, minerals, and fossils of the Harlech Dome match those of the Avalon Peninsula in Newfoundland. Both these regions and many others now scattered around North Atlantic margins were originally a part of "Avalonia" (named after its Newfoundland element), an archipelago calved off the coast of Africa near the South Pole.

2 Ocean Cycles

Tuzo Wilson developed his idea of an ancestral Atlantic Ocean into a model of ocean cycles now known as "Wilson Cycles." These explain the interactions of the continents with the ocean basins over geological time. Wilson proposed that there are six stages in the life of an ocean basin, as illustrated.

From the first diagram we can visualize an uplifting stage that produces a rift valley, which sinks to form a depression called a graben and fills with shallow sea. Secondly, continued rifting and then seafloor spreading results in the formation of a mature ocean like the present Atlantic (the third diagram). This is followed by a fourth stage, a period of decline—in effect the failure of the spreading-center system that leads to the compression of the ocean crust by encroaching continents. The ocean crust is destroyed by subduction beneath the continents. Occasionally, by some little-understood quirk of tectonics, the heavier ocean floor overrides the edges of the lighter continents. This phenomenon is exemplified by the panorama of Lark Harbour, where remnants of the Iapetus Ocean floor, called "ophiolites," overrode the edge of the ancient continent of Laurentia about 450 MYA. This process is called *ob*duction, the opposite of *sub*duction.

The penultimate and fifth stage in an ocean cycle is the narrowing of the ocean to the point that continental margins collide, leading to a prolonged period of mountain building such as that in the Mediterranean Alpine belt today. The final act is the joining of continents into one landmass—as seen today in India with Asia, where the suturing has caused the formation of the Himalayas.

Wilson Ocean Cycles

[*continued from page 55*]

near the South Pole, wedged between African Gondwana and South American Gondwana. Between these northern and southern extremities of the megacontinent lay yet another sector of the modern North Atlantic realm, a volcanic archipelago called "Avalonia."

The name Avalonia was derived from the Avalon Peninsula in southern Newfoundland, called after a mystical isle of Celtic mythology and Arthurian legend. In the language of plate tectonics, "Avalonia" defines certain modern North Atlantic regions when they were parts of an "island arc." This arc took the form of a substantial curving chain of volcanic islands rather like Japan in size, shape and volcanicity. Just as Japan today is volcanically active off the coast of Asia, in its time Avalonia was active off the coast of African Gondwana. Ultimately, Avalonia became a displaced terrane, a phenomenal drifter with a life of its own.

The Avalon Peninsula in Newfoundland is the most significant remnant of Avalonia today. Like those of its counterpart, the Iapetus Ocean floor, other remnants are scattered around the modern North Atlantic's shores. They form parts of southern Scandinavia and northern Germany, the Low Countries, much of central England and Wales. They also include large parts of New Brunswick and Nova Scotia, the complete Boston Basin, parts of Rhode Island, and pieces of both North and South Carolina, and Georgia.

CLEARLY, THERE HAVE BEEN extraordinary geographical transformations in the North Atlantic region during the past 750 MY. There were correspondingly momentous changes in both the evolution of the region's biosphere and in the diversity of its biota.

The evolution of eukaryotic multicellular organisms [see pages 48-49 "Life Begins" and pages 50-51 "Crisis and Diversity"] was in its early stages around 2,000 MYA. Subsequently, whatever continental assemblies (if any) existed between the two dates, the supercontinent Rodinia was fully assembled by 1,100 MYA and disassembled into widely separated megacontinents and individual continents by 750 MYA. From this summary it can be seen that the global distribution of very early forms of eukaryotic life was profoundly influenced by such continental breakup and reassembly in a startling variety of environments from polar to equatorial. Following the breakup, the hundreds, possibly thousands, of eukaryotic species evolved in a variety of ways, and in varying environments, as individual continents and megacontinents were reassembled into supercontinental Pangea. In turn, the break-up of Pangea into the modern continents and oceans resulted in the global redistribution of original Pangean species. These evolved into modern species, resulting in the present distribution pattern of flora-fauna.

Of course, this broad statement is a gross oversimplification of a very complex process. Its intention is simply to highlight the role evolving continents and oceans played in the evolution of life. This and the three following chapters will portray some of the main events that characterize the incredible 400 MY of evolution from 650 MYA to 250 MYA: a period that commenced with the appearance of the first known marine animals and ended with a planet that teemed with both marine and terrestrial life, during the final assembly of Pangea.

THE STORY BEGINS IN WALES in the early 19th century. At that time, natural science was the vogue and geology and paleontology (the study of fossils) were blossoming new sciences. In 1840 the Swiss naturalist Louis Agassiz and others had engendered ridicule for the then preposterous idea that some of the northern continents had been largely covered with ice in the not-so-distant past; and in 1858 Charles Darwin and Alfred Russel Wallace would provoke fierce debate about the evolution of species with their joint papers to the Linnean Society in London. It was an exciting and highly contentious time for natural science.

In 1831 the rocks of North Wales and the borderlands between Wales and England had attracted the attention of two geologists,

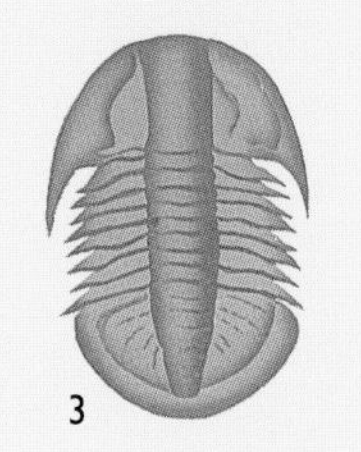

3

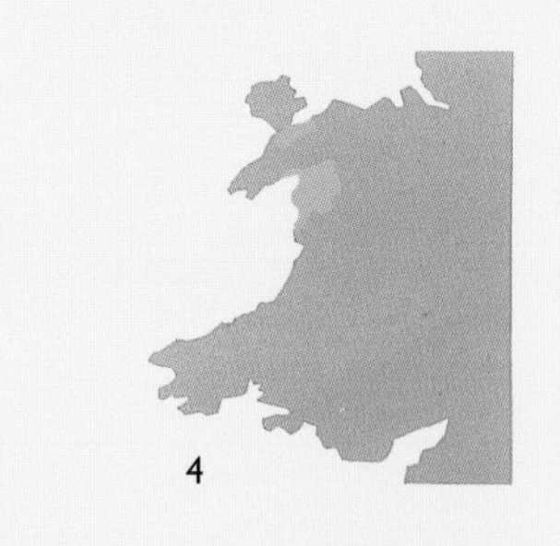

4

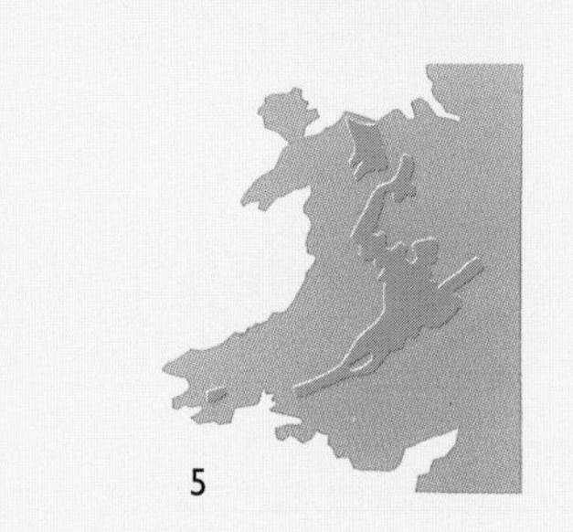

5

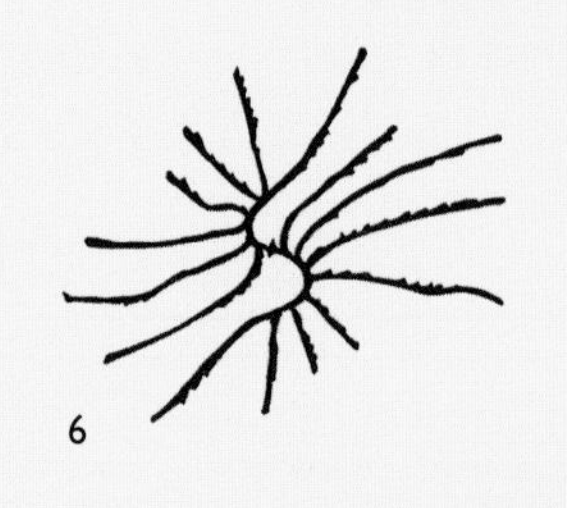

6

7

Adam Sedgwick and Roderick Impey Murchison. These borderland rocks were recognized to be very old. Some of their shale formations (shale is mud and silt compressed into rock) contained a great variety of the fossils of a small marine invertebrate with a triple-lobed body and many legs. It was called a "trilobite"—the oldest form of life known at that time (**3**). No fossils of any form of life had been found in rocks beneath those that contained trilobites.

Sedgwick studied the rocks in a mountainous region of North Wales known as the Harlech Dome [Essay 1]. Sedgwick's main interest was "structural geology"; the form, character, and continuity of the rock formations across the landscape. After years of work he was able to describe an evidently related sequence of successive rock strata and their deformation into huge folds. His two-dimensional sectional drawings clearly showed that strata of slates, shales, grits, and other "soft" rocks in the Harlech Dome region seemed to have a base on a "hard" rock platform (rock such as granite or basalt). In 1835 Sedgwick used the word "Cambrian" to describe the series of "soft" rocks (rocks formed from sediments) that rested on the platform (**4**); he coined the term from a 12th-century Latinized version (*cambriae*) of the Welsh word for "people" (*cymry*). The platform beneath the Cambrian formations that created so evident a boundary was termed "Precambrian." It was impossible to judge in Sedgwick's time that the platform on which the Harlech Dome stood represented only a fraction of the "Precambrian Period." In fact it represents intervals of time greater by far than all the subsequent geological periods from the Cambrian to the present put together.

Murchison had avoided intruding on Sedgwick's territory and worked in adjoining areas—the Border country between England and Wales. Murchison's emphasis was on the fossil content of his rock formations rather than on their structural relationship. He too identified a system of succession (**5**), and in 1835 coined the term "Silurian" to describe it. This was an adaptation of the name of an ancient Welsh Border tribe, the Silures, guerillas of Roman times.

During the 1840s other geologists found that when traced in other parts of Britain, Sedgwick's Cambrian succession and Murchison's Silurian had many rock formations and fossils in common. The two systems overlapped to a very significant degree, and some rock formations and their trilobites and graptolites (**6**) were identical. The scene was set for one of the most furious, vitriolic, and prolonged disputes in the history of science. Sedgwick claimed that Silurian structures were mainly Cambrian, which Murchison hotly contested. By the 1850s fossils, mainly of trilobites of different ages, had been widely found in successive sedimentary rocks of Cambrian/Silurian affinity in America and in Europe. These established that the lower half of Murchison's Silurian really belonged neither to his system nor to Sedgwick's (**7**). So the name "Ordovician" was coined from the name of another Welsh tribe, the Ordovices, contemporaries of the Silures, and used to describe the rock formations in dispute. This controversy went on for more than 70 years, only being settled when the name Ordovician Period was officially adopted.

Ultimately, eleven hotly contested but nonetheless unequivocally individual geological periods were identified (twelve in the U.S.). These began with the Cambrian Period and progressed through the Ordovician, Silurian, Devonian, and other periods to the present. To these, subdivisions of "epochs" and "ages" were added. And to these, literally thousands of further names describing particular rock formations of particular ages in particular localities. Today these geologic "time scales"—expressed in relative terms—are the basic reference tools for geologists all over the world. Sometimes, however, there is disagreement about their "absolute" dates (see below). After the brief historical account that follows we will avoid all but the most necessary references and stick to absolute dating where possible. (See page 330 for an abbreviated version of the 1999 GSA time scale.)

THROUGH THIS BEWILDERING array of names one factor stood out above all others: the eleven geologic "periods" were distinct from each other in the life forms that had evolved and extinguished in their time. Yet it was also clear that the periods could themselves be divided into evolutionary stages: a stage of early life, a stage of much more advanced life, and a stage that corresponds to life on Earth today. These three divisions were called the Paleozoic Era (early life), Mesozoic Era (middle life), and Cenozoic Era (recent life)—all derived from the Greek language. The three eras are together called the Phanerozoic Eon, and span 543 MY.

We know today that Precambrian times lasted over three billion years. This represents the length of time it took for the first prokaryotic bacteria to evolve into eukaryotic organisms and for the first simple eukaryotic animals to appear [see pages 48–51]. Yet, during the Phanerozoic Eon, in just under one-sixth of the time, the evolution of eukaryotes advanced from the first creatures with hard parts to the astonishing diversity of life that exists on the crowded planet we inhabit today.

This biological explosion seems primarily to have been due to the strong tendency of all eukaryotes to overpopulate. Unless restricted in numbers by predation, limited by living space or lack of nourishment, or reduced by intolerable changes in their specific environments, eukaryotes increase in numbers at a geometrically progressive rate. Thus, dominant species tend naturally to self-destruct by overpopulation. This process is one of many contributors to mass extinctions that permits underlying species to emerge and evolve, sometimes to achieve dominant roles.

The establishment of eons, eras, periods, epochs, and ages was and still is crucially important as a method of "relative" dating in which "time" is related to the progressive appearances and disappearances of "index fossils." Two very good examples are trilobites named *Olenellus* (from the Greek for "bent elbow": **8**) and *Paradoxides* (Greek for "self-contradictory": **9**). The fossil index provides a means of "dating" sedimentary rocks by referring their fossil content to stages of evolutionary succession. With a series of exact matches it can be said that the rocks in which the fossils were found are "Middle Cambrian" or "Late Ordovician," and so on. But it was not until methods for measuring the half-life decay of radioactive isotopes were developed that it became possible to express the record in "absolute" time. In this case age is expressed in terms of the number of years that have elapsed since a particular type of rock was formed, within a defined margin of error. Then, most importantly, one dating method is used to check the other. The absolute numeric dating used throughout this book is expressed in millions of years (abbreviated to "MY" and "MYA" for millions of years ago). Several slightly different chronological timescales are published by authoritative institutions in different parts of the world; opinion is divided about absolute dates for particular intervals of time. However, since this book's original draft was based on the Geological Society of America's (GSA) absolute scale at the time, the book was updated before publication to conform with the 1999 GAS scale.

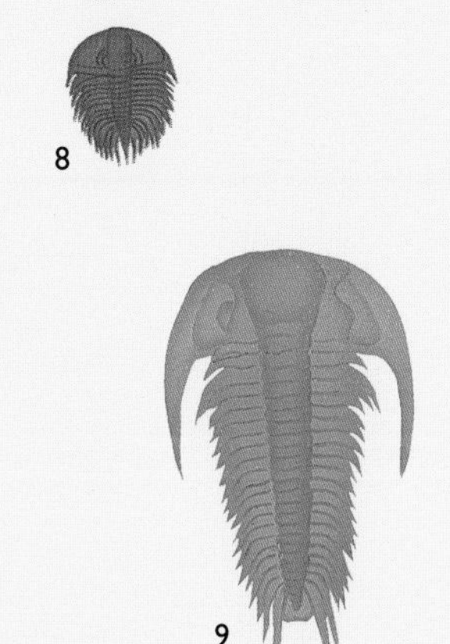

8

9

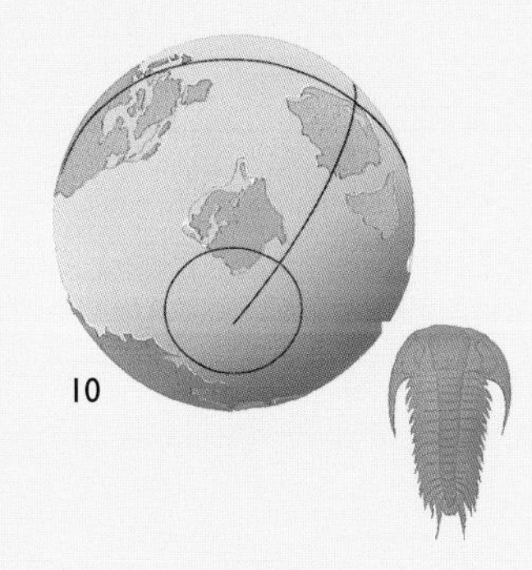

10

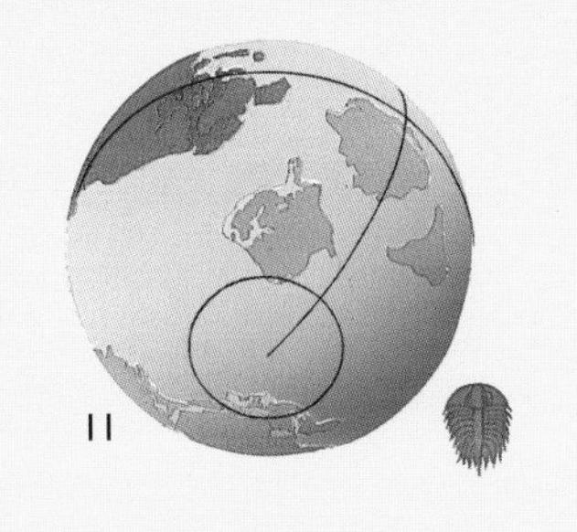

11

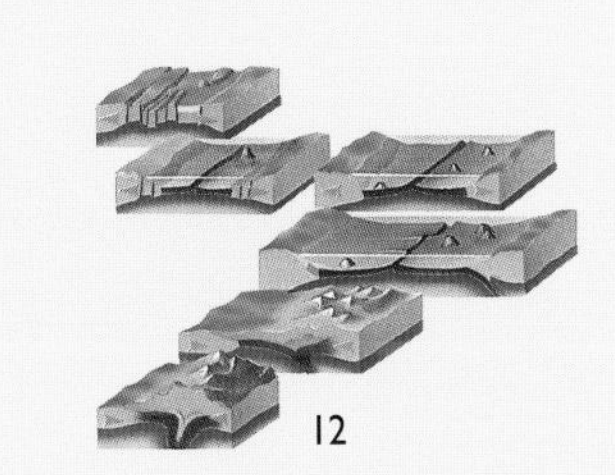

12

Olenellus, Paradoxides, and other trilobites made a second important contribution to science, and to the first steps in the settlement of a controversy that raged on both sides of the Atlantic. This time it was a matter of paleogeography rather than age. In 1966 J. Tuzo Wilson (the Canadian who named the Wegener Fault) published a benchmark paper in the British science journal *Nature*. This was before the proof of continental drift had been established, and while Wegener's notion of a Pangea was still considered a fantasy by many geoscientists. Wilson listed a series of contradictory facts that had been observed by a number of other geologists. These included the point that locations of two "faunal realms," of which *Olenellus* and *Paradoxides* were important members, were plainly in contradiction. Wilson offered a solution that is considered to be one of the key contributions to the advance of tectonic science [Essay 1].

Because of its frequent discovery in several localities in Britain and elsewhere in Northern Europe, *Paradoxides* (**10**) was attributed to the "Atlantic realm." On the American side of the ocean *Olenellus* (**11**) appeared with equal frequency and even wider distribution: it was therefore attributed to the "Pacific realm." *Olenellus* is anatomically different from *Paradoxides* and is found in somewhat older Cambrian rocks. Yet, in spite of the intervening width of the Atlantic Ocean and the difference in age-of-occurrence between the two trilobites, occasionally both were found in separate rock formations in matching locations on opposing sides of the ocean. One trilobite species or the other was on the wrong side of the Atlantic.

Before Wilson's paper was published, it had already been noticed that these awkward-to-explain fossil regions coincided in an extraordinary fashion if maps of the continental margins were fitted together. It had also been suggested that this coincidence could be explained if Europe and North America had once been joined before being separated by the present North Atlantic Ocean. Following this line of reasoning, Wilson now proposed that there had been a "proto-Atlantic Ocean" (since named the Iapetus Ocean) in Cambrian times. He suggested that this had closed to form Wegener's Pangea, and that the ocean had subsequently reopened along the general line of the original suture but in a slightly different configuration. This would account for the mixed faunal realms on both sides of the new ocean (and various other geological contradictions).

The decade that followed this crucial paper and its proofs was an exciting one in which the evidence of seafloor spreading and continental drift was established. Wilson further developed his concept of a proto-Atlantic by showing that there were six stages in what he had now concluded to be an "ocean cycle" [Essay 2].

The first stage, Wilson suggested, is "embryonic" in character; it takes the form of an uplifting and subsequent "extension" (stretching) of part of a continent's crust (**12**), like that seen in today's East African Rift Valley. Next is the "young ocean" stage of rifting and spreading, during which a sunken valley forms with a volcanic rift at its center: this fills with a shallow sea—like the present Red Sea and the Gulf of Aden. The "mature" stage of ocean formation is like that which has been reached in the present North Atlantic. This has great continental shelves on either side of the ocean carrying a heavy load of sediments deposited from continental rivers. The sediments lithify and form layered sedimentary rocks.

After the mature stage there follows a period of steady "decline" in which the ocean floor is destroyed by encroaching continents, like the Pacific Ocean floor beneath parts of the West Coast of America today. Continued compression of the ocean floor between continents also results in the creation of mountain ranges. These are often formed from marine sedimentary rocks on opposing landmasses, such as the Alps and other mountains associated with the diminishing Mediterranean Sea—a remnant ocean. Such mountain building marks the "terminal" stage in the cycle. The very last act is the fusing of colliding continents, like the fusion between India and Asia which has resulted in the uplift of the Tibetan Plateau and the formation of the Himalayas.

Wilson also argued that if there had been a Pangea then the supercontinent must have had predecessors. But how did this concept evolve into today's understanding of the prior existence of Rodinia and possibly other Precambrian supercontinents? Esoteric clues to events in the remote past have been pieced together and interpreted, using both the most intricate techniques of satellite measurement and delicate instruments whose readings can measure the slightest tremor deep within the Earth. Results are analyzed in conjunction with the 19th century's index-fossil record and its encyclopedic 20th-century supplement.

Paleogeographic reconstruction is complicated; two of the results of tectonics add immeasurably to the difficulty. The first is that the fossil record is by no means complete, mainly because of the massive destruction of ocean floors and continental margins, and the distortion of continental sedimentary rock into mountain ranges.

[*continued on page 66*]

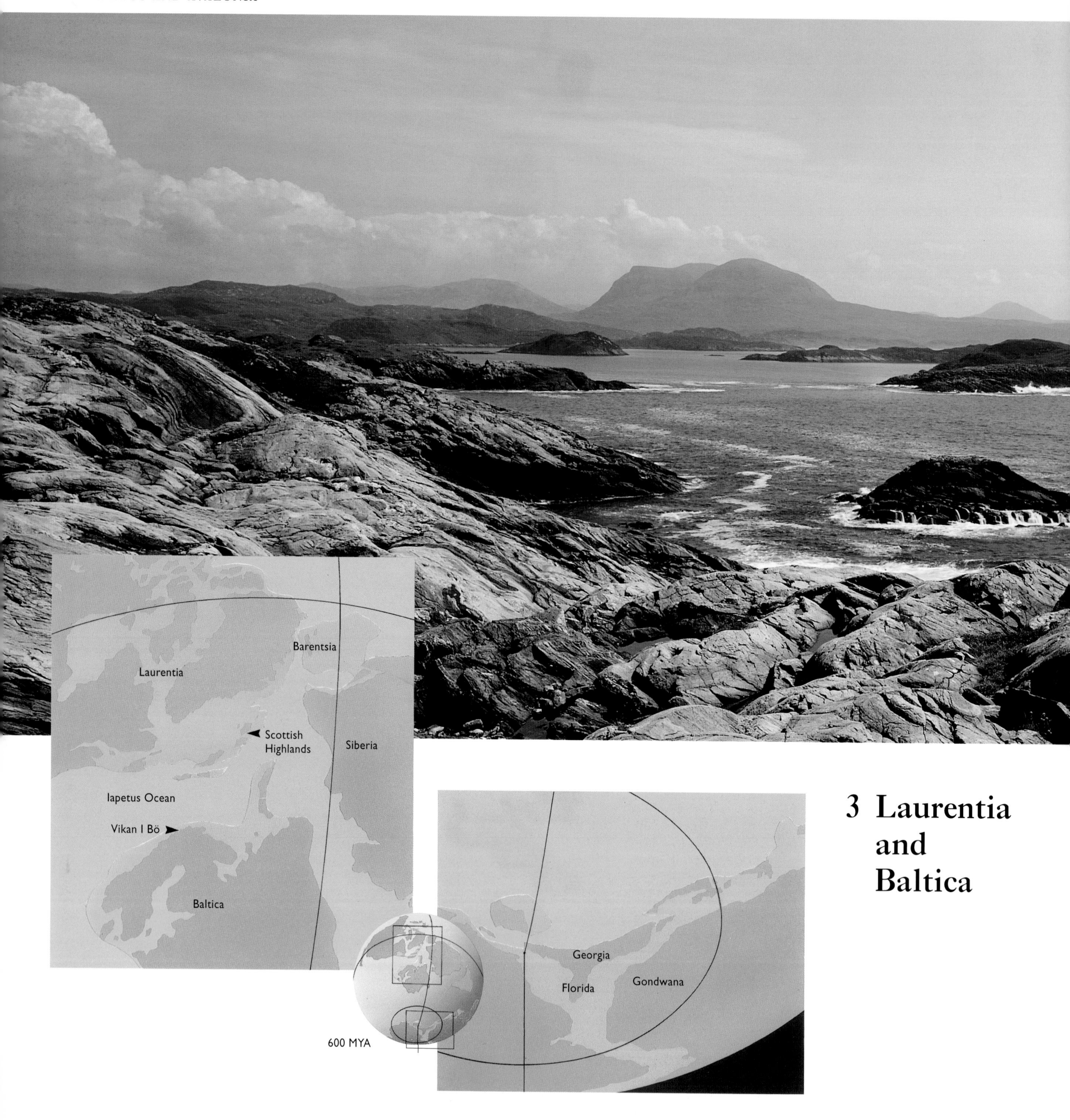

3 Laurentia and Baltica

Vikan I Bö, Westeralen, northern Norway

During the assembly and breakup of the loose-knit pre-Pangean supercontinent of Rodinia in the period 1,100–650 MYA, the continents of Baltica, Siberia, and Laurentia, collided to form a megacontinent called "Laurussia." Baltica included northern Europe and European Russia; Laurentia included North America and Greenland. The Iapetus Ocean formed during the breakup of Laurussia about 650 MYA, when Baltica separated from Laurentia.

At that time Laurentia was well south of the equator and included Greenland, Spitsbergen, and northern Newfoundland in its makeup. In fact, most of the basement rocks of Scotland and northern Ireland were part of southeast Greenland. But the Laurentian continent excluded some well-known North American elements such as the Boston Basin, parts of the Carolinas, peninsular Florida, and southern Georgia, whose basement rocks were then part of the southwestern coast of Gondwana.

The main panorama is part of Laurentia. It shows the Torridon Hills seen from Badcall Bay in northwest Scotland. The rocks in the foreground are some of the oldest basement rocks in the British Isles, Lewisian schists 2,900–2,700 MY old: they match their counterparts in Greenland quite perfectly in age and style. The Torridon Hills are relict sandstone mountains formed near the Laurentian coast about 600 MYA on the margin of the Iapetus Ocean. The inset picture is of a part of Baltica that lay on the opposing shore of the Iapetus Ocean. These Norwegian rocks are the oldest known formations in continental Europe and were formed about 3,400 MYA.

Trevezel, Brittany, France

Tillites, Granville, Normandy, France

4 Armorica

We turn now to the southwestern margin of Gondwana, to a region next to Avalonia called Armorica. Between 650 MYA and 543 MYA there was a major mountain-building episode in this region, called the "Cadomian orogeny" (the names *Cadomia* and *Armorica* are Latin adaptations derived from the Celtic language).

The Cadomian orogeny contorted and disrupted the southwestern margin of Gondwana, which then included the basement rocks of many other elements of modern Europe. The main picture here is of Cadomian granite. This is best exposed in the Channel Islands off the coasts of Brittany and neighboring Normandy. Elsewhere the remnants of the Cadomian orogeny are less easily distinguished because of "overprinting" by later orogenies that occurred during the final assembly of Pangea.

Corbiere, Jersey, Channel Islands, U.K.

The inset picture shows an outcrop of rock on a misty moor in Brittany, all that is now visible of a once majestic Armorican mountain range. The second inset picture is a closeup of lithified "tillites" that include a rounded cobble of Cadomian age found in the coastal cliffs at Granville in Normandy (not to be confused with Grenville in Quebec). Tillites are the result of the grinding action of ice against rock, and they vary in size from flour-sized grains to large boulders. As the orogeny progressed, the Avalonian and Armorican regions were in the South Polar region at a time when there was an intense ice age.

[*continued from page 61*]

The second problem is that if rocks of any type are heated beyond their Curie point, and then cool down again, their magnetic compasses and isotopic clocks reset to the time and place of that event. The record of previous events is "overprinted," and in some cases their fossil records are also destroyed. Such resetting can occur as a consequence of deep burial, miles beneath the surface, by contact with heat from volcanic events, by coal burning underground, or even by overlying forest fires [Essay 5].

SOME PALEOGEOGRAPHERS specialize in the overall paleogeography of the past. The reconstruction of such ancient geography is a highly specialized field of research; individual reconstructions vary and are subject to revision as research continues; the older the period of reconstruction, the more controversial the result. All paleogeographers are mountain-men whose work largely depends on "orogenies" (Gr. *oros genes*, mountain-born)—episodes of mountain-building that cause chains or belts of mountains to form, commonly thousands of miles long. Each orogeny results in many mountain ranges that often involve regions on opposite sides of a sea or an ocean. The common factor is that the mountains in the orogenic belt were produced by the same tectonic events. For instance, the orogenic event that has produced the Alps (as a result of Africa colliding with Eurasia) also caused other mountain ranges to form on either side of the Mediterranean from Morocco and Spain in the west, to the Balkans in the east (**13**). All these ranges are considered to be part of the "Alpine orogeny." Deformed structures in this extensive region are simply referred to as "Alpine." Previously existing mountains such as the ancestral Pyrenees, were "overprinted" by the Alpine orogeny. Thus one would say that the ancestral Pyrenees were formed during the "Variscan" orogeny, an episode that occurred during the formation of Pangea, and that they were overprinted by the formation of the present mountains during the "Alpine" orogeny. Such is the language of the tectonic scientist.

Every orogenic belt has a distinctive and recognizable tectonic "style," characterized by remnants of its mountain structures, volcanoes, and heat-altered "metamorphic" rocks, as well as by its sedimentary rocks and their fossil content, by remnants of its glaciations, and by odd bits of associated seafloor. The latter are sometimes pushed up onto continental surfaces instead of being destroyed beneath opposing continents ("obduction" instead of "subduction").

Of the many orogenies that occurred during the assembly of Pangea and breakup of Rodinia, two in particular figure in the reconstruction of the paleogeographic story of the North Atlantic world of today. They also help to explain those Pacific realm and Atlantic realm trilobites that led Tuzo Wilson to his hypothesis about the prior existence of an ancestral Atlantic Ocean.

The first orogeny occurred when Laurentia, Baltica, and Siberia formed "Laurussia" 1,100–800 MYA; it is called the "Grenville" orogeny (**14**) after Grenville in the Province of Quebec, Canada [Essay 3]. The second and younger orogeny occurred 650–543 MYA during the breakup of the assemblage, and is called the "Cadomian" orogeny. *Cadomia* was the Roman name for Caen, in Normandy, chosen because it is in Normandy, neighboring Brittany, and in the Channel Islands off the coast of Brittany that Cadomian orogenic relics have been most thoroughly studied [Essay 4].

The Quebec region of Precambrian Laurentia rifted apart to form the Iapetus Ocean. As a consequence, by 600 MYA Quebec (in Laurentia) and Norway (in Baltica) had become separated on the opposing shores of the near-equatorial Iapetus. At this time Brittany (called Armorica, the Latin name for Celtic Brittany) was half a world away, adjacent to Avalonia in a region of Gondwana that was approaching the South Pole. Laurentia and Gondwana were separated by the continent of Baltica, which had an ocean on either side.

At this time there was an intense ice age, called the Varangian Glaciation, after glaciated rocks of this age found in Norway, sometimes referred to as the "Snowball Earth." So, the Cadomian

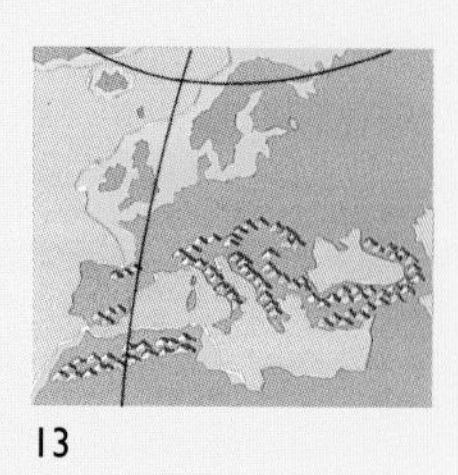

13

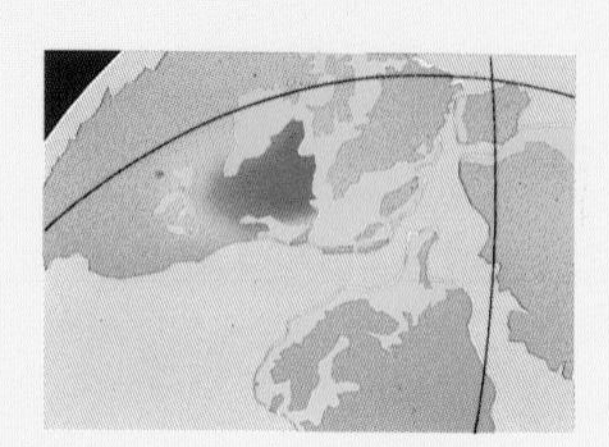

14

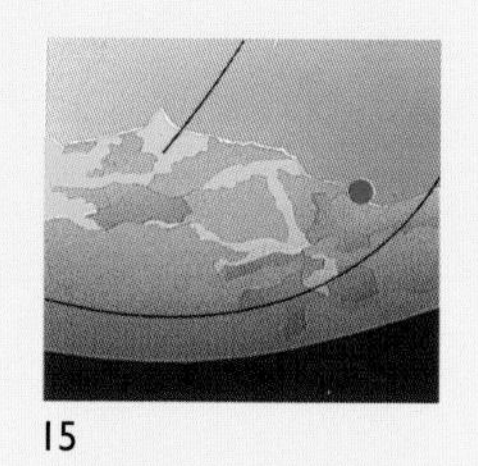

15

16

ranges were frequently snowbound and glaciated as the ice age waxed and waned. Armorican peaks were carved by glaciers, and their lower slopes were reduced to tillites, a rubble of boulders, cobbles, gravel, sand, silt, and mud that was borne down to the Gondwanan continental margins by ephemeral rivers. What remains of those very ancient Armorican mountains and the products of their erosion forms the spectacular cliffs and beaches of Brittany and of the Channel Islands today (**15**). One can still stand on the remains of that incredibly old mountain range and imagine a cold and misty Armorican landscape.

This broad-brush picture of the paleogeography of the Precambrian and Cambrian world puts those enigmatic "Pacific" *Olenellus* and the "Atlantic" *Paradoxides* trilobites into a clearer perspective. *Olenellus* lived on Laurentian shores in the near-tropical Iapetus Ocean. Millions of years later *Paradoxides* and other members of that family lived on the sediments and swam in the shallow seas off the Avalonian region of Gondwana well south of the Equator. By this time the Varangian Ice Age had ended, sea level had risen very appreciably, and shallow seas were beginning to flood the continents.

It seems that *Olenellus* (which some paleontologists consider to be from a line of ancestral trilobites) and the younger *Paradoxides* were only distantly related, but sparked a revolution in natural science.

TO VICTORIAN GEOLOGISTS in Britain the boundary between the Precambrian and the Cambrian periods was very distinct, as it proved to be in many localities around the world where Cambrian rocks are found. There appeared to be no fossils in any Precambrian rocks. Then suddenly, at one moment in geological time and in successive formations from Cambrian rocks upward, there were trilobites and other fossils in ever-increasing numbers. From this evidence some Victorian scientists concluded that the beginning of the Cambrian Period must also mark the beginning of life on Earth.

As we have seen, today's continents were in distant places in Late Precambrian and Cambrian times. There were several individual ocean cycles during the course of the assembly and breakup of Rodinia, involving prolonged periods of mountain building and erosion. The products of that erosion accumulated into 15,000 meter-thick (50,000 ft) sedimentary deposits at the crustal margins, whose cumulative weight caused crustal depressions on the edges of such "passive" continental margins.

These thick deposits—now transformed into sedimentary rock—were partially destroyed and deformed as the ocean later closed again and continents collided, at what now became "active" continental margins. This sequence of formation and subsequent destruction of rock is known as the "rock cycle" [Essay 5]. As mentioned previously, the recycling of rock is the main reason why fossils of ancient organisms are comparatively rare. But other significant processes also add to the difficulties of reading the rock record.

As we have seen, the Victorians did not quarrel over the unequivocal division between Precambrian and Cambrian rocks; their argument was over the ambiguous division between Cambrian and Silurian formations. The reason for the clear and widespread division between the Precambrian and Cambrian rocks in many regions of the world is that global sea level in the Late Precambrian fluctuated very considerably, probably by 300 m (1,000 ft) or more. When sea level fell, more continental rocks became exposed and were subject to erosion. The rock that was worn away created a significant gap in the subsequent stratigraphic and fossil record. When sea level rose, margins and their continents were flooded once more and new sediments were deposited on worn-away older surfaces. The interface between new sedimentary rock and the old eroded surface beneath it (**16**) is called an "unconformity" [see page 138].

TWO BASIC phenomena control major changes in global sea level: short-term waxing and waning of ice ages, and long-term swelling or shrinking of the ocean floor near ocean spreading-centers. A change from ice age conditions to temperate conditions, or to greenhouse conditions with little or no ice at the poles, causes fluctuations of up to about 120 m (400 ft) within periods counted in thousands of years. But variations in the amount of volcanic activity at seafloor spreading-centers can account for semi-permanent fluctuations of up to an *additional* 240 m (800 ft) or more in global sea level lasting millions of years. Thus if spreading-centers are extremely active, as they were during the breakup of Rodinia, the regions of hot, swollen rock on either side of spreading-centers are extended and so displace a proportionate volume of seawater in the global ocean.

When increased rates of seafloor spreading coincide with greenhouse conditions and the melting of ice at the poles, the low-lying regions of continental platforms can flood to a depth of 365 m (1,200 ft) or more. Such shallow "epicontinental seas" and coincidental greenhouse climates have prevailed on Earth through much of geological time. Ice ages are comparatively rare but are usually prolonged events—up to 60 MY in duration. The current ice age started about 1.64 MYA, and should continue, with interglacial interludes as at present, until the current interpolar ocean circulation is replaced by an equatorial circulation as the modern continents drift and reassemble [see Chapter XI: *Fountains of Youth*].

Whether or not the continents were flooded by epicontinental seas (**17**) determined whether or not sediments and their fossils were deposited on continental surfaces or on continental shelves. That sudden appearance of trilobites in Cambrian rocks that misled the Victorians was because there was a considerable fall in sea level in Precambrian times. This had resulted in the deep erosion of Precambrian rock surfaces. Sea level rose dramatically in Cambrian times and shallow-sea sediments were deposited on now lowlying eroded surfaces. So it was not that the trilobites marked the beginning of life, but that their predecessors had evolved on low-lying continental shelves; these were mostly destroyed by future continental collisions.

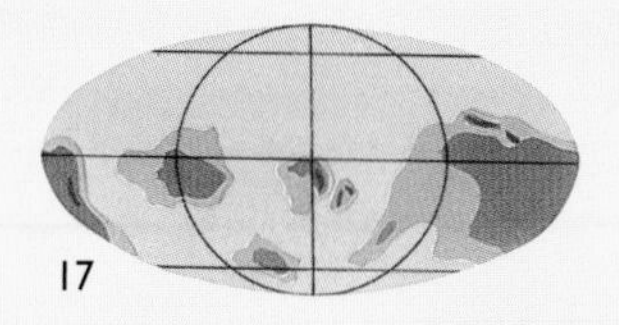
17

18

19

20

THE OLDEST-KNOWN forms of structured eukaryotic organisms are called "acritarchs." They are usually spherical in shape and vary in size from microscopic to merely minute. No one knows exactly what role they played in the evolution of eukaryotic life but they are thought by many to be representative of the transition from single cellular to multicellular life forms. Apart from the acritarchs, the first major discovery of ancient Precambrian animal fossils was made in 1947 near some old lead mines in the Ediacara Hills, a desolate part of South Australia. Subsequently it was found that an extensive area of fine-grained sandstones in that region contained the fossils of a variety of previously unknown soft-bodied animals without any hard parts, now dated at about 650 MYA.

The Ediacara fossils fall into four major groups (**18**): jellyfish-like animals that could swim: frond-like animals that could not swim; trace fossils of worm-like creatures leaving tracks and burrows; and "trilobitomorphs," bottom-living creatures that may or may not have had affinities with later Cambrian animals. All these creatures lived on intertidal flats and in shallow seas off the shores of Australia—at that time part of an equatorial Precambrian megacontinent that incorporated prototype Antarctica, India, and several elements of Africa. Not surprisingly such fossil faunas, wherever found, are called Ediacaran assemblages. Such communities not only provide a glimpse into the early evolution of animals, but also show that complex life forms existed very much earlier than was previously suspected. Other assemblages were later discovered at Mistaken Point on the Avalon Peninsula, and elsewhere on the Newfoundland peninsula. Subsequently they have been discovered on every continent on Earth except Antarctica—today enveloped by ice.

The Mistaken Point fossils are beautifully preserved "casts" of the original animals. These casts are found on the surface of very fine-grained rocks (**19**) called "turbidites." The latter were formed from the residue of an avalanche of silt, mud, and rock debris that had poured down a steep continental slope and settled at the foot of the slope. The Mistaken Point faunas had lived on the surface of such turbidite deposits. When volcanic eruptions took place in Avalonia, then in the mid-latitudes of the Southern Hemisphere, fine ash from the eruptions had settled on the surface of the sea some distance away from the coast. The ash sank to the ocean floor, where it smothered the fauna. In due course the ash consolidated into a "volcanic tuff." This formed the matrix of a mold in which replicas of the original creatures were formed out of lithified silt. Later sediments were deposited on top of the tuff. A new generation of Avalonian animals established itself on the uppermost surface, only to be buried by a subsequent ash fall (**20**)—and so on [Essays 6 and 7].

By definition turbidites are not shallow-water rocks. In fact Newfoundland geologists are of the view that the Mistaken Point turbidites were formed in a "moderately deep" ocean environment thousands of feet deep—sufficiently deep to be inky black and devoid of sunlight. Most Australian Ediacaran communities lived in the shallow sunlit waters of equatorial Australia where photosynthesis (producing carbohydrates and releasing oxygen by the action of sunlight) was an important key to their survival. How then did the Avalonian fauna survive at the bottom of a "moderately deep" ocean environment generally hostile to community life?

TODAY, DEEP-SEA ORGANISMS depend on oxygen dissolved in seawater to "breath" and the scavenging of organic debris to "feed." Alternatively they depend upon chemosynthesis at thermal vents (**21**) located at spreading ridges or at deep cold-water seeps at the foot of steep continental shelves—such as those off Florida. Animals living near such vents depend upon their symbiotic relationship with "autotrophic" (Gr. *auto trophe*, self-nourishment) prokaryotic bacteria to survive. Autotrophs are anoxic bacteria that can metabolize sulphur and other "noxious" substances. Their digestive systems contain enzymes that reduce dissolved sulfide compounds and produce oxygen and carbon compounds as "waste" products—to the benefit of the vent-animals in which they live.

Organic debris in Precambrian seas was primarily prokaryotic in origin—the oceans were simply alive with bacteria. Also the dissolved oxygen in Precambrian oceans was perhaps only a hundredth of today's level, while the presence of toxic chemicals was much higher than today. Ediacara were eukaryotes and therefore oxygen-dependent. Could the deep-sea species at Mistaken Point have depended upon chemosynthesis?

The Ediacarans are the first-known soft-bodied creatures. Indeed, it seems that there were no animals with hard parts in the global ocean at that time. Neither shelled nor chitinous animals yet existed (chitin is an organic material much like a human fingernail). But around 543 MYA at the beginning of the Cambrian Period, Ediacara gradually disappear from the fossil record and the first shelled animals, the Tommotian fauna, appeared.

The gradual disappearance of the Ediacara fauna after about 100 MY of dominance is thought to represent the first major mass extinction of eukaryotic life on Earth. Just as the extinction of dinosaurs is considered to mark the end of Cretaceous time, the extinction of the Ediacara is considered by some geologists today to mark the boundary between the Precambrian world and the Cambrian world.

But why did the Ediacara fauna suffer extinction? The arrival of the first shelled animals, shortly supplemented by the appearance of chitinous animals (the first trilobites for example) suggests an answer. It could be that, along with a radical warming of the climate, the chemistry of the global ocean and atmosphere was undergoing crucial change, with increased levels of dissolved oxygen in seawater and free oxygen in the atmosphere. The change must have been fundamental although its cause is not known. However, the changing environment simply did not support the continuity of

[*continued on page 72*]

Delta fan, Baillarge Bay

5 The Rock Cycle

Baillarge Bay, north Baffin Island, Canadian High Arctic

The rock cycle

The castellated arctic cliffs in the aerial panorama and inset photograph are an accumulation of lithified sediments originally deposited in a shallow sea during the assembly of an ancient supercontinent over a billion years ago. These sediments were the products of erosion from mountains formed from the remains of even older mountains. Together the pictures show several stages in the "rock cycle"—a byproduct of ocean cycles. As Tuzo Wilson showed, ocean cycles produce mountain-building episodes, which in turn initiate rock cycles.

The inset picture is of a classic "deltaic fan," a delta that has accumulated at the foot of a stream-cut valley. As the cliffs above the fan erode they contribute boulders, cobbles, pebbles, gravel, sand, and silt to the formation of the delta. After these sediments transported downhill to the sea, the larger rocks formed the visible subaerial fan, while the finer-grained sediments accumulated to build a submarine skirt. As the overall size of the underwater skirt increased, the gravels and silt began to lithify into sedimentary rock by the pressure of their own accumulating weight.

About half the mass of sedimentary rock on Earth today was formed from rock that has already undergone part or the whole of the cycle illustrated here. Each cycle takes about 600 MY. From this it is reasonable to assume that five such rock cycles have taken place in the last 3,500 MY and that the product of each cycle lost about half its mass by being subducted into the Earth's interior. It follows that much of the fossil record of life on Earth has been destroyed during this process.

6 First Life

The main picture shows part of a fossilized stromatolite reef dated 1,950 MYA. This was planed by ice during the last glacial advance. The inset picture of Prismatic Spring depicts an even earlier stage in life's evolution—the kind of "warm little pond" (in Charles Darwin's phrase) in which biogenesis is thought to have taken place around 3.5 billion years ago (BYA). It is now thought by some biologists that biogenesis could also have taken place at superheated hydrothermal vents on the ocean floors and other inhospitable places with abundant very hot water and minerals.

Each stromatolite, a mat-like structure, was formed by a colony of single-celled unnucleated bacteria, micro-organisms that replicated perfect clones of themselves. Such "prokaryotes" (once known as blue-green algae but now called "cyanobacteria"), converted the carbon dioxide-based atmosphere into carbohydrates by photosynthesis: they lived on the carbohydrates and exhaled oxygen as a waste product. These bacterial reefs became so prolific worldwide that over many millions of years they contributed to the conversion of the Earth's atmosphere. The proportion of carbon dioxide was vastly reduced and that of the oxygen greatly increased—a development that caused gradual and increasing environmental stress on the world's stromatolite community. Oxygen was poisonous to most early forms of life. The gradual adaptation of micro-organisms to an oxygen-based atmosphere was a key step in the evolution of the nucleated cell—the "eukaryotic" cell. This led to the evolution of multicellular macro-organisms and an explosion of new forms of life on Earth.

Bacterial growth, Prismatic Spring, Yellowstone National Park, Wyoming

Stromatolites, East Arm, Great Slave Lake, Northwest Territories, Canada

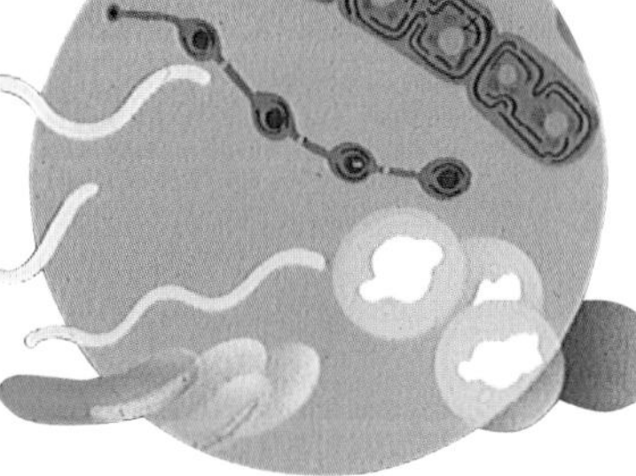

Earliest prokaryotic life

[*continued from page 67*]

Ediacaran communities—and since these communities could not adapt to the new conditions for life, they became extinct.

THE TOMMOTIAN FAUNA—named after the Siberian rocks in which they were first found—were the first known animals with hard parts. They were a diverse group of tiny invertebrates, miniature shelled animals that measured a fraction of an inch in length—although there are exceptions to this limitation in size. Some had crude mollusk-like coiled shells (**22**); others had tubes that are presumed to have contained worm-like animals; and others had glass-like spicules. These tiny creatures were widespread by 543 MYA but in a few million years seemed quickly and universally to vanish from the fossil record. This was just as the first animals with chitinous exoskeletons, the trilobites, and larger shelled animals began to appear. And by about 525 MYA a greater diversity of invertebrates existed in shallow Cambrian seas than there is in the modern oceans. This frantic evolutionary pace is termed the "Cambrian Explosion" and it could only have taken place if global environmental conditions had changed to favor such an event. So it is likely that the chemical and physical conditions for life in the oceans had changed radically.

The fact of a "Cambrian Explosion" was first uncovered by a discovery made by Charles D. Walcott, then Secretary of the Smithsonian Institution, in 1909. Walcott excavated a site high in the Rocky Mountains of British Columbia, where the so-called Burgess Shales, dated at 515 MYA, proved to contain a quite astonishing number of fossils of previously unknown Cambrian animals. In total it has been estimated that over 40,000 live animals were buried in the mud that accumulated to form the shales at the foot of a steep cliff in a shallow sea. About 11 percent of the fossils are of different types of algae (seaweed): the balance is made up of varieties of sponges, sea pens, mollusks, worms, crustaceans, trilobites, sea urchins, chordates and others. Most of these creatures were bizarre in form and alien in character to those already known. They were mostly small but sometimes quite large—up to 50 cm (20 in) in length.

Subsequently, other sites similar in nature to the Burgess Shale site were discovered in British Columbia and in other parts of the world. A site in Utah is younger in age but those in China, northern Greenland, and South Australia are older: all were equatorial in disposition in Cambrian times. But were they the result of a wild and extravagant evolutionary experiment that had little eventual significance? Were they an overreaction, perhaps, to the development of the correct chemical and physical formula that supported burgeoning life on Earth? Or, as many paleontologists think probable, did the Burgess community in fact contain all the attributes that survived to trigger the evolution of modern species—ultimately including humans?

At the end of Cambrian time (490 MYA), there were four successive waves of extinction leading to a mass extinction of many species, including the Burgess communities. This pattern of almost cyclic extinction suggests environmental instability as the possible cause. But while many varieties of animals were decimated in late Cambrian times, including trilobites, sufficient numbers survived the waves of extinction not just to continue the process of evolution but to develop and exploit the circumstance of their survival. There were animals with legs and exoskeletons, soft-bodied animals—some with teeth, some with shells; animals with delicate glassy structures like sponges, and some animals that were little more than swimming tubes fitted with fins (**23**). The latter were the most primitive of the jawless fish—the first fish in the sea.

All these extraordinary changes occurred in the aftermath of several intense and long-lasting ice ages during a period of vigorous seafloor-spreading activity, swollen spreading ridges, and epicontinental seas. The global climate was now warm and balmy. The Earth was briefly in the doldrums ... a lull between storms. The assembly of Pangea was about to begin.

22

23

Turbidite formations, Mistaken Point, Avalon Peninsula, Newfoundland, Canada

7 First Animals

The oldest known animal fossils on Earth are called "Ediacara" after a locality in Australia where they were first discovered. These, and the Avalonian Ediacara, pictured here at Mistaken Point in Newfoundland, were eukaryotes: their multiple cells had nuclei, and they reproduced sexually. The Avalonian species of Ediacara lived on the flanks of a continental shelf off Avalonia about 550 MYA. They took many exotic forms but their natural history is little understood, for they have no living counterparts.

Avalonian fauna most probably lived in the pitch black of the deep sea. This is thought likely because their fossils are found on successive surfaces of deep-water "turbidites," shown in the inset picture. These are rocks formed from clouds of fine sediment that settle after a submarine avalanche. But how did these strange animals survive? One can only conjecture. Perhaps the sea at that depth was naturally rich in organic remains raining down from above? Or perhaps this fauna was dependent for survival upon bacteria capable of breaking down toxic chemicals into compounds that could be metabolized: a process called "chemosynthesis."

Avalonian Ediacara, Mistaken Point

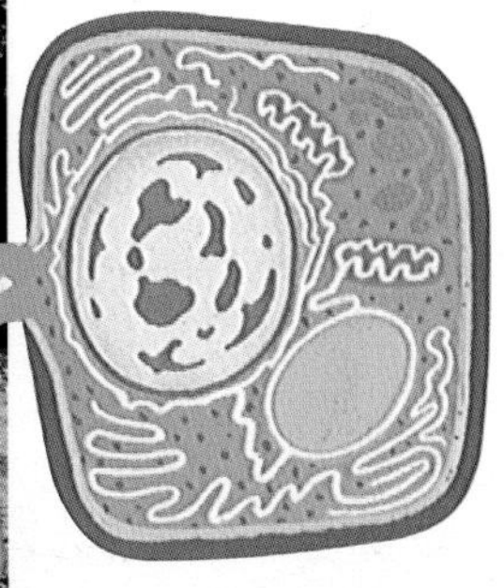

Earliest eukaryotic life

Magddlenefjord and the Losvik Mountains, northwestern Spitsbergen

During the general assembly of Central Pangea, ranges of mountains formed part of a continuous chain, mostly Alpine in stature but sometimes of Himalayan scale, that stretched for 9,650 km (6,000 miles) from the northern tip of Greenland to present-day Mexico. The mountains of Spitsbergen pictured here are remnants of the most northerly of these ranges, the Caledonides. They were among the first mountain ranges to be formed as the Iapetus Ocean and Tornquist's Sea closed during the collision of Laurentia, Baltica, and Avalonia.

TORNQUIST'S SEA
490–360 MYA

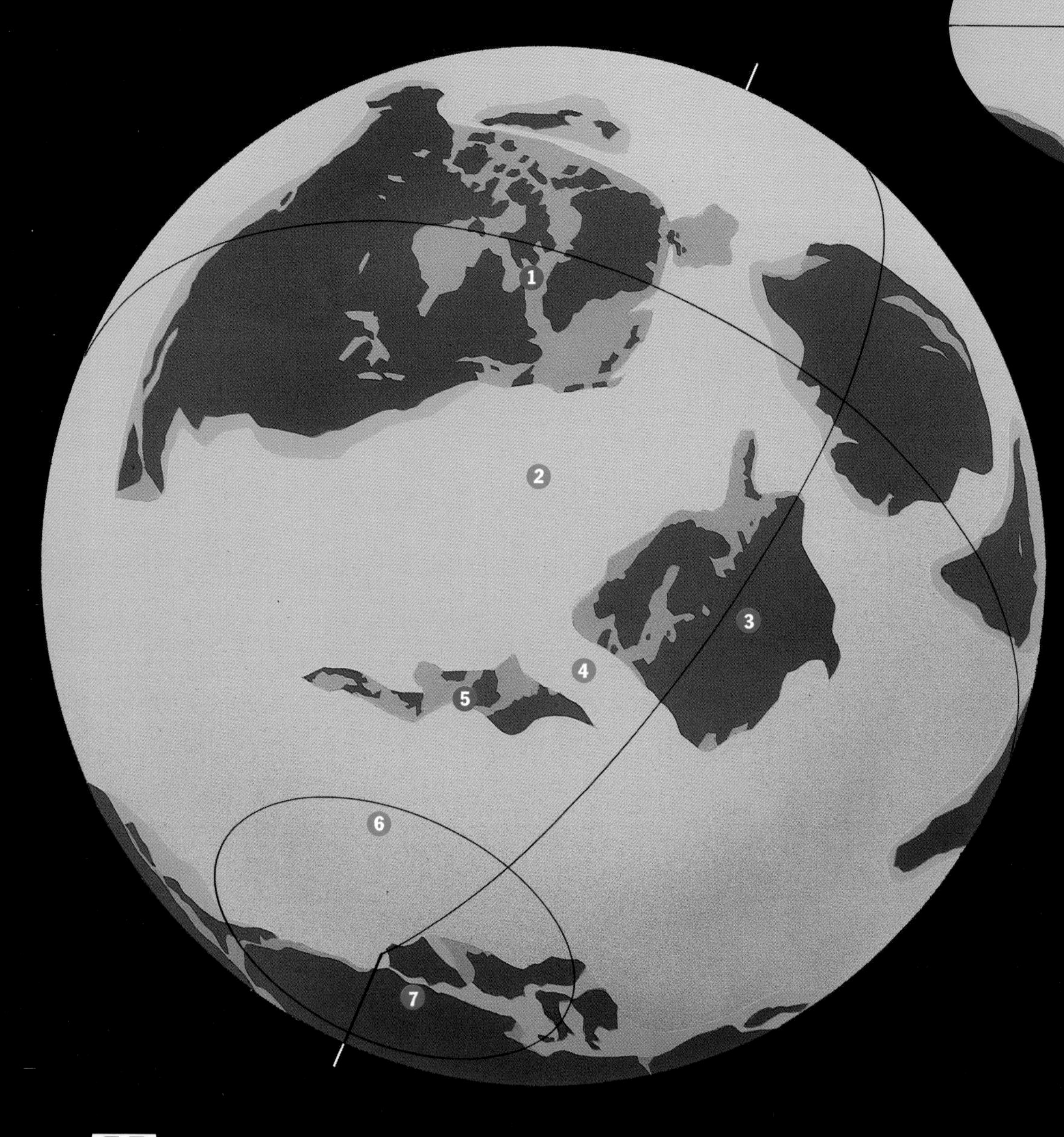

1. LAURENTIA
   Northern Newfoundland
   Northern Scotland
   Svalbard (Spitsbergen)
2. IAPETUS OCEAN
3. BALTICA
   Scandinavia and Baltic Countries
   Northern Denmark
   Part of Northern Germany
   European Russia
4. TORNQUIST'S SEA
5. AVALONIA
   Low Countries and Southern Denmark
   Southern Ireland, Wales, England
   Southern Newfoundland
6. RHEIC OCEAN
7. GONDWANA
   Iberia
   Armorica
   (Brittany, parts of France and Germany)

Tornquist's Sea of the chapter title (shown here at 460 MYA) is the narrowing ocean between northward-drifting Avalonia that included parts of England, Wales and southern Ireland, and Baltica—Norway, Sweden, east Baltic countries, and European Russia. The Iapetus Ocean is caught between Avalonia and Baltica, which are moving northward as they converge upon the southeast coast of Laurentia. This movement will lead to the eventual destruction of the Iapetus Ocean and the assembly of Newfoundland, the British Isles (apart from southwest England), and the docking of the Low Countries with Scandinavia.

The Rheic Ocean at this time was still continuing to widen between the southern coast of Avalonia and the coast of Gondwana near the South Pole. It began to form when Avalonia calved off Gondwana. However, by this chapter's end it too will be in the process of being destroyed during the general assembly of Central Pangea.

The oval projection shows the whole surface of the Earth as it is thought to have appeared at the time of these events. Mountain ranges were beginning to form in the north—the Caledonides shown on the chapter title page—and sea level was generally up to 300 m (1,000 ft) higher than it is today.

The key to reconstructing the story to this point has been Tuzo Wilson's theory that there was once an ancestral North Atlantic Ocean—the Iapetus Ocean. When this claim was made in 1966, the geoscience community was deeply split over Wegener's concept of continental drift and the previous existence of the supercontinent of Pangea. Yet here was a onetime antidrifter and distinguished member of the geoscience establishment arguing for the existence of a previous ocean. He was making a serious proposition based on a few anomalous fossil trilobites and remnant island arcs. He was also admitting by inference that there had indeed been a Pangea that had formed from continents that had collided during the closure of an ocean.

Judging from Wegener's drawings and text, he had visualized Pangea as the archetype continent from which present global geography had evolved. Wilson was now proposing an elegant theorem about what had happened to cause this geological equivalent of the "Big Bang," while other scientists were still arguing about even the possibility of continental drift, let alone assuming its probability! Furthermore, Wilson inferred that if continents drifted, then they and their predecessors had drifted since time immemorial. Wilson's proposition was indeed a major advance in the tectonic revolution.

Today we know not only that Pangea existed, but that a Precambrian supercontinent existed about 1,100 MYA—Rodinia. It is also thought possible that Rodinia had at least one predecessor. Moreover, it is now theorized that Rodinia itself had rifted into three elements, one over each pole and one smaller equatorial continent, and that it had then reassembled into other continental clusters. And finally this assembly was torn apart by the opening of new oceans—the Iapetus Ocean among others—to result in the formation of Pangea. In fact the assembly, breakup, and reassembly of supercontinental clusters may have characterized the general evolution of the Earth's surface since primordial times.

It is evident that the breakup of Rodinia ultimately resulted in the formation of the megacontinents of Laurentia and Gondwana and several lesser continents such as "Baltica," or proto-Northern Europe. Later stages in the assembly of Pangea are identified by the age of the mountain ranges formed during the collision of Laurentia and Baltica, and subsequently the collision of Baltica-Laurentia with Gondwana [Essay 1]. There were many other continental collisions during this period, such as Baltica with Siberia and other Eurasian continental elements such as an island-area called Kipchak, and North China.

Many familiar mountain belts were formed on or near the margins of the modern North Atlantic during the assembly of Pangea. These included the mountains of Norway and Greenland, the Scottish Highlands, the English Lake District, the ancestral ranges of Brittany, Iberia, northwest Africa, and the East Coast of North America. This general process was complete about 280 MYA. However, some mountains, the Alleghenies and Ouachitas, for instance, continued to form because of the jostling of Gondwana and Laurasia (**1**), as they assumed their final jigsaw fit. The completed panoply of Pangean mountains formed an almost continuous belt over a stretch of 9,650 km (6,000 miles) from Spitsbergen to Mexico. Here the Spitsbergen-Mexico axis linked with other belts in the global system at that time, including the ancestral Andes, the Cape Folds of South Africa, the Transantarctic belt, and the Australian east coast fold-belt (**2**).

The architecture of the mountain ranges around the North Atlantic was closely but separately studied by European and American scientists for more than a century before the prime cause of their formation was understood. Naturally, the European and Greenland coastal ranges and the North American coastal ranges are named differently: the Caledonides and the Appalachians. But it is now known that the mountain-building events that led to the formation of these ranges, their orogenies, dovetail both in time and cause. Meanwhile, different terms had been coined for virtually the same orogenies: "Variscan" and "Caledonian" (and other names) to describe European mountain-building episodes of the time, "Taconian," "Acadian," and "Alleghenian" (and other names) to describe corresponding American coastal orogenies. To avoid what would otherwise prove to be a confusing array of nomenclature, the story of the assembly of Pangea that follows will be presented in terms of stages in the destruction of oceans and seas.

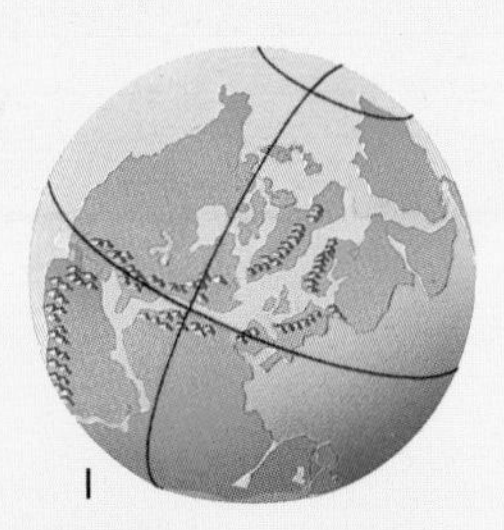

1

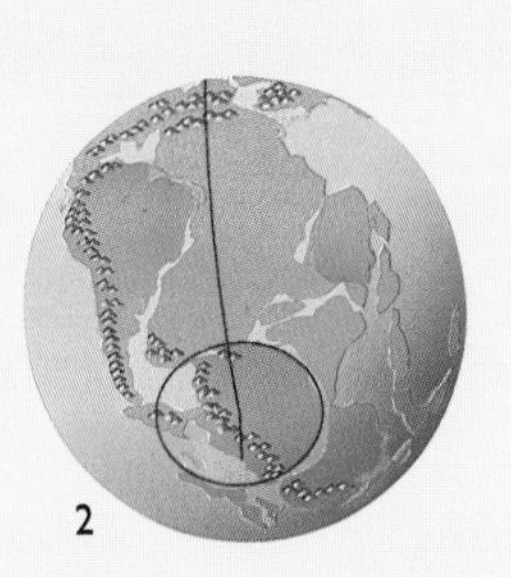

2

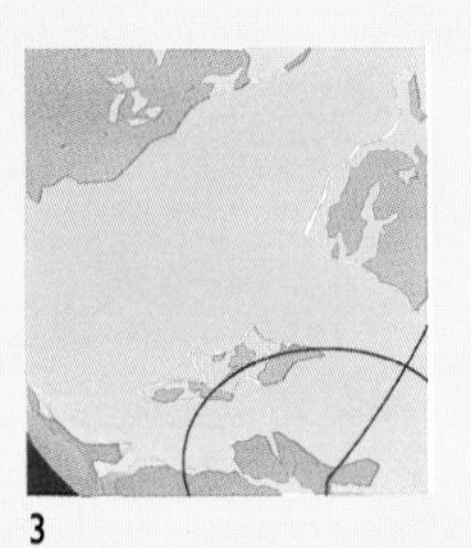

3

4

The first ocean in the proto-Atlantic region to meet its *coup de grâce* in the Pangean saga was the ocean that separated the Avalonian region of coastal Gondwana from the subcontinent of Baltica. Nearing the end of its cycle, this southern ocean had been reduced in extent to a sea called "Tornquist's Sea" (named after a Polish geologist). The second ocean to be destroyed was the Iapetus (**3**), which is at its widest as we pick up its story 490 MYA but is destroyed as the chapter progresses. The third is Iapetus' sister ocean, the Rheic Ocean, which opened as the Iapetus closed and was destroyed during the final assembly of Pangea. The name "Rheic" is derived from Greek mythology. (Rhea and Iapetus were brother and sister, the children of Gaea, the Earth goddess. The sibling relationship of Rhea and Iapetus explains the personification.) But first we turn to the destruction of Tornquist's Sea.

In 1910 a Polish geologist named Tornquist who lived and worked in Denmark, published a paper in a German scientific journal. The paper drew attention to a line of structural features in the natural boundary between Scandinavia and Poland, to the north of the line, and northern Germany to the south. In 1982 two British paleontologists, L.R.M. Cocks and R.A. Fortey, deduced from a wealth of fossil evidence, that around 480 MYA a narrowing sea, which they called "Tornquist's Sea," had separated Baltica from Avalonia. During the closure, the northern end of Avalonia collided with the Danish region of Baltica. Cocks and Fortey proposed that Tornquist's "line" of structural features (**4**) marks the line of suture that formed between the southern shores of Baltica and the northern shores of Avalonia when the two collided 440 MYA [Essay 2]. This collision and the destruction of Tornquist's Sea were thus the first of many steps in the making of the North Atlantic realm as we know it. For instance, the rocks of northern Avalonia underlie much of present England, southern Ireland, and the Low Countries.

There is a remarkable modern analogue for the separation of Avalonia from Gondwana: the cleaving of part of California from the West Coast of North America. If we indulge in reasonable but still speculative prediction about the future of California (based on recent computer modeling), we can project a clearer image of the Baltica/Avalonian event.

The coast of California today includes a 1,600-km (1,000-mile) block of mixed continental and volcanic rock. This *terrane* (the spelling is that of D. Howell who made the discovery) is a term adopted to describe all "displaced" or "suspect" terrain. In this case the "suspect terrane" is west of the notorious San Andreas Fault. The fault runs from the Gulf of California to Cape Arena on the coast of northern California. The "Baja-Arena" block is analogous to ancient Avalonia, in that Avalonia was also a mix of continental and volcanic displaced terranes once accreted onto the edge of Gondwana, just as the Baja-Arena block is now attached to North America. However, there was a considerable difference in size: Avalonia is categorized as a "microcontinent" while the Baja-Arena block is not—although it is a sizable chunk!

At present the Baja-Arena block is being slowly transported up the western edge of North America by the northward motion of the Pacific Ocean floor on which it is riding. This explains why the region is prone to severe earthquakes. Scientists who study this movement predict that in tens of millions of years the Gulf of California will extend to become an open-sea strait. As the Baja-Arena

[*continued on page 82*]

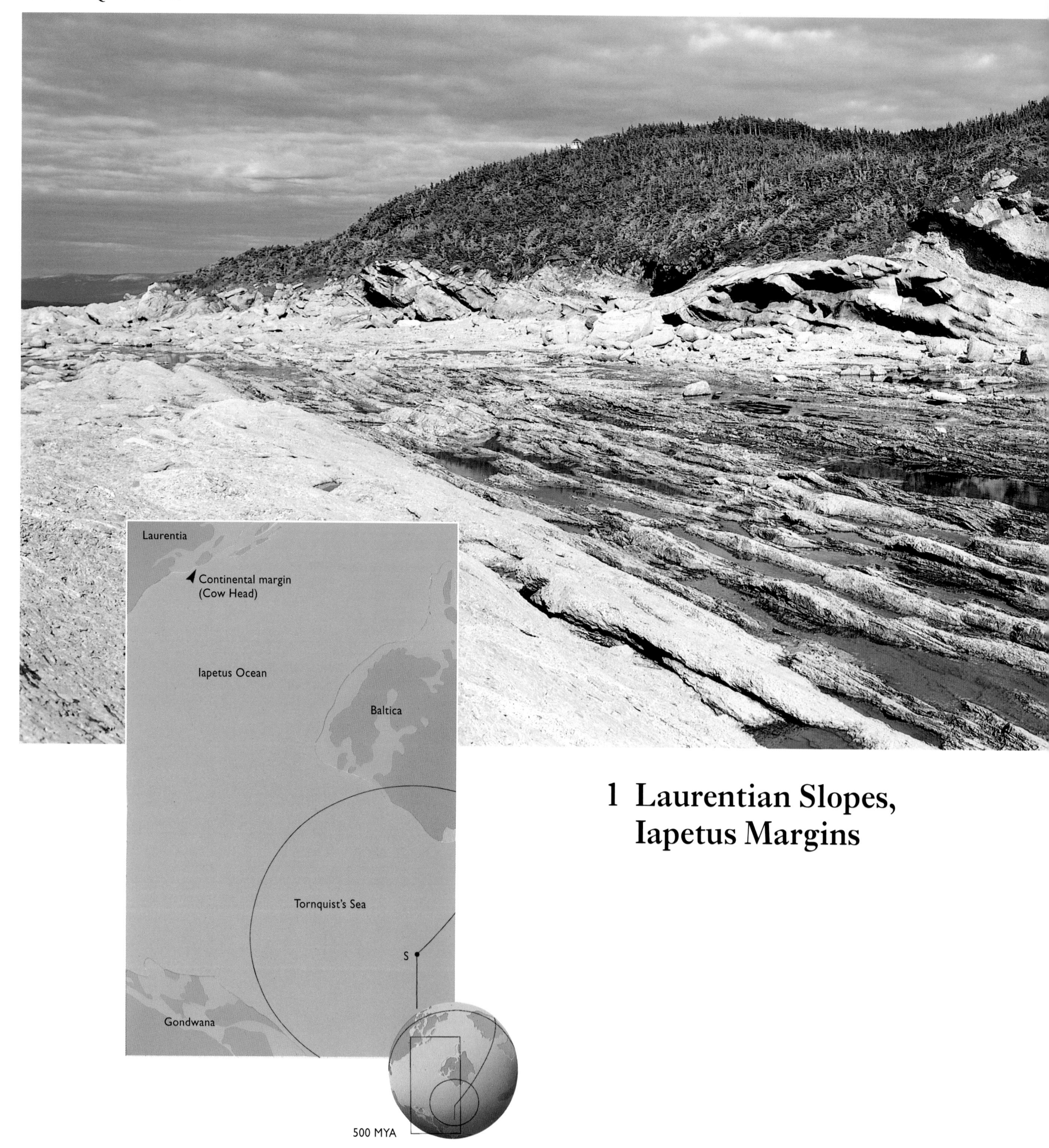

1 Laurentian Slopes, Iapetus Margins

Between 510 and 500 MYA Laurentia straddled the Equator from present Mexico to northern Greenland. The megacontinent began both to rotate and to drift eastwards towards the subcontinent of Baltica. Because of this movement, the Iapetus Ocean started to close. Meanwhile, southernmost Gondwana was driving towards the South Pole from the Eastern Hemisphere.

This panorama shows part of the Laurentian continental shelf on the Iapetus seaboard at that time. The shelf was tilted and obducted (pushed up) onto the continent's surface —now a part of the northern peninsula of Newfoundland—as the Iapetus Ocean closed during later stages in the assembly of Pangea. We are looking at the extreme edge of the shelf at the point at which it once plunged steeply into the Iapetus Ocean: the original angle of the continental slope can be judged by turning the picture counterclockwise through 45°.

The house-sized blocks of limestone that form the low-lying cliffs at the center-right of the picture are about to slide or tumble down the continental slope into the depths of the Iapetus. The rocks in the foreground are formed from "breccia," then part of an unstable submarine slope. Breccia is small, sharp, angular pieces of rock like those found at the base of cliffs on a "talus slope."

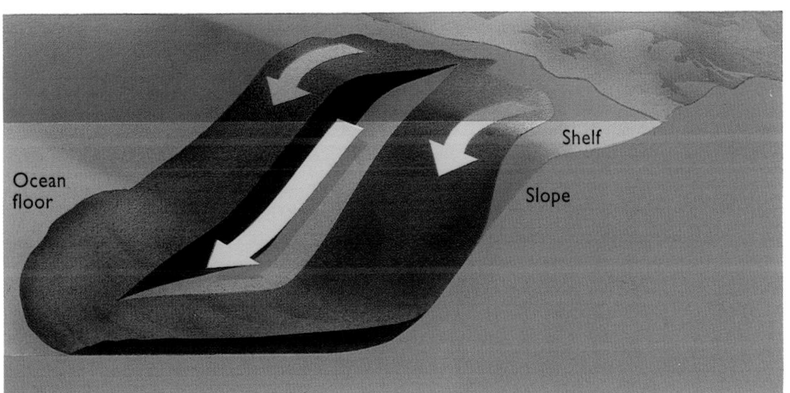

The Rias Altas, Galicia, Spain

2 Rheic Genesis, Tornquist Demise

At around 500 MYA a huge island archipelago, Avalonia, was wrenched away from the edge of the megacontinent of Gondwana. A narrow strait formed between the two margins as Avalonia was carried northwards towards Baltica. The strait widened into the Rheic Ocean (Rhea and Iapetus are the names of a brother and sister in Greek mythology).

At least part of the rifting process occurred between the Avalonian and Iberian regions of Gondwana, now the Rias Altas region of Galicia on the Atlantic coast of northern Spain (inset). The main picture, of cloud-covered Cabo Ortegal, shows one of five volcanic structures that have surfaced from great depth in the Rias Altas region. These were formed near the Antarctic Circle above a "hot spot" in the Earth's mantle 500–440 MYA—indicators of continental rifting and subsequent seafloor spreading. When active, each hot spot was 30 km (20 miles) or more beneath the surface and perhaps 100–200 km (60–120 miles) wide .

Avalonia moved north from Gondwana towards Baltica, with which it "docked" 440 MYA. The suture runs beneath the North Sea and divides Baltic Europe from the Avalonian regions of the British Isles and the Low Countries. The northern end of Avalonia

was separated from Baltica by "Tornquist's Sea," named after the geologist who first identified the probable suture line in 1910, but without knowing its significance.

The paleomaps show stages in this process. The trilobite and graptolite fossils *Opipeuter* and *Monograptus* are among those used to reconstruct this paleogeography.

[*continued from page 77*]

block moves out into the Pacific Ocean, it will become a 1,600-km-long (1,000-mile) archipelago. If the Pacific Ocean floor continues its present northerly direction of movement, the northern extremity of the Baja-Arena archipelago will eventually "dock" with Alaska. This startling suggestion is not as novel as it may seem, for it has been shown that many other continental displaced terranes similar to the Baja-Arena block have done this very thing in the past; much of Alaska, and other parts of the West Coast, consists of such accreted elements, some of which are believed to have traveled significant distances, perhaps a thousand kilometers or more, before reaching their present port of call.

This San Andreas model of strike-slip movement of displaced terranes, and continental accretion of the West Coast, prompted geoscientists to look again at the tectonic history of the East Coast of North America. They found overwhelming evidence that there is indeed a parallel: the East Coast is also largely made up of an accretion of ancient displaced terranes that include parts of Avalonia.

So today's northeast Pacific Ocean is closing, just as its much smaller counterpart, the Iapetus Ocean, was closing from about 460 MYA. The Gulf of California's tendency to propagate to the north and open a gulf between the Baja-Arena block (**5**) and the mainland, corresponds to the early stages of formation of the Rheic Ocean. The northern end of what will become the Baja-Arena archipelago will eventually dock with Alaska, just as the Avalonian microcontinent docked with Baltica with the closure of Tornquist's Sea. If the Pacific Ocean continues to close, the ultimate fate of the Baja-Arena archipelago might be its destruction between converging continents. This would be analogous to the fate of the Avalonian microcontinent.

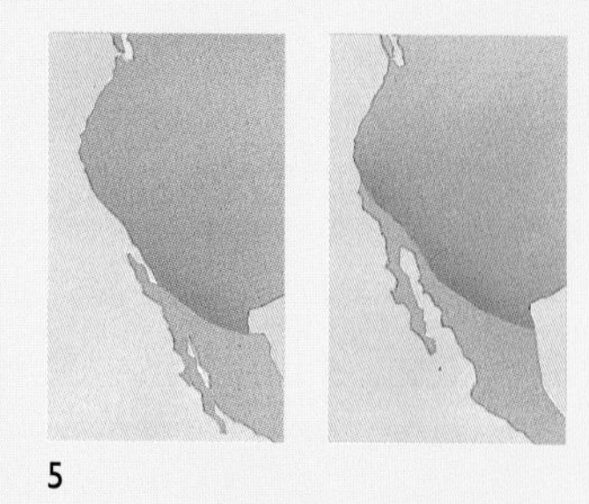

5

6

7

8

AS WE SAW in the previous chapter, the continental motions that had contributed to the closing of first the Iapetus and then the Rheic Oceans were in fact the rotation of Laurentia and the drift of Gondwana across the South Pole and north towards the Equator. These movements began as the Iapetus Ocean opened in late Precambrian times. The early Cambrian Equator crossed Laurentia from what is now Mexico through Hudson Bay and Greenland, almost at right angles to the present orientation of North America. During the opening of the Iapetus Ocean to its widest point (around 500 MYA) Laurentia drifted westward along this equatorial line while it rotated in a counterclockwise direction about an axis centered in the region of Hudson Bay.

The net result of Laurentia's gyration and Gondwana's extraordinary northward drive across the South Pole was that Baltica/Avalonia was caught between the two. The subcontinent of Baltica and its Avalonian attachment were large enough to survive the crush almost intact, but the rest of the long tenuous Avalonian arc was trapped between opposing continents—squeezed like a slippery pip between the fingers. Avalonia was crushed, and its pieces were scattered and accreted onto the East Coast of North America.

The fragments of the Avalonian microcontinent consisted of continental elements and volcanic islands. Today these fragments form the Avalon Peninsula of southern Newfoundland (**6**), the northeastern half of Nova Scotia, parts of Maine and Rhode Island, the Boston Basin, and regions of the Carolina states—together with central parts of the British Isles and the Low Countries that are now attached to Baltica. Avalonia stretched southwest from Baltica, separating Laurentia from Gondwana. The closing Iapetus Ocean lapped its northeastern shores and the widening Rheic Ocean lapped its southwestern shores.

At that time the non-Avalonian elements of the future Western nations were still scattered. For instance, Scotland and Northern Ireland were part of Laurentia on the far side of the Iapetus Ocean, along with the northern peninsula of Newfoundland. The basement rocks of Florida, Iberia (and other Mediterranean countries such as Italy and Greece), and northern France (Armorica) were all still a part of, or near to, the shores of Gondwana.

IT WAS THE DESTRUCTION of the Iapetus Ocean, and later the Rheic Ocean, that created those very Pangean landforms that eroded and later rifted apart to form the present North Atlantic Ocean bordered by its present spectacular landscapes and mountain vistas.

For example, a glance at any relief map of the British Isles is enough to give the sense of a distinctly NE/SW rake to the general topography of Ireland, Scotland, the north of England, and North Wales. This includes the borderlands between England and Scotland from Berwick-on-Tweed on the northeast coast of Britain to the Solway Firth on the west coast (**7**). And if this line is extended across Ireland with a distinct dip to the south, its length will mark the approximate line of suture between what was once the southwestern margin of Laurentia and the northeastern margin of Avalonia. Although the historic English/Scottish border was defined because it was the best line of defense, and bloody wars were therefore fought to defend it, the fact remains that the "border" and its extension across Ireland form the natural geological frontier between a part of North America and the rest of the British Isles. The Romans even immortalized part of its length by building Hadrian's Wall, completed in 136 AD.

As opposing continental margins converged (460–410 MYA) the Iapetus Ocean floor was assaulted from both north and south. Part of it was destroyed beneath Laurentian sea-covered shelves, with the consequence that volcanoes formed both offshore and onshore [Essay 4]. Millions of years of volcanic activity combined with erosion, sedimentation, and uplift to produce the weathered igneous and sedimentary rocks that we know today by such evocative names as Connemara in the Republic of Ireland and the Grampian Highlands in Scotland. To the south of the Iapetus suture zone similar volcanism resulted from the subduction of the Iapetus floor beneath Avalonia. This produced the volcanic and sedimentary rocks that now form the mountains of the English Lake District, of Snowdonia in North Wales, and of Leinster and Tyrone in Ireland [Essay 3].

In addition to much of the British Isles, and the basement rocks of Holland and Belgium, the northeastern element of the Avalonian microcontinent included the Avalon Peninsula of Newfoundland, part of Nova Scotia, and the now-submerged Grand Bank of Newfoundland—which because of its fisheries became one of the key factors in the development of the Western World. Britain and Newfoundland, now separated by several thousand miles of open ocean, were for most of their existence literally neighbors.

UNLIKE THE IAPETUS subduction beneath Avalonian Britain and Ireland, slivers of the ocean floor abutting Avalonian Newfoundland and Nova Scotia were "obducted" instead of "subducted" during their collision with Laurentia; they were pushed up onto the Laurentian continental margin. Indeed a large region of central Newfoundland, the Dunnage Zone, contains vestiges of the Iapetus. Here a large variety of "ophiolites," ocean-floor rocks and sediments, form the landscape: mountains formed from pillow lavas, vertical dikes, and gabbro. The corresponding region of Britain, the Southern Uplands of Scotland, is formed from an "accretionary prism," a wedge of Iapetus Ocean ophiolites accumulated as the seafloor plowed up onto the continental margin. (The island of Barbados is a modern accretionary prism: **8**.)

Newfoundland is structured from tectonic elements that have their counterparts in the makeup of the British Isles. The most obvious NE/SW diagonal features that segregate some of these elements in both Scotland and Newfoundland are the Great Glen in the Highlands of Scotland and the Grand Lake region separating the northern peninsula of Newfoundland from the Dunnage Zone. The first marks the line of the "Great Glen Fault" and the second the line of the "Cabot Fault" [Essay 5].

The Great Glen Fault is the site of perhaps the best known of many enchanting tourist haunts in Scotland. It nestles between the rugged bastions of the Highlands of northwest Scotland—on the northern shores of Loch Ness and Loch Lochy—and the more gently undulating Grampian Highlands, which rise like billowing heather-covered waves from the lochs' southern shores, with Ben Nevis and Glencoe marking their southwestern extremity. What is perhaps not generally known is that the Cabot Fault nestles in a remote and inaccessible but equally enchanting lake-filled glen remarkably similar to the Great Glen, a similarity that goes beyond casual coincidence of the form of the lakes and mountains; these places are closely related. Both faults have been long dormant and are now offset and separated by about 3,200 km (2,000 miles) of North Atlantic Ocean, but 380–340 MYA they formed sections of the same curving complex fault system that followed the margin of northeastern Greenland.

It is very remarkable that the movement along strike-slip faults, the drift of continents, the opening and closing of oceans, and other long-ago events can be reconstructed in some cases so precisely in both time and place that paleogeographers can draw a general overall picture of the Earth's geographical history. They make these interpretations from "paleolatitudinal indicators" in the form of geomagnetic, fossil, and environmental data. They also compare both the tectonic style and the geology of rocks on opposing continental margins such as those on either side of the Great Glen and Cabot Faults (**9**). But the clinching confirmation for such reconstructions, or even the rationale for proposing them in advance of the conclusive geophysical data, is the form and occurrence of contemporary life found in the fossil record. Deducing the life cycle of an ocean, the proof of a Tornquist's Line or of an Iapetus suture, or of the existence of a Rheic Ocean, depends as much on the interpretation of fossils and deducing the environments in which the fossilized animals once lived, as upon the style and evolution of the rocks themselves. But before we can discuss this interplay of paleontology and geology we must first get a perspective on our human perception of "time." It would be misleading to discuss organic evolution and geological evolution in the same unqualified time frame. We need an analogy to provide a perspective. Why not make the number of words in this chapter represent the period of time the chapter covers?

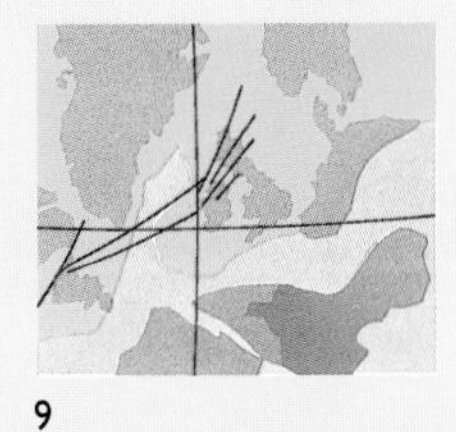

9

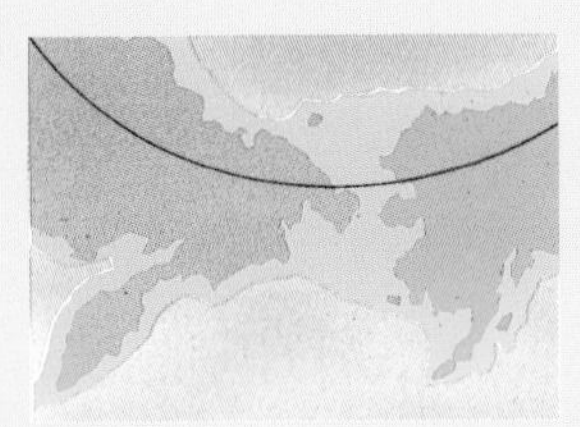

10

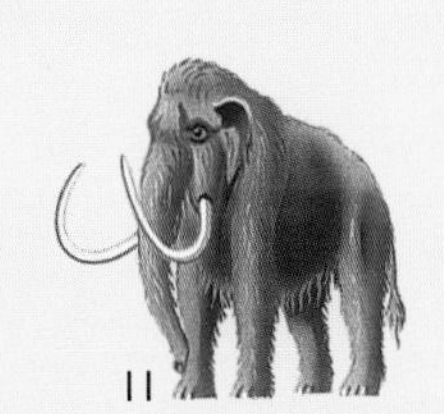

11

APPROXIMATELY 6,500 WORDS will be needed to describe the events of the 145 MY covered in this chapter. Each one word therefore represents about 25,000 years. Although in human terms 25,000 is an impressive number of years, it becomes insignificant if expressed in geological terms: 0.025 MY. Yet as we shall see in the following paragraphs, 25,000 years can be a very significant period in both evolutionary and environmental terms but not long in tectonic terms. To make this point we need to return to the geological present from the geological past, for a few paragraphs.

THE RECENT PLEISTOCENE EPOCH, a major stage in the present ice age, lasted about 1.6 MY, and during this time there were a number of glacial advances, glacial retreats, and interglacial periods (we are living during an interglacial period in the continuing Cenozoic Ice Age at the moment). The last major glacial advance reached its peak 18,000 years before the present (BP), when much of North America and Europe were covered in ice sheets two miles thick. By about 10,000 BP these had already begun to retreat, and today in the Northern Hemisphere only the Greenland ice sheet and a few relatively minor ice caps in Iceland remain.

Now let's consider just a few of the evolutionary and environmental events of that 18,000-year period—remembering that tectonic movement in that time accounts for an extra mile or so on the width of the North Atlantic Ocean.

During the glacial advance global sea level fell by 100–140 m (300–400 ft) or more below the present level because of the great volume of water taken up by continental ice. This caused continental shelves between islands and mainlands to be uncovered around the world and these, in turn, allowed animals and humans to migrate and to populate virgin territory. The best-known such "landbridge" is the at-times 1,000-km (600-mile) wide Bering Landbridge between Asia and North America (**10**), because it was this subcontinental-sized region that enabled humans and animals to migrate to the North American continent from Asia. As global climate warmed from about 18,000 BP, the ice sheets retreated, so that sea level rose again and isolated humans and animals in their new localities.

As recently as 10,000 BP many species of long-haired ice age animals that are now extinct (**11**) still roamed the tundra and plains of the Northern Hemisphere in vast numbers. In that 10,000 years the human species has advanced from hunting and butchering with stone tools to practicing husbandry, building computers, and manipulating genes. And perhaps even more significantly, in just 10,000 years human global population has grown from a few tens of millions at most to almost six billion—a growth rate that seriously threatens overpopulation of the Earth by humankind and the extinction of many species of plants and animals and possibly ourselves.

The stage for this environmental and evolutionary drama has been set and played out in the space of less than one word's length in our chapter covering 145 MY. But you and I, whose individual life spans would not warrant a fraction of the space taken by one comma, would judge that the environmental and evolutionary changes that have either taken place or have been triggered in the course of that "one word" have been brought about by "gradual" change.

THIS PERSPECTIVE OF TIME allows us to think about what is meant by words like "gradual" and "rapid" when arguments arise about the respective merits or demerits of Darwin's "gradualistic" theory, and the newer theory of "punctuated equilibrium." Gradualism claims that evolution of species takes place steadily by natural selection, while punctuated equilibrium, as proposed by N. Eldredge and S.J. Gould in 1972, claims that evolution only takes place after long intervals of standing still, or stasis, "punctuated here and there by rapid events of speciation." So vehemently did the advocates of the Eldredge-Gould concept of punctuated equilibrium express their views that the issue became a media event. Press articles and television programs suggested that there was a great row going on in the academic world (which there was) because Darwinian theory had proved fallacious and was about to be replaced with a new theory (which it was not). What was really happening was a healthy, if somewhat heated and overdue, reappraisal of Darwin's century-old ideas in the light of present-day knowledge.

Given that the environment of the Earth can be reasonably stable for long periods (millions of years) and that it can also change to an ice age extreme "rapidly," how can we judge the speed of evolutionary change? Many specialists now agree with the view expressed by C. Emiliani: that living things in the geological past evolved "in a continuous spectrum ranging from lethargic to spasmodic." As global environmental conditions waxed and waned during the assembly of Pangea and after, speciation proceeded either slowly, when ecological stress was applied gradually over millions of years, or with alacrity, when ecological stress was "rapidly" applied as a result of "sudden" global environmental stress. The former resulted in selective speciation and the latter in multiple speciation in a single stage—the processes differed from each other only by a matter of degree.

SMALL BUT EVIDENT increments of evolutionary change in the same families of animals over long periods allow paleontologists to make judgments about the disposition of continents and oceans hundreds of millions of years ago. The present, very complex geographic and stratigraphic location of fossil animals is plotted from the results of years of patient fieldwork, library research, and reference to existing fossil records; then the possible original location of the sedimentary rocks in which the fossils were found is reconstructed.

[*continued on page 90*]

3 Iapetus Destruction

In this panorama of Borrowdale and Derwent Water in the English Lake District, we are looking across what was once a level coastal region of northern Avalonia and, figuratively, in the direction of an approaching Laurentian margin. The subsequent collision and fusion were the cause of massive deformation. The Avalonia-Laurentia suture is just over the horizon beyond Skiddaw (the mountain right of center), and follows a diagonal path across the British Isles from Berwick-on-Tweed on the east coast, through the Solway Firth to the west coast of Ireland. Skiddaw is structured

from folded slates (and some volcanic rocks) that were first deposited as mud and silt on the Avalonian margin to a thickness of several miles. The pressure and heat resulting from the destruction of the Iapetus seafloor turned the mudstone into slate.

As the Greenland region of Laurentia approached Avalonia (448–438 MYA) the Iapetus Ocean floor was subducted beneath the Avalonian margin, causing a chain of volcanoes to form from Derwent Fells above Borrowdale (the hills on the left horizon) to Snowdonia in North Wales, and Leinster in Ireland. As a result, a deposit of volcanic ash and sediments piled up a mile deep above the slates, and onto this pile yet another mile-thick deposit of limestone—the product of warm, shallow, coral seas that flooded the continents at that time.

When the last of the Iapetus Ocean floor was destroyed beneath this enormous sedimentary basin, the collision with Laurentia caused this edge of Avalonia to be uplifted, folded, and fractured. The *coup de grâce* came around 408 MYA, with the injection of vast amounts of granite beneath Shap Fell, to the east of this picture.

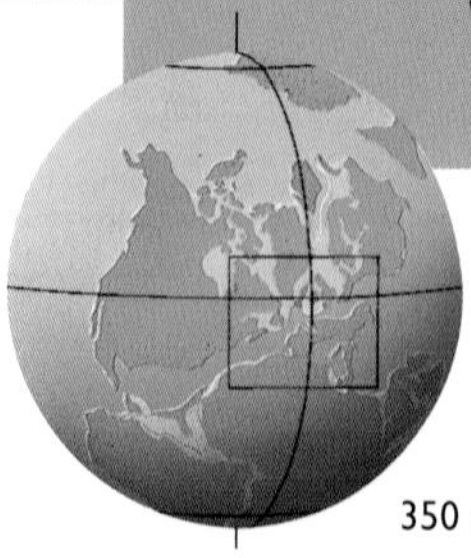

4 Subduction and Orogeny

The assembly of Pangea continued as Avalonia/Baltica collided with Laurentia. This collision was along a southerly front that stretched from Newfoundland to New York on the Laurentian coast of North America.

As remnants of the Iapetus Ocean floor were being destroyed between Avalonia and Laurentia, the Rheic Ocean floor was itself gradually being subducted beneath the south coast of Avalonia (see diagram above). This section of the Maine coast in Acadia National Park was a part of that region. It was once the roof of a huge chamber of softened and partially melted rock that formed above a mass of intruded granite known as a pluton. This and many other plutons developed beneath the newly formed Laurentian continental margin around 350 MYA.

Plutons form at great depth during the subduction of an ocean plate, and when continents collide. In either case, water driven off from the descending rocks acts as a flux that causes adjacent crustal rocks to melt. As the rocks melt they form huge balloon-like globules of molten magma. Because they are lighter than surrounding rock they rise towards the surface. As they do so, they melt overlying and adjacent "country" rocks that shatter, soften, and fall into the molten granite.

The panorama shows the upper part of a plutonic chamber in which granite from the pluton has penetrated every nook and cranny in the surrounding country rock. The inset diagram (right) shows a small piece of sedimentary country rock that has peeled away from the roof of the chamber and fallen into the granite brew.

Hunter's Head, detail

The Great Glen Fault, Loch Lochy (foreground) and Loch Ness (mid-distance), northern Scotland

5 Relative Faults

The loch-filled Great Glen of Scotland is well known to tourists throughout the Western World: the Northern Highlands are on the left and the Grampian Highlands are on the right of the inset picture. But Grand Lake (the main picture), roughly 3,000 km (2,000 miles) away across the North Atlantic in a remote region of Newfoundland, is relatively unknown. Yet the rocks of both regions are Laurentian, and both mark the once-continuous general line of the same fault system, the Great Glen and Cabot Faults, which were active 380–340 MYA during the assembly of Pangea. While all the basement rocks on both sides of the fault system are Laurentian, the regions to the left and right of both pictures came from different parts of Laurentia.

The Cabot Fault, Grand Lake, northern peninsula of Newfoundland, Canada

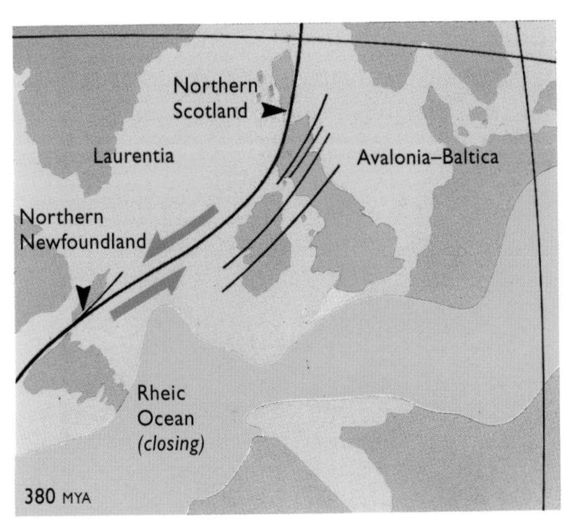

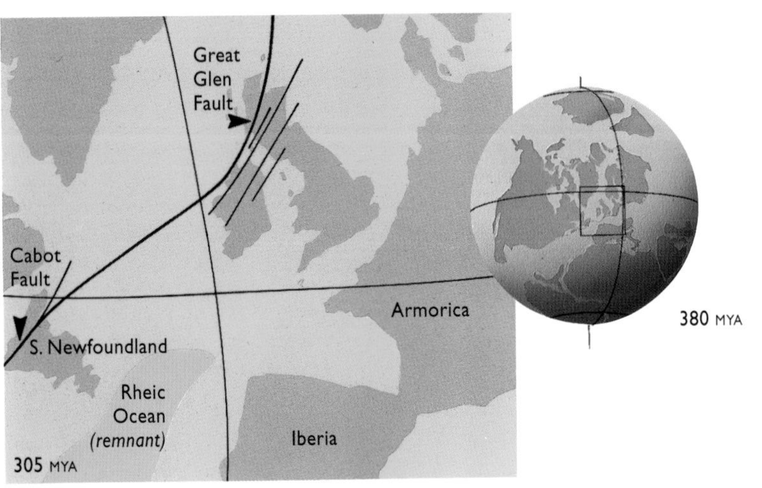

[*continued from page 83*]

It was in this way that Cocks and Fortey reached their conclusions about the paleogeography of the disposition of Tornquist's Sea. They drew on the fossil record and the evolutionary succession of marine invertebrates within particular groups such as trilobites, brachiopods, graptolites, conodonts—soft-bodied animals with hard parts (**12**)—and mollusks, to name a few. Equatorial faunas were the most diverse and occupied a broad band, while an ocean apart and near to the South Pole off the shores of Gondwana there were fewer species, if any, of the same kind. There were differences between the same "genera" (of trilobites for example) of animals that lived off the shores of Laurentia, Avalonia, Baltica, and Gondwana, for they had evolved in different environments affected by latitude and continent–continent proximity. But as continental elements jostled, rotated, drifted, and eventually collided, so the faunal differences on either side of Tornquist's Sea became less pronounced. After continental suturing there was little or no difference between members of the same genera of neighboring faunas [Essay 2].

THE FOSSIL RECORD contains minor glitches: "background extinctions," the occasional permanent disappearance of one or more species during relatively stable periods. There are also major glitches that provide paleontologists with global boundaries in the fossil record. These glitches are so extraordinary when judged by the numbers of species or families or even "orders" that permanently disappeared in a brief period of geological time during a "mass extinction," that they raise large questions about cause and effect.

There are many and often interrelated factors in a variety of scenarios that are thought to have caused mass extinctions. These include cycles of tectonic activity and geochemical changes of the kind we have discussed here and in previous chapters, changes in sea level, climatic change, extraterrestrial influences such as changes in the orbits of the Earth and Moon, and bolide impacts on Earth's surface ("bolide" is a general term used to describe any projectile hitting the Earth from outer space).

Whatever the cause of a mass extinction, the fossil record of the aftermath of that extinction demonstrates an unequivocal response. There have been five or more major periods of extinction during the past 700 MY. (Some paleontologists suggest as many as sixteen, depending upon how the word "major" is defined.) Each extinction event appears to have been followed by a period of renewed eukaryotic diversity. Each period of renewal has commenced after the decimation or even complete elimination of a broad spectrum of species and has resulted in an increase in the diversity of surviving eukaryotes. Of these, and by way of example, four minor mass extinctions between 540 and 490 MYA occurred when the Precambrian megacontinents were breaking up and Gondwana was approaching the South Pole from the Eastern Hemisphere. These extinctions decimated the early invertebrates, notably the trilobites, and were followed by both the resurgence of the surviving species of trilobites and by the evolution of other animals. These included starfish, corals, bryozoans and the jawless fish (510–460 MYA: **13**).

THE NEXT MASS EXTINCTION, the Ordovician event (450–440 MYA), when Gondwana was near the South Pole and Avalonia was about to collide with Baltica, affected a much wider variety of marine invertebrates, including the resurgent trilobites and the early reef-building animals that had appeared in the interim. It is thought that this mass extinction was caused by a general cooling of the Earth's climate, resulting in both an ice age and a radical change in the pattern of ocean circulation. The ice age began after a 40 MY period of high sea level (around 300 m or 1,000 ft higher than now) that had caused the flooding of low-lying continental regions by epicontinental seas. Global climate fluctuated during the preglacial period, but was always warmer than the global average today. One can imagine that marine life flourished during these periods of shallow seas that remained equably warm for millions of years at a time, suffering only

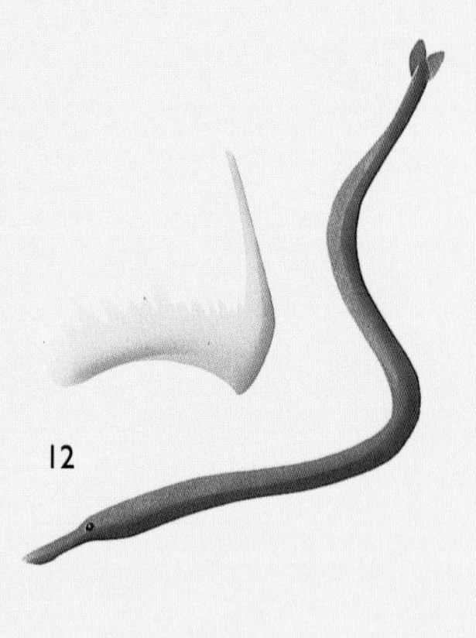
12

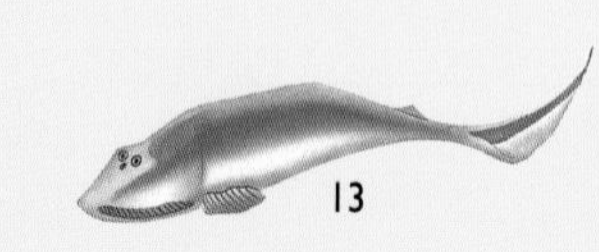
13

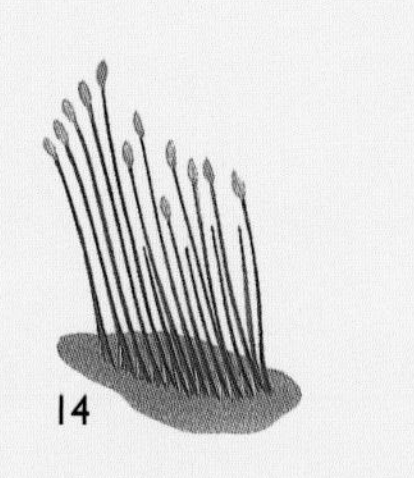
14

15

16

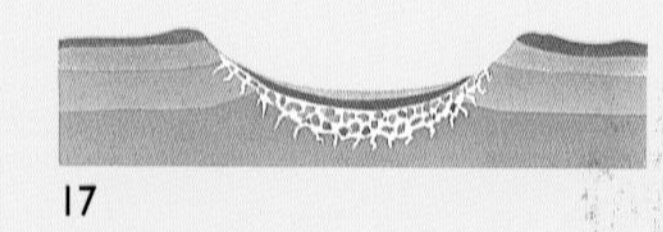
17

background extinctions induced perhaps by comparatively small changes in epicontinental sea level and climate.

This was an extremely active time in the Earth's tectonic history: Laurentia was rotating; Gondwana was drifting across the South Pole and northwards towards the Equator; Tornquist's Sea and the Iapetus Ocean were closing and the Rheic Ocean was opening. The swelling of ocean floors at the many mid-ocean spreading-centers had caused volume-for-volume displacement of the world ocean, and this displacement induced continental flooding—such displacement is thought to be the prime cause of epicontinental flooding.

These environmental conditions were the opposite of "normal" in the sense that ice ages "normally" result in a lowering of sea level below continental surfaces. But "sudden" change from balmy temperatures to colder conditions seems to have led to mass extinctions on a global scale and to the replacement of the biomass with those species of invertebrates that proved capable of adapting to such stressful conditions. When the ice age ended about 430 MYA epicontinental seas continued to flood the continents, but world climate conditions became balmy again, leading to a great surge and diversification of marine life.

THE NEXT EPISODE, the Frasnian event, occurred 367 MYA, when both Tornquist's Sea and the Iapetus Ocean had closed and the Rheic Ocean was nearly closed. It saw the end of many marine invertebrates, particularly reef-builders and families of primitive fish. Meanwhile the first plants had begun to flourish (around 420 MYA; **14**) and the first invertebrates (scorpions, arthropods) had appeared on land [Essay 6]. (The word "Frasnian" is the geological term for one of about 60 "ages" in the geologic time scale; it was followed by the "Famennian" Age. Together the Frasnian [374–367 MYA] and Famennian [367–360 MYA] Ages serve to comprise the "Late Devonian Period.")

In 1957 a Canadian scientist, Digby J. McLaren, now a distinguished elder-statesman of his profession, was mapping rocks in the Northwest Territories when he found (according to his own account) that he could trace the boundary between Frasnian and Famennian rock formations (**15**) by a well-marked feature: in the upper Frasnian formation there were abundant index fossils, particularly the brachiopod *Atrypa*, certain types of rugose corals, and bottom-dwelling immobile creatures called stromatoporoids (**16**); all were shallow-water organisms that lived on reefs or on sandy or muddy seabeds in epicontinental seas. But McLaren could not find one of these fossils in the overlying rocks of the Famennian formation. From this and similar observations elsewhere, McLaren deduced that there had been a major worldwide mass extinction of shallow-water fauna that had ended the Frasnian Age. He later suggested (in 1970) that "in view of the kinds of organisms that had disappeared, the most likely cause was extreme turbidity prolonged over many weeks and months." He went on to suggest that a possible explanation of such turbidity could be that a bolide of significant size had hit the Earth at the time of the Frasnian/Famennian boundary (**17**). Some additional supporting evidence for this conclusion has been found in southeastern Nevada (J.E. Warme et al. 1991). Here, in Late Devonian times, about 13,000 sq. km (5,000 square miles) of surface rock was torn up and disrupted by catastrophic waves or currents of the kind that could have been generated by a bolide impact in the ocean.

Nevertheless, the prime cause of the Frasnian event is still a matter of sharp debate. In a sense it has been overshadowed by a separate and much publicized theory that the dinosaurs may have been extinguished by a bolide impact around 65 MYA. However, immediately before the Frasnian mass extinction, global temperature appears to have been much higher than today and epicontinental seas still flooded the continents: there is no record of an ice age in Laurentia, although the Brazilian region of Gondwana was then in the region of the South Pole. Regardless, there was an undeniable and sudden startling change in this environment, and there appears also to have been an almost instant annihilation of many species of both marine invertebrates and vertebrates on a global scale.

A NUMBER OF BOLIDES hit the Earth around this time, as supported by the evidence from Nevada. Judging from the size of their impact craters none were small, and at least one was very large. It made a crater in the region of Siljan, Sweden, which measures 51 km (32 miles) across. This means that the object that made the crater could have measured over 5 km (3 miles) in diameter (the rule of thumb for guesstimating the size of a bolide is to divide the crater diameter by ten). The bolide might have been a stray asteroid, an icy comet with a rocky core (like Halley's Comet), or an object from outside the solar system. But whatever the nature of this bolide, the crater it made on impact is dated 372–358 MYA—the Frasnian Age was 374–367 MYA. The Siljan impact is not necessarily the one that caused the Frasnian event, but it is at least a serious candidate, although some scientists do not agree that a single bolide could cause a mass extinction. However, not all bolide impacts are known. Averages of six out of every seven are likely to have landed in the ocean, so there would be no trace of any impacts in the ocean more than 180 MY old, because ocean floor up to that age has been destroyed.

IN FRASNIAN SEAS, the early jawless fish that had first appeared about 510 MYA (the precursors of today's flesh-boring and blood-sucking lampreys and hagfish) coexisted with the sometimes giant armored fish, known as arthrodires (**18**), and the cartilaginous fish. Some of the first cartilaginous fish, which had appeared about 410 MYA (ancestors of the modern shark, skate, and ray) may have given birth to live young. Although cartilaginous fish generally were restricted to salt water, they had the world ocean at their disposal and were therefore under little or no pressure to improve their extraordinarily successful anatomical and exterior design: their modern counterparts are presumed to be relatively unchanged from their ancestors. Female sharks today give birth to live pups that struggle to free themselves from a placental-type envelope, and then to break their placental umbilical cords. Yet in spite of this early evolutionary advance, which presumably applied to the extraordinarily successful ancestral sharks, it was the lobefin fish, the Rhipidistians, that were the most successful in terms of ultimate evolution: they led to amphibians, reptiles, and mammals.

Rhipidistians (Gr. *rhipidos*, fan: a reference to fan-like fins) had bony skeletons, probably fertilized their eggs outside their body, and had adapted to freshwater environments (**19**). They all had fins with hand-like bone structures, as seen in the modern coelacanths occasionally caught off Madagascar (coelacanths appeared about 390 MYA). By the end of the Devonian Period (360 MYA) some freshwater Rhipidistians (**20**) had evolved into tetrapodal amphibians—the primary ancestors of all terrestrial vertebrates. By this time most jawless and armored fish had become extinct, along with a great variety of species of shallow-marine invertebrates: victims of the Frasnian mass extinction. The Rhipidistians survived until about 268 MYA, but were extinct long before the greatest of all known mass extinctions, the Permian mass extinction (255–250 MYA) that followed the final assembly of Pangea.

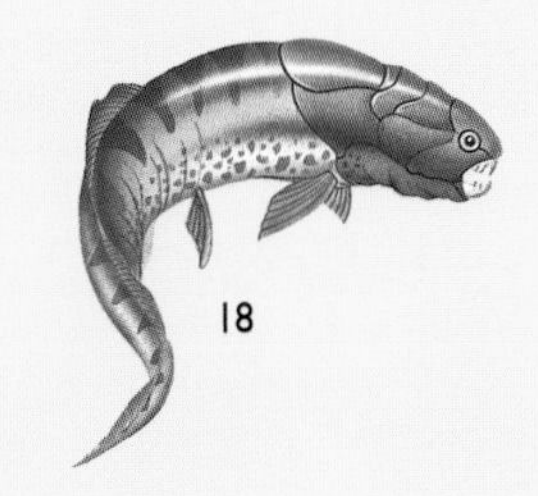
18

19

20

21

THE EARLIEST AMPHIBIANS appeared 374–360 MYA and were to some extent intermediate between the Rhipidistians and the early reptiles. They were clumsy-looking animals over one meter (three feet) long with comparatively large skulls and stocky limbs. Their skeletal pattern was the paleontological equivalent of a Model T Ford, because all subsequent terrestrial vertebrates are built on similar lines. Reproduction, like that of their Rhipidistian forebears, probably depended on external fertilization. They laid their eggs in water and their offspring went through an aquatic larval stage.

The oldest known amphibians include *Ichthyostega*, a carnivorous creature (**21**) that could open its jaws wide enough to attack, kill, and feed off large victims—including other *Ichthyostega*. They fed in shallow water or on fish stranded at the margins of lakes and swamps. Large terrestrial invertebrates (such as crunchy millipedes around a meter long) also provided at least part of their food supply. Being land animals (although restricted to a swampy environment) gave the amphibians an advantage over the predatory shallow-water Rhipidistians. By competing with *Ichthyostega* and its kindred species for food in shallow-water swamps, the Rhipidistians contributed to the pressures of a post-mass extinction era. But the situation now favored animals with tetrapod limbs, which allowed them to move from pond to pond—which Rhipidistians themselves could not.

The amphibians took full advantage of their terrestrial mobility. They lived in the warm, humid world of land plants and swamps that predominated at the continental margins, river banks, and lakes of late pre-Pangean times. With these advantages the extraordinary rate of reproduction, radiation, and diversification of early amphibians is not surprising.

Meteor Crater, Canyon Diablo, Winslow, Arizona

6 Marine Catastrophe, Terrestrial Providence

The remote volcanic region of Iceland shown in the main picture here may resemble the terrestrial conditions in which the first primitive land plants began to clothe swampy areas of the Earth around 420 MYA.

Towards the end of the Devonian Period (408–360 MYA), known as "the Age of Fish," early jawless fish were evolving simultaneously with gigantic armored fish, but none of these animals appear in the fossil record after the period ended. Jawless and armored fish, along with many species of marine invertebrates, were victims of a mass extinction that took place around 367 MYA.

Several large bolides hit the Earth about this time, and one or more of these might have contributed to the Devonian extinction. One bolide dated between 372–358 MYA made a crater in Sweden that measures 52 km (32 miles) across, and may have been formed by a bolide measuring 5–6 km (3–4 miles) in diameter. This impact is at least a serious candidate for contributing to the Late Devonian mass extinction. There is also evidence of a gigantic tsunami in Nevada in Late Devonian times that suggests an ocean impact at the time. The picture on the previous page is of a small impact crater formed when an extraterrestrial object hit the Earth over 5,000 years ago.

Cartilaginous fish (ancestors of today's sharks and rays) survived the catastrophe, as did some coelacanth-like lobefin fish (the Rhipidistians). These creatures were perhaps the most important product of evolution in the "Age of Fish" because shortly after its end lobefins evolved into tetrapodal amphibians—they were therefore the prime ancestors of all four-limbed terrestrial animals, including *Homo sapiens.*

Rhipidistian (lobefin)

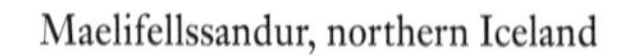
Maelifellssandur, northern Iceland

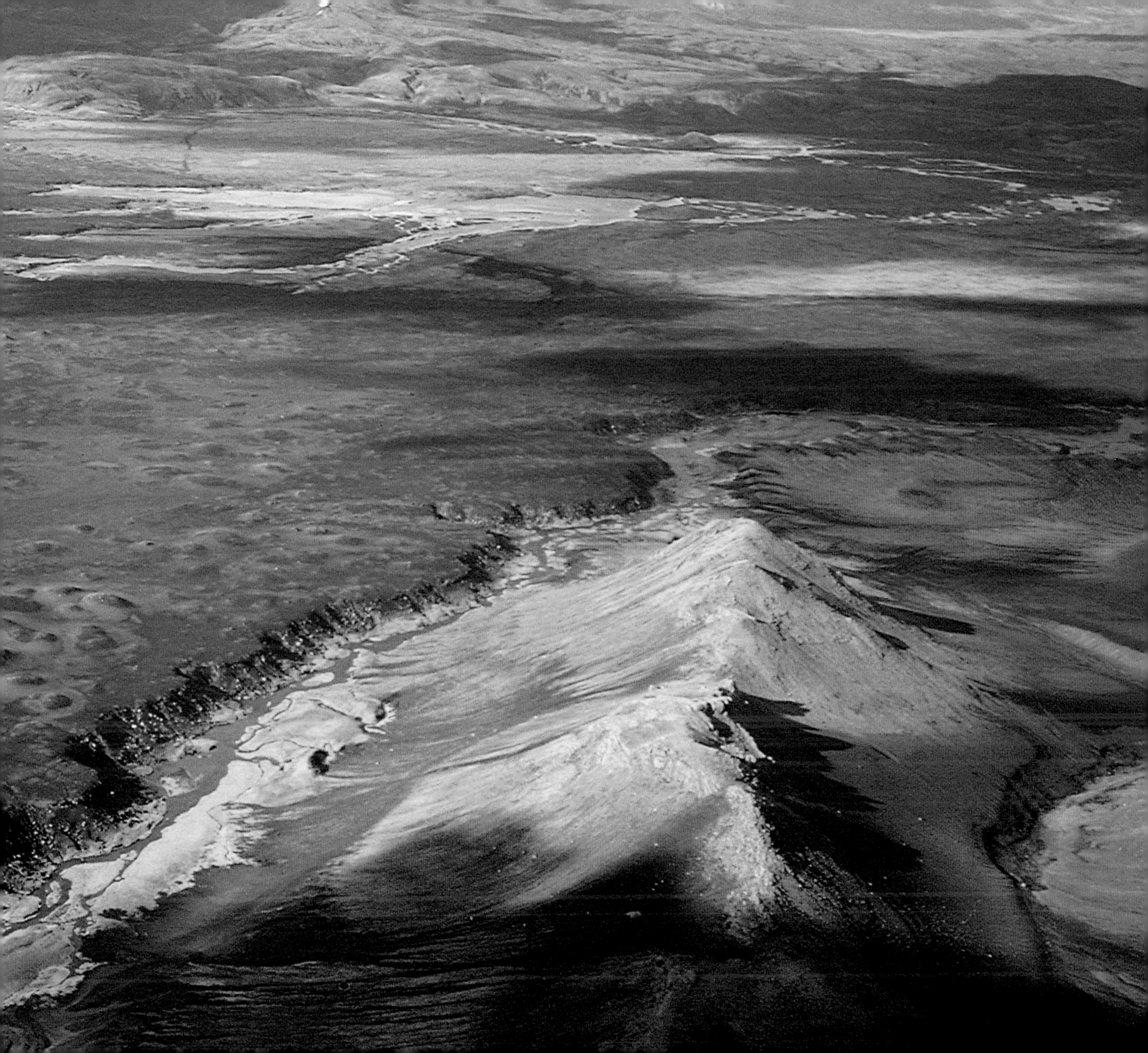

V THE THIRD AGE

Boscastle Harbour, North Cornwall, England

About 450 MYA Avalonia and Armorica, two neighboring regions on the leading edge of Gondwana near the South Pole, were rifted apart by the formation of a new ocean, the Rheic Ocean. A hundred million years or so later the two micro-continents were rejoined as they collided near the Equator during the destruction of the self-same ocean that had caused their initial separation. That collision, during the assembly of East Central Pangea, caused spectacular deformation of the kind we see in the foreground in southwest England. The picturesque Cornish harbor in the background and the valley beyond were the site of a collision between two minor elements of Avalonia jammed together during the

THE THIRD AGE
360–286 MYA

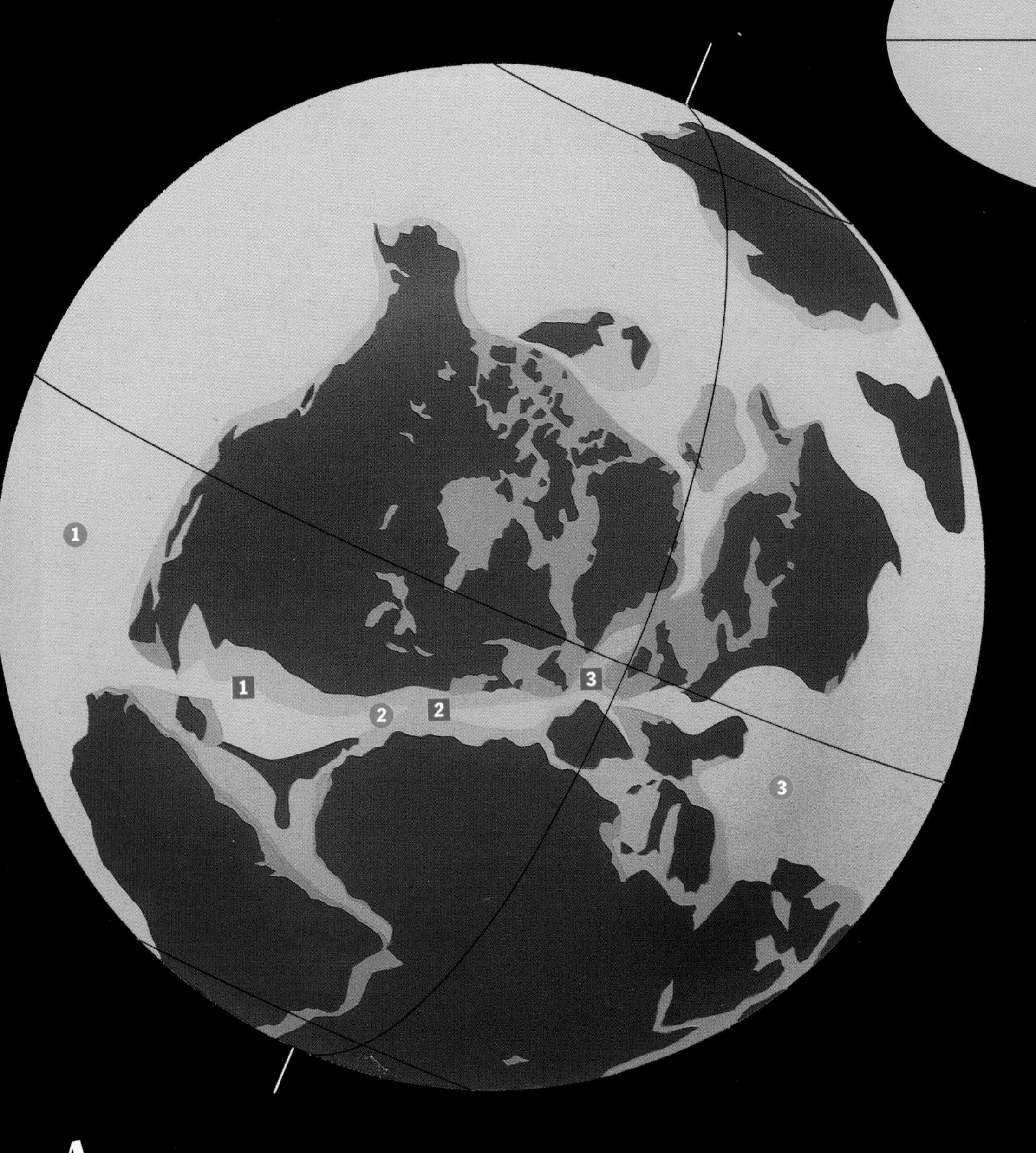

SUPERCONTINENTAL PANGEA

1 West Central Pangea

2 Mid Central Pangea

3 East Central Pangea

OCEANS

1 Panthalassa (future Pacific Ocean)

2 Rheic Ocean (in course of destruction)

3 Tethys Ocean (future Mediterranean)

All that remained of the Rheic Ocean at the time of this global reconstruction at 340 MYA—during the final assembly of Pangea—was a narrowing seaway between two interacting megacontinents. In East Central Pangea the Rheic seafloor was being destroyed during the collision of Avalonia (the British Isles and the Low Countries) with Armorica (Brittany and the Rhineland region of Western Europe). Southwest of this region a stray piece of Gondwana called Meguma Terrane (southern Nova Scotia) collided with northern Nova Scotia (a part of Avalonia already sutured to Laurentia). Meanwhile, east of the Meguma fragment the Rheic Ocean was being destroyed beneath Iberia. The broader picture depicted on the oval projection makes clear the difference in size between Laurentia, which had now incorporated both Avalonia and Baltica (but not Siberia), and the enormous Gondwanan assembly that dominated the Southern Hemisphere.

The Lord of the Rings is a novel, a tale of heroic dimension written by J.R.R. Tolkien (1892–1973), an Oxford University professor of English language and literature. Tolkien's characters lived in the good or evil kingdoms of Middle-earth, which were separated by mountain ranges stretching from the Northern Waste to Far Harad. There is an incidental but astonishing resemblance between Tolkien's imaginary Middle-earth, and the mountainous central region of Pangea with its fantastic plants and fearsome creatures. For instance, in a summary of Hobbit history (in *The Fellowship of the Ring*) Tolkien wrote,

"Those days, the Third Age of Middle-earth, are now long past, and the shape of lands has been changed; but the regions in which Hobbits then lived were doubtless the same as those in which they still linger: North West of the Old World and East of the Sea."

The reader may readily understand why Tolkien's phrase "the Third Age of Middle-earth" has been adopted for the titles of two chapters in this book. This and the following chapter will tell the story of the "Elder Days that are now lost and forgotten," the story of the incredible but *real* world of Wegener's supercontinental Pangea. They will also tell of the evolution and catastrophic end of many of its extraordinary forms of life.

Central Pangea was not only an intriguing part of the Ancient World in the Tolkienian sense. It was an assembly of the once-contiguous continental bits and pieces of Central Pangea that now rim the North Atlantic Ocean and form the sovereign states of the present Western World.

At around 350 MYA, the margins of the present continents that surround the North Atlantic Ocean were in close proximity (**1**). They straddled the Equator and stretched from the shores of an ocean-sized embayment to the east called the Tethys Ocean, to the very edge of Wegener's world ocean, Panthalassa, in the far west. During its assembly, the future Central Pangea was at first geographically similar in appearance to the Sea of Japan today. It had a marginal island-arc complex similr in tectonic style to modern Japan, open to the ocean. Later it was Mediterranean-like, a closed sea. Ultimately (around 250 MYA) it was a contiguous if restless whole (**2**). But throughout it was broadly divided northeast to southwest by almost unbroken chains of coastal or continental mountain ranges, almost certainly snowcapped at times, and in the later stages of formation perhaps Himalayan in stature. Fast-flowing rivers scoured mostly barren ruddy-colored landscapes [see pages 116–17], but from 300 MYA some regions were clothed by brilliant green tropical swamp forests.

We left the first stage of the assembly of Pangea at about 360 MYA, when Laurentia and Baltica/Avalonia were firmly welded together to form parts of the megacontinent of Laurentia-Baltica. Only the Rheic Ocean, the ailing sister of the demised Iapetus Ocean, now separated the northern group of continents, Laurentia-Baltica, from Gondwana in the south. During the destruction of the Iapetus, systematic strike-slip faults—long, deep, almost parallel and gently curved fractures—had formed along the northeastern edge of Laurentia, between the two rotating blocks of Laurentia and Baltica/Avalonia as they collided. The continued slippage between the opposing surfaces of these faults caused previously unrelated terranes to become juxtaposed: opposing shores of Loch Ness in the Great Glen and Grand Lake in the northern peninsula of Newfoundland are notable examples [see pages 88–89].

Such events were a rehearsal for the tectonic trauma that led to the final assembly of Pangea, when Gondwana and Laurentia-Baltica collided. While the scenario was similar—continent–continent collision—the scale was colossal in comparison. According to C.R. Scotese (the author's mentor for paleogeography) and others, a minor clockwise rotation of Laurentia-Baltica (**3**) (about 2°) and a simultaneous major clockwise gyration of Gondwana (about 20°),

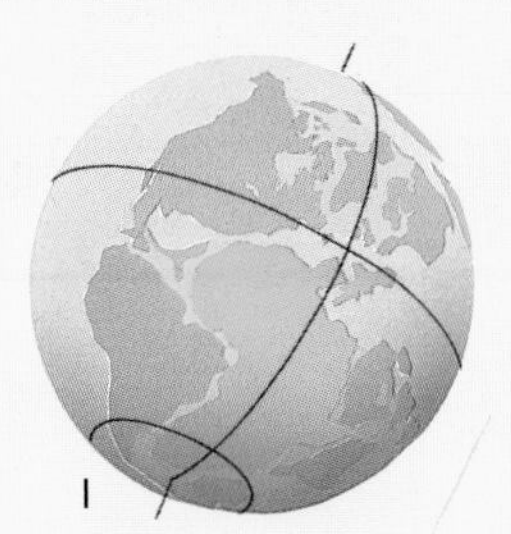
1

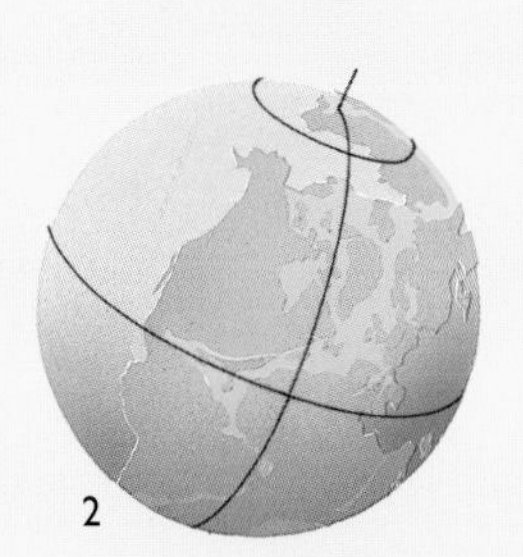
2

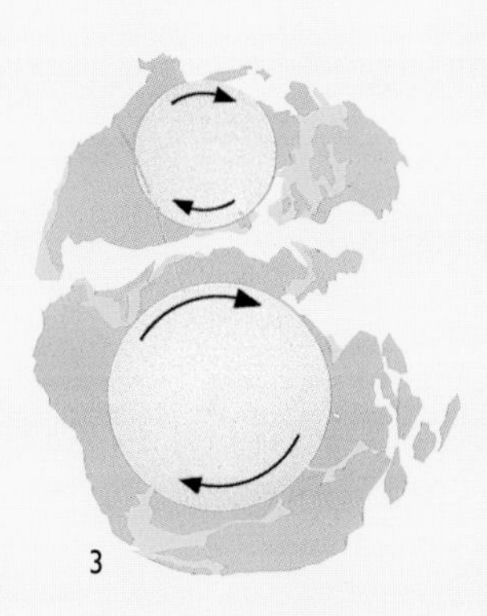
3

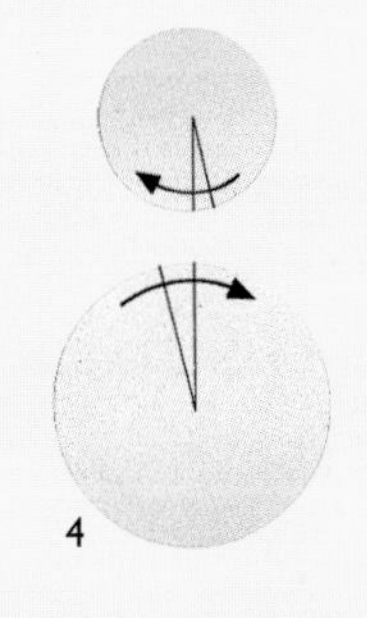
4

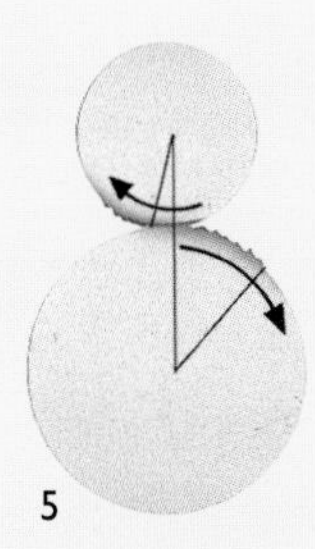
5

while both were converging 350–286 MYA, resulted in strenuous strike-slip faulting between the colliding megacontinents. In the process the opposing margins of Laurentia-Baltica and Gondwana were carried some 2,560 km (1,600 miles) past each other. As a consequence, the margins were greatly disrupted, distorted, and partly destroyed. This sliding and grinding displacement determined the final juxtaposition of the present Atlantic realm. But how could two megacontinents rotating in the *same* direction have suffered a 2,560-km (1,600-mile) relative displacement of their margins?

Imagine two bicycle wheels of different sizes mounted on an adjustable jig in the same horizontal plane: the larger represents Gondwana and the smaller Laurentia-Baltica. Both wheels are almost imperceptibly rotating clockwise (**4**), with their tire treads a fraction of an inch apart. While from a distance it is difficult to detect movement of the wheels, if we look closely at the point where the tires are nearly touching we will see that the tread surfaces are moving quite rapidly relative to each other. This is because the treads are moving in *opposite* directions at the point of near-contact, although the wheels are rotating in the *same* direction. The speed of this relative movement is analogous to the situation between the margins of Gondwana and Laurentia-Baltica.

To understand the great destruction during the assembly of Central Pangea, we must now adjust our bicycle wheels so that the tire treads are slowly but relentlessly jammed hard against each other (**5**). Instead of coming to a grinding halt, suppose that a powerful mechanical drive is applied to them, and that the larger Gondwana-wheel is driven at more than twice the rate of the Laurentia-Baltica-wheel. If we rotate the Gondwana-wheel through about 20°, we will find that both tire treads have been severely damaged by friction and shear, but the different rates of rotation have caused damage in different ways. Due to the faster peripheral speed and the greater distance traveled by the larger Gondwana-wheel's tire, its tread will have sustained lighter but more extensive damage. The damage to the smaller Laurentia-Baltica-wheel will be proportionately less extensive but much more concentrated, and there will be a churning effect on its surface. We will also find that the common points of initial contact on the treads are now separated by a distance equivalent to the combined peripheral movement of *both* wheels away from the common starting point.

The same principle accounts for the relative displacement of the margins of Laurentia-Baltica and Gondwana as they collided 350–286 MYA—about 2,500 km (1,600 miles) of displacement!

Of course, Laurentia-Baltica and Gondwana were far from being circular. They did not rotate smoothly for 64 MY (their movements were erratic), nor were they fixed on the Earth's surface: Gondwana was drifting due north while Laurentia-Baltica was drifting more slowly to the north-northeast. But the bicycle wheel model does demonstrate the interaction of tectonic forces in the region of Central Pangea during its final assembly. The model accounts for the complete destruction of the remnant Rheic ocean-floor; and the intense deformation and "megashear" faulting of Laurentia-Baltica caused by the churning effect as Gondwana collided with it. The result was the compression of all the elements that became constituents of the North Atlantic realm into a close-knit geological society—Central Pangea.

This chapter examines the main consequences of this collision: the assembly of Nova Scotia and Western Europe, and the formation of the southern Appalachians. The next chapter discusses the overlapping series of events that resulted from the collision of South American and West African Gondwana with Laurentia-Baltica in the region that was to become the Gulf of Mexico.

An early victim of these megacontinental gyrations when rotations began about 350 MYA was a large chunk of Gondwana called

[*continued on page 104*]

The Ovens region, near Lunenburg, Nova Scotia

1 Gondwana Fragment

A detached "fragment" of the Gondwanan continental margin now called Meguma Terrane became an early victim of the megacontinent's northward drift and rotation. Around 350 MYA the Meguma fragment collided at an oblique angle with the Avalonian margin of Laurasia. This "fragment" is now the club-shaped southern section of Nova Scotia and extends beneath the Gulf of Maine and the North Atlantic Ocean. It forms the submarine Scotian Bank next to Grand Bank of Newfoundland.

If one projects the line of a fault in the Bay of Fundy between Nova Scotia and New Brunswick (shown in the diagram) across Nova Scotia and out into the Atlantic, most of the elements to the north of the line are Avalonian in origin. They were sutured to Laurasia during the closure of the Iapetus Ocean. Everything to the south of this fault, as far as the second and lower fault in the illustration, is Meguma—*displaced terrane*. However, experts cannot agree exactly where on the Gondwanan margin Meguma Terrane originated.

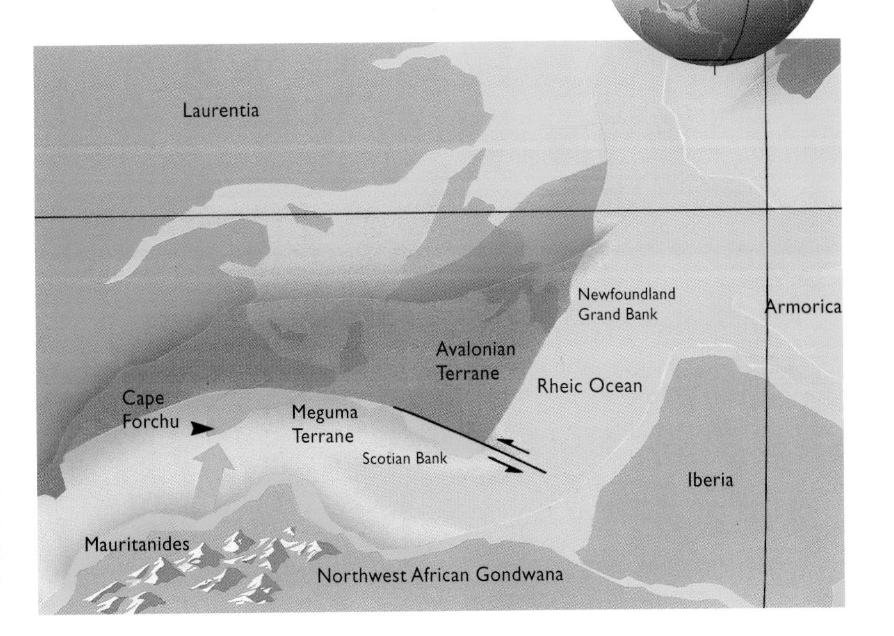

The most favored of several theories is that the terrane formed in a deltaic environment—that it accumulated from sediments washed down from the Mauritanide mountains of northwest Africa and was then wrenched from the Gondwanan margin off Morocco by movements of the Rheic Ocean floor. Ultimately, part of that floor was subducted beneath the Meguma element to form the granite plutons that underlie most of southern Nova Scotia today.

The pictures show the after-effects of this tectonic event. The panorama is of Cape Forchu East, on the southern tip of Nova Scotia. The rock in the picture was formed from ignimbrites—compacted volcanic ash from one of the most violent forms of volcanic eruption. The smaller picture of the Ovens region near Lunenburg on the Atlantic coast shows Meguma Terrane sedimentary rocks altered by heat. The original sedimentary rocks are believed to have formed off the coast of West Africa about 500 MYA, when this region of Gondwana was within the Antarctic Circle.

2 Last of the Rheic

The Rheic Ocean floor was destroyed or disrupted because of the violent megacontinental collision and interaction in the region of East Central Pangea during the period 350–286 MYA, an episode of European mountain-building called the Variscan orogeny. During the Variscan, parts of the Rheic seafloor were subducted beneath Armorica and destroyed as they descended into the asthenosphere, while other parts were deeply buried and deformed beneath the newly formed Variscan mountain belt.

Kynance Cove, pictured here, is a holiday beauty spot on the Lizard Peninsula, a remnant of that Rheic seafloor. In Britain the Variscan front runs in a direct line that crosses southern Ireland. It then follows the South Wales coastline to the Mendip Hills, and continues beneath the south of England and the English Channel. It finishes beneath the Low Countries and the valley of the River Meuse in Belgium. Other relics of the Variscan orogeny and the Rheic Ocean mark the grossly deformed Armorican zone of confrontation with Laurentia-Baltica, such as the Ardennes, Vosges, and the Black Forest regions of the lower and upper Rhine Valley.

During this alpine mountain-building event, remnants of the Rheic Ocean floor were thrust over the Laurasian foreland, then buried by Variscan mountain debris. Eventually, as the surface above this contorted region was eroded, a complex of Rheic Ocean floor debris and sedimentary coverings was exposed—an "ophiolitic complex." This complex now forms the Lizard Peninsula of Cornwall and the bedrock of the English Channel between Lizard Point and Start Point in Devon.

3 Folding Front

The tremendous energy exerted during the Variscan orogeny is well expressed in these chevron-folded cliffs at Hartland Quay in North Devon. The force of Armorica's collision with the Laurentian foreland was such that gently folded formations were distorted into chevron folds—like the closing bellows of an accordion. The shore at low tide has exposed the stumps of previous cliffs undermined and eroded by wave action.

Warren Cliff, in the inset picture, shows a detail of the Crackington Formation. This is a classic example of sharply folded sandstones and shales that were presumably formed on the Avalonian margin of Laurentia-Baltica before the collision with Armorica gained momentum. The trend of the folds—from east to west—records the direction of the collision.

The Devon coast in this region is exposed to the wildest and most damaging waves that can be produced by an Atlantic gale, and has over the centuries been the scene of countless shipwrecks. Nevertheless, Hartland Quay was built in the late 16th century, as were many other quays on the north coast of Devon and Cornwall, by the local lord of the manor to give his otherwise isolated community access to Elizabethan maritime trade. Remains of the quay are scattered on the shore at the bottom left of the panorama.

Warren Cliff, detail

[*continued from page 97*]

"Meguma Terrane"—possibly a huge offshore island in the Rheic Ocean (like Madagascar off East Africa today), or land detached from the mainland because of the "megashear" between Gondwana and Laurentia-Baltica (**6**). Most paleogeographers believe it calved off Morocco in northeast Gondwana [Essay 1]. Whatever its origins, the huge Meguma block came to rest against an Avalonian part of Laurentia-Baltica, and today forms the southern section of Nova Scotia. The present Bay of Fundy marks the line of suture between Meguma Terrane and Avalonia. If one follows the narrowing Bay of Fundy and extends its line across Nova Scotia and well out into the North Atlantic, then within reasonable limits submarine shelves north of it are Avalonian and those south of it (including the Scotian Bank) are Meguma Terrane.

As Meguma Terrane was docking with its present Avalonian counterpart in Laurentia-Baltica, the Armorican (Brittany) and Iberian (Spain and Portugal) assembly was encroaching on the Rheic Ocean floor [see paleoglobe p. 96]. The subsequent megacontinental collision resulted in the destruction of the Rheic Ocean floor beneath Armorica and Portuguese Iberia, and the marriage of Armorica and Iberia to Laurentia-Baltica (**7**) (i.e., the suturing of southwest and northwest Europe). The sequence of these events paralleled the current continent-continent collision of Africa with Eurasia, which has resulted in subduction of the Mediterranean seafloor beneath Italy and Sicily, producing Mount Vesuvius and Mount Etna. The present collision has also resulted in the formation of the Swiss Alps and other alpine ranges from North Africa to the Balkans.

THE ARMORICAN COLLISION with Laurentia-Baltica resulted in the formation of huge "batholiths," extensive groups of individual plutons such as those illustrated in the previous chapter [see Essay, pages 86–87]. Perhaps the best known of the Armorica/Laurentia batholiths are those in southwest England (the Cornubian batholith of Devon and Cornwall) and in Iberia (Portugal and Spain). The granite batholiths were formed in two distinctly different ways: through continental collision on the one hand and subduction on the other. Their granites are different in composition and so in mineral content—impacting on the technological development and wealth of nations in the North Atlantic realm.

The massive flooding of the Armorican continental surface by andesite resulted in the formation of the Massif Central of France. Andesite, which is named after the South America Andes, is a volcanic rock that is intermediate in composition between basalt and rhyolite. It is formed during subduction when seafloor basalt is thought to intermix in varying proportions with melted continental rocks (**8**). Seafloor basalt is derived from melted mantle rock deep beneath the surface, while rhyolite is formed from melted continental rock. The common factor between andesite, basalt, and rhyolite is that they all reach the surface of the Earth through volcanic vents. However, if their individual magmas are *intruded* into other rock formations at depth instead of being *extruded* onto the surface through volcanic vents, they form different types of rock: an andesitic mix forms "diorite," a basaltic mix forms "gabbro," and a rhyolitic mix forms "granite." In this context, the Massif Central is another consequence of the rotation and collision of Gondwana and Laurentia-Baltica. The andesitic flooding was caused by the interaction of the Rheic Ocean floor with the underside of Armorica as the ocean floor was subducted beneath the continental margin.

MUCH OF THE CORNUBIAN batholith lies beneath sea-covered continental shelf. The visible parts of it form the rocky outcrops of Dartmoor and Bodmin moors, St. Austell, Carnmenellis, Land's End, and the Isles of Scilly off Land's End. Each of these localities is perched on the top of an enormous granite pluton shaped like an inverted but deformed pear [Essay 4]. The Cornubian plutons were formed from melted continental rocks of *sedimentary* origin, called "S-type" granites. It is possible to define the tectonic environment in which a

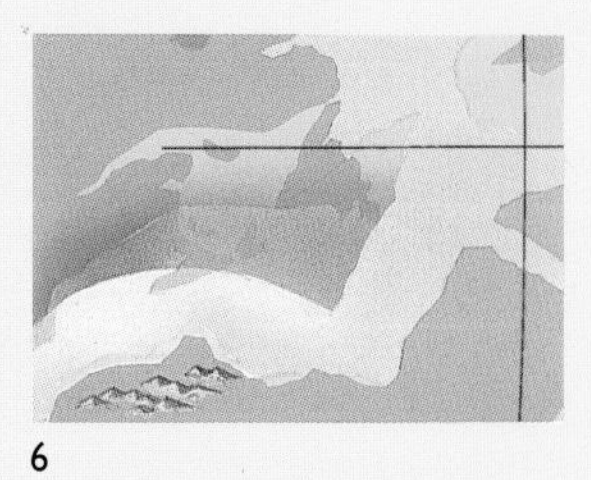

6

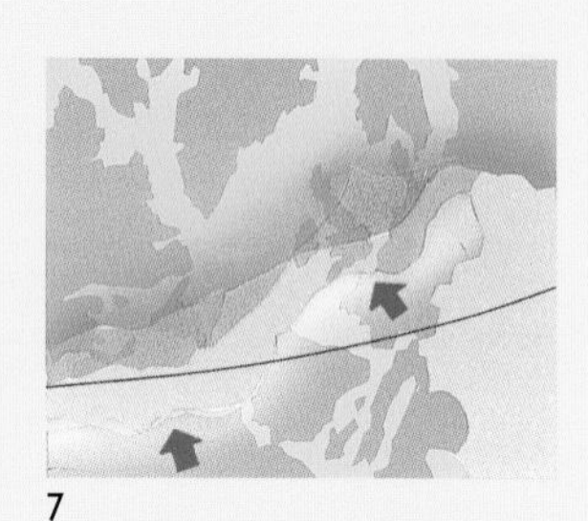

7

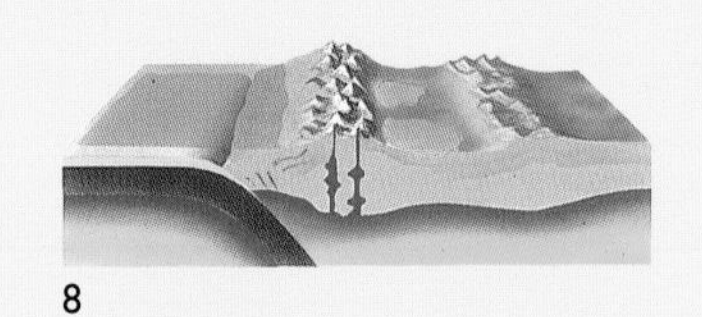

8

9

batholith was formed by its granite type and to some extent by its plutons' mineral content. One characteristic of Cornubian S-type granites is that they typically contain a high proportion of tin in their metal ores [Essay 5]. On the other hand, batholiths formed in Iberia are generally rich in sulphides (metallic compounds of sulphur), associated with their formation during the subduction of an ocean crust beneath a continental margin, and the subsequent formation of andean-type mountains. Here, subduction of the Rheic Ocean floor beneath western Iberia resulted in the formation of a natural mountain barrier between Portugal and Spain.

In the case of the Cornubian batholith (formed at the same time as a similar batholith in Czechoslovakia called the Bohemian massif) continent–continent collision resulted in the formation of alpine-type mountains with folded and overthrust rocks of mainly sedimentary origin (**9**). This led to a thickening of the continental crust in that region—it caused rocks formed on the surface to either be thrust upwards or become deeply buried in the collision zone. The higher the mountains became the deeper the burial of associated rocks, until conductive and radioactive heating inside the Earth caused them to melt and form granites. Most of the metal ores in these granites were derived from continental material and had received a relatively small infusion of mantle rock (basalt).

Granites containing a high proportion of tin, like the Cornubian batholith, enabled the technology of the Copper Age (about 5,000 BP) to advance to the Bronze Age (about 4,000 bp). Copper alone is a soft metal with limited use for making tools; but an alloy of copper with about 10 percent tin produces bronze, a material hard enough to make quite a range of effective tools. Tin opened an era of technology for Western populations that did not change until the discovery of ways to smelt iron from its ore marked the beginning of the Iron Age (about 3,000 BP).

IN CONTRAST, IBERIAN granites were formed through heating of the descending Rheic Ocean floor beneath the Iberian continental crust. The ocean floor acted like a gigantic conveyor belt feeding an assortment of minerals and lithified sediments that had accumulated on its surface into this interactive zone. From here the process of pluton formation and "diapirism" (the tendency for materials lighter than surrounding rocks to rise towards the surface) in the continental crust above the subducting plate was similar to that already described, but with different consequences.

First, andean-type mountains were formed on the Iberian continental margin, in contrast to the midcontinental alpine-type mountains that result from the destruction of mainly sedimentary rocks in continent–continent collisions. Second, the principal source of metal ores for the Iberian batholiths was the subducting Rheic Ocean seafloor: This affected both the process of granite production and the types of ores produced, which were richer, more varied, and more prolific. This wealth of minerals is exemplified by the multicolored, almost grotesque landscape of an ancient mining area in the Sierra Morena and Rio Tinto regions of Andalusia. For thousands of years streams carrying red, yellow, green, purple, and black colored mineral deposits have fed the Rio Tinto as it flows from the Sierra Morena to the Gulf of Cádiz in southern Spain [Essay 6].

The "Rio Tinto ore body" (as it is known), and its associated mineral belts in Portugal, are so rich in copper, lead, zinc, manganese, silver, and gold and many others, that together these regions are considered one of the richest sources of metal ores on Earth. There is evidence that Copper Age humans mined here 5,000 BP, but that once the easily accessible copper minerals (malachite) near the surface had been exhausted, these people could not dig deeper because they were limited to stone and copper tools. However, archaeologists believe that once the Bronze Age had commenced, mining activities in the Rio Tinto region proceeded almost without interruption; the ores that were worked, and are still being worked, reflect both human technical progress and changes in ideas of value.

For example, around 2,900 BP the Phoenicians (who had by then invaded Spain) are said to have mined and smelted so much silver from the Rio Tinto mines that, when their ships were loaded to the gunwales for transport back to the eastern Mediterranean, they replaced their iron anchors with anchors cast in silver! The wealth of Mediterranean empires over the millennia, like that of the Carthaginians (2,500 BP) and others later, was largely dependent on Rio Tinto silver. And judging by the volume of slag left by the Romans after *their* smelting activities, they too benefited because of the subduction of the Rheic Ocean floor beneath Iberia (**10**): Roman slag dated around 2,200 BP is estimated to weigh more than nine million tonnes (10 million tons).

EXTENSIVE AREAS of East Central Pangea (now Western Europe) suffered side-effects from the combined tectonic assault of simultaneous collision and subduction. This resulted from the aggressive northward movement and rotation of Gondwana and the more modest coincidental rotation of Laurentia-Baltica. This period of mountain building is known as the "Variscan" orogeny (some geologists call it the "Hercynian" orogeny: **11**): it lasted from 345 to 280 MYA [Essay 2]. Armorica and Iberia and their retinues were in the middle of it all, caught between the two megacontinents. Even rocks well away from the more violently active zones were "gently" folded: regions that had been almost flat were contorted into undulating hills interspersed with occasional, more radical uplifting, refolding, and overthrusting. The general direction of the folding was decided by the existing lie of the land and its structural weaknesses, as well as the direction of the compressional force acting on it. Resulting valleys and hills in southwestern Britain, particularly the coal field regions of South Wales, are haphazard and disordered.

It was the same in mid-Central Pangea, where the Gondwanan "wheel" had an even more destructive effect on the edge of the smaller Laurentian one. The deformed regions are now known as the Mauritanide fold-belt of West Africa (around Dakar) and the Southern Appalachian province of North America, particularly the coal field regions of Pennsylvania and West Virginia. Again, a period of mountain building called the "Alleghenian" orogeny overlapped the Variscan orogeny between 330 and 300 MYA, when the first effects of the angular impact of the continental rotations were felt. In addition, the Alleghenian orogeny had a second pulse, from 300 to 250 MYA. During this period the trauma of the main collision led to the complete destruction of the Rheic Ocean floor beneath the converging and rotating megacontinents. Subsequent overthrusting led to chaos on the Laurentian margin, with a rippling effect that impacted far from the confrontation zone into modern Pennsyl-vania and West Virginia.

Vast areas of coal-bearing rocks in these compressed and undulating regions were eroded so that coal seams were preserved in downfolds (synclines), while the upfolds (anticlines) were worn away and later covered by new sedimentary rocks—sometimes then folded yet again (**12**). Refolding produced extra pressure and heat, which turned some coal into "anthracite," a very hard, almost pure-carbon form of coal—in effect, charcoal. These now-scattered and buried coal- and occasionally anthracite-bearing synclines north and south of the Western European suture became the coal-mining areas of Britain, France, the Ruhr, and northern Spain. Coal and anthracite ultimately unleashed the forces of the Industrial Revolution in Britain where, due to deforestation, there was a shortage of trees and therefore of charcoal for iron smelting. This revolution spread to Western Europe and ultimately to regions of North America, where rich coal fields of the same age as their European counterparts had formed in Pennsylvania. In the 19th century, Pittsburgh, set in the haphazard undulations of Pennsylvania, became the focal point of a transformation that was to see the end of the Industrial Revolution and, with the mass production of steel, the beginning of the Technological Revolution.

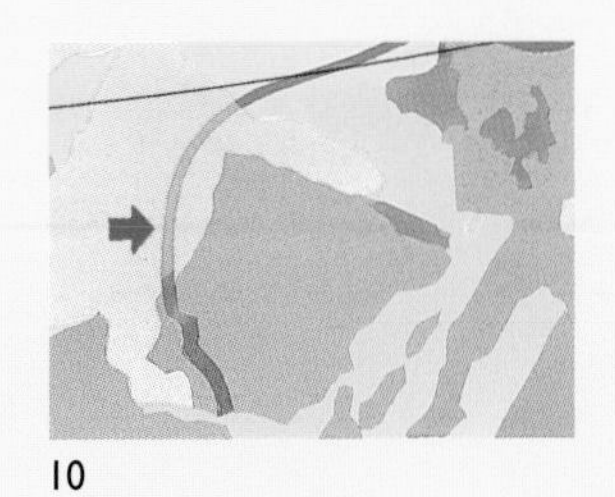

10

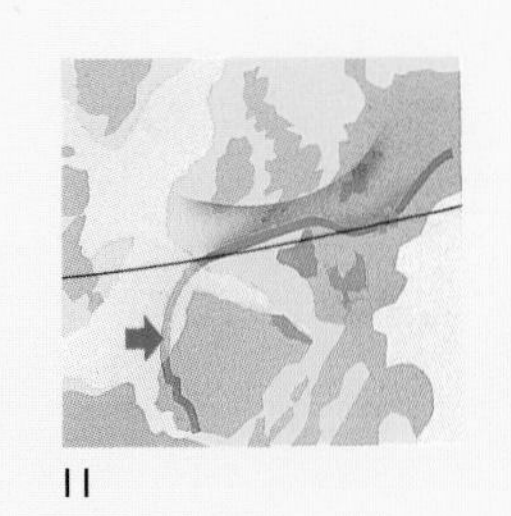

11

12

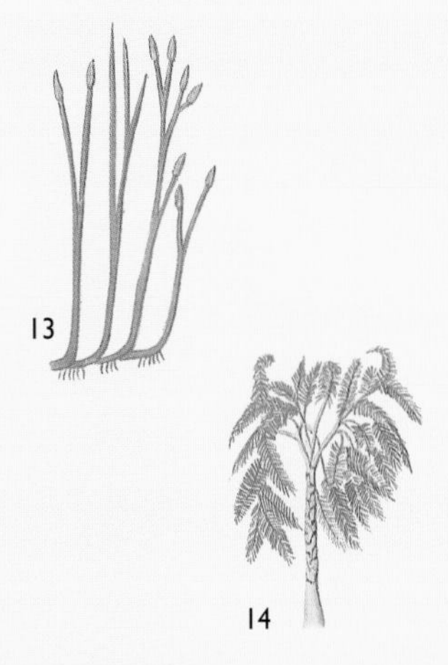

13

14

THE GEOLOGICAL PERIOD in which coal was formed on a prolific scale was an obvious benchmark, and was named the "Carboniferous Period" by British geologists. Before this coal-forming era, plants had a tenuous hold, but after it began they had suddenly and unequivocally taken possession of the habitable landscape. The Victorians observed that the Carboniferous in Britain had two distinct halves: "Early" (360–320 MYA) and "Late" (320–286 MYA). American geologists refer to the same intervals as the "Mississippian Period," characterized by limestone formation and little coal; and "Pennsylvanian Period," characterized by major coal deposits. The divisions arise from the quite different environments in which the limestones and coals of these regions were formed.

The Carboniferous Period, and Gondwana's period of rotation and northward drift, are virtually of the same duration (360–286 MYA and 350–286 MYA, respectively): not a superficial coincidence. The "carboniferous" nature of the geological period was radically influenced by tectonic events and the idiosyncracies of global climate. This connection requires discussion of the evolution of the coal forest plants, one of several key factors that together decided the form of the natural world today.

ACCORDING TO THE FOSSIL RECORD, land plants first appeared around 420 MYA. They were rootless and leafless, with creeping stems with short branches—like barbed wire lying on the ground. They fell into two basic groups, nonvascular plants (bryophytes) and vascular plants (tracheophytes): the bryophytes (Gr. *bryon phyton*, moss plant) are thought to have appeared first. It is broadly agreed that eukaryotic green algae were the common ancestor of both types, and that they invaded wetlands at a very early date, certainly over 420 MYA. The reasons for this conclusion are mainly that plants and green algae both contain the same types of chlorophyll and accessory pigments. The chlorophylls and pigments accept different wavelengths of light, which enable photosynthesis and the production of carbohydrates such as starch, the main food reserve of plants. Green algae and plants have other characteristics in common: the presence of cellulose, a common pattern of sexual reproduction, and minute details of cell division that are identical.

Of the early land plants a group characterized by the genus *Rhynia* is by far the most important (**13**); plants in this group were the ancestors of all tracheophytes (Gr. *tracheia*, artery), which are distinguished from bryophytes by the presence of water-conducting tissue in their stems. This vascular system takes the form of a bundle of tubes that transports water, carbohydrates, and trace elements throughout the plant. The bryophytes (mosses, liverworts, and their kind) have, at most, a rudimentary system of fluid transport—therefore the term "nonvascular." This weakness means that nonvascular plants are inevitably small and are tied to a moist environment in which to live and reproduce.

Vascular plants are broadly divided into two major groups, *seed* plants and *seedless* plants. And in turn each of these can be further and extensively categorized. But in general we will restrict ourselves to the central evolutionary theme of this book: the progress and biogeography of plants and animals, as the continents and oceans themselves were undergoing evolutionary change.

Two types of plants with fern-like foliage grew in the coal forest: the often huge spore-bearing fern trees (true ferns) and the smaller "seed-ferns" (**14**). This is unfortunate nomenclature. "Seed-ferns" were not ferns, they were gymnosperms—plants with "naked seeds" that led to the evolution of flowering plants around 140 MYA. The "angiosperms," as flowering plants are called (Gr. *angos*, capsule, seed-case), account for the vast majority of plants on Earth today. As we shall see in later chapters, angiosperms literally changed the face of the living world. Along with the shifting continents, angiosperms determined the direction of evolution by creating circumstances favorable to the explosive evolution of insects and, much later, to the coevolution of flowering plants and mammals. The angiosperm

[*continued on page 110*]

4 Plutonic Tors

Like Yosemite in California and Rio de Janeiro in Brazil, southwest England is a small and exclusive world with its own special character. The common factor between these widely separated places is that all three depend on a group of plutons as the prime source of their natural beauty. But unlike their counterparts in California and Brazil, the counties of Devon and Cornwall in southwest England are on top of a series of virtually intact plutons, collectively called the "Cornubian batholith." The batholith was formed at the very roots of alpine mountains during the collision of Armorica with Laurentia-Baltica.

During their genesis most granites are thoroughly mixed and pressure-cooked. Melted continental basement rocks receive an infusion of basalt (melted mantle rocks in the case of the Cornubian batholith) and mineral-laden water that is under terrific pressure. The mineralized water acts as a flux, which aids in making granites less dense and more fluid than surrounding rock. Consequently, molten balloon-like plutons form, and these stream towards the surface.

As they rise, plutons cause uplift and erosion of the continental surface above them. In the case of the Cornubian batholith, such uplift caused the formation of Dartmoor—portrayed in the panorama—Bodmin Moor, and the Isles of Scilly. It also forced parts of the deeply buried Rheic Ocean floor to the surface, where it forms the Lizard Peninsula.

5 Cornubian Lodes

Cornubian granites were formed from melted rock of mainly sedimentary origin—a direct consequence of the intercontinental collision between Armorica and Laurentia-Baltica. The metal ores they contain are therefore quite different from those formed because of the subduction of the Rheic Ocean beneath Iberia [Essay 6]. In fact, the tectonic environment in which any batholith formed can be deduced from its type of granite and to a lesser extent by its mineral content.

Granites are infused with high-pressure mineralized water during their genesis. As rising plutons cool, these solutions crystalize to form "stocks" of metallic ores. Because cooling granite naturally forms rectilinear blocks and fractures—as seen in this picture of Land's End—stocks form a huge matrix encompassing the blocks like an old-fashioned egg box. It is the walls of the matrix that are either quarried or mined for tin and other metal ores.

The Cornubian batholith has been an important source of tin since Bronze Age times. To a lesser degree, it has also been the source of other base metals like copper and lead, and precious metals like silver and gold. Its most prized mineral asset today was not discovered until 1746. At this time a Plymouth chemist named William Cooksworthy recognized that Cornish clays, which form near the surface of a pluton as the result of weathering and chemical alteration of its granite, were of the same composition as kaolin,

St. Austell china-clay pits, Cornwall

Botallack tin mines, Cornwall

a key mineral first discovered by the Chinese. They used it to produce "china," the thin-walled translucent ceramics we call porcelain. In recent times, however, the principal use of Cornubian china-clay has changed from porcelain to high-quality paper-making. The china-clay content of paper largely determines its weight, texture, and glossiness. The paper used to produce this book probably originated as granite formed deep beneath the colliding margins of Armorica and Laurentia-Baltica.

[*continued from page 105*]

flower was the key factor in this dramatic change, a change that is arguably as important as the evolution of eukaryotes in a previously prokaryotic world.

Flowers are designed, among other functions, to attract the right kind of flying insect, and sometimes to repel the wrong kind. Where flying insects were the catalysts of angiosperm reproduction, they ensured maximum pollination with the least waste of pollen production and of subsequent formation of fruits and/or seeds. The latter were the source of the plants' continuity.

Although this evolutionary breakthrough took over 140 MY to complete, there is no doubt that the first steps were taken in the now nearly assembled Pangea. The main players evolved in the coal forests: the gymnosperms (the seed-ferns) already mentioned, and the flying insects to which we will now turn.

UNLIKE LATER FLYING ANIMALS, such as flying reptiles and birds that developed wings from modified forelegs, the wings of flying insects developed from modified armored plates that protected the tops of their delicate and therefore vulnerable leg joints. These plates were useful for grasshopper-style gliding and eventually developed into powerful flying mechanisms. As far as is known, the dragonflies (**15**) and mayflies of the coal forests were the first to develop such mechanisms. They were "exopterygotes," (Gr. *exo pteron*, outside wing) insects that hatch straight from the parental egg as miniature adults minus wings. Their descendants include their modern and far smaller counterparts, as well as animals like grasshoppers, locusts, and earwigs.

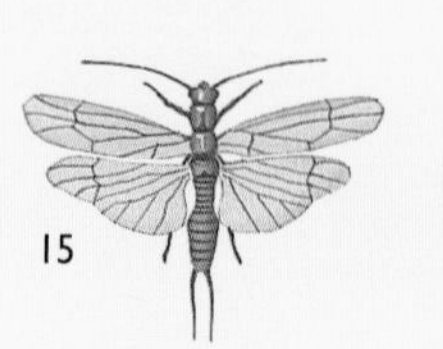
15

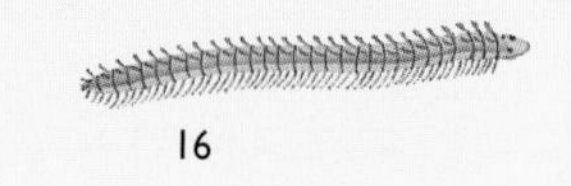
16

At some time during the Carboniferous, an important divergence from the exopterygotes resulted in the evolution of the "endopterygotes" (Gr. *endo*, inside): insects that lay eggs that hatch into grubs, not miniature adults. The infant creatures undergo a dormant pupal stage, often within a plant, from which they emerge as adults, complete with wings and reproductive organs. Today this category of flying insect, the largest by far, includes bees, wasps, ants, butterflies, and moths—animals of the class Insecta. the characteristics of these small air-breathing insects reflect the very creatures that contributed vastly to the evolution of angiosperms.

The giant *wingless* insects, and millipedes, centipedes, and their class (Myriapoda: **16**), which also lived in the coal forests, were among the first land animals. Somewhere down the evolutionary line in Precambrian times, a division occurred in the mainstream evolution of segmented worms (Annelida), the class which included the first marine worms, ultimately resulting in insect evolution. As plants invaded the land, some burrowing marine worms adapted to terrestrial conditions and became burrowing "earth" worms—perhaps not too difficult a transition. These creatures played a key role in the conversion of humus and granite-based clays into the rich swampy soils that allowed land plants to develop into the giants of the coal forests.

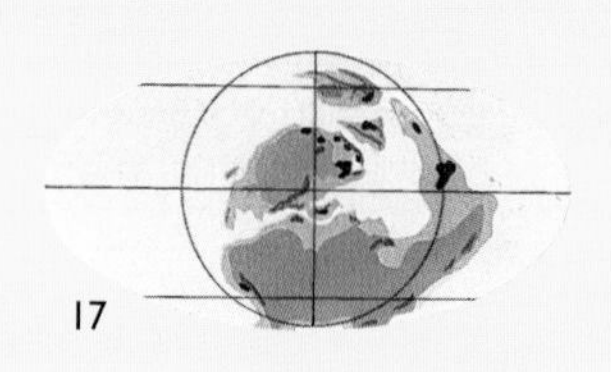
17

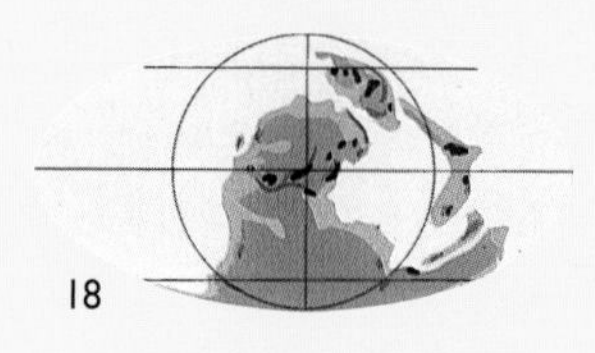
18

WE HAVE MADE THE POINT that global climate, fluctuating sea level, and tectonic events were irrevocably linked with coal formation. The great tectonic event of the time was the general assembly of Pangea, particularly of Central Pangea. As we return to physical rather than organic evolution, the main geographical focus shifts from Euro-American Pangea to Afro-American Pangea. As mentioned above, American geologists refer to the European term "Carboniferous Period" as the Mississippian and Pennsylvanian Periods (360–320 and 320–286 MYA). The Americans recognize that two very different geological styles were at work in Laurentia during these periods.

The Mississippian Period is largely dominated by marine limestones (**17**), formed when limy mud was laid down in the shallow seas that covered much of what is now central and western North America. This included the area of mid-America known as the Mississippi Valley, the region above the confluence of the Mississippi and Ohio Rivers, which gives its name to the Mississippian Period.

The marine fossils of Mississippian age found in this region suggest that a very warm climate and high sea level existed in this region. Marine fossils found there show that the continental shelves of Laurentia-Baltica were contiguous by this time, allowing the migration of shelf faunas from one area to another. Creatures, such as mollusks, which lived in East Laurentia-Baltica (northern Europe) match those in West Laurentia-Baltica (mid-North America). This suggests to paleontologists that there had been a migration of fauna across the Arctic region. The latter must have been free from ice then to allow such migration. Although many of these same marine fossils are found in the Mississippi Valley, there are other "localized" species not found outside the region. Because of the presence of local species the shallow continental seas in the Mississippi Valley region must have been in semi-isolation.

UNLIKE THE MISSISSIPPIAN PERIOD, the Pennsylvanian Period (**18**) is rich in coal, the product of lush tropical swamps subject to frequent periods of heavy rain—like today's Amazon rain forests. Typically, such forests were underlain by a "terrigenous" sandstone (a rock type made up of fragments of pre-existing rocks eroded from

Tree ferns near Mount Pelée, Martinique

the surface of the land), covered by shale. As the forests died, their partially decomposed debris accumulated in extensive areas of thick peat. The vegetation did not rot completely because tree trunks, branches, leaves, and animal remains were not only smothered by their own accumulating weight but also submerged in tannin-laden swamp water, an "anoxic" (oxygen-free) environment, and an essential condition for subsequent formation of coal [Essay 7]. Episodically over millions of years, cyclic inundation by shallow seas drowned repeated forest growths, so that the peat of each forest was covered by sediments. The sediments lithified into thick deposits of shales and *marine* limestones, which eventually gave way to sandstone and shale formations again. The whole cycle suggests that, in accord with the glacial rhythms of an ice age, a shallow sea had first advanced and then retreated from a deltaic environment. The rock sequence produced in this cycle is called a "cyclothem." It consists of sandstone, shale, and coal, followed by sandstone, shale, marine limestone, and shale—then back to the sandstone and shale sequence beneath the next coal seam.

The Pennsylvanian pattern of coal-forest growth on a terrigenous basement (**19**), followed by marine invasion and a return to a cover of nonmarine conditions, was repeated at least a hundred times in some localities during the Pennsylvanian Period. Each of these coal-forming periods resulted from repeated invasions and withdrawals of shallow seas over a flattish landscape, where a change of 100 m (300 ft) in sea level could result in either the advance or the retreat of a shoreline by 160 km (100 miles) or more. But why were there so many of these transgressions and regressions during the Pennsylvanian Period?

19

The rhythmic formation of coal, along with more extensive glaciation in Southern Gondwana during Pennsylvanian—but not Mississippian—time, provides an answer. There was a severe ice age during the Pennsylvanian half of the Carboniferous Period but *not* so severe, or as prolonged, as during the Mississippian half (the so-called Carboniferous Ice Age overall lasted from about 330 to 260 MYA). Each shallow-sea transgression is thought to represent an interglacial period.

Nobody knows why the Carboniferous Ice Age was triggered, but there are clues—particularly the fact that much of Gondwana was over the South Pole at this time, while the North Pole was an open ocean. This geography would have allowed an interpolar ocean circulation, considered by some to be an essential ingredient of an ice age. Additionally, large amounts of carbon dioxide were being absorbed by vast areas of coal forest, and this may be a secondary factor—it could have produced a "refrigerator" effect (the opposite of the present "greenhouse" tendency).

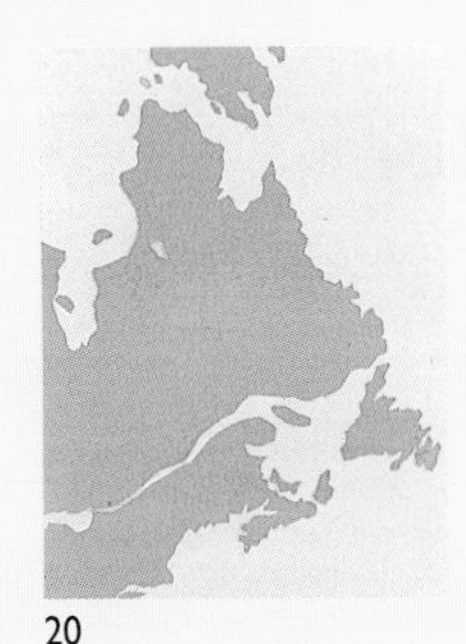

20

As discussed in Chapter III: *Iapetus and Avalonia*, the advance and retreat of ice sheets during ice ages causes variations in sea levels in the range of 180 m (600 feet) from one extreme to the other. One could argue that at present we are living during an interglacial period of a prolonged ice age, the Cenozoic Ice Age. which commenced in the Southern Hemisphere 20–15 MYA; and yet the continents are not flooded by epicontinental seas. Why then should the continents have been flooded in Pennsylvanian times?

One suggestion is that the Carboniferous was a time when unusually high rates of spreading activity caused extra swelling on the world ocean floor. This in turn caused an increase in the displacement of the world ocean. Sea levels rose and low-lying continental regions were flooded. But the prolonged pre-Pangean ice coincided with the final assembly of the supercontinent and the effect of accumulating ice at the poles tended to reduce global sea level. The shallow continental seas caused by displacement must have fluctuated in depth according to glacial advances and retreats.

Such geologically frequent variations in sea level, and corresponding fluctuations in global climate, caused ecological stress. Some species became extinct as a consequence, while others responded to the challenge; they flourished and diversified. This was gradual evolution of the Darwinian kind, and continued for thirty million years after the end of the Carboniferous. It was not until Pangea was fully formed and sutured, and the great southern ice age had ended, that this Darwinian progress was punctuated by another spasm of mass extinction. When it came (255–250 MYA), it was the greatest extinction that there has ever been.

There are unexplained contradictions about this steady-state period of evolution. Many factors usually attributed to the cause of mass extinctions at other times in the Earth's history were also present during the Carboniferous Period. Even so, life ambled steadily onward, and no mass extinctions were registered in the fossil record during this period. Yet there were radical changes in global climate, unprecedented tectonic trauma, and a reduction of available living space on drastically reduced continental margins.

Besides these perhaps less dramatic events, a significant number of large extraterrestrial objects hit the Earth around 300 MYA. Although occasional "bolide" impacts are part of the planet's day-to-day history in the geological sense, it is rare that such impacts can be linked directly to major extinctions. In putting the case for mass extinctions claimed to be caused by bolide impacts it is wise also to look at examples of floral/faunal survival of such impacts—such as those that occurred in the Carboniferous Period. At least nine terrestrial impact craters were made by bolides towards the end of the Carboniferous—eight in North America and one in Brazil. It is probable that these craters represent only a small proportion of the bolides that actually hit the Earth between 300 and 286 MYA. Statistically, if there had been nine impacts on Pangea, the odds are that a far greater number struck the Panthalassa Ocean floor, which was an open space roughly six times the area of the supercontinent. Traces of such ocean-floor impacts were destroyed along with the seafloor itself during subsequent subduction beneath continental margins.

Existing Carboniferous craters vary in size from about 5 to 30 km (3 to 20 miles) in diameter. The largest bolides in the cluster that hit the Earth during a 20 million-year span struck Laurentia-Baltica simultaneously around 290 MYA. Their twin lake-filled craters in the Canadian province of Quebec are called Clearwater West, which has a diameter of 32 km (20 miles), and Clearwater East, with a diameter of 22 km (13.75 miles: **20**). The size of these crater lakes suggests that the largest bolide was 3–5 km (2–3 miles) across and the smaller one between 1.5–3 km (1–2 miles).

In calculating the damage of a bolide impact, scientists consider estimates of its speed of entry, where it hit and at what angle, and the character of its composition—anything from a ball of ice to a chunk of iron. It has been calculated that if an object about 10 km (6 miles) in diameter were to hit the Earth today, the subsequent vaporization, pressure wave, and ejecta would immediately destroy life within a radius of hundreds of miles. Its impact would also cause the formation of a global dust cloud in the stratosphere. Some scientists propose that this would absorb all but 1.0 percent of solar radiation for three to six months, sufficient to cause the equivalent of a "nuclear winter" that could have a disastrous effect upon all life on Earth. But many scientists do not agree that a bolide impact would, on its own, be sufficient to cause mass extinction, as this event appears to show.

The Carboniferous bolide impacts are real enough. Their effect must have been devastating. Yet there appears not to have been a mass extinction as a consequence. Maybe the ice age already in progress was intensified by the occurrence of repeated "nuclear winters": it certainly continued well into the next stage of geological time—the Permian Period. Or maybe a bolide has to be of a special composition to wreak havoc in the form of mass extinction? K.J. Hsü has suggested that a high proportion of deadly prussic (hydrocyanic) acid contained in the mass of an icy comet would do the trick: indeed, prussic acid is a common compound on some outer planet moons that are themselves subject to impact from other bodies.

But, like Tolkein's Hobbits of Middle-earth at the end of their *Third Age*, the denizens of Central Pangea lived to see another day.

For this is the Ring of Fire, and with it you may rekindle hearts in a world that grows chill.

6 Iberian Deformation

The panorama illustrates one of the many dramatic effects of Variscan folding and uplifting as Iberia was assaulted from the north by waves of deformation. These resulted from the collision of Gondwanan Armorica with Laurentia-Baltica around 300 MYA. The result of the orogeny was to cause the uplift and overthrusting of a gigantic limestone platform from where it had formed below sea level, to become the Picos de Europa mountains of northern Spain. The same Variscan event also caused neighboring regions to form immense synclines (downfolds) at or near sea level. Coal swamps thrived in these basins, which later became the low-lying crescent-shaped coal fields of Basque country bordering the Picos de Europa.

Simultaneously, the subducting Rheic Ocean seafloor southwest of Iberia acted like a conveyor belt for both the minerals it had contained since its formation, and the minerals deposited as sediments on its surface. This rich mix of basalt and metallic minerals was blended with molten continental rocks as the ocean plate descended beneath Iberia. The granites in the plutons formed above the descending plate as a result were of a type different from

the contemporary Cornubian granites, and much richer in metallic minerals. They contained large quantities of sulphides (metallic sulphur compounds) of copper, silver, gold, lead, zinc, manganese, and many other elements.

The mining and smelting of sulphides, from Phoenician and Roman times to the present, has reduced the Sierra Morena and Rio Tinto regions of Andalusia in southern Spain to a stark and varicolored landscape—one which contains the largest known reserves of metallic minerals on Earth.

7 Quivering Earth

Okefenokee, the Choctaw Indian word for "quivering earth," is the largest and most primitive swamp in North America today. It is also a coal swamp, in which peat formed from dead vegetation has already accumulated to a thickness of 6 m (20 feet) or more. As in Carboniferous times, we are in an interglacial period of an ice age now. If global warming continues, and sea levels continue to rise, the swamp could become a shallow marine environment, where sandstone, shale, and perhaps limestone will accumulate on top of the peat. In a future glacial period sea level will fall as ice begins to accumulate at the poles once more. In this event the Okefenokee region will again be exposed above the sea and perhaps a new coal-forming swamp will develop above the old.

Repeated cycles of this sort over hundreds of thousands of years result in the accumulation of a specific series of sedimentary rocks and peat in layer upon layer. Each complete sequence of rocks is called a "cyclothem." Pressure from above causes the peat to metamorphose into seams of coal. Additional compression and heat during an intense period of folding and compression can improve the quality of coal by driving out further impurities. The top grade, "anthracite," is almost pure carbon—once the key fuel used to achieve the high temperature required for manufacturing steel. Anthracite was formed in Pennsylvania (inset picture) during the Alleghenian orogeny 300–250 MYA, and in Europe during the Variscan orogeny, 345–280 MYA.

During the Pennsylvanian Period of coal formation (320–286 MYA), hundreds of millions of acres of coal swamps clothed flat low-lying regions of Central Pangea. The appearance on Earth of vast numbers of giant land plants may have been one of several factors that triggered the Carboniferous Ice Age, which created the cyclothems. Carboniferous forests absorbed huge volumes of carbon dioxide, contributing to a "refrigerator effect." The ice age lasted 70 million years (330–260 MYA) and produced over 100 glacial cycles.

The prairies (open water) and hammocks (floating islands) of Okefenokee

Okefenokee Swamp, Georgia

Seneca Rocks, Potomac River headwaters, Pennsylvania

VI

MIDDLE EARTH

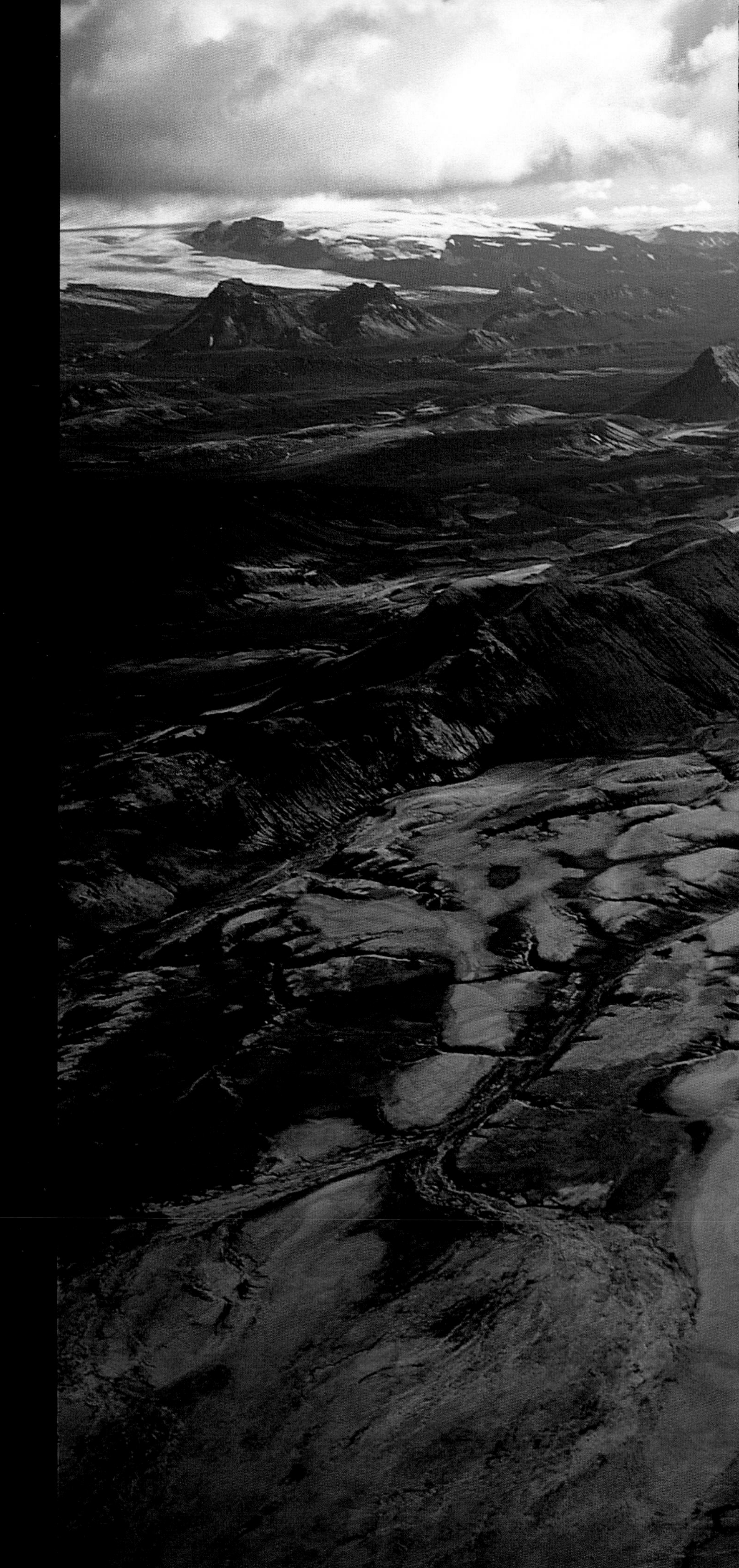

THIS PRIMITIVE SCENE, PHOTOGRAPHED NEAR THE EDGE OF AN ICELANDIC ICE CAP, MAY RESEMBLE THE CHARACTER OF A MID-LATITUDE PANGEAN LANDSCAPE 286 MYA. A PROLONGED ICE AGE HAD REACHED ITS MAXIMUM INTENSITY AT THIS TIME. AS IN THE PRESENT ICE AGE, PERIODS OF GLACIAL-MAXIMUM CLIMATE CONDITIONS WERE FOLLOWED BY INTERGLACIAL INTERLUDES AND THEN A PLUNGE INTO THE NEXT GLACIAL ADVANCE. HOWEVER, BY THE TIME PANGEA REACHED THE ZENITH OF ITS ASSEMBLY AROUND 250 MYA, GLOBAL CLIMATE HAD CHANGED FROM ICE AGE TO GREENHOUSE AGE—WITH NO ICE AT THE POLES—AND THE EARTH HAD SUFFERED THE MOST DEVASTATING MASS EXTINCTION IN ITS EVOLUTIONARY HISTORY.

Maelifellssandur, Iceland

MIDDLE EARTH
286–206 MYA

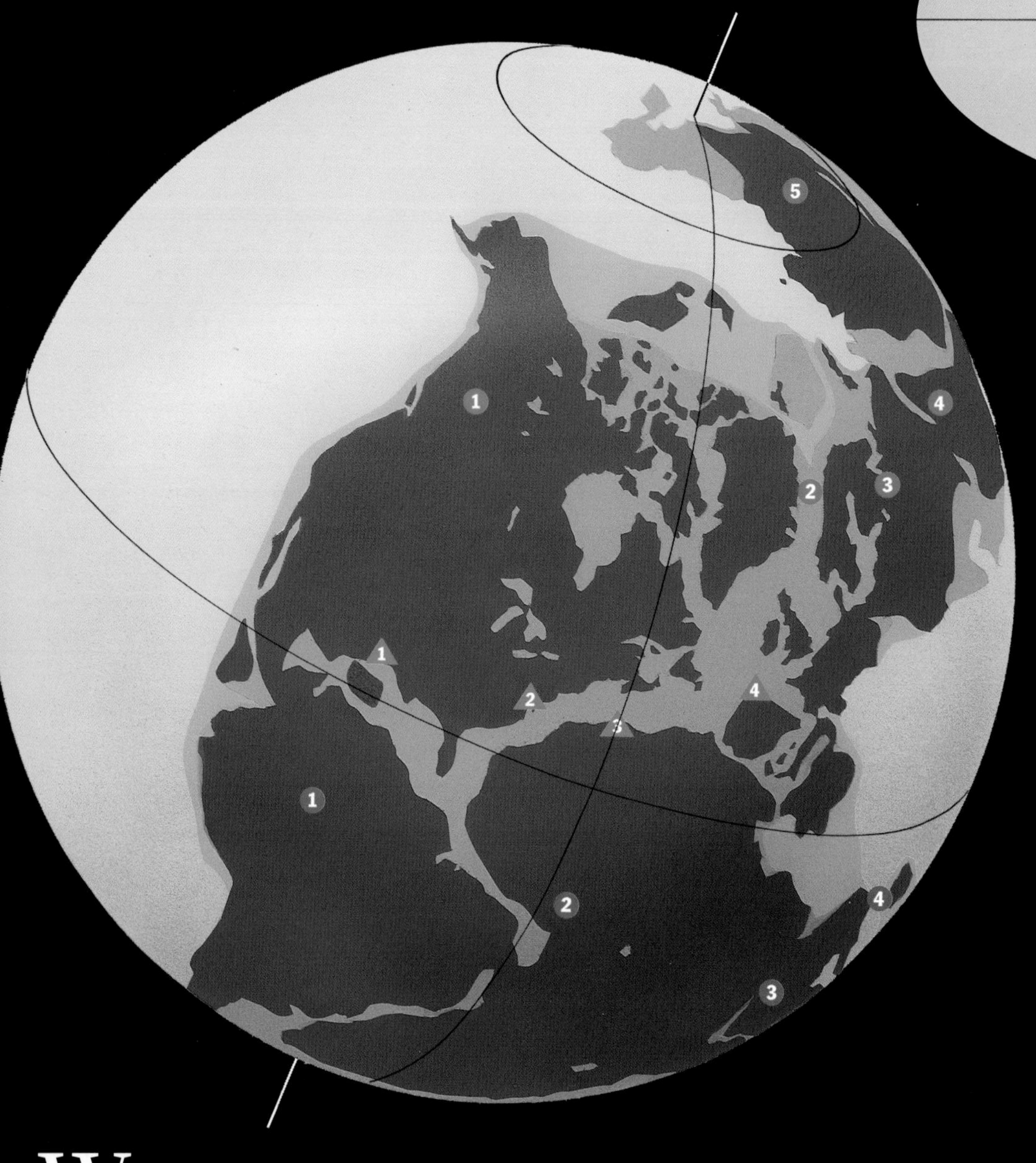

NORTHERN PANGEA: LAURASIA

1. Laurentia (North America, Greenland, Svalbard, Avalonian Europe)
2. Caledonides
3. Baltica (European Russia, Scandinavia, and Baltic Countries)
4. Urals
5. Siberia

CENTRAL PANGEA

1. Ouachita Ranges (Mexico-Ouachita Mountains)
2. Alleghenian Ranges (Appalachians)
3. Mauritanides (West African coastal ranges)
4. Variscides (Iberia, Armorica)

SOUTHERN PANGEA: GONDWANA

1. South America
2. Africa
3. Arabia
4. Turkey and Iran

Wegener visualized Pangea as a single supercontinental mass, but today Pangea and its predecessors are thought to have been ephemeral montages of megacontinents. The Pangean mix of continents existed from at least 360 to 200 MYA, and possibly from 387 to 180 MYA. Central Pangea was firmly established by 250 MYA—as seen here. At this point West African and South American Gondwana were sutured to North American Laurasia, forming the extremely mountainous regions of West Central Pangea and Mid Central Pangea.

As the oval projection shows, a string of continental elements enclosed an equatorial ocean east of Central Pangea. These included (from north to south) North China and South China, Vietnam, Laos: Cambodia in one block, and Burma, Thailand, Sumatra, and Borneo in another. The string also included the microcontinents of Tibet, Turkey, and Iran, plus legions of oceanic islands. At that time there were two oceans on Earth, not one as Wegener had supposed. These were an eastern equatorial ocean, the Tethys Ocean, and the all-embracing Panthalassa. The Mediterranean Sea is all that now remains of the Tethys. The Pacific Ocean is all that remains of the Panthalassa.

In his foreword to the second edition of the three-volume *The Lord of the Rings*, J.R.R. Tolkien explained that that saga had begun as a sequel to *The Hobbit*, published in 1937. The sequel had been "drawn irresistibly towards the older world, and became an account, as it were, of its end and passing away before its beginning and middle had been told." In *The Hobbit* the author provided a detailed map of a region beyond the Misty Mountains of Middle-earth, but he only set out the geography of Middle-earth as a whole when the first volume of *The Lord of the Rings* was published in 1954. This included a map of Middle-earth drawn by Christopher Tolkien, the author's son.

The geography of Middle-earth was conceived more than a decade before the proof of seafloor spreading and continental drift had put the controversial subject of Wegener's Pangea in a more positive light. Yet the Tolkien map of Middle-earth bears an uncanny resemblance in both topography and scale to our now quite detailed knowledge of the topography of Central Pangea. Such similarity inspired the use of *The Third Age* and *Middle Earth* for the titles of this and the previous chapter—chapters that together relate the story of the end of the "older world" and the evolution of the embryonic new world, our Western World.

Like Central Pangea, Middle-earth was divided in two (**1**) by mountain ranges. According to the scale of the Tolkien map, the Misty Mountains stretched roughly a thousand miles from Forodwaith to Dunland, the equivalent, shall we say, of the Appalachian mountain chain in Central Pangea, which stretched some 2,400 km (1,500 miles), or about the distance from Maine to Georgia.

Tolkien's Middle-earth has the River Isen flowing through the Gap of Rohan. The gap separated the Misty Mountains from their natural extension, the 800 km (500 miles) of the White Mountains of Gondor. In similar fashion an ancestral Mississippi River in Central Pangea once flowed, as its modern counterpart still does flow, through a gap between the Southern Appalachians and *their* natural extension, the Ouachita Mountains (both *Ouachita* and *Appalachian* are Indian tribal names). In Pangean times this river was part of a substantial mountain system that extended about 2,080 km (1,300 miles) into the Mexican region of Central Pangea.

One only has to rotate Tolkien's map of Middle-earth clockwise through ninety degrees to see that one could indeed be looking at a rough topographical map of equatorial Central Pangea (**2**). The rotated map is reminiscent of the West African and South American elements of Gondwana colliding with the southwestern region of Laurasia to form the Southern Appalachian and Ouachita ranges (pronounced *wach-it-aw*) around 286–245 MYA. And that is precisely the place and time at which we start this chapter.

The collision of West African Gondwana with North American Laurasia resulted in a continuous process of mountain building that progressed from East to West Central Pangea. The Variscan orogeny commenced in European East Central Pangea 345 MYA and was complete by 280 MYA. The Alleghenian orogeny in Mid Central Pangea commenced 330 MYA and the Ouachita orogeny in West Central Pangea began around 320 MYA. Both ended about 250 MYA when Pangea had reached its closest fit. They were a result of the collision of first West African and then South American Gondwana with Laurasia, and were the last act in the final assembly of Pangea.

These events were sequential because the megacontinents were rotating as they collided. The Southern Appalachian region of Laurasia and the Mauritanide Mountain region of West African Gondwana were the regions of closest contact between the rotating megacontinents. They therefore suffered the greatest destruction by overthrusting, folding, and general contortion, a process that resulted in possibly the highest mountains formed in Central Pangea [Essay 4]. The subsequent Ouachita phase of collision resulted in the destruction of what was left of the Rheic Ocean floor in the region of the present Caribbean, the Gulf of Mexico, Florida, Texas, and neighboring states. But before we begin a review of this final event in the assembly of Central Pangea, we must first put the supercontinent as a whole into perspective.

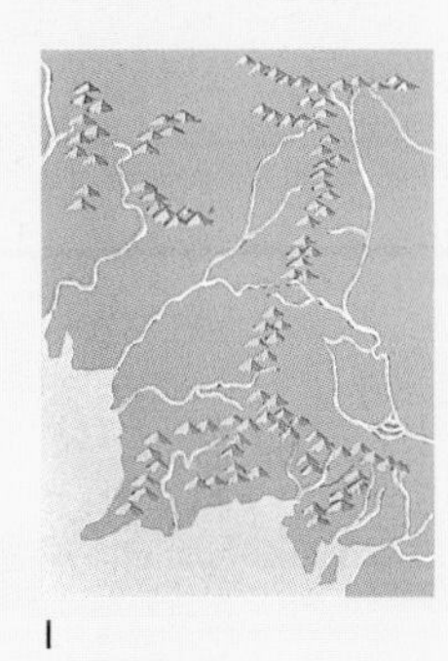

1

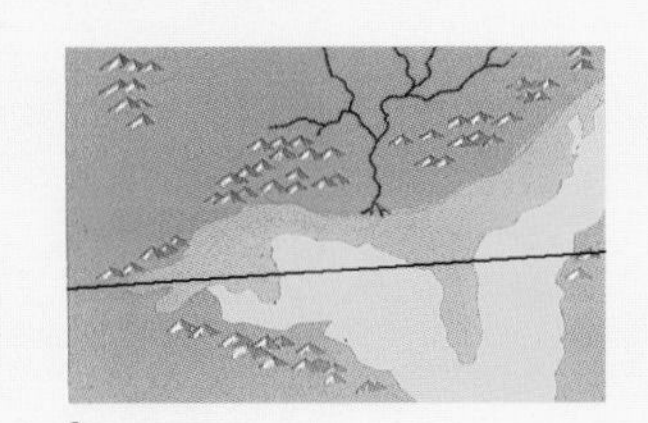

2

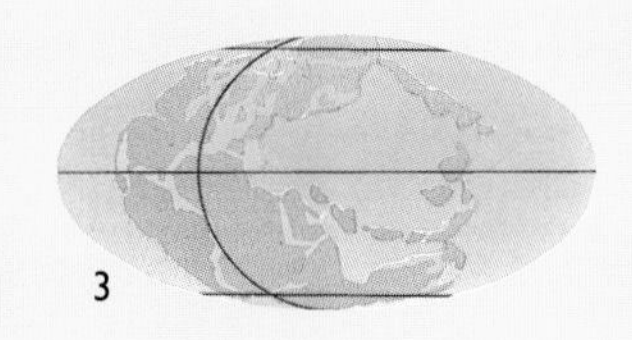

3

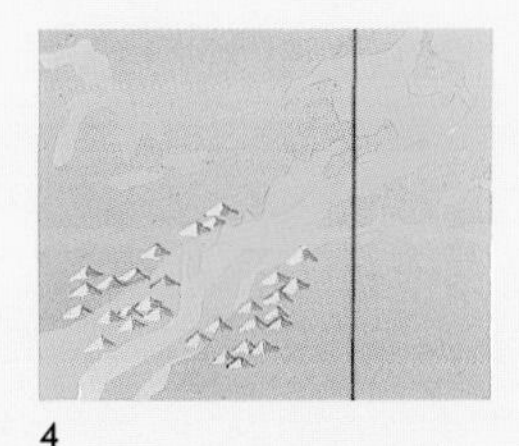

4

The still-emerging picture of Pangea is much more complex than Wegener's comparatively simplistic idea of an Earth with one primordial continent and ocean. Pangea was dynamic, not static, and one's mental image of Pangea must reflect the period of its existence that one is considering. It is now believed to have been made up of an ever-shifting aggregation of two megacontinents, Gondwana and Laurasia, several subcontinents and microcontinents, mostly Asian, and a legion of volcanic ocean islands similar to those of today. Pangea was never welded firmly together in one solid piece at any one moment in time, as Wegener had supposed. It is now thought to have been an ephemeral montage of these pieces from at least 360 to 200 MYA, and according to the latest reconstruction, possibly from 387 to 180 MYA.

To this point we have been looking almost exclusively at the history of the Western Hemisphere of the Earth. If we now look briefly at the Eastern Hemisphere around 250 MYA we will find a number of continental blocks within the world ocean. We will also see that these form a loop of tectonic elements north and south of the equator (**3**). North and South China are north of the Equator. South of the equator there are strings of other very mountainous blocks, one consisting of Vietnam, Laos, and Cambodia, and another of Burma, Thailand, Sumatra, and Borneo. To this string we can add the microcontinents of Tibet, Iran, and Turkey immersed in shallow seas, and note that the Iranian microcontinent is separated from Arabian Gondwana by a deep rift—the future Persian Gulf.

All these elements are of course differently arranged today, but when they formed the borderlands between the Panthalassa Ocean and an interior ocean called the Paleo-Tethys, 250 MYA (the reader is reminded that *Panthalassa* was Wegener's term for "all-ocean," the complement of Pangea, "all-land"), much of the modern Far Eastern world was strung across the Eastern Hemisphere like a necklace of oversized beads. The necklace extended from Siberia in the north to Australia in the south.

In Greek mythology Tethys was a sea goddess, who with her brother Iapetus—of *Iapetus and Avalonia*—was descended from Gaea, mother of the Earth. At its maximum, around 250–230 MYA, the Tethys Ocean was about the size of today's Indian Ocean. It was defined to the east by a mountainous Asian archipelago, and to the west by vast, shallow, sea-covered continental shelves off the shores of East Central Pangea—the future oil fields of the Arab world. The Tethys has since been reduced through stages to the comparatively minuscule, almost landlocked Mediterranean Sea.

During this time, East Central Pangea, the region of present-day Western Europe and Northwest Africa, was riven by rift valleys and shallow seas. In contrast to these first signs of the breakup of the supercontinent, Mid Central Pangea, the present-day eastern coast of North America and the western coast of Africa (**4**) were about as closely bound as India, the Himalayas, and Tibet are today, with similar topographic results. But the westernmost part of Central Pangea, the region of the present Mesoamerica, was still being compressed into the andean-type margin that typified the coasts of South American Gondwana and North American Laurasia.

All paleoglobes in this book show the modern geographic outlines of familiar places. If one was to show the continental and other shapes as they appeared at different geological times, they would be unrecognizable. The oval projections show the landmasses more realistically. It follows that our paleoglobes can be unintentionally misleading. We meet one such instance as we resume the story of the Ouachita orogeny and the destruction of the Rheic Ocean floor.

[*continued on page 124*]

1 Ancestral Rockies

The last act in the assembly of Pangea was the collision of South American Gondwana with southwestern Laurasia. The resulting Ouachita orogeny (320–250 MYA) caused the formation of the Ancestral Rocky Mountains and other striking features described in the following essays. The *Ancestral* Rockies are characterized by the red rock formations in the inset picture. They were "ancestral" in the sense that after they had been largely reduced by erosion another range, the *Laramide* Rockies, formed in their place (represented by the ridge at the top right of the inset). When that range had also been reduced it was followed by yet another—the modern *Southern* Rocky Mountains, typified by Long's Peak (main picture), a structure of granite basement rocks capped by sedimentary rocks in Precambrian times.

This succession of mountain ranges occurred for quite different tectonic reasons, but with one common factor. All three ranges were formed in a region of inherent structural weakness in the "stable platform." This is the very thick and otherwise geologically undisturbed continental region represented by the Great Plains and their underlying rocks, stretching east for a thousand miles or more beyond the ridge at the top right of the inset picture.

The Ancestral Rockies were formed by the flexing of the stable platform during the Ouachita orogeny, when South American Gondwana collided with the edge of Southwestern Laurasia. This caused ancient faults to open in the Laurentian stable platform (Laurentia was the North American region of Laurasia). The faulting allowed part of the Precambrian basement on which the platform stands to be thrust towards the surface, forming mountains. In scientific language this phenomenon is called "tectonic escape."

Red Rocks Park, Morrison, Colorado

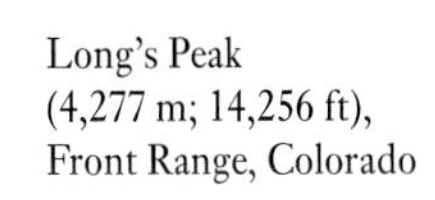

Long's Peak
(4,277 m; 14,256 ft),
Front Range, Colorado

2 West Central Pangea

As the Ancestral Rocky Mountains were being formed around 250 MYA, the last of the Rheic Ocean floor between southwestern Laurasia and South American Gondwana was subducted and destroyed beneath the colliding megacontinents. The leading edge of South American Gondwana then began to overthrust the Laurentian stable platform, at the same time as the weight of accumulating mountains caused the edge of the platform to arch downward. This in turn produced "foredeep basins"—troughs at the base of the mountains on their stable platform side. One of the best known and best preserved of these foredeep basins, called the Permian Basin of West Texas, is illustrated here in the Guadalupe Mountains of present-day Texas and New Mexico.

This basin, now the Midland & Delaware oil and gas basins of Texas, was open to the ebb and flow of the Panthalassa Ocean from the west. As the orogeny progressed and the foredeep grew deeper, reefs and carbonate banks at or near sea level stood correspondingly higher above the seafloor, the structures that now form mountains. But the flow of fresh seawater into the basin was restricted and so could not deliver the dissolved oxygen required by an expanding population of bottom-dwelling benthic fauna, or replenish the oxygen consumed during the decomposition of dead marine organisms on the seabed. The bottom of the basin thus became anoxic and the teeming multitrillions of organisms that had once populated the basin were simply pickled in brine. Over millions of years, they were ultimately transmuted into hydrocarbons by the pressure of accumulated sediments above, the heat generated with burial, and the natural heat of the Earth. This Guadalupe region of the foredeep has so far generated a hundred billion barrels of oil and a hundred trillion cubic feet of natural gas from such marine organisms.

[*continued from page 119*]

The V-shaped notch region (the "Piedmont microplate") between South American and West African Gondwana—which appears in all the paleoglobes in this book from 600 MYA to the point we have now reached (286–206 MYA)—includes an outline of modern Florida. The basement on which Florida's carbonate rock has accumulated is Gondwanan displaced terrane, but the paleoglobe reconstruction is unintentionally misleading: the peninsula's distinctive modern outline did not exist in this form then (**5**).

ALTHOUGH THERE ARE other interpretations of the genesis of the Gulf of Mexico region (including Florida, the Bahamas, and the Caribbean basin), here we are following reconstructions made by M.I. Ross and C.R. Scotese. According to these scientists, the basement rocks of Florida are composed of two segments that were offset until about 150 MYA (**6**). That part of Florida now to the south of a rough line drawn from Tampa to Palm Springs was simply driven along a strike-slip fault into its present conjunction with the northern section. The southern section included the basement on which the future Bahama Islands were to form, and had affinities with the Guinea Plateau region of West African Gondwana. The northern section was associated with Senegal and southern Mauritania, near modern Dakar in West Africa. On the other hand, the Cuban and the Greater Antilles basement rocks next to the southern Florida-Bahama elements had a close association with the Venezuelan region of South American Gondwana.

What is now the North Atlantic margin of South America was far more extensive in Gondwanan time. Besides the notch region it included the continental rocks that form the base of Yucatán and other lesser fragments of today's Caribbean region. Yucatán was next to or part of the Colombian region of South American Gondwana. Like Florida and the Bahamas, the Yucatán basement later became the foundation of an enormous coral edifice. This was compressed by its own weight into a carbonate structure, a vital component of the very remarkable structure of the modern Gulf of Mexico and the Caribbean—a story chronicled in the next chapter.

THE LAURASIAN side of the confrontation zone bore the brunt of the Gondwanan onslaught in the Ouachita orogeny. This zone stretched from Mexico in the west to present-day Tennessee and parts of Alabama and Georgia. At various stages during this intercontinental collision the region resembled different phases in the present Alpine orogeny (itself the final stage in the closure of the Tethys Ocean). At a late stage the Ouachita orogeny may have resembled the Aegean Sea, with its many islands and neighboring Balkan and Anatolian mountain ranges. Still later, it could have resembled the Black and Caspian inland sea regions of Europe in their present Carpathian setting. These include both the highest and lowest topographical features in Europe, Mount El'brus in the Caucasus, 4,570 m (15,233 ft) high, and a point near Baku in the Caspian Sea , 829 m (2,764 ft) below sea level.

In the final stages of the Ouachita orogeny, the leading edge of South American Gondwana was literally thrust up and over the edge of Laurasia. The alpine-type ranges that formed as a result are thought to have been about the stature of the modern Swiss Alps, but three to four times their extent. The Ouachitas ran from the Mexican volcanic belt 2,080 km (1,300 miles) east to Mid Central Pangea, where they met the extremity of the Himalayan-style giants that had formed during the Alleghenian orogeny. As the Ouachita orogeny was in progress, further chaos was added by the descent of part of the Panthalassa Ocean floor beneath the edge of West Central Pangea, a subduction process that resulted in the formation of coastal volcanic ranges like the modern Andes (**7**).

The Ouachita and Alleghenian orogenies met in a region now known as the Mississippi Embayment. What remains of the eastern extremity of the Ouachita alpine range is buried deep in the embayment beneath the accumulated sediments of the geologically very young Mississippi River, and its *many* predecessors. Other remnants of the Ouachita Alps have been reduced to modest rolling hills in central Arkansas and southeastern Oklahoma, and to some scattered outliers in the Mexican volcanic belt.

5

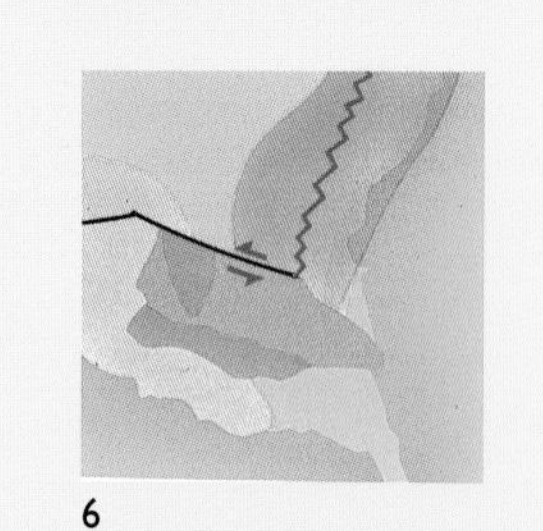

6

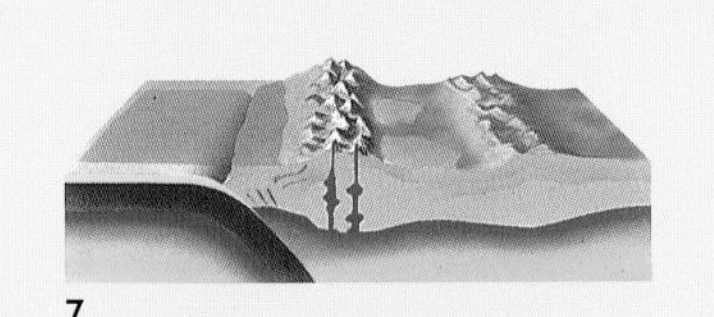

7

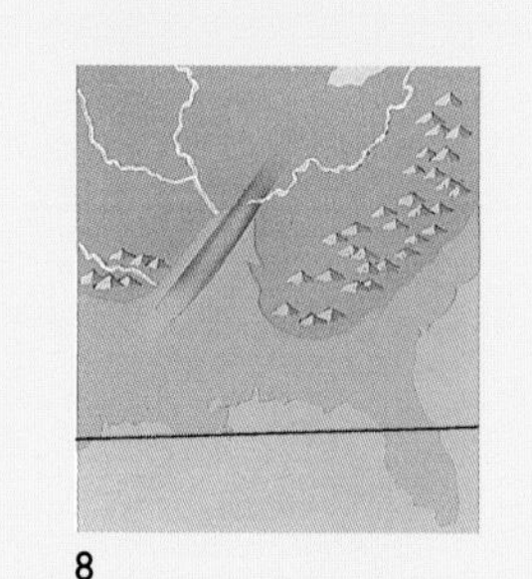

8

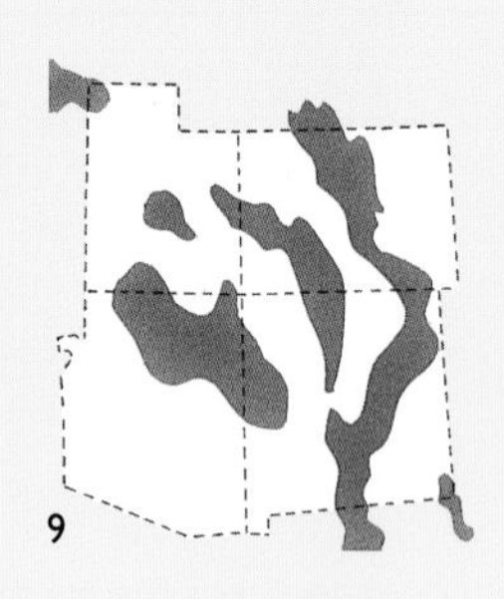

9

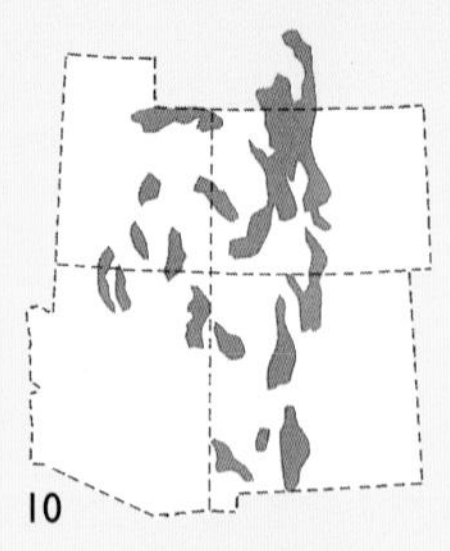

10

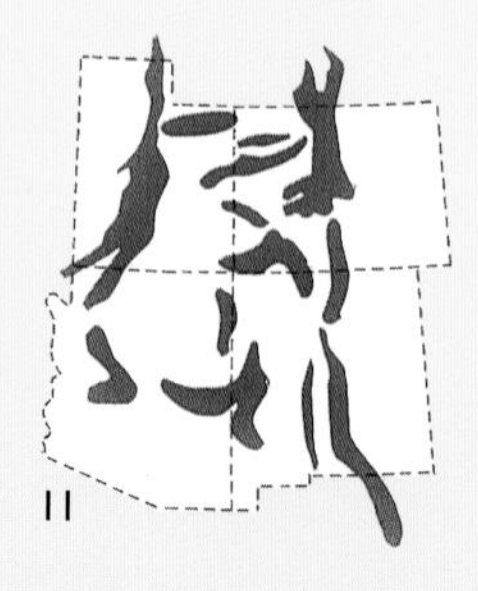

11

The burial of the eastern flank of the Ouachitas beneath the Mississippi Embayment was long suspected (**8**). It was described by K.D. Nelson and others during a seismic profiling project that also identified the probable line of suture between Laurasia and Gondwana, deep beneath the sediments of the modern coastal plain between Georgia and Florida. But why the eastern Ouachitas subsided into the embayment region is not known.

Like most great rivers, the modern Mississippi flows for at least a part of its length above the site of an ancient rift. This New Madrid Rift is believed to have formed at least 600 MYA. It now lies far beneath the neck of the Mississippi Embayment, still seismically active and a dangerous source of major earthquakes—potentially the most damaging in North America. The New Madrid series of major earthquakes in 1811–12 are considered the most severe in historic times. The neck of the embayment above the rift today takes the form of a "graben," a shallow undulation in the landscape that channels water just as the waist of an hourglass channels sand. The neck constricts and guides silt-laden water from the confluence of the Mississippi and Ohio Rivers to the north of the embayment, and then on to the vast Mississippi flood plain to the south [Essay 3].

The Mississippi, the Ohio, and their tributaries today drain most of the North American Midwest. This region is called the "stable platform"—it is comparatively undisturbed by tectonic trauma. The stricture between the Ouachitas and the Southern Appalachians that overlies the region of the New Madrid Rift has frequently been the focus of continental river-drainage systems. Conversely, it has also been the entry point of epicontinental seas that have spilled onto the stable platform as sea level has fluctuated through time. Indeed, the New Madrid Rift formed a stricture that is reminiscent of Tolkien's notion of the River Isen. Tolkien had the Isen flowing through the Gap of Rohan [see page 119].

WE MUST NOW ADD a new and important factor to this scenario. The mountain building that resulted from the Ouachita and Alleghenian orogenies had more than a locally destructive effect on Laurasian margins. Overthrusting had a cumulative effect that can be likened to the response of an outstretched hand to a sudden weight: the heavier the weight the more the hand, the wrist, and finally the arm, are forced to bend. In a similar way, the stable platform was arched downward in response to the accumulating overthrust rock. At the same time the platform was being compressed by the lateral force of the Gondwanan impact. The combination of weight and lateral compression resulted in the formation of the Ancestral Rocky Mountains in southwestern Laurasia (**9**).

The *Ancestral* Rockies are so named because after they had been largely worn away by erosion, another range, the *Laramide* Rockies (**10**), formed in their place. When that range had been reduced it was followed by yet another, the modern *Southern* Rocky Mountains (**11**). This succession of mountain ranges formed in different ages because of separate tectonic events. Nevertheless, all three systems have one factor in common: they developed in a region of the stable platform that has inherent structural weaknesses [Essay 1].

In the case of the Ancestral Rockies, the flexing of the platform added to the force of South American Gondwana's impact on Laurasia. This caused ancient faults to open—just as the fingers of that metaphorical outstretched hand (this time palm downward) would be forced to bend and to open if subjected to steady lateral force from fingertips to wrist. The gaps between the "opening fingers" of the Laurasian stable platform were the ancient fault systems that reactivated. They allowed the Precambrian basement on which the platform stood to be thrust to the surface and form mountains—a phenomenon called "tectonic escape."

THE MOST IMMEDIATE effect of the accumulating weight of mountain ranges on the edge of the Laurasian stable platform was to cause "foredeep basins" (elongated troughs that filled with sediment) to form at the base of the mountains on their stable platform side. What happened to the marine organisms that lived in those basins, and ultimately to the basins themselves, provides a perfect model of the interaction of life, climate, and tectonics.

For example, as we saw in the last chapter, during the Pennsylvanian coal-forming period 320–286 MYA, basins gradually filled with shallow river water and terrestrial sediments upon which coal forests took root. Following a rise in sea level, these swampy areas were repeatedly drowned by shallow seas and smothered by marine sediments. Such coal-forming cycles continued for as long as warm and humid conditions lasted and as often as epicontinental seas could flood the basins, which were foredeep basins associated with the Alleghenian orogeny. As the Alleghenian orogeny reached its climax and was overtaken by the Ouachita orogeny during the Permian Period (286–245 MYA), seas could no longer enter Mid Central Pangea and coal-forming cyclothems ceased to develop.

In West Central Pangea the Ouachita orogeny (called the "Marathon" orogeny in Texas) was also causing foredeep basins that *were* subject to flooding by shallow seas. In the Southern Rocky Mountains near the New Mexico and Texas border, one can actually stand in a foredeep basin on a West Central Pangean seabed with carbonate banks and reefs, once the margin of a Permian sea, rising far above one's head [Essay 2]. The reef is now a mountain range, the Guadalupe Mountains, and includes Carlsbad Caverns, one of the world's most spectacular cave systems, which burrows deeply into the ancient reef.

This system of mountains and basins is one of North America's most prolific petroleum provinces, called the Permian Basin. The two basins that developed at the foot of the Guadalupe Mountains, the Delaware and Midland Basins, are highly productive. They have so far produced around a hundred billion barrels of oil and a hundred trillion cubic feet of natural gas—hydrocarbons transmuted into their present form from the remains of marine organisms. Picture for a moment the teeming marine life that those figures represent. Consider the countless trillions of organisms, mostly minute, that must have existed in this inland sea to form such enormous volumes of fossil fuels. Generation upon brief generation of organisms succeeded each other year after year for tens of millions of years, living and dying in a sea trapped in the rain shadow of the Ouachita Mountains near the Permian equator. At this time a similar scenario applied to the Arabian continental shelves of the Tethys Ocean and other Pangean localities, as we shall see later.

FROM THE TIME it began to form 260 MYA, the Guadalupe foredeep was filled and replenished by seawater through a narrow inlet called the Hovey Channel, which connected the basin to the Panthalassa Ocean. The channel remained open though constricted through much of Permian time. At first, when low reefs were beginning to build on the stable platform edge of the foredeep (**12**), there was an adequate supply of oxygen-rich seawater flowing through the channel to support a rich bottom-dwelling "benthic" fauna.

As the orogeny progressed and the foredeep grew deeper, the reefs at or near sea level grew correspondingly higher above the seafloor, but the flow of fresh seawater into the foredeep remained about the same. The flow was not sufficient to maintain an expanding population of benthic fauna nor, more importantly, to replenish the oxygen consumed during decomposition of the steady rain of dead organisms from above. The bottom of the shallow sea gradually became anoxic: benthic fauna could not survive, and organic material on the seabed was preserved.

As the reefs built higher, limestone rubble and silt fell down the fore-reef slopes in occasional submarine avalanches. Some boulder-sized reef fragments were swept onto the basin floor by these turbidity flows; one can still see the debris today. Over many millions of years sediments from hundreds of thousands, possibly millions of avalanches covered, and ultimately deeply buried, layer upon layer of preserved faunal remains. At some point during this process the foredeep was divided into two basins by block faulting: one basin (the Midland) silted up completely, but the other (the Delaware) remained linked to the sea via the Hovey Channel.

As the Permian global climate grew steadily warmer, the back-reef lagoons in the Delaware Basin evaporated, producing crystallized salts. Then the sea in the basin itself began to evaporate until it too precipitated into salts, which slowly accumulated and filled the basin. The accumulated thickness of these lithified sediments produced terrific pressures on the preserved organic mix of proteins, carbohydrates, and fats. Heat generated by radioactive isotopes of uranium, thorium, and potassium in the surrounding rocks also acted on the mixture, reducing it to hydrocarbons. The transformation of the organic mix into the equivalent of billions of barrels of oil and trillions of cubic feet of natural gas (the "maturation period"), took millions of years. In more recent times rainwater has dissolved and washed away the soluble minerals in many parts of the foredeep, leaving the magnificent Permian reef and basin structures exposed.

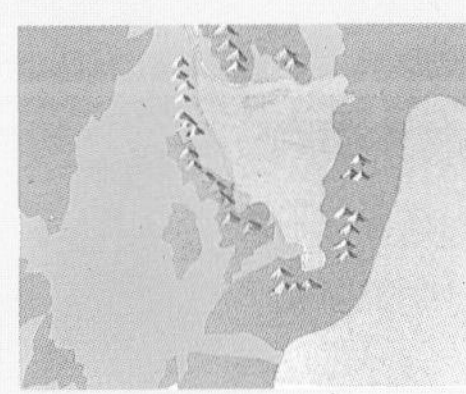

13

AT THE TIME that the Guadalupe foredeep basins were steadily filling with evaporites (rocks formed from residues after evaporation), thousands of miles away a similar process was taking place in two basins in East Central Pangea. These vast basins, known in Europe as the Northern and Southern Permian Basins, were occasionally flooded by the arctic waters of the Panthalassa Ocean through a broad but shallow channel. The vast, flooded basins formed an epicontinental sea called the Zechstein Sea (**13**) [Essay 5]. The Zechstein formed and evaporated five times in a period of less than twenty million years near the end of Permian times (263–245 MYA). The channel to the sea flowed through a graben over an active rift between the Pangean regions of Greenland/Scotland and Norway—a rift that ultimately became part of the seafloor-spreading system in the northern Atlantic.

14

Zechstein is the place in northern Germany where evaporite deposits were first described by geologists early in the 19th century. The British geologist Roderick Murchison (who had quarreled with Adam Sedgwick over the classification of rocks in Wales) put these deposits into their correct geological context after discovering (in 1841–42) a "new" series of rock formations in the Russian province of Perm (therefore the term "Permian"). Murchison's Permian formations overlie Carboniferous rocks on the western slopes of the Urals, mountains formed during the collision of present-day south central Russia with Europe 310–270 MYA.

The Permian formations in the Urals proved to contain sedimentary sequences corresponding to the Zechstein deposits in northern Germany. After many years of exploration it was discovered that the Permian-Zechstein series extended beneath the surface of Denmark, The Netherlands, beneath the North Sea to northern England, and beneath the Irish Sea north of the Isle of Man. As with Murchison's other discoveries, his classification of the Permian Period as a geological system in its own right was controversial. It was not finally accepted by all geologists until 1948.

The Zechstein Sea varied in depth from literally zero, when the sea level fell below that of its connecting graben, to several hundred feet when sea level rose again. The Northern and Southern Permian Basins were contained to the south by alpine ranges, stretching from the Black Forest region in modern southern Germany (**14**) and the Vosges in northeastern France, via the French Massif Central to the Cornubian mountains of southwest England. All these ranges were products of the Variscan orogeny that took place during the collision of Gondwana with European Laurasia (described in the last chapter).

[*continued on page 130*]

3 Ouachita Connection

It was the Laurasian side of the zone of confrontation with South American Gondwana that took the brunt of the megacontinental collision during the Ouachita orogeny (called the "Marathon" orogeny in Texas). The alpine-type ranges that resulted stretched about 2,080 km (1,300 miles), from the shores of the Panthalassa Ocean in West Central Pangea to the base of the Himalayan-sized giants in the Alleghenian Appalachians of Mid Central Pangea. In modern terms this would be from present-day Mexico to Tennessee, northern Alabama, and Georgia. The ranges met in the region of what is now the Mississippi Embayment. What is left of the eastern end of the Ouachita Alps is buried deep in this embayment beneath the accumulated sediments of the recent Mississippi River and its many predecessors.

The neck of the Mississippi Embayment today takes the form of a gentle undulation in the landscape, a shallow valley, or graben, which has formed above the site of the ancient and still seismically active New Madrid Rift. The graben serves to channel water drained from the stable platform just as the waist of an hourglass channels sand. As shown in this panoramic photograph, the New Madrid neck constricts and guides silt-laden water from the confluence of the Mississippi and Ohio Rivers to the north of the embayment, and empties onto the vast, near-flat Mississippi flood plain to the south. The stricture has also been the entry point of many epicontinental seas that have spilled onto the stable platform as sea level has fluctuated. Vestiges of the Ouachita alpine range today take the form of comparatively modest rolling hills in central Arkansas and southeastern Oklahoma, with scattered outliers in the Mexican volcanic belt.

4 Mid Central Pangea

The mountains formed during the Ouachita and Alleghenian orogenies were alpine in character; they were the folded, refolded, overthrust, and uplifted edges of Laurasia and Gondwana that collided in what is now the Mississippi Embayment. The Alleghenian range to the east was far greater in stature than the Ouachita ranges. The former, since reduced to the modern Southern Appalachians, is thought to have included giants of Himalayan proportions. They were the highest mountains in Central Pangea and perhaps twice the elevation of their counterparts in West Central Pangea. Their spectacular height resulted from proximity to the actual point of contact between the rotating and colliding Laurasia and the Mauritanide region of West African Gondwana.

Perhaps this twilight scene of the Blue Ridge Mountains above Shenandoah Valley, Virginia, is a fitting memorial to the snow-clad Everests, Nuptses, and Kangchenjungas that once stood here.

[*continued from page 125*]

These European Permian basins, both northern and southern, were the result of tectonic processes quite different from the ones that created the Guadalupe foredeep. The northern basin is associated with a rift that was active throughout Permian time, and caused a major depression in the basin floor, into which sediments and evaporites accumulated. This feature is named the Oslo Graben, after the Norwegian capital, which overlies part of the rift today. The tectonic and climatic events that led to the formation of the Southern Permian basin and its minerals were different again, resembling events that are taking place today in the Great Basin and Range Province of Utah, Nevada, and southeast California. If one wants to see the conditions in which the Zechstein basins were formed, and experience Central Pangea at around 250 MYA, the Great Basin is about as close as one can get.

THE GREAT BASIN is landlocked and confined by mountains, notably the Sierra Nevada of California to the west. As we have seen, the Zechstein basins were also landlocked and confined by the high Variscan alpine ranges to the south. Sediments washed down fromthe Variscides had accumulated on the Avalonia-Baltica margins, weighing them down and so causing the original Zechstein basins to develop. The floor of the Great Basin is being stretched by tectonic activity, the result of past ocean-floor subduction beneath California in a process called "crustal extension." As a result, the Great Basin is now twice its original size, and has the thinnest known continental crust on Earth today. Similarly, it is thought that the Zechstein basin also experienced such crustal extension in Pangean times. Other parallels include the great influence of the glacial and interglacial periods of the present ice age; the environment of the Zechstein basin was greatly affected by fluctuations in sea level at the end of a prolonged ice age.

As the Great Basin floor is being stretched, it also sags and twists—just as a sheet of paper with parallel slots cut into it will "sag and twist" if the sheet is "extended" by pulling its opposing edges apart. The geological result is "block faulting," in which the crust breaks into peaks and troughs (**15**). Over a period of tens of millions of years the structurally weaker troughs and faults in the basin floor have been punctured by volcanic intrusions and eruptions, the source of many economically important minerals such as gold, silver, copper, and lead. The basin has also been etched by many low-lying valleys like Death Valley, 85 m (282 feet) below sea level and about 224 km (140 miles long), flanked by parallel ranges like the Panamint Range and the Funeral and Black Ranges near Death Valley. This pattern is the reason geoscientists call the region the "Basin and Range Province."

Although the Great Basin is well north of the Equator, roughly at the same northern latitude as the Zechstein Sea in Permian times, it is still one of the hottest desert regions on Earth. It lies in the rain shadow of the Sierra Nevada (**16**), with peaks ranging from 2,400–4,200 m (8–14,000 ft), a striking contrast to the adjacent basin floor, which, due to thinning and block faulting, is either near or below sea level. The Sierra Nevada wrings the moisture out of warm west winds blowing across California from the Pacific Ocean, leaving the basin to the east almost completely desiccated [Essay 6].

The perpetual drought in the Great Basin is relieved by violent thunderstorms that sometimes result in short bursts of extremely heavy rain—light rain simply evaporates as it falls. Such cloudbursts occur over the interior ranges in the basin, causing torrents of muddy water to roar down the mountainsides, scouring minerals out of the rocks as they go. If the supply of mineral-laden flash-flood water, or mountain streams from surrounding snowcapped peaks is insufficient to make good the losses by evaporation, any standing water left in the basin leaves a deposit of crystallized salt like those on the floor of Death Valley [Essay 5].

If, as happened during the last glacial advance, the water supply is greater than the loss, free-standing water accumulates until it floods most of the Great Basin. The lake that formed is called Lake Bonneville (**17**) and has since been reduced by evaporation and catastrophic discharge to the size and hypersaline state of the present Great Salt Lake. The rest of the Great Basin has become a desert of the kind typified by Death Valley—where the highest temperature recorded in the shade is 57.1°C (134°F), and the highest ground temperature 88.5°C (190°F).

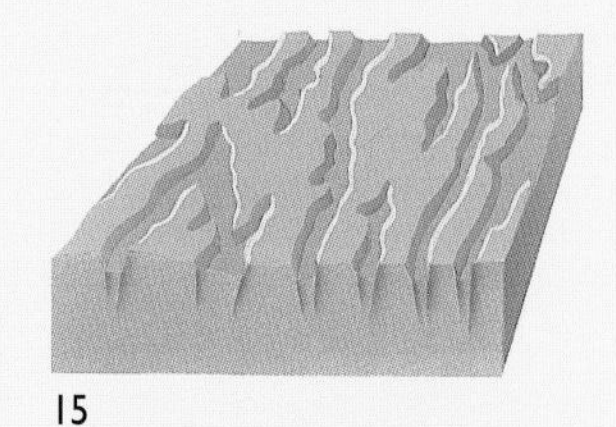

15

16

17

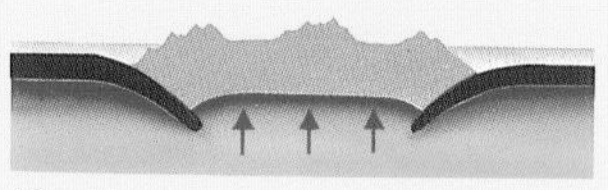

18

THIS IS HOW IT might have been in the southern regions of the Zechstein basin. Transitory shallow seas and ephemeral freshwater lakes were fed by flash-flood water from the flanks of the Variscan ranges and block-faulted mountains in the thin-floored basin. These seas and lakes were the source of immense deposits of salt, potash, gypsum, and other minerals—deposits that sometimes exceed 900 m (3,000 ft) in thickness. Between cycles of epicontinental seas, wind-blown sand dunes accumulated in places on the bone-dry basin floor, and petrified into thick sequences of permeable "eolian sandstone" (named after Aeolus, Greek god of the winds), which were in turn covered by other deposits [Essay 5].

As Murchison first discovered in the Urals, such sandstone formations overlie Carboniferous rocks that include coal deposits. When the Southern Basin was subjected to more tectonic pressure, the Carboniferous coal seams beneath the Permian sandstones were further compressed, forcing gas out of them and into the overlying eolian sandstone, which acted as a reservoir, efficiently sealed at the top by Zechstein evaporite minerals deep beneath the bed of a modern epicontinental sea—the stark and often frigid North Sea. Evaporites are similarly important sealing agents in the Permian Basin of West Texas.

The first of the five Zechstein Sea cycles took place about 265 MYA, towards the end of a 60-million-year ice age. The global climate was remarkably similar to that of today, although low-lying regions of Central Pangea became desert-like through Permian time because they straddled the Equator and were dominated by ranges of high mountains that cast rain shadows. As the Zechstein cycles progressed, global climate continued to warm appreciably, a trend that continued for 20 MYA after their end. As we shall see in future chapters, high global mean temperatures then predominated for a further 200 MY, and remained generally above today's average until the onset of our Cenozoic Ice Age.

THE IMMEDIATE RESPONSE of the average global sea level to increases in global temperature in Late Carboniferous and Permian time was surprising. At the conclusion of an ice age, global sea level should rise as the ocean accommodates the melting ice—yet (according to P.R. Vail of Rice University, Texas) global sea level *fell*. It seems that *average* sea level dwindled to below today's level when the Carboniferous Period ended—and stayed that way throughout the existence of Pangea. The Zechstein cycles were caused by relatively minor increases in sea level during a period of general decline. Only when the supercontinent was well into the process of rifting and breaking up did sea level again begin to rise, around 180 MYA. It then increased well above the present level and remained high until the onset of the present ice age. As we shall see, these conditions had a truly profound effect upon the evolution of life on Earth, so the explanation of this paradox is critical. The answer may lie in the special nature of supercontinental clusters—their very size and concentration in a specific region of the Earth.

ASSUME FOR A MOMENT that all the continents on the planet were once wedge-shaped—like the cut segments of a round gateau—and that they were driven together pointed end first by the movement of the seafloor surrounding them. The most obvious results of such perfect symmetry would be the compaction of the supercontinent into a circular shape (**18**). This would inevitably result in the formation of alpine-type mountains with deep-seated roots at the center where the wedge points had collided. On the perimeter of this imaginary supercontinent, where the edge is still thin enough to allow subduction of the surrounding ocean floor, we would expect to see an unbroken chain of Andean-type mountains.

The "footprint" area of the imaginary supercontinent would have been reduced by the process of collision, and the crust would have thickened with the formation of both central and peripheral mountains. The circular supercontinent would therefore have a high profile and a footprint less than the total area of the individual wedge-shaped continents that contributed to its makeup. At its zenith, the continent would take the form of a towering circular edifice rising far above its surrounding seafloor, with roots plunging deep into a hotter than usual and therefore swollen region of the asthenosphere beneath—an induced hot spot.

With the addition of a warming climate and melting ice caps (**19**), huge volumes of water would be added to this model world ocean, even as the circular supercontinent was being compressed and uplifted. From this we can readily imagine that, over the long term, although the volume of the ocean has increased, to an observer standing on the supercontinent, sea level would appear to have fallen. The warming trend in our model is a short-term factor operating in fits and starts. Tectonic changes are slow and measured. This difference in behavior would account for the apparent contradiction of cyclic episodes of basin flooding in a period of falling relative sea levels—such as those in the Zechstein region.

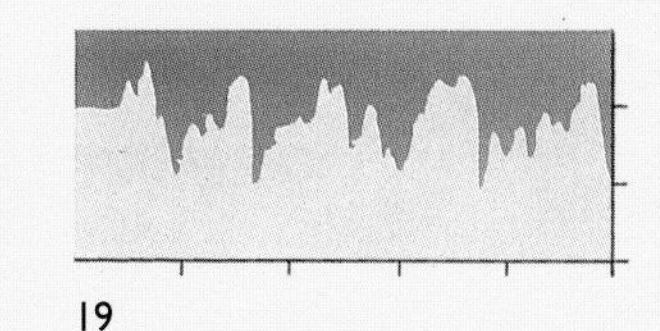

19

During breakup the imaginary supercontinent would increase in area and reduce in thickness (**20**) as its mountains eroded and its crust was stretched by the effect of the hot spot in the asthenosphere beneath it. The supercontinental crust would therefore begin to extend, as in the real world the Great Basin is being extended and the Zechstein basin was extended. From this it seems likely that all supercontinental assemblies are predestined to self-destruct and breakup because they cause overheating of the asthenosphere beneath them.

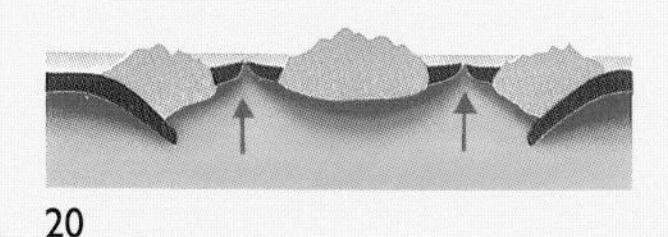

20

The end of Permian time, 245 MYA—near the climax of Pangea's existence, the ultimate stage of supercontinental compression described above—proved to be a turning point in the Earth's history. This was the end, so to speak, of the "Elder Days of Middle-earth." Pangea now entered a period of supercontinental aging and decline called the Triassic Period. ("Triassic" refers to three stages of deposition first recognized in German rock formations dated 245 to 206 MYA). The Triassic was the time in which the initial rifting of Central Pangea began [Essay 8].

As these tectonic events were progressing, significant bio-evolutionary steps were also taking place. These developments both preceded and followed the greatest known mass extinction of life that appears in the fossil record. This was the great Permian extinction that brought to an end the opening stages in the evolution of multicellular forms of life, the Paleozoic Era, and marked the beginning of the "middle ages" of evolution, the Mesozoic Era [Essay 6].

Near the end of Permian time more than half the known families of animals on Earth were extinguished or seriously depleted—and 75–95 percent of marine invertebrates were permanently extinguished. The latter included the last trilobites, all the horn corals, and most of the brachiopods. Among the terrestrial animals most amphibians and many early reptiles were extinguished. Also, over a period of 15 MY, spore-bearing plants gave way to conifers and their kind—seed-bearing plants, the gymnosperms.

Even so, it would be wrong to suggest that this mass extinction was a *dramatic* event. Contrary to popular belief, few if any mass extinctions are "instant" catastrophes. This mass extinction, like most others, was spread over a period of five million years, from 255 to 250 MYA. The probable causes are manifold and perhaps the result of a domino effect—as one cause of stress was superimposed upon yet another cause.

A deep and prolonged ice age had now ended in the Southern Hemisphere. Since this had no apparent effect upon sea level, as demonstrated above, the supercontinent had not been flooded by epicontinental seas. In fact, the creation of Pangea had the opposite effect—the total length of coastline available for shallow-sea marine animals had been greatly reduced. Significantly, Pangea had been underpinned by subducting seafloors, which had led to an excess of volcanism. This factor is highlighted in a paper by D.H. Erwin published in *Nature* in 1994. It details a series of overlapping events that coincided with the ultimate formation of Pangea 250 MYA. These events include an extended period of excessive volcanism in South China (259-242 MYA), excessive flows of flood basalt lava—called "traps"—in Siberia (253-241 MYA), and a global reduction in oxygen levels arising out of the other two events (258–243 MYA).

The formation of the Siberian traps was a devastating event: thousands of cubic miles of flood basalts (lavas similar to those in Hawaii today) erupted and spread over tens of thousands of square miles in a period of 12 million years (these figures translate to hundreds of thousands of cubic kilometers and tens of thousands of square kilometers). This prolonged and widespread volcanicity, and the effect of Andean-type volcanoes dotted around Pangean shores, would have had a noxious, perhaps profound effect upon the global atmosphere, and must surely have contributed to a Permian greenhouse effect which in turn helped bring an end to the 60 MY Permo-Carboniferous Ice Age.

The unification of the supercontinent had drastically changed the pattern of ocean circulation; had greatly decreased the gross area of its interconnected elements; and had increased the stature and extent of its equatorial mountain ranges. For all these reasons, the pattern and character of the supercontinental climate had also changed radically. The final assembly of Pangea was itself the probable root cause of the Permian marine extinction, through the radical changes in climate, sea level, and atmospheric toxicity that it produced over millions of years.

The existence of coal-producing swamps in Central Pangea had depended upon fluctuating sea level and warm humid equatorial climates during the Permo-Carboniferous Ice Age. In some regions (Pennsylvania for example) coal-forming cyclothems ended 286 MYA. This was not because the ice age had finished—on the contrary, it was at its height at this time. However, continental collision and mountain building had restricted the direct access of shallow seas to the coal-forest basins in Central Pangea (**21**).

21

As the Permo-Carboniferous Ice Age waned during the final assembly of Pangea (286-245 MYA), West Central Pangea had continued to come together in the Ouachita orogeny. Meanwhile, East Central Pangea had begun to extend—the first signs of continental breakup, exemplified by the extension that was taking place simultaneously in the region of the Zechstein basin (discussed earlier). Concurrently, the equatorial climate also began to change in an ever-broadening band from equator to poles. The climate had fluctuated between interglacial and glacial maximum conditions for 75 MY during the Permo-Carboniferous Ice Age. Now it had changed to the variable conditions of a humid greenhouse age with little or no ice at the poles, a climate regime that lasted for 200 MY. So it was that by the middle of the Triassic Period (245-206 MYA) and the breakup of Pangea, the planet had suffered a fundamental change in the conditions for life.

Every form of eukaryotic life is to a greater or lesser extent water-dependent. Consequently, all terrestrial eukaryotes had to evolve special mechanisms to perpetuate their existence. The plants and animals of the Carboniferous swamps in Central Pangea evolved in warm and humid wetlands that proved to be evolutionary cul-de-sacs. Specialized for life in swamps, they could not move from one

[*continued on page 136*]

5 East Central Pangea

Potash evaporites near Moab, Utah

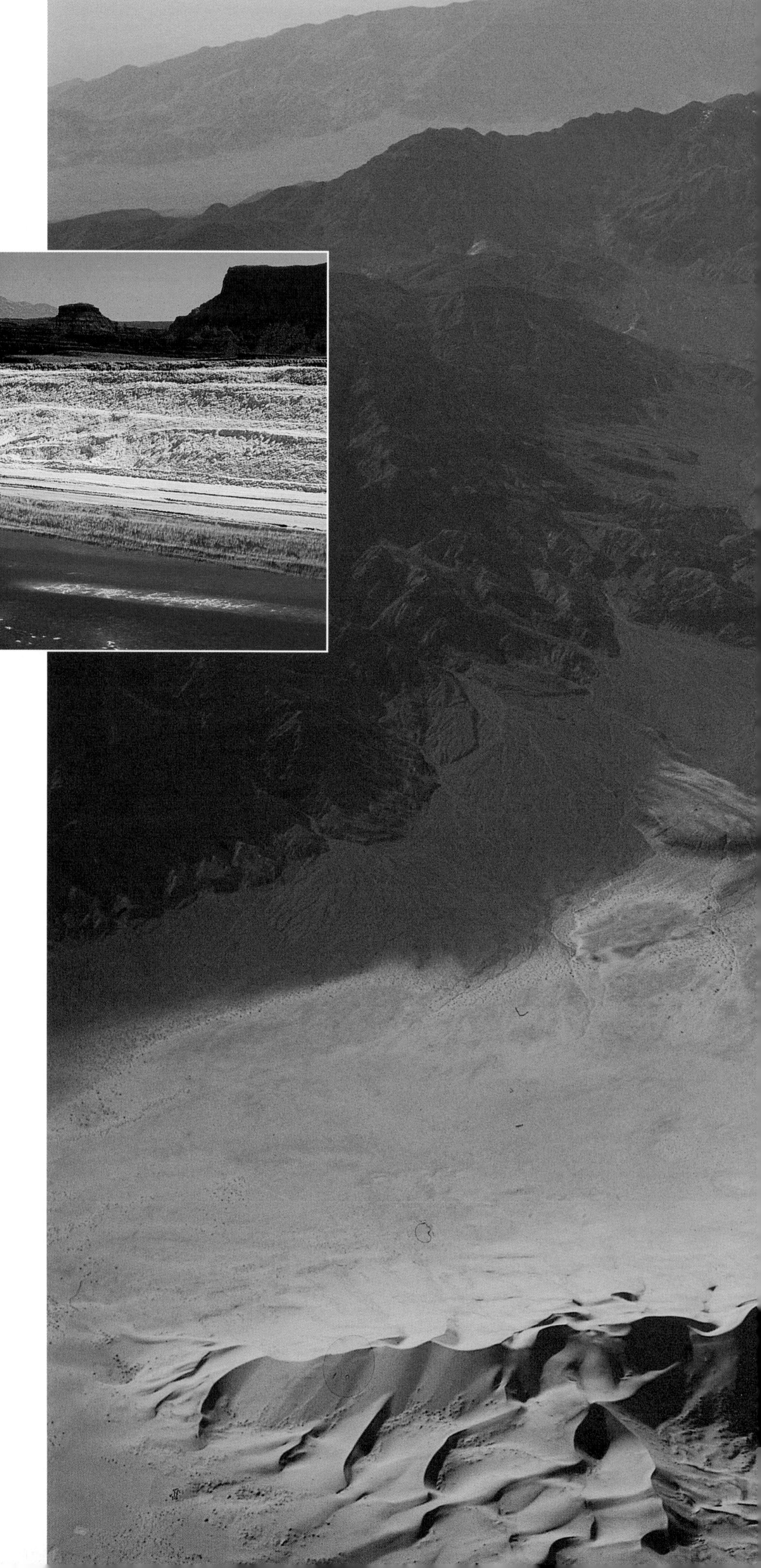

Almost 9,600 km (6,000 miles) to the east of the Guadalupe basins, with the Himalaya-like Alleghenian Mountains between the two, there existed another, even more extensive, basin. This too was frequently filled with a shallow epicontinental sea—the Zechstein Sea. The modern site of the Zechstein Sea is the stark and often frigid but shallow North Sea, which has flooded the Northern and Southern Permian Basins. 250 MYA these basins were extremely hot and arid desert environments similar to the modern Great Basin and Range Province of Utah, Nevada, and California. Today's Great Basin lies in the "rain shadow" of the Sierra Nevada. The Zechstein basins were in the rain shadow of similarly high mountains.

The floor of the Great Basin is subject to stretching, thinning, and block faulting—"crustal extension" caused by tectonic forces at work beneath its surface. The Zechstein basins were also subjected to crustal extension and block faulting. It is therefore likely that residual lakes in the Zechstein basins were fed by snow-melt from high mountains, and by flash floods that scoured the flanks of the mineral-laden block-faulted mountains within the basin. This scenario would account for the immense accumulation of common salt, potash (inset picture), gypsum, and other mineral deposits that are found throughout the greater Zechstein basin in Western Europe today.

The picture of a rare "Chinese wall" formation of star-shaped wind-blown dunes was taken from above the Great Basin. The dunes lie at the base of an alluvial fan, where deposits from a mountain stream are deposited on the basin floor. Dunes like these may have formed porous wind-blown sandstones above Carboniferous coal seams. These sandstones were later capped by deposits of mineral salts in the Zechstein basin, and acted as reservoirs for gas released by the compressed, heated, and devolatilized coal beneath them. They are now the source of the huge volumes of natural gas trapped in the North Sea's Southern Permian Basin.

Eureka Dunes,
The Great Basin,
Southern California

Bentonite formed from chemically modified clay, Death Valley

6 Extinction

Between 255 and 250 MYA more than half the species of animals on Earth, including 75–95 percent of marine species, were permanently extinguished. This mass extinction is the greatest such event found in the fossil record. It is thought that the main contributory cause was the fundamental change in the Earth's climate from an ice age to a desert age, which took place during Permian times (286–245 MYA)—but there is an additional explanation.

During the Permian Period continental shelves were destroyed and reduced during the final assembly of Pangea. On land, conditions for life in Central Pangea, where equatorial mountain ranges

caused rain shadows to form, are thought to have been analogous to conditions for life in the American Southwest today.

Death Valley (inset), one the best-known features of the Great Basin, is one of the hottest desert regions on Earth: the highest recorded temperature there is 57.1°C (134°F) in the shade. This basin, and many hundreds of others to the far left of main picture, lie east of the Sierra Nevada, which has many peaks in the 2,400–4,200 m (8,000–14,000 ft) range. Prevailing moisture-laden winds from the Pacific Ocean precipitate most of their moisture as heavy snow on the Sierra, causing a rain shadow to form over the Great Basin to the east. As a result, both the basin and the neighboring Colorado Plateau are extremely arid, and this aridity largely dictates the nature and diversity of life in these regions. The high elevation, great expanse, and profound ecological effect of the Californian Sierra must be modest in comparison to the height, extent, and devastating aridity caused in Central Pangea by the Late Permian mountain ranges. In addition to arid conditions, excessive volcanic eruptions in China and India (which released thousands of cubic miles of lava) also produced anoxic gases into the atmosphere throughout the period of the Permian extinction.

[continued from page 131]

damp region to another in response to drought. Terrestrial life faced a new problem, the change from prolonged moist conditions to alternating dry and moist conditions.

As the Permo-Carboniferous Ice Age ended, such conditions spread progressively from equatorial and subtropical regions to regions that were subject to more significant seasonal change. The result was a ripple effect as ancient floras and faunas were replaced by species that could survive the new conditions.

For example, plant spores that required moist conditions to germinate, and amphibian eggs that could only be deposited and hatched in water, were no longer the most effective means for perpetuating life across broad regions of the planet. As time went by, the swamp communities of the Paleozoic world gave way to forms of life whose reproductive methods were viable in both moist and dry climates.

THE KEY PLAYERS in this crucial trend of terrestrial evolution were unrelated members of plant and animal families that had lived in drier habitats along river banks and on hillsides. These flora-faunas had one important feature in common. During their reproductive processes they spawned embryos enclosed in nearly waterproof casings that were resistant to desiccation but permeable to oxygen. The plants that had this facility are called "gymnosperms" because they produced "naked seeds" (the literal meaning of gymnosperm). The animals were primitive reptilians called "cotylosaurs" (Gr. *kotyle sauros*, socketed lizard), the stem-reptiles from which all others are descended. The first fossil of this character was classified as a cotylosaur, a name that we will continue to use here for simplicity, although some paleontologists have reclassified these animals into several groups. Cotylosaurs were responsible for one of the most significant advances in the evolution of animal species: they produced the first "amniote" eggs, enclosed in leathery shells. The word amniote is derived from Gr. *amnos*, lamb, a reference to the fetal membrane, *amnion* [Essay 7].

Plants that had seeds and animals that produced amniote eggs were in an excellent position to face the environmental stresses in Permian times. They were able to reproduce, to radiate, and then to evolve, adapt, and diversify in widely separated regions of Pangea. At that time the bulk of the supercontinent was a unified landmass, so apart from vast mountain ranges in Central Pangea there were few natural barriers to inhibit plant and animal migration.

The first true seed plants, the pro-gymnosperms, appeared about 363 MYA and it was from these that the gymnosperms—seed ferns, conifers, yews, ginkgos, and cycads—now evolved (**22**). Because of their adaptability to a variety of ecosystems, these orders gradually gained ascendancy over the more specialized spore-bearing plants. They remained the dominant flora until they in turn were superseded by angiosperms (Gr. *angeion sperma*, capsule seed)—another crucial step in the evolution of life as we know it. Flowering plants, which are the dominant plants on Earth today, are thought to have evolved from the gymnosperms. [This revolution, and the coevolution of insects that made it possible, are discussed further in Chapter IX: *Maritime West*].

The cotylosaurs first appeared about 310 MYA (**23**). They were small, about 15–30 cm (6–12 in) long. Their sprawling gait and appearance generally resembled that of modern lizards. Their fossils are distinguishable from contemporary amphibian fossils by the changes in the structure of their reptilian skulls and jaws, and other skeletal changes. It is speculated that perhaps the shape of cotylosaur skulls allowed the development of stronger jaw muscles, and the resultant improvement in the articulation of their jaws led them to eat animals and plants that amphibians found difficult or impossible to eat. The cotylosaurs could snap at and crush their prey.

Relatively early in Permian time the cotylosaurs diverged into several orders, including what are called synapsids. These animals are commonly called "mammal-like reptiles," because their skulls had evolved "holes"—apertures characteristic of mammal skulls, even today (the name synapsid is a reference to a double cranial opening). The number and size of cranial openings that evolved in early reptilian skulls may also have contributed to the evolution of their jaws, teeth, the range of their hearing, the effectiveness of their vision, and the size and shape of their brain cases.

In Pangea at its zenith the mammal-like reptiles were represented by two main groups, the pelycosaurs (Gr. *pelykos*, pelvis, simply another anatomical reference: **24**), some of which had sails on their backs, and the therapsids (Gr. *ther*, wild animal; *hapsis*, arch). The therapsids evolved from the pelycosaurs, and both groups were very diverse until the Permian extinction 255–250 MYA. After that time the pelycosaurs disappear from the fossil record and therapsid diversity was severely reduced.

22

23

24

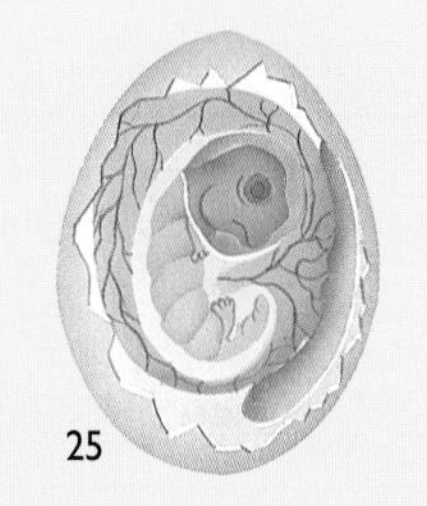

25

THE DEVELOPMENT OF AMNIOTE EGGS, and their superiority over amphibian eggs, had a profound influence on the course of terrestrial evolution. The success of the amniotic system of reproduction changed the main thrust of evolution completely and exponentially. Its effect on the course of life on Earth was perhaps as profound as the development of eukaryotic forms of life from simple prokaryotes had been in Precambrian times.

It is thought that amphibian eggs in Pangean times were similar to those of modern amphibians that live in warm climates and inhabit near-anoxic swamp water. The eggs were probably plate-like so that they floated on or near the surface. In this way the embryo could breathe air as it developed, rather than having to remove dissolved oxygen from water (as the more familiar globular cold-water amphibian egg casings of today permit). As now, the Pangean amphibian embryo probably went through several larval stages of development after it emerged from its original egg casing. Much earlier amphibians, found in Czechoslovakia, had gills that are evident in their fossil larvae. After the larval stage the adult animal was free to leave its watery environment except during mating.

In contrast, the amniote egg (whether contained in an oxygen-permeable shell (**25**) or kept within a female, which supplies oxygenated blood to the embryo) has three enveloping membranes that are outgrowths of the embryo itself. There is the amnion (which protects the embryo), the allantois (which collects its waste), and the chorion (which encloses yolk sac, embryo, and allantois). This triple-membrane system encloses the embryo, which develops without an aquatic larval stage, in its own self-sufficient but watery world. The earliest known fossil of such a reptilian egg has been found in Lower Permian rocks in Texas.

Because the cotylosaurs, the stem-reptiles, were the originators of the amniote egg, they and their countless descendants collectively are called "amniotes." This distinguishes them from all other vertebrates; the fish and the amphibians remain tied to an aquatic or semiaquatic environment.

The amniotes of late Pangean times included marine, terrestrial, and flying reptiles (the early dinosaurs and the first birds), as well as the therapsid group of mammal-like reptiles, which survived the Permian extinction to produce the first true mammals during the final days of the Pangean supercontinent about 210 MYA.

From this it is clear that the amniote method of reproduction became all-embracing. It is a fact of life that all cold-blooded and warm-blooded reptiles and birds, and all mammals—monotreme, marsupial, and placental alike—are descended from the cotylosaurs. So, in Middle Earth (as Tolkien put it in *The Lord of the Rings*), there was indeed

> *One Ring to bring them all and in the darkness bind them*
> *In the Land of Mordor where the Shadows lie.*

Tree ferns near Mount Pelée, Martinique, Lesser Antilles INSET: Cycads, Andromeda Gardens, Barbados, Lesser Antilles

7 Resurgence

During the final assembly of Pangea, as the Permo-Carboniferous Ice Age waned, humidity gave way to arid conditions in equatorial regions of Pangea. Plant communities were forced to adapt from wetlands to dry land. They did so in ever-widening zones radiating from the tropics to the temperate and high-latitude regions of Pangea—where conditions favored reptiles over amphibians and seed-bearing plants over spore-bearing plants.

Because they were dependent upon water-bound eggs, many varieties of amphibians were extinguished. The survivors lived on the river banks and associated wetlands where they flourished and even challenged reptiles for dominance. But it was inevitable that the reptiles would predominate in the end. They produced oxygen-permeable eggs in a leathery shell, "amniote" eggs that were self-sufficient and could be laid out of water on dry land. In evolutionary terms these eggs were more important than the reptiles that first produced them. Their development led to a dramatic change in the direction of animal life and particularly in the evolution of the mammal-like reptiles—the precursors of all mammals.

Reptiles had evolved similarly to plants. Climate conditions no longer favored spore-bearing plants (main picture), which preferred moist conditions. The seed-bearing plants now came into their own. Their primitive enveloped seeds, some in cone-like structures, were the equivalent of the reptilian egg and its shell—they readily adapted to warmer and drier conditions. These plants, of which conifers and their cones are an example, are called "gymnosperms" and they reproduced by distributing wind-blown pollen. The evolution of yews, ginkgos, and cycads (inset) followed, and such plants now began to dominate the land. Gymnosperms were the botanical equivalent of the mammal-like reptiles, for they led to the evolution of angiosperms, the flowering plants that dominate the planet today and on which much of life on Earth depends.

8 The Great Unconformity

The panorama of the Grand Canyon of the Colorado River overleaf provides a rock-based overview of this book's storyline up to the time of the final assembly of Pangea 250 MYA—the average age of the Canyon's north and south rim-rocks. But the Canyon still has many secrets to be unravelled—for instance, the tilted strata at the center of the left-hand page of the panorama are formed from Precambrian rocks of the Unkar Series and are surmounted by an unbroken line of formidable cliffs. The Unkar Series are part of a once-mountainous landscape in which the peaks were reduced to a level plain by erosion, a "peneplane." The cliffs above are part of the Tapeats Sandstone Formation that surrounds the Canyon at this point [see key diagram below and block diagram opposite]. The formation starts in the far distance to the left, and ends at the eroded escarpment in the foreground.

Obviously, the peneplane was formed before the Tapeats Sandstone was deposited on top of it in early Cambrian times, around 530 MYA. But the youngest rocks in the Unkar Series were formed long before, about 650 MYA. Because the peneplaned surface of the tilted strata does not match up in age with the Tapeats Sandstone above, it is called an "unconformity." This is just part of the Great Unconformity of Grand Canyon, one of the world's most outstanding geological features.

The age difference between individual steps in the sloping Unkar Series and the Tapeats varies from 80 to 200 MY depending upon which part of the angled series is making contact with the sandstone cliffs above (called an angular unconformity). In effect, these cliffs form a geological window-frame enclosing a series of much older and noticeably dissimilar rocks. The latter are the remains of a mountainous, volcanic landscape of great age that may well have been part of a previous supercontinent called Rodinia, thought to have existed from about 1,100 MYA.

The Tapeats and the underlying tilted Unkar Series continue out of the picture to the left, down Grand Canyon into Upper Granite Gorge. Here, at Hance Rapid, the Tapeats continues but the underlying rocks are intersected first by much older Precambrian sedimentary rocks from the Unkar Group, and later by even older rocks, the peneplaned roots of an alpine-type mountain range. The latter, still capped by the Tapeats, are crystalline granites and gneisses, the oldest rocks in Grand Canyon. Some of these are dated *c.*1.7 billion years ago and are over a billion years older than the Tapeats deposited upon them. This older section of the Great Unconformity, together with its overlying Tapeats, continues down the Canyon for another 32 km (20 miles) before gradually disappearing beneath the surface of the Colorado River.

But how was such a vast and varied Precambrian mountainous landscape reduced to a level plain studded with occasional resistant features like the two that can be seen in the panorama: Kwagunt and Chuar buttes? (Chuar is to the left of center and rises above the far horizon. Kwagunt is to the left of Chuar, farther away and more difficult to identify.) How was the scene set for the deposition of vast volumes of sand to form the Tapeats, in the balmy early Cambrian climate, as the first trilobites appeared in tropical seas?

In 1963, J. Tuzo Wilson published a benchmark paper in support of Wegener's supercontinental hypothesis. He added a comment to the effect that if there had once been a Pangea as Wegener had proposed, then the supercontinent could have had predecessors. Some scientists support this view, but others do not.

Those who argue for the existence of supercontinents before Pangea, claim that another such assembly formed during the period 1,100 to 650 MYA. Their opponents say that this claim is invalid because the available data for the supercontinent is tenuous and contradictory. Even so, there is a growing if somewhat guarded consensus between the two viewpoints for the existence of Rodinia, a supercontinental clustering that may have existed 1,100 MYA [see page 55]. Even so, there is disagreement about the disposition of Rodinia's continent-sized elements, particularly after its breakup and for the period 750 to 580 MYA—a timeframe that includes the youngest part of the Great Unconformity.

The pro-supercontinental school supports the notion that by at least 650 MYA parts of Rodinia had reassembled into an antarctic-based supercontinent with a major northern arm extending to a point north of the Equator (C.R. Scotese: 1997). The opposing view claims a general scattering of continents around and to the south of the Equator at this time, and does not support an earlier or coincidental polar-based assembly (P.E. Hoffman and D.P. Schrag: *Sci.American*: Jan. 2000). However, both schools accept the reality of an ongoing ice age, severe but spasmodic, with two to four pulses in the period 750 to 580 MYA—the time of the so-called "Snowball Earth." It follows that those who agree with the notion of a contiguous antarctic-based supercontinent for this period of time have little difficulty in explaining a heavily glaciated supercontinent. However, those who promote an equatorial distribution of scattered continents need to explain how these continents could have become heavily glaciated—as evidence shows they were at the time. They have developed an ingenious theory that produces a self-perpetuating and extremely cold global climate with abrupt and very warm intervals, within the 750-580 MYA time frame.

For our purpose, the important common factor between these conflicting interpretations of the paleogeography is that both schools support the concept of a heavily glaciated Laurentia (proto-North America) around 600 MYA—Hoffman and Schrag, for instance, suggest global average temperatures of -50°C (-58°F) for a period of 10 MY during this time. From this we can assume that whatever Laurentia's latitude at *c.*600 MYA, sub-antarctic or sub-equatorial, there might well have been heavy glaciation in the Grand Canyon region of Arizona between 650–580 MYA.

This period covers the time frame for the levelling of the Precambrian rock formations beneath the Tapeats. In fact, the forces of erosion produced a peneplaned surface reminiscent of that produced by ice sheets, but this idea was discounted when put forward years ago because no evidence for glaciation of the Grand

Canyon region in Precambrian times then existed. However, evidence of glaciation in the region during this Snowball Earth period, if it exists, may be found adjacent to the canyon, rather than within. For instance, tillites of the right age—products of the glacial grinding of rock surfaces—are found in both Nevada and California's Death Valley.

Various other causes of the peneplanation have been postulated since the 1800s. For example, E.D. McKee (in the Four Corners Geological Society Guidebook, 1966) cited "intense chemical weathering" and "transgression of Late Precambrian seas" as factors. But if these were secondary erosional forces rather than primary causes, they could have completely destroyed evidence of previous glacial erosion. Perhaps the time has come to take another look at the origins of the Great Unconformity of Grand Canyon?

Sea levels, which had fallen dramatically as continental ice locked up fresh water during the Snowball Earth period, were restored to well above present-day levels as each glacial advance ended. Shallow seas invaded low-lying parts of Laurentia at these times. After the last advance sand transported by rivers began to accumulate on peneplaned surfaces—the Tapeats Sandstone Formation was the result of one such period of prolonged deposition.

The distinctive Tapeats cliffs mark the boundary between the Precambrian and Cambrian time and the beginning of the Paleozoic Era—the era of ancient life. The sedimentary rocks on the right and left horizons of the panorama were formed just before and just after the zenith of Pangea's existence, 250 MYA. The Kaibab Limestone on the North Rim to the left of the panorama was formed *c.*260 MYA. The Moenkope Formation on the South Rim to the right is dated *c.*230 MYA. It follows that all the rock formations from the Tapeats to the rim-rocks are the products of Pangea's assembly—the rotation, collision, and re-assembly of Rodinia's breakaway continents.

On the diagram below, the graph to the left shows the number of plant and animal families that evolved and became extinct during this same period of time. The Great Unconformity and other post-Cambrian unconformities in the Canyon are represented by blocks on the right representing missing sedimentary rock formations or periods when no rocks were formed because the region was above sea level, as it is today. The blue transparent boxes beside these blocks denote the approximate time span of ice ages throughout the period. These changes contributed to the incidence of unconformities in rock formations worldwide.

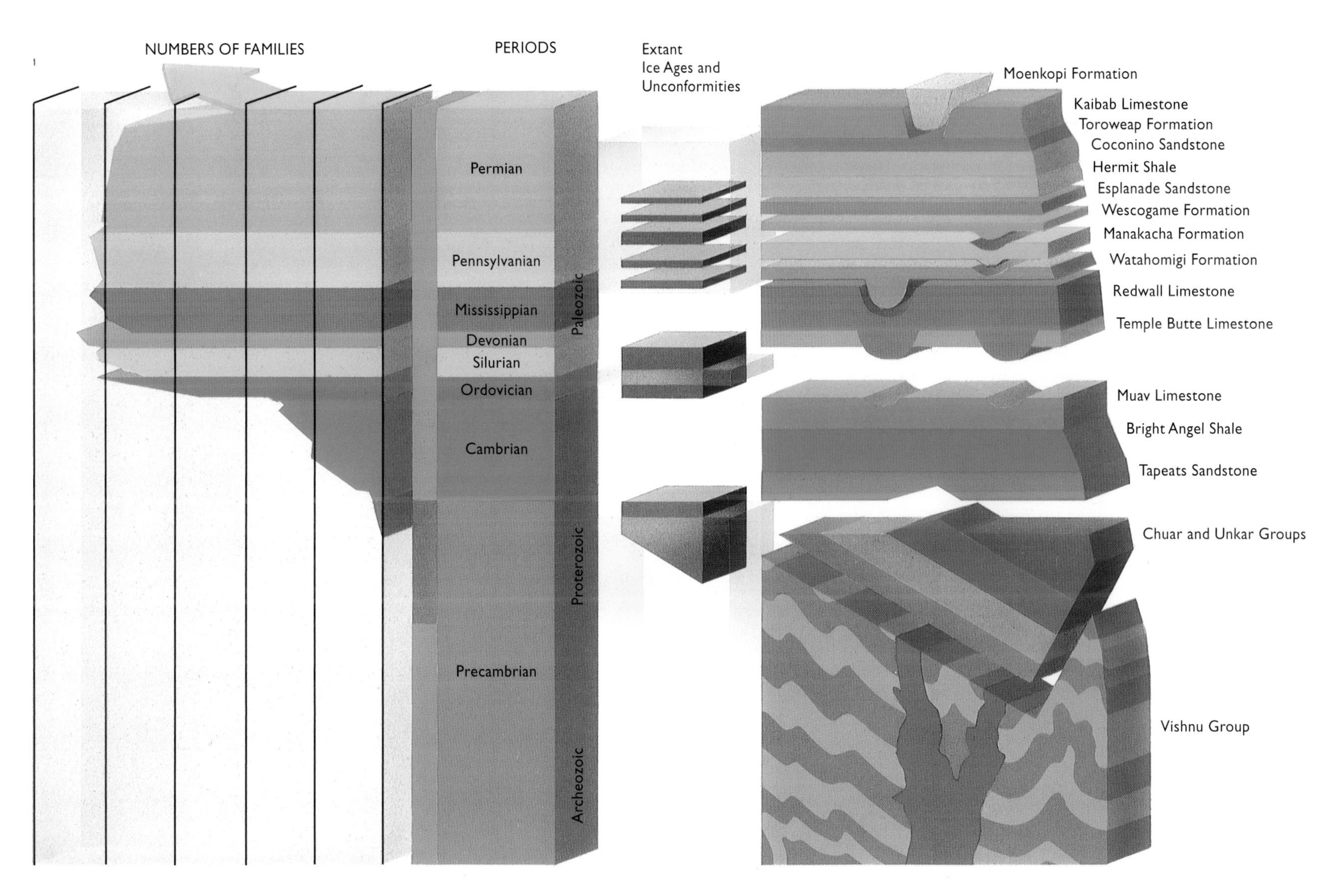

In the panorama we are looking northwards at the Unkar and Marble Gorge regions at the eastern end of the Grand Canyon of the Colorado. The South Rim is to the right of the picture and the North Rim to the left. The panorama was taken from a point on Escalante Butte about 600 m (2,000 ft) below Desert View (top, far right) and 900 m (3,000 ft) above Unkar Rapid (center left).

VII THE NEW WORLD

The supercontinent remained a contiguous but restless whole until about 220 MYA but by 206 MYA Central Pangea began literally to be torn apart. Here the Hudson River flows between New York City and the Palisades—cliffs that were a volcanic expression of that breakup, and part of a system of active volcanic rifts that stretched from Nova Scotia to the Carolinas. North America was then connected to West Africa, with the New York region opposite West Africa between Cap Bojador, south of the Canary Islands, and Cap Blanc in Mauritania. Today New York City is on the same latitude as Lisbon in Portugal, and separated by 5,425 km (3,370 miles) of ocean.

THE NEW WORLD 206–144 MYA

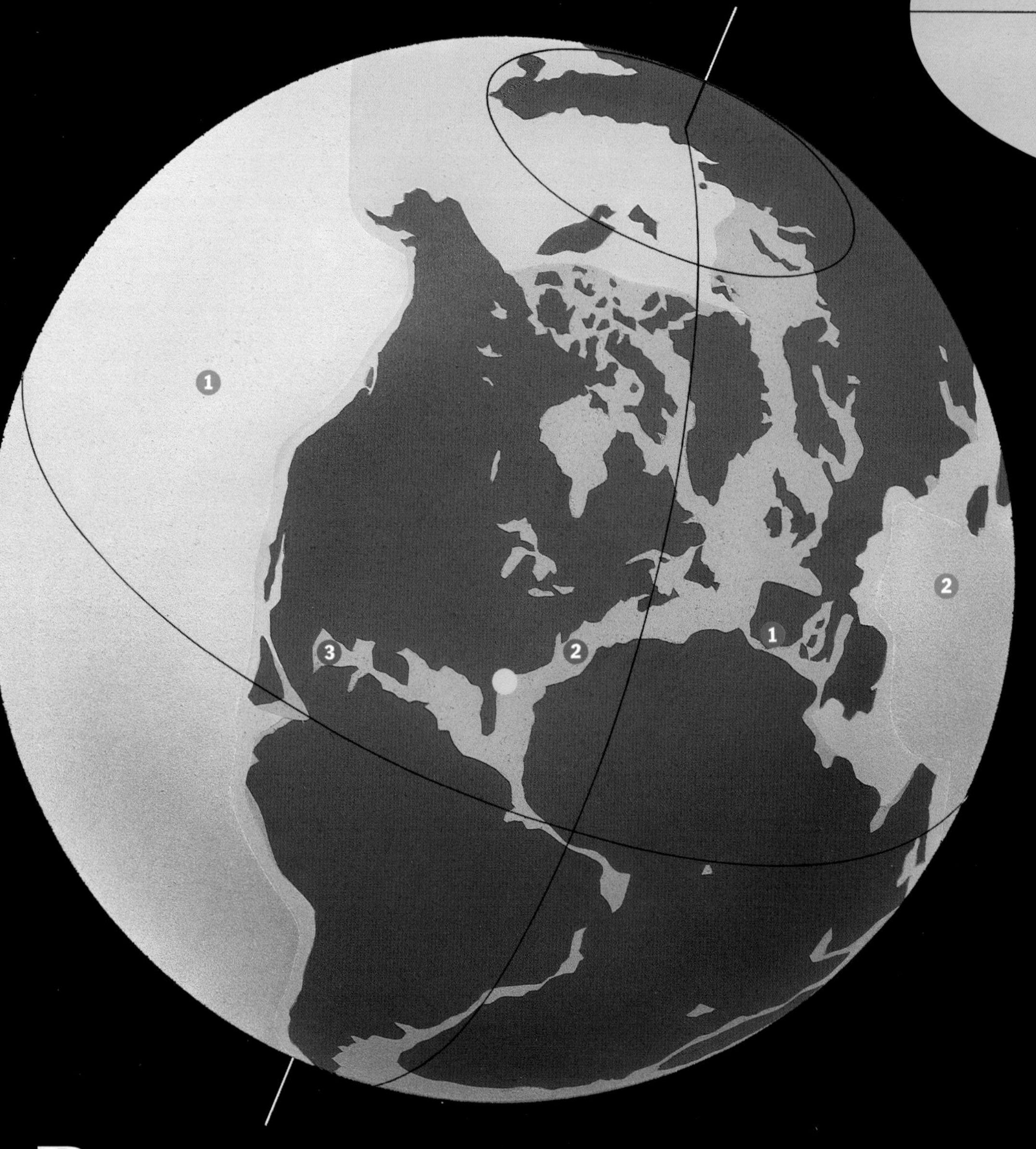

OCEANS

1. Panthalassa
2. Tethys Ocean

1 GIBRALTAR-RIF SEAWAY
Morocco (Casablanca)/Spain (Gibraltar)
via continental shelves
to Tethys Ocean

2 PROTO-CENTRAL ATLANTIC
Florida-Bahama block
via Palisades (NY)/Cap Bojador (Morocco)
to Nova Scotia/Morocco (Agadir)

● MID CENTRAL PANGEA
Basement rocks of future Piedmont
(East Coast of North America),
Florida-Bahama block and Cuba

3 AMERICAS SEAWAY
Florida-Bahama block
via continental shelves
to Panthalassa Ocean
Proto-Gulf of Mexico

Pangea as a whole had been drifting northwards during its assembly. By 195 MYA the northern elements of the supercontinent had clustered around the North Pole, as is strikingly evident from the oval projection. But even as Pangea reached and held the stage of closest fit, it began to fracture. The structure of Central Pangea had become so thick that overheating resulted, and rupturing inevitably followed. In this chapter we will see the initial division of Central Pangea into post-Pangean Laurasia and post-Pangean Gondwana, and the first stage in the formation of the Central Atlantic Ocean.

The paleoglobe shows Central Pangea at 195 MYA, about 30 MY before a shallow seaway began to flood the spreading region between North America and West Africa in Mid Central Pangea. This Proto-Central Atlantic was an extension of the Tethys Ocean through Gibraltar. Meanwhile, a second shallow sea flooded into West Central Pangea from the Pacific Ocean—the Americas Seaway. The Florida-Bahama carbonate platform inhibited development of a connection between the two seaways until about 142 MYA, but once that connection had been made, the Central Atlantic Ocean was established.

THE FACT THAT there was once a Pangean supercontinent, a Panthalassa Ocean, and a Tethys Ocean, has profound implications for the evolution of multicellular life on Earth. These considerations were unknown to the scientists of the 19th century—making their scientific deductions even more remarkable.

Quite independently of each other, Charles Darwin (1809–82) and his young contemporary Alfred Russel Wallace (1823–1913) reached the conclusion that life had evolved by natural selection. Wallace later wrote in *My Life* (1905) of his own inspiration:

Why do some [species] die and some live? The answer was clearly that on the whole the best fitted lived. From the effects of disease the most healthy escaped; from enemies the strongest, the swiftest or the most cunning; from famine the best hunters ... then it suddenly flashed on me that this self-acting process would improve the race, because in every generation the inferior would inevitably be killed off and the superior would remain, that is, the fittest would survive.

Both Darwin's and Wallace's ideas about natural selection had been influenced by the essays of Thomas Malthus (1766–1834) in his *Principles of Population*. Their conclusions, however, had been the direct result of their personal observation of animals and plants in widely separated geographic locations: Darwin from his experiences during the voyage of the *Beagle*, and particularly during the ship's visit to the Galapagos Islands in the East Pacific in 1835: Wallace during his years of travel in the Amazon Basin and in the Indonesia-Australian Archipelago in the 1850s.

Darwin had been documenting his ideas on natural selection for many years when he received a paper on this selfsame subject from Wallace, who asked for Darwin's opinion and help in getting it published. In July 1858, Charles Lyell and J.D. Hooker, close friends of Darwin, pressed Darwin to present his conclusions so that he would not lose priority to an unknown naturalist. Presiding over the hastily called but now historic meeting of the Linnean Society in London, Lyell and Hooker explained to the distinguished members how "these two gentlemen" (who were absent: Wallace was abroad and Darwin chose not to attend), had "independently and unknown to one another, conceived the same very ingenious theory."

Both Darwin and Wallace had realized that the anomalous distribution of species in particular regions had profound evolutionary significance. Subsequently, Darwin spent the rest of his days in almost total seclusion thinking and writing mainly about the origin of species. In contrast, Wallace applied himself to the science of biogeography, the study of the pattern and distribution of species, and its significance, resulting in the publication of a massive two-volume work, *The Geographical Distribution of Animals*, in 1876.

WALLACE WAS A GENTLE and modest man, but also persistent and quietly courageous. He spent years working in the most arduous possible climates and terrains, particularly in the Malay archipelago. He made patient and detailed zoological observations and collected huge numbers of specimens for museums and collectors—which is how he made a living. One result of his work was the conclusion that there is a distinct faunal boundary (**1**), called "Wallace's Line," between an Asian realm of animals in Java, Borneo and the Philippines and an Australian realm in New Guinea and Australia. In essence this boundary posed a difficult question: how on Earth did plants and animals with a clear affinity to the Northern Hemisphere meet with their Southern Hemispheric counterparts along such a distinct Malaysian demarcation zone? Wallace was uncertain about demarcation on one particular island—Celebes, a curiously shaped place that is midway between the two groups. Initially he assigned its flora-fauna to the Australian side of the line, but later he transferred it to the Asian side. Today we know the reason for his dilemma. 200 MYA East and West Celebes were islands with their own natural history lying on opposite sides of the Tethys Ocean. They did not collide until about 15 MYA. The answer to the main question is that Wallace's Line categorizes Laurasia-derived flora-fauna (the Asian) and Gondwana-derived flora-fauna (the Australian), fauna that had evolved on opposing shores of the Tethys. The closure of the Tethys Ocean today is manifested by the ongoing collision of Australia/New Guinea with Indochina/Indonesia (**2**) and the continuing closure of the Mediterranean Sea—a remnant of the Western Tethys Ocean.

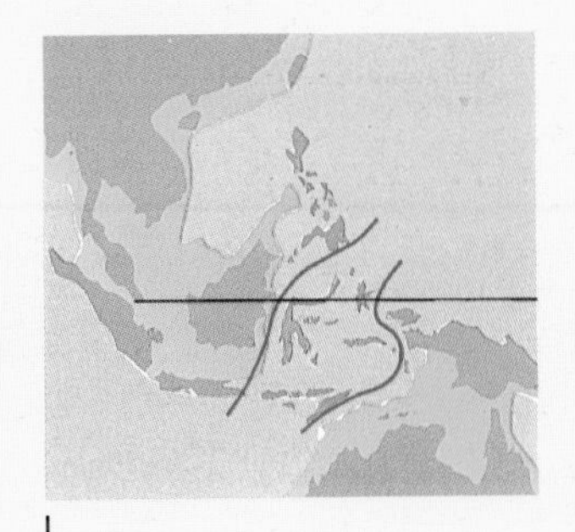
1

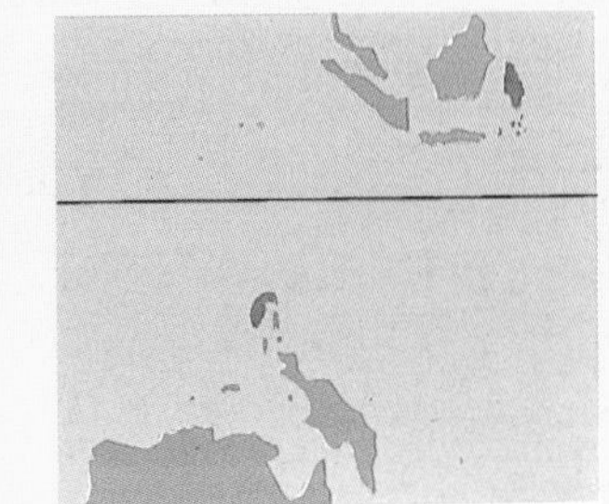
2

IN HIS *ORIGIN OF CONTINENTS AND OCEANS*, Wegener quoted at length from Wallace's *Geographical Distribution of Animals*. According to Wegener's reading, Wallace had identified three clear divisions of Australian animals, which supported his own theory of continental displacement. Wallace had shown that animals long established in southwestern Australia had an affinity with animals in South Africa, Madagascar, India, and Ceylon, but did not have an affinity with those in Asia. Wallace also showed that Australian marsupials and monotremes are clearly related to those in South America, the Moluccas, and various Pacific islands, and that none are found in neighboring Indonesia. From this and related data, Wegener concluded that the then broadly accepted "landbridge" theory could not account for this distribution of animals and that only his theory of continental drift could explain it.

The theory that Wegener dismissed in preference to his own proposed that plants and animals had once migrated across now-submerged intercontinental landbridges. In 1885, one of Europe's leading geologists, Eduard Suess, theorized that as the rigid Earth cools, its upper crust shrinks and wrinkles like the withering skin of an aging apple. He suggested that the planet's seas and oceans now fill the wrinkles between once-contiguous plateaus.

Today, we know that we live on a dynamic Earth with shifting, colliding and separating tectonic plates, not a "withering skin," and the main debate in the field of biogeography has shifted. The discussion now concerns "dispersalism" versus "vicarianism": unrestricted radiation of species on the one hand and the development of barriers to migration on the other. Dispersion is a short-term phenomenon— the daily or seasonal migration of species and their radiation to the limits of their natural environment on an extensive and continuous landmass. Vicarian evolution, however, depends upon the separation and isolation of a variety of species within the confines of natural barriers in the form of islands, lakes, or shallow seas—topographical features that take a long time to develop.

ON A CONTINENTAL SCALE, barriers can take a variety of forms. There could be an intolerable warming of the climate at the equator, or the development of a mountain range, or the opening of an ocean basin, or the formation of an ice sheet. This type of barrier can take thousands or millions of years to develop. The resulting isolation causes pressures to develop in indigenous populations, which promote competition for food and space and forces adaptation to a new environment. Such adaptation often results in specialization, creating new species within indigenous populations, and therefore promoting diversity.

During the Jurassic Period, the time frame of this chapter (206-144 MYA), terrestrial populations were widespread but modest in diversity. As Pangea began to rift apart, populations were separated and isolated from each other. At the height of Pangea's existence sea level was relatively low (about the same as today), but as breakup progressed sea levels rose, increasing the isolation of terrestrial populations on the southern megacontinent of Gondwana and the northern megacontinent of Laurasia. Isolation grew with the subsequent formation of individual continents, subcontinents, and islands in a progressive *vicarian* process. It led to the speciation of innumerable animals and plants, increasing the diversity of species that had arisen from common roots after the Permian mass extinction.

That extinction (250-245 MYA) had decimated marine life to about 5 percent of its original species. It had at least halved the

[*continued on page 152*]

1 Pangean Breakup

About 200 MYA New York's Manhattan Island and the now tree-hidden Palisades on the New Jersey shore of the Hudson River were opposite the Cap Bojador-Cap Blanc region of West Africa. At that time, the present continental margins of North America and Northwest Africa, from Long Island to Nova Scotia, and from Agadir to Rabat, were also in close proximity. Both margins were separated by a tropical zone littered with vast areas of salt deposits, and cut by wide rivers burdened with silt from the foothills of the fast-eroding mountains on either side of the old suture.

Between 220 and 200 MYA innumerable rift valleys had formed from eastern Greenland to Guinea in West Africa at the start of the breakup of Pangea. Along what is now the East Coast of North America, the "Newark Rift system" stretched from the Carolinas to Nova Scotia. When the supercontinent finally rifted apart between the East Coast and Northwest Africa, it separated along the approximate line of a previous suture. Water flooded into the breach from the Tethys Ocean in the east to form the proto-Central Atlantic, from Gibraltar to the Bahamas. As seafloor spreading took hold, the tension pulling the neighboring margins apart eased, and the now-inactive rift valleys were filled by lakes and sediments.

During the active lifetime of the East Coast rifts basalt flooded parts of what are now the states of Connecticut and New Jersey. Some magma that could not reach the surface intruded beneath existing rocks to form volcanic "sills"—so named because of their elongated, horizontal windowsill form. One such sill has since been uplifted and uncovered by erosion to form a high escarpment, the Palisades cliffs of the Hudson River—a memorial to the origins of Pangea's disintegration.

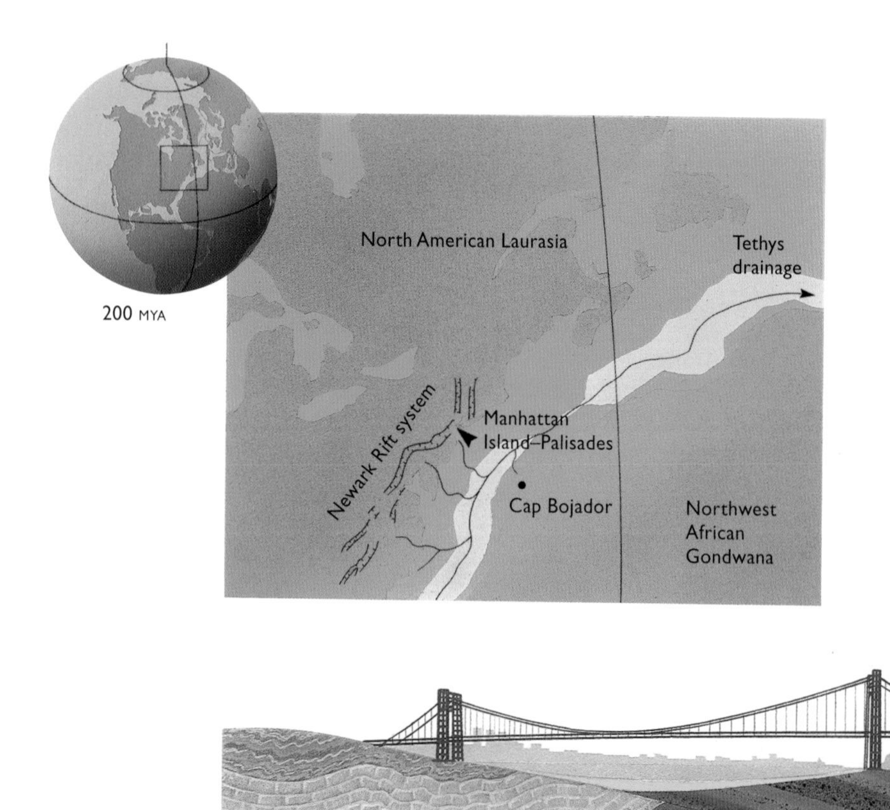

Manhattan Schist–Inwood limestone

Palisadian sill

Palisades of the Hudson River, New Jersey

Post-Pangean North American Laurasia
Gibraltar-Rif Seaway
Ziz Valley
Tethys Ocean
Cap Rhir
Proto-Central Atlantic
Newark Rift system
Northwest African Gondwana
Florida-Bahama Block
180 MYA
180 MYA
Nova Scotia
Morocco Meseta
Rift valleys
200 MYA
Proto-Central Atlantic
Morocco
Sea incursion
180 MYA
Proto-Central Atlantic
Morocco
Central spreading
160 MYA

2 Gibraltar Seaway

At the time the Newark Rift system was becoming volcanically extinct, a major system of rift valleys was forming across the opposing corners of Northwest Africa and Iberia, in a line parallel to the Strait of Gibraltar. Unlike the Newark Rifts, the Iberia-Gibraltar-Morocco system, called the Gibraltar-Rif Seaway, became a budding ocean—but an ocean that ultimately failed to mature. During the period 206–165 MYA some Moroccan rift valleys acted as troughs for seawater flowing from the Tethys Ocean into the separating region between Africa and North America. In fact the cliffs of Cap Rhir near Agadir match similar but submerged rock formations off the coast of Nova Scotia. Cap Rhir marks the edge of the present passive margin of Morocco, where the Gibraltar-Rif Seaway once flowed into the region that ultimately became the North Atlantic Ocean.

The panorama is of the Ziz Valley in Morocco, a rift valley that cut its way into the Moroccan Meseta but failed to propagate. It became a sea-filled backwater, and filled with sediments. About 175 MY later, the valley was exhumed and raised to its present elevation by the uplift of the High Atlas Mountains during the current collision of North Africa with Eurasia.

Amazingly, this ancient failed-rift valley still retains its original salt deposits, sun-baked mud flats, carbonate platforms, sponge mounds, algal reefs, huge fossil ammonites, and debris from submarine avalanches called turbidites. Some larger features that are characteristic of the Mid-Jurassic marine environment of the Gibraltar-Rif Seaway dominate the central horizon in the main picture; the base of Jebel Assaureur, a Ziz Valley atoll structure, can be seen in the right-hand foreground.

The map to the left shows a reconstruction of the present Central Atlantic region during the breakup of Pangea. The cross-sections illustrate stages in the formation of the proto-Central Atlantic—the precursor of the North Atlantic Ocean.

Thingvellir, Iceland

Krafla caldera, Iceland

3 Proto-Central Atlantic

About 165 MYA the Mid Central Pangean crust had been stretched to its maximum. It could extend no more, so it fractured and collapsed in a jagged line from Nova Scotia/Morocco to the Bahamas. The division roughly corresponded to the zone where Gondwana had originally collided with Laurentia-Baltica. The now-collapsed basin filled with a shallow sea called the "proto-Central Atlantic—a seaway that served to link the Tethys Ocean through the Strait of Gibraltar to the region of the Bahama Banks.

The dots on the adjacent map mark the locations of volcanic regions in the spreading-center that crosses modern Iceland—they indicate regional rifting processes analogous to those that occurred in Mid Central Pangea. The line stemming from each dot shows the extent and direction of volcanic networks, called "fissure swarms," which develop beneath each caldera.

In Iceland only a few volcanic centers, in the complete length of the spreading ridge which crosses the island from north to south, are active at any one time. The inactive spreading-centers between them, like the quiet Thingvellir rift valley in the inset picture, remain dormant for centuries before reactivating. Currently active regions, such as the Krafla caldera in the panorama, are hot and swollen. As they become dormant after a period of violent activity, their upper parts solidify to form an unyielding wedge between separating margins. The wedge causes stresses to build up in neighboring dormant regions, making the latter vulnerable to renewed volcanic activity.

During such wedging processes the North American plate, to the left of the main picture, is slowly inched apart from the Eurasian plate, to the right. It does so at a fingernail rate of growth, about an inch each year—roughly the rate that post-Pangean Laurasia separated from post-Pangean Gondwana to form the proto-Central Atlantic.

[*continued from page 145*]

number of terrestrial families—perhaps more: nobody knows the correct percentage. The pelycosaurs, widely dispersed mammal-like reptiles (**3**), some with sails on their backs, were among the terrestrial victims. However, some therapsid descendants of the pelycosaurs survived the mass extinction event. Although greatly reduced in diversity they were prolific in number. Towards the end of the Triassic Period (210 MYA), for unknown reasons they became smaller and more mammal-like in anatomy, leading to the evolution of the first true mammals during the final days of the contiguous Pangean supercontinent (210-200 MYA).

3

ALTHOUGH THERAPSIDS were the paramount reptiles of their time, many other orders also stemmed from the first amniote-egg laying reptiles. The result was a bewildering variety of terrestrial creatures in the predominantly reptilian world that existed until the demise of the dinosaurs. Accordingly, before discussing the isolation and evolution of animals and plants in post-Pangean Gondwana here [post-Pangean Laurasia is discussed in the next chapter], we will put the main players of this primarily reptilian world into perspective.

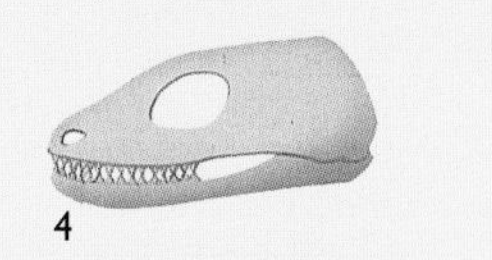

4

APART FROM THEIR NOSTRILS and eye sockets, the earliest reptiles, the lizard-like cotylosaurs that first appeared about 310 MYA, did *not* have cranial apertures (**4**): this physical characteristic is termed "anapsid." The most direct living descendants of the cotylosaurs, though not lizard-like, are nevertheless the aristocrats of today's reptilian world. They are heavily armored, carnivorous, herbivorous, and omnivorous reptiles, the amphibious anapsid turtles (**5**) and the terrestrial tortoises. The oldest known fossil of this order of reptiles is dated 220 MYA. Subsequently, they evolved beaks for teeth, and while terrestrial varieties retained clawed feet, their amphibious counterparts evolved paddles for legs. The sprawling gait, heavy shells, and slow movement of modern turtles and tortoises on land, are reminders that these animals are very primitive—yet extraordinarily successful.

5

The pelycosaurs, the first mammal-like reptiles, appeared about 280 MYA. They and their descendants, the therapsids, had *two* openings in the skull, one on either side in addition to eye and nostril apertures (**6**), and because of this they and their descendants are termed "synapsid." The dual openings made synapsid skulls lighter, which allowed them to hold their heads higher. Previously splayed legs were now superseded by legs that supported the animal's weight from directly below. This in turn allowed these animals, whose ancestors were quadrupedal, to become capable of faster movement to evade predators or to catch prey. The synapsid openings also allowed jaw muscles to develop inside the skull, and permitted changes in the jaw itself and in the arrangement of its teeth. Fossils of therapsids are distinguished from those of pelycosaurs by differences in dentition: therapsid teeth were sometimes canine-like—differentiated with nipping incisors, piercing canines, and slicing or crushing back teeth. In evolutionary terms the most important of the therapsids are therefore called "cynodonts" (Gr. *kyon odon*, dog tooth: **7**).

6

During Permian and Triassic time cynodonts became the dominant carnivorous therapsids (**8**), and dicynodonts the dominant herbivores (Gr. *di*, two: they had no teeth except two long canines in the upper jaw). As they evolved, cynodonts became more erect in posture; and because of their longer legs they could lie down and get up again. Some paleobiologists interpret this development to mean that the most advanced cynodonts may have suckled their young. Some larger predators had saber-like canine teeth. The largest grew to a size proportionate to their largest herbivore victims—some of which were up to 4.8 m (16 ft) in length and weighed as much as a small elephant. Some therapsid carnivores were dog-like in tooth style and size: still others were very small and fleet of foot.

7

During their evolution, the number of bones in the cynodont jaw, and the way in which the jaw hinged to the skull, had changed. This change resulted in three small subsidiary jawbones next to the skull, adapting to form a primitive hearing apparatus similar to the middle ear of modern mammals. In these advanced reptiles the quadrate, articular, and stapes bones were equivalent to the incus, malleus, and stapes in mammals—the hammer, anvil, and stirrup of the middle ear. Most of the early reptiles probably felt vibrations, but were hard of hearing; the advanced therapsids, however, could hear. Most also had an acute sense of smell, for their olfactory organs were well developed. It is suggested, on the evidence of the comparative anatomy of living tetrapods, that the addition of hearing to smell and vision increased the volume and complexity of information flowing to the brain so much that it helped trigger an increase in brain size.

8

The change in jaw design and dentition also enabled more complex movements. Unlike the anapsid reptiles, cynodonts could chew and finely divide their food with cheek teeth; eating ground-up food improved the metabolic rate of the animal and the amount of energy available to it. Synapsids had also developed a double palate, enabling them to breathe while eating, a facility that further improved metabolic rate.

AFTER CHEWING AND GRINDING up food in the mouth, where the process of digestion really begins, the biochemical process of breaking down food continues in the stomach. Such "catabolism" is followed by "anabolism," the process of absorbing the digested food and transferring its latent energy into the bloodstream through the walls of the upper intestine. The whole process of breakdown, buildup, and absorption is called "metabolism." Smaller animals need a faster metabolism than larger ones—they lose heat faster because they have a larger surface area relative to their mass. In fact, they need to eat almost continuously and to grind their food as finely as possible before swallowing. A high metabolic rate also allows quick reflex action, and sustained fast movement—typical attributes of warm-blooded animals.

Warm-bloodedness means having the ability to regulate blood temperature automatically at a level that can supply enough energy to activate the brain and body irrespective of ambient conditions—day or night, winter or summer. Although warm-bloodedness vastly extended the territorial and climatic range of animals, the attribute had the disadvantage of requiring the consumption of large amounts of food to "stoke the furnace."

Conversely, "cold-blooded" terrestrial animals, which often have blood temperatures similar to their warm-blooded counterparts, do not require a high metabolic rate. Their bodies both absorb and store the sun's heat—the bigger the animal the bigger the reservoir of heat. Consequently, cold-blooded animals need *much* less food per pound of body weight than warm-blooded animals and therefore do not need to convert consumed food into energy quickly.

In a sense, cold-blooded terrestrial animals in Late Triassic times were driven by unregulated solar batteries supplemented by a modest intake of food calories. Warm-blooded animals were controlled by thermostats and driven by their intake of food calories exclusively. There might have been several interim stages between absolute cold-bloodedness and absolute warm-bloodedness in reptiles, as there are many such intermediate stages among modern animals. In addition, one should remember that heat exchange between an animal and its environment is a two-way process.

From this one can see that "cold-blooded" (originally meaning "cold to the human touch") and "warm-blooded" (meaning the opposite) are subjective and quite misleading terms for describing animals. Biologists use terms that are much more relevant: "ectothermic," meaning dependent on external heat (Gr. *ektos therme*, outside heat), "endothermic," meaning dependent on controlled inner heat (Gr. *endos*, within), and "heterothermic," indicating intermediate animals with imperfectly controlled blood heat (Gr. *heteros*, other). Heterotherms include some insects, fish, and other animals. These terms accurately describe the very complex nature of living creatures and their physical relationship to the natural world.

It follows that advanced forms of endothermy requiring complex physiological thermostats are more likely to have evolved first in smaller animals that depended on efficient metabolism. The smaller the animal the more vital it was for its survival to counteract a cold nighttime environment quickly—large reptiles could depend on a proportionately larger reserve of bodily heat. There is also a probability (some believe "certainty") that medium-sized fast-moving reptiles, which required a higher-than-usual reptilian metabolic rate, were heterothermic, if not necessarily endothermic.

IT WAS THE SMALL CYNODONTS that proved to be the ultimate products of therapsid evolution. About the size of a modern badger, one creature was called *tritylodont* (from the Gr. *tria tyl odont*), because of its highly differentiated "three-knobbed" cheek teeth. Tritylodonts had small pits and grooves in their snouts, implying that they had touch-sensitive whiskers, and therefore also fur to insulate their bodies on cold nights in seasonal climates (**9**). They were probably endothermic, although their body temperatures are thought to have been low, possibly below that of animals of similar size today. They were shrew-like in appearance and probably nocturnal, feeding at night when the early dinosaurs were supposedly inactive.

Experts find it difficult to separate fossils of the last of the endothermic therapsids, the *tritheledonts* (Gr. *tria thele odon*, three-nipple teeth), from the earliest true mammals, endothermic animals with milk-secreting glands. Mammals of this period are identified with certainty by their characteristic single-boned lower jaw (or a fragment of it). It is that one-piece lower jaw, and the two cranial openings, which are still the hallmarks of mammals 200 MY or so after their first appearance.

The largest group of all reptiles were "diapsids," which also date from Pangean times to the present. They had four cranial openings (**10**)—two on either side of the head, besides eye sockets and nostrils. The first diapsids appear in the fossil record about 285 MYA, but the most important, the "archosaurs" (Gr. *archos*, ruling), evolved about 245 MYA. They are characterized by a broadly expanded skull, sometimes with double rows of teeth, plus ankle bones that helped some of their early descendants, the dinosaurs and the pterosaurs, to assume an upright posture, which in turn led to bipedalism. Besides being diapsid, all archosaurs, including their immediate and modern descendants (crocodiles, birds, lizards, and snakes), had an extra pair of openings behind the nostrils and in their lower jaws (**11**).

Archosaurs were both herbivorous and carnivorous in habit: the carnivorous animals fed on fish, early amphibians, and smaller mammal-like reptiles. They diversified around 230–225 MYA when they split into several lines, some of which led to the pterosaurs (flying reptiles), others to the dinosaurs (**12**). Yet others led to the phytosaurs, aquatic predators more fearsome than modern crocodiles. The fantastic diversity of these terrestrial groups during the post-Pangean era began only after the mass extinction that marked the end of Triassic time (206 MYA) and is described later.

All the other mammal-like reptiles, several reptilian orders, and many amphibians were either already in decline and were finally decimated during the Triassic event, or were eliminated by it altogether. Only the early pterosaurs and dinosaurs, along with the tritylodonts and the earliest mammals, survived the mass extinction. With the extinction of many other animals, the way was now clear for the ascendancy of the dinosaurs during the breakup of Pangea.

IN THE LAST CHAPTER Pangea was portrayed as a compressed aggregate of previously separate continents. It was implied that, among other contributory causes, the very fact of aggregation in the region of Central Pangea had predetermined the supercontinent's tectonic fate. Pangea was likened to a hypothetical circular mass formed from continent-sized gateau-like wedges driven together to

9

10

11

12

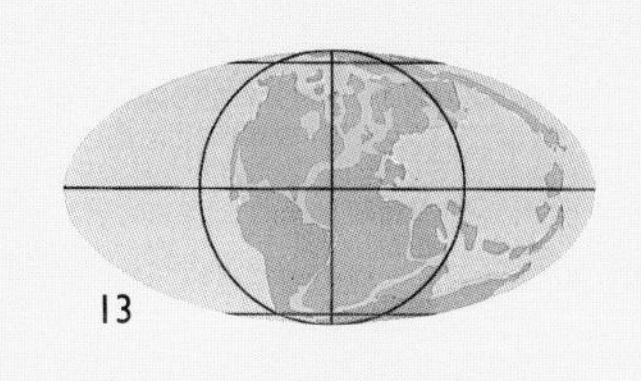

13

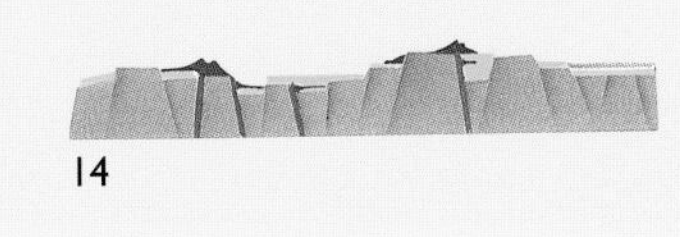

14

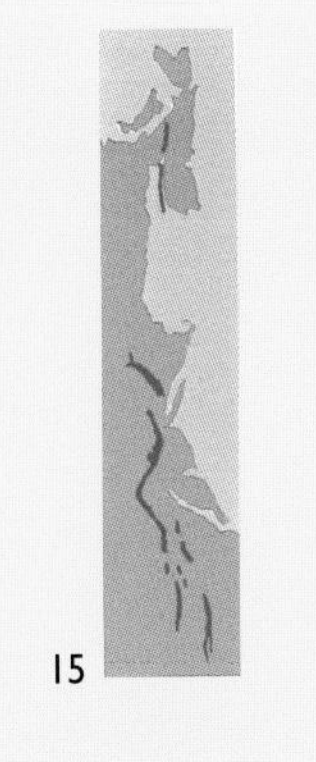

15

form a very thick structure at the point of collision. The roots of the model structure penetrate into the very hot regions of the asthenosphere, even as parts of the surface reach for the sky as high alpine mountains. It was concluded that the net effect of compression and increased thickness resulted in the formation of a hot spot in the mantle below it, causing internal melting, weakening, and eventually breakup of the model supercontinent.

Pangea, of course, was not circular; it was an elongated, lopsided amalgam (**13**) resulting from the conjunction of Laurasia with Gondwana. It was also an immense consolidation of granitic and sedimentary rock concentrated into one-sixth of the total area of the Earth's surface. This thermal blanket caused overheating of the Earth's asthenosphere, and therefore swelling and distortion of the "geoid" (the term for the imperfect surface shape of the planet). By Triassic times (245–206 MYA) the heat generated beneath Pangea had caused melting and weakening below its thickest parts. Meanwhile, the continued slow and independent rotation of the Laurasian and Gondwanan elements caused twisting and wrenching of the suture from East Central Pangea to West Central Pangea. So it is most likely that a combination of subterranean melting, wrench-faulting, and weakening of the suture caused the breakup of the supercontinent and its division into post-Pangean Laurasia and Gondwana.

THE INITIAL BREAKUP OF PANGEA, and then its fragmentation into the modern continents, progressed in stages. At first there were prolonged periods of rifting that involved crustal extension (**14**) and the formation of valleys (grabens) and ridges (horsts), like those discussed in the previous chapter [see Essay, pages 132–33]. Further rifting forced actual continental separation and eventually the development of seafloor spreading-centers between continents. [To avoid confusion over the use of the general term "rifting," which applies to *both* extension and spreading-center processes, we will use "crustal extension" and "continental separation" when describing the breakup of Pangea into the modern continents.]

The prolonged period of crustal extension is generally believed to have been caused by a period of "doming" in the lithosphere beneath Central Pangea. Upwelling and accumulation of molten mantle material deep beneath the supercontinent raised individual domes and then extended them. This process caused a three-armed rift system to form, particularly in structurally weak parts of Pangea, such as the suture between Laurasia and Gondwana [Essay 3]. Although the doming and crustal extension were accompanied by volcanic eruption, actual continental separation did not occur until grabens had sunk and thinned immediately above the hottest central regions of their parent domes. Volcanic rifts then ruptured the grabens, eventually propagating and joining to form a continuous spreading axis. But for every rift that ultimately became a mid-ocean spreading-center there were many hundreds of rifts that were either inhibited for many millions of years, or failed to form an ocean and filled with sediments.

The Newark Rift system and the Palisades of the Hudson River (**15**) are examples of such "failed" rift systems [Essay 1]. Around 210–200 MYA the many rift valleys that existed along what is now the East Coast of North America, then part of the suture between Laurasia and Gondwana, extended at least 2,240 km (1,400 miles) from Nova Scotia to the Carolinas. They may once have extended as far as the North Sea region in East Central Pangea and have reached as far south as the present Gulf of Mexico in West Central Pangea. As rifting failed, some grabens became channels for river systems, or were filled by lakes and later drowned by shallow seas.

The valleys, originally up to 1,350 m (4,500 ft) deep, filled with sediments that accumulated to form rock up to 3,900–4,200 m (13,000–14,000 ft) in thickness. This apparent paradox is explained by a phenomenon called "isostatic compensation." This is the creation and maintenance of an equilibrium by adjustment of the

[*continued on page 158*]

4 Florida-Bahama Block

This aerial picture portrays a part of the vast Bahama Platform that began to form about 165 MYA at the southern end of the rift between Laurentia and Gondwana as the proto-Central Atlantic opened. The platform now consists of the Great and the Little Bahama Banks and others, on which the comparatively minuscule Bahama Islands are encrusted. It has grown so massive that it rises from the seafloor at the edge of one of the deepest regions of the North Atlantic, the Hatteras Abyssal Plain, and breaks the surface of the sea 5,445 m (18,150 ft) above the abyss. The total height of the structure from its basement rocks (which penetrate the seafloor) to its surface is thought to be well over 5 km (3 miles).

It is generally believed that the Bahama Banks developed from atolls that formed on inactive volcanic islands. According to C.R. Scotese, whose paleo-reconstructions we are following through the book, these volcanoes formed where the southeastern margin of a strike-slip fault met the proto-Central Atlantic spreading-center (see diagram). Scotese terms this region the North Bahama Fracture Zone, and concludes that the fault's sliding interface extended

from what is now the Gulf of Mexico, across southern Florida to the submarine Guinea Plateau off West Africa. In Scotese's view, the fault was the product of a wrenching movement, a "megashear" in West Central Pangea as post-Pangean Laurasia separated from post-Pangean Gondwana. A side-effect of the shear would have been to correct an offset of southern Florida from northern Florida. However, this is a controversial interpretation, and some paleogeographers take a different view. They think that there is little evidence for such a megashear, and that there was not an offset of the Florida peninsula.

Whatever the correct interpretation may be, it is generally agreed that as the original volcanoes became extinct, they cooled, contracted, and settled deeper into the shallow tropical sea that now flooded the region. The reefs that grew upon them gradually built up to form atolls. As the first volcanic structures cooled and sank, sediments accumulated within and between the atolls to establish the foundations of an ever-growing carbonate edifice built by marine organisms.

5 Americas Seaway

The basement rocks of the Florida-Bahama block inhibited the extension of the proto-Central Atlantic. But this did not prevent the fragmentation of the future Gulf of Mexico and Caribbean regions, or the development of the Americas Seaway to the west of the block.

As the Caribbean events proceeded (165–142 MYA), rotation of the megacontinents caused a 480-km (300-mile) displacement between North American Laurasia and South American Gondwana. This megashear created the trans-Mexican volcanic belt, while the Florida-Bahama block rotated in concert with Gondwana. To further complicate this chaotic tectonic scenario, the Pacific Ocean crust continued to subduct beneath the nearby western margin of Pangea.

Meanwhile, the continental crust in the future Gulf of Mexico stretched and subsided. An intermittent supply of seawater from the Americas Seaway flooded the basin, and caused vast quantities of salt thousands of feet thick to accumulate as the seawater gradually evaporated. Continued crustal extension in the Gulf led to seafloor spreading. Ultimately, spreading was inhibited by a second spreading-center that opened in the Yucatán region of Gondwana. This second center resulted in the complete separation of Yucatán from Gondwana and the formation of the proto-Caribbean Sea.

As seafloor spreading in the Gulf of Mexico ended, movement along the North Bahamas Fracture Zone [Essay 4] also stopped. The Florida-Bahama block, including the basement rocks of Cuba, now became permanent parts of the North American plate. Seafloor spreading commenced between the southeastern tip of the huge Florida-Bahama block and the Guinea Coast of Africa. This separation allowed the Americas Seaway spreading-center to connect to the proto-Central Atlantic spreading-center, and the two seaways to join the Pacific Ocean to the Tethys Ocean. At this point, the proto-Central Atlantic and the Americas Seaway ceased to exist—they had become the Central Atlantic Ocean.

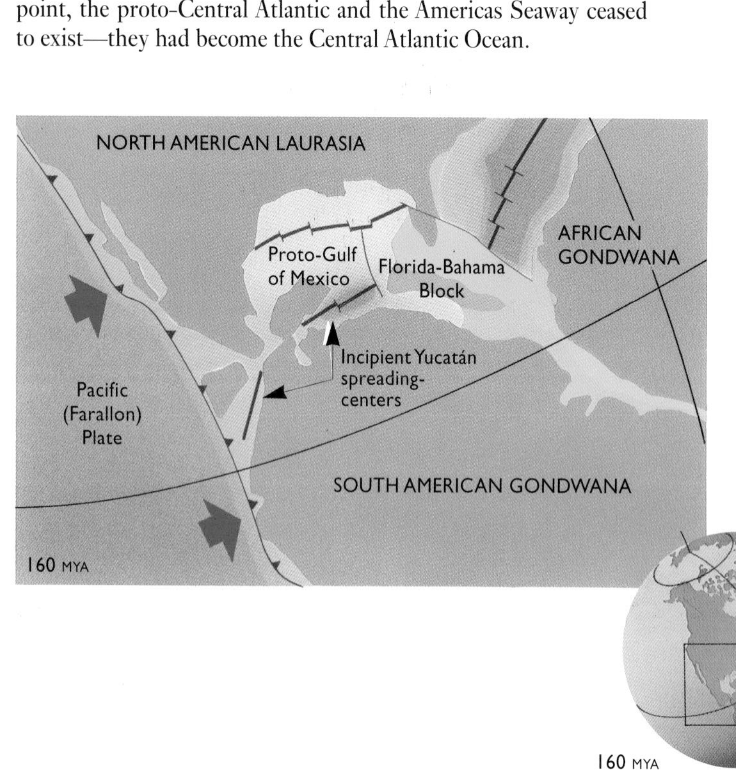

Gulf of Mexico, back berm
near Pensacola, Florida

[*continued from page 153*]

Earth's lithosphere to weight on its surface. For example, water 1,350 m (4,500 ft) deep can cause a basin floor to sink by 540 m (1,800 ft). A similar depth of sediment alone can cause a subsidence of 2,700 m (9,000 ft). So when "isostatic depression" has been accounted for, 4,050 m (13,500 ft) of sediments *can* accumulate in a basin with an initial water depth of only 1,350 m (4,500 ft). Subsequently, as the sedimentary rock is reduced by erosion, the basin floor is relieved of weight and rebounds ("isostatic rebound").

IT WAS DURING THIS PHASE of increasing volcanism in Central Pangea, around 206 MYA, that another great mass extinction took place. This was equal to, if not greater than, that most publicized of all mass extinctions, the "K/T event," which coincided with the end of the dinosaurs [a focal point of Chapter IX: *Maritime West*]. According to the paleontologist P.E. Olsen, between 40 and 50 percent of all terrestrial tetrapod families were extinguished 206 MYA. The victims included most of the remaining mammal-like reptiles and the amphibians. In the marine realm both invertebrate and vertebrate families were hugely affected and some were eliminated.

This Triassic extinction may have been caused by excessive volcanism at the time, or by a bolide impact on the Earth's surface—as is also claimed for the K/T incident. In the case of the Triassic event there is strong circumstantial evidence that links it to a particular bolide crater at a particular time: the Manicouagan impact structure of Quebec in Canada. But the Triassic event was not accompanied by the appearance of an iridium anomaly anywhere in the rock record (which is considered proof of the K/T bolide impact).

In the early 1990s, a rich assemblage of fossil animals was discovered in a Newark Rift-type basin in the region of the Bay of Fundy, Nova Scotia. The rock in which the fossils were found dated to half a million years *after* the Triassic event, in the Jurassic Period (206–144 MYA), which is named after the Jura Mountains between Switzerland and France. The Nova Scotian rocks contain fossils of dinosaurs, crocodilians, mammal-like reptiles, and some amphibians, but no fossil bones of animals that were extinguished at the end of Triassic time. In paleontological terms this deficiency, the complete absence of all the previously extant animals from this assemblage, strongly supports the idea that the Triassic mass extinction was "sudden." This is in marked contrast to a "prolonged" event like the great Permian extinction, which is thought to have been caused by changes in climate and sea level over a period of about 5 MY.

The Manicouagan crater (**16**), one of the world's best-known impact craters, is about 480 km (300 miles) to the north of the Bay of Fundy. The structure is 70 km (44 miles) in diameter and was produced by a bolide that was at least 9.6 km (6 miles) in width. The date of the impact is 210 MYA (plus or minus 4 MY). Was this impact the cause of global calamity? Or was the mass extinction caused by the excessive volcanism in Central Pangea? Or perhaps a combination of the two?

DINOSAUR BONES FIRST APPEAR in the fossil record about 235 MYA. At that time these animals were small (chicken-sized) to medium-sized carnivores. They are distinguished from other diapsid reptiles by their ability to stand erect and run on their hind legs (**17**), counterbalanced by their elongated tails while in motion. They also used their tails for tripod-like support while stationary. Dinosaurs were bipedal at an early stage in their evolution, the first animals that made this advance. This factor, plus their agility and their short, clawed forelimbs, made dinosaurs highly effective predators. Some later dinosaurs, mainly the very large or the heavily armored herbivores, reverted to quadrupedalism (**18**) but retained hindlegs that were about twice the length of their forelegs.

Mainly because of their bipedalism, the early dinosaurs enjoyed great competitive advantage over the slower-moving, less agile mammal-like reptiles. Some paleontologists believe that this predatory advantage enabled dinosaurs to reduce the therapsids to the

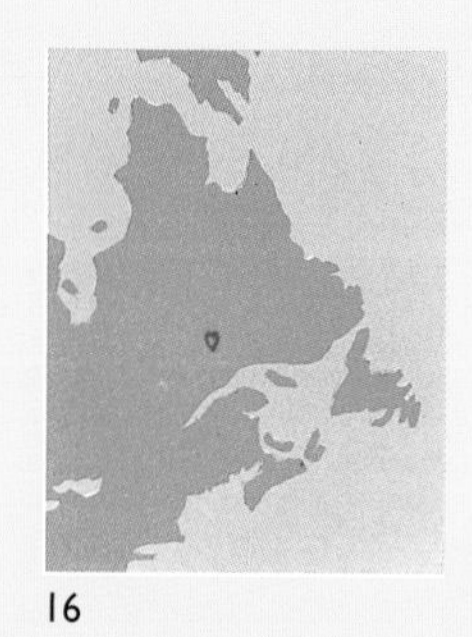

16

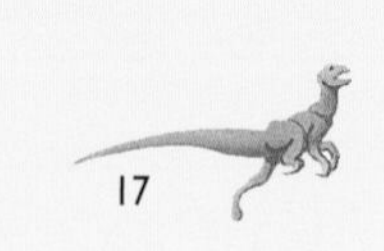

17

18

19

20

point of near-extinction before the Triassic event. Their bipedal gait, the reasoning goes, gave the carnivorous dinosaurs an ability to move at great speed over the ground, and they were able therefore to outrace the quadrupedal therapsids and devour them. On the other hand, at least part of the reason for the disappearance of Triassic therapsids was because they were evolving into mammals.

The early dinosaurs' ability to move quickly to chase other animals, or to fight similar-sized animals for prolonged periods, or to flee from large predators such as phytosaurs, raises the issue of whether dinosaurs were endothermic. This is a highly controversial subject because by definition modern reptiles are "cold-blooded"—although a few rare living species are heterothermic. In view of their need for considerable reserves of power, it seems likely that many species of carnivorous dinosaurs were at least heterothermic (like the modern armadillo, a heterothermic mammal), while really large herbivores, like the well-known *Diplodocus*, were probably homeothermic (their body temperatures controlled by their great size). With one exception [page 159], no fossils have been found to promote the mammalian notion that any of the dinosaurs had hairy or fur-covered bodies to insulate them against heat loss.

ONE TYPE OF ANIMAL OF THIS PERIOD (*c.*150 MYA) that was insulated was the feathered variety—namely the flying animal known as *Archaeopteryx* (Gr. *archaeo pteryx*, ancient wing), assumed by many to have been the first birds. Coincidentally, the other flying animals of the time, the pterosaurs, are also thought to have been insulated. This is conjectured from a fossil of what appears to have been a once hair-covered pterosaur discovered in Kazakhstan. Both the feathered birds and bat-like pterosaurs needed hollow bones for lightness, coupled to a high metabolic rate and a high-pressure heart-lung system to sustain flight. The consensus view is that both were endothermic [Essay 6].

The first appearance of pterosaurs in the fossil record (215 MYA) shows that by that time these reptiles had become highly specialized for flight. Accordingly, they had diverged from their archosaur lineage before or coincident with the dinosaurs. Although pterosaur bones were hollow and bird-like, the bone structure of their membrane-covered wings shows that they did not evolve into birds.

Pterosaurs were particularly numerous in Central Pangea, and their fossils are found in large numbers in Kansas, in Britain, and in the Solnhofen limestone of Bavaria, where *Archaeopteryx* fossils are also found. They became widely dispersed and highly diversified in all regions of the Pangean and post-Pangean world (except Antarctica, as far as is known) during the 150 MY of their existence.

The early species generally had short faces and necks but long tails. The later species that replaced them from about 150 MYA often had quite large and elongated crests or fins on their heads. They had long faces and necks to support the crests. These animals are called *Pterodactyls* (Gr. *pteron daktulos*, wing finger: **19**) and they had short tails. Pterosaurs (**20**) varied from sparrow-sized to the size of an overlarge, man-made hang glider complete with pilot.

In fact the larger pterodactyls resembled hang gliders in many ways: they shared design features, performed in a similar way, and were both endangered by high or gusty winds. The largest known pterodactyl wingspan was 10.8 m (about 36 feet), and the weight of that particular animal is estimated to have been about 81 kg (180 pounds). Pterosaurs' brains were larger than other reptiles of comparable size. This suggests that they had developed the high level of muscular coordination that would have been required for maneuverability in flight and for takeoff and landing. The larger animals had only to spread their wings in a steady breeze and give a push with their feet to be airborne. They could then stay aloft for extended periods—just as hang gliders do today. They probably used their crests as front-end rudders—this is a speculative idea, but there is no doubt that the large crests had an aerodynamic effect, and therefore it is highly likely that they evolved with a deliberate aerodynamic purpose. The smaller pterosaurs were the most active fliers—they probably had to flap their wings energetically to fly and to maintain flight.

THERE IS GREATER MYSTERY about the origin and habit of *Archaeopteryx*—the earliest known birds that appeared about 150 MYA (**21**). Why and how did *Archaeopteryx* learn to fly? Were its ancestors tree-dwelling dinosaurs that learned to flutter down from tree branches, or running dinosaurs that gained speed on the ground by flapping their wings like frightened chickens? Why did they have three claws extending from the center of both wings at the joint? Were these used to help them clamber out of reach of forest predators? One clue to their use may arise from a modern but primitive bird. The chicks of the extraordinary South American hoatzin bird, found in the upper Amazon Basin, have almost identical claws in infancy, which are used for just this purpose.

Much is known about the pterosaurs because their fossils are both many and diverse. But there are only six fossils of *Archaeopteryx* in existence: the first was found in 1861, the second in 1867, and the rest in the 20th century. The most recent and largest specimen lay unrecognized in a private collection of fossils until it was rediscovered in 1988. Because of differences in sizes between specimens, it is now thought that perhaps two species of *Archaeopteryx* are represented in the series of six. All the specimens originated from the exceptionally fine-grained Late Jurassic Solnhofen limestone of Bavaria—East Central Pangea. So clear and articulated are the bones, the impressions of feathers and the shape of wings, that these features are judged to be very like those of flying birds today. In addition, *Archaeopteryx* had an avian wishbone, which in living birds forms one of the attachment sites for flight muscles. One of the main differences between modern birds and their predecessors, is that *Archaeopteryx* had a long bony tail and teeth—like a reptile.

Most paleontologists today think that *Archaeopteryx* evolved from a small, bipedal, fast-running dinosaur which had evolved feathers from its reptilian scales. Indeed, modern birds (the chicken, for example) have scaly legs that graduate into body feathers. Advocates argue that such an evolutionary change would have improved the animal's forward motion and therefore its rate of survival in a very competitive world. Feathers would also have conserved body heat if the animal was ectothermic or heterothermic. The *Archaeopteryx*, the reasoning goes, fed its young on an insect diet. Young and old alike may have climbed into trees and bushes, using clawed wings, in search of insects to supplement their diet. They became arboreal by habit, using flapping and gliding flight to get from tree to tree. When they landed on the ground, their sharp-clawed wings would have helped them to scramble back up—like the hoatzin bird today.

In recent times a bird of the same age as *Archaeopteryx* (*c.*150 MYA) was discovered in northeastern China and named *Confuciusornis.* It has wing claws but lacks teeth. Perhaps more startlingly, in 1986, an unarticulated avian fossil was found in West Texas—West Central Pangea. Like *Archaeopteryx* and *Confuciusornis,* this Central Pangean animal had a reptilian brain case and a wishbone. It was named *Protoavis* (Gr. *protos*, L. *avis*, original bird: **22**) by its discoverer, S. Chatterjee, who asserted that the fossil is that of a bird that lived 225 MYA—long before the dinosaurs. This highly controversial claim was eclipsed in 1998 by the discovery by J. Qiang, P.J. Currie *et al* (*Nature*: June 1998), also in China, of several species of feathered theropods dated *c.*120 MYA. These animals were two-legged, fast-running dinosaurs about the size of a turkey. One of several species found is called *Caudipteryx,* a reference to its tail-like feathers. Another is *Protoarchaeopteryx robusta*—a reference to the transitional creature it resembles. None of these animals could fly and they all lived 30 MYA after *Archaeopteryx*. so, while strengthening the relationship between dinosaurs and birds, their significance has yet to be fully understood.

21

22

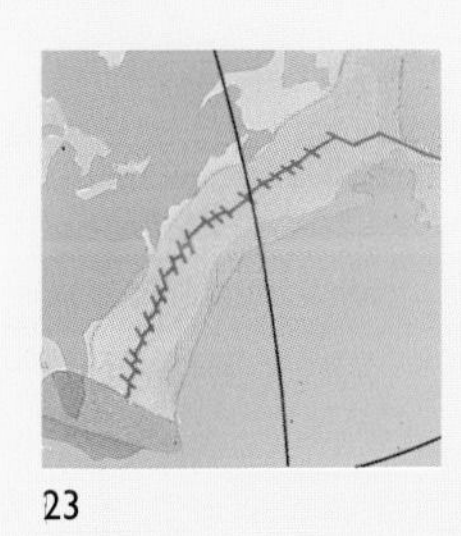

23

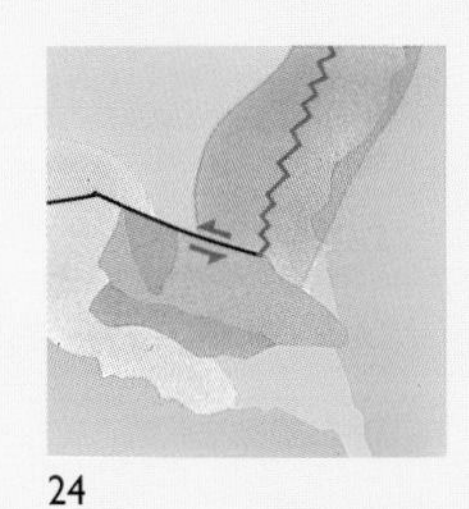

24

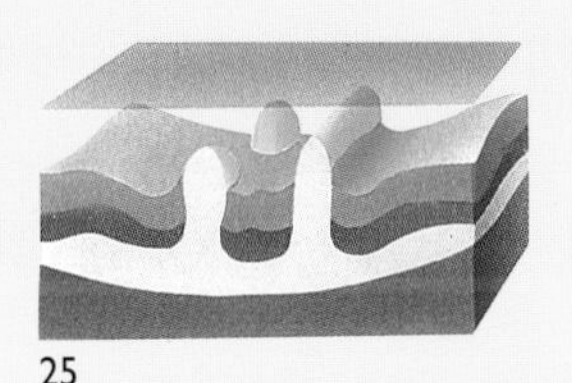

25

RETURNING NOW to the tectonic environment of these evolutionary events, we recall that there was intense crustal extension in Central Pangea between 230 and 187 MYA, in Late Triassic and Early Jurassic times. At the end of that period the East Coast of North American Laurasia, from Manhattan to Nova Scotia, was still quite close to the Moroccan region of northwest African Gondwana. Ultimately, rifts on either side of the Florida-Bahama block began to propagate into spreading-centers about 180 MYA and the numerous tectonic movements that had been taking place throughout Central Pangea, including the East Coast region, reduced in activity.

By 165 MYA the propagating rift system in the Bahama region had split the supercontinent into what now became post-Pangean Laurasia and post-Pangean Gondwana. Today we call this rift system the Mid-Atlantic Ridge. At the time of its formation, the ocean-to-be was the narrow proto-Central Atlantic (**23**), stretching from Nova Scotia and Morocco to Florida-Bahama and Senegal. It was fed by seawater flowing from the Tethys Ocean through the Gibraltar system via a short-lived rift valley in northern Morocco and the coastal region of Iberia [Essay 2].

THE GEOMETRY OF THE PROTO-CENTRAL ATLANTIC was decided by the ancient and therefore relatively cold suture between West African Gondwana and North American Laurasia, which reopened along its entire length in a a single event. Although the break was clean-cut it was jagged in shape. Parts of Gondwana were left on the Laurasian side of the seaway and vice versa [Essay 3]. West of the Florida-Bahama block the supercontinental suture was still warm and malleable after the comparatively recent (in geological terms) conclusion of the Ouachita orogeny. The Pacific Ocean floor was continuously being subducted beneath the west coasts of both North and South America. As a result, the breakup of western Central Pangea was fragmented and tectonically disorganized.

Spreading activity in the proto-Central Atlantic caused stress, but this was inhibited by a formidable piece of old Gondwana, the "notch region" (the Piedmont plate) between Africa and South America, described in the last chapter. This plate included Cuba, Florida, and the Bahamas. According to C.R. Scotese and his associates, the Central Atlantic spreading-center made no headway into the Piedmont block of continental basement rock. Instead, the northern edge of the block sheared at right angles to the rift and slipped along the break at the "North Bahama Fracture Zone" (**24**), which then propagated into the Gulf of Mexico [Essay 5]. Volcanoes formed on the surface where the fault met the Central Atlantic spreading-center, and these volcanoes that became the foundation of today's Bahama Banks, on which the present Bahama Islands are superimposed. Even so, many tectonic scientists disagree with this interpretation. They suggest that South Florida's basement rocks are in fact volcanic rocks produced by an old hot spot, and that the Bahamas and Cuba overlie a basement formed by that hot spot's track.

Sparked by movement along the fracture zone, a prolonged period of crustal extension that lasted from 180 to 160 MYA stretched and thinned the future central "Gulf of Mexico" region. The resulting basin was sporadically connected to the Pacific Ocean by the Americas Seaway, and because of this tenuous marine connection vast salt deposits several thousand feet thick accumulated in this region. These, and similar salt deposits in the coastal regions of the proto-Central Atlantic and off the shores of the Tethys Ocean, were later to play an important role in the recovery of oil and gas.

As the thick deposits of Jurassic salt became covered with younger sedimentary rock, the salt was pressurized to a point where it behaved like a liquid. The "liquefied" salt pierced weaknesses in overlying formations, creating "diapirs"—huge salt domes (**25**) that often resemble long-stalked, capped mushrooms. As they rose, the diapirs deformed the surrounding hydrocarbon-bearing rock, folding it upwards towards the salt column. This process permitted oil and gas in the rocks to migrate in the direction of the diapir. Here

[*continued on page 162*]

6 First to Fly

In order to fly, an animal must be built like a small aircraft: it must be light, winged, aerodynamic, supercharged, and powered by adequate fuel. Birds have all these characteristics. They have hollow bones, feathered wings, powerful muscles, a high-pressure heart and lung system, and a steady supply of calorific fuel, derived from a high rate of metabolism, to enable them to sustain flight. Heat produced by this rapid metabolism allows birds to maintain a *controlled* high body temperature, so they are endothermic. Birds are more than just "warm-blooded"—they generate their own heat (Gr. *endos therme*, within heat).

The first birds, the *Archaeopteryx* (Gr. *archaeo pteryx*, ancient wing), appear in the fossil record of 150 MYA. But they were preceded in the skies by the pterosaurs (Gr. *pteron sauros*, winged lizard), which date from 215 MYA. Experts agree that pterosaurs were probably endothermic.

Egrets and blue herons, Lake Windsor, Great Inagua, Bahama Islands

Pterosaurs diverged from their reptilian lineage before or coincident with the dinosaurs—to which they were only distantly related. The pterosaurs were prolific and ranged far and wide. In size they were diverse; some could be as small as sparrows or as large as hang gliders: the small animals flapped their wings vigorously, as bats do today, while the larger pterosaurs used air currents for gliding. Although their bones were hollow and bird-like, the bone structure of their membrane-covered wings suggests that they did not evolve into birds.

Unlike pterosaur fossils, *Archaeopteryx* fossils are rare (only six specimens exist) but in recent times a bird of the same age as *Archaeopteryx*, with the same wing-claws, but lacking teeth, was discovered in northeastern China and named *Confuciusornis*. In 1986 another, highly controversial new species of avian fossil was found. It dates from 225 MYA and is named *Protoavis*. If its affinity with birds is accepted (which is unlikely until a fully articulated example is found), then birds may have to be reclassified as an earlier order of hollow-boned flying animals predating the pterosaurs. Whatever their ancestry, by 140 MYA birds had already diverged along several adaptive pathways, and two very different kinds had emerged. There were certain points of resemblance; both had reptilian teeth: some were gull-like birds with strongly keeled sternums, a mark of proficient flying; others were flightless cormorant-like diving birds. However, to add to the mystery of bird/dinosaur evolution, a turkey-sized dinosaur with feathers, dated 120 MYA, has been found in China [see page 159].

The Bahamian salina scene pictured here features the modern blue heron and egret inhabitants. Fom our contemporary perspective, such a scene is perhaps the nearest we can get to the ambience of Early Cretaceous times.

[*continued from page 159*]

the hydrocarbons became trapped in porous reservoir rocks at the edges of the salt domes, or beneath the mushroom cap sealed by impermeable salt. Today reservoir rocks like these are the focal points of oil and gas exploration in the Gulf of Mexico, which has several hundred salt diapirs. This scenario is repeated wherever remnant margins of the Americas Seaway (**26**), the proto-Central Atlantic, and the Tethys Ocean can be found today.

By 140 MYA seafloor spreading had begun to the southeast of the apparently indestructible Florida-Bahama block. At last, there was a continuous spreading-center that bypassed the block, and the proto-Central Atlantic and the Americas Seaway were joined. An ocean link was now established between the Tethys Ocean and the Pacific Ocean. Post-Pangean Laurasia and Gondwana were at last separate entities. The Central Atlantic Ocean was born, and the *New World* became a reality.

Today's Bahama carbonate platform and Florida peninsula are memorials to that significant moment in geological time when a great ocean opened. We have only recently understood that the Bahama Banks have grown in stature as the ocean has widened and deepened, and that they are probably more than 4.8 km (3 miles) thick [Essay 4]. The explanation for this extraordinary height has been available from the early 19th century. Charles Darwin was the first to work out the principles of how such structures are built.

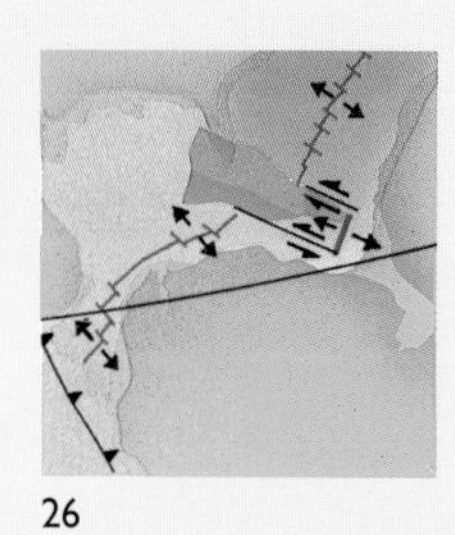

26

27

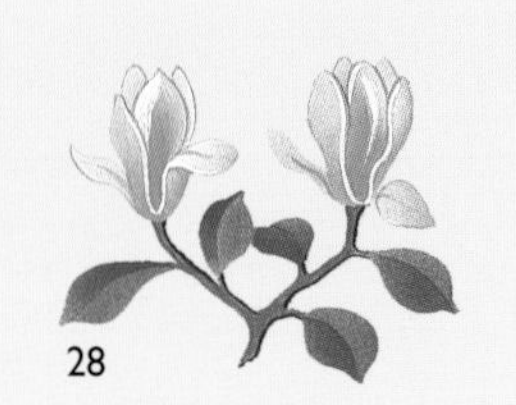

28

IN THE SPRING OF 1836 Darwin spent ten days visiting and exploring the Cocos Islands in the Indian Ocean—a few hundred miles southeast of Wallace's future happy hunting grounds in Indonesia. Darwin was by then near the end of his five-year voyage as a naturalist aboard the British ten-gun brig HMS *Beagle.* Like many others before and after him, Darwin was overwhelmed by the beauty of the Cocos and other coral islands he had visited. He wondered how such islands could possibly have formed in a deep oceanic region. His analysis of the available facts led him to a series of conclusions that still hold good. Darwin presented them as a straightforward process of thought in his notes dated April 12, 1836. They were later included in his book *The Voyage of the Beagle*, published in 1839.

Darwin reasoned that in tropical seas a coral reef forms just below the sea surface around a partly submerged volcano. As the volcano becomes extinct it cools and subsides. As the level of the sea rises on the flanks of the volcano, the corals grow higher (**27**), building on those that have died in the deeper and cooler waters out of reach of sunlight. The process of building continues until only the peak of the now-extinct volcano shows above the sea. This peak becomes encircled by a coral reef pounded by oxygen-rich surf. Ultimately, Darwin continued, the remnant volcano will sink out of sight, leaving an atoll that will continue to grow, surrounding a shallow sandy-bottomed lagoon without an island at its center. And in principle this is how it is believed the Bahama carbonate bank began to form, but on a huge scale. It began as a cluster of innumerable reefs partially surrounding the volcanoes that, according to the Scotese model, formed at leaky joints between the Florida-Bahama transform fault and the proto-Central Atlantic spreading-center.

Darwin's elegant explanation of the formation of atolls was not repeated in the case of another 19th-century mystery—the origin of angiosperms, or flowering plants. These plants were evolving when the Bahama carbonate bank was in its infancy [Essay 7]. The mystery troubled Darwin, Wallace, and others throughout their lives, but it may be nearer to a solution today.

FLOWERING PLANTS produce nearly all the foodstuffs that most populations of mammals and insects depend upon—a huge variety of leaves, grasses, vegetables, pollen, seeds, and fruit. As we shall see in later chapters, angiosperms were important factors in the coevolution of insects and herbivorous and omnivorous reptiles and mammals—from beetles and dinosaurs to honey bees and humans. The appearance of the first true angiosperm thus represented a giant step forward in the evolution of terrestrial life. The advance from gymnosperms (naked seeds) to angiosperms (covered seeds: **28**) was the plant kingdom's equivalent of the advance from amniotes (hatched from a shelled egg) to placental amniotes (born from a womb) in animals. Therefore, to understand the world as we know it today, we need to understand the mechanisms of the first appearance and subsequent evolution of angiosperms.

THE TWO MAIN CLASSES OF ANGIOSPERM, "dicots" and "monocots," are terms derived from *dicotyledon* and *monocotyledon*, which in turn describe the distinctly different character of their "embryos." The "cotyledon" part of the two words (Gr. *kotyledon*, cup-shaped) refers to the embryos' leaf-like organs, of which the dicots have two and most monocots have one. The cotyledon in the dicot seed stores food, while the cotyledon in the monocot seed absorbs food. When the seeds are "hydrolyzed" (activated by moisture) they split and later discard their seed-coat (the skin), sprouting roots, stems, and leaves. The dicot cotyledons become the plant's two primary leaves and the monocot cotyledon becomes the plant's single primary leaf. Of the two the dicots are long considered to have been the more primitive—but there is still some debate in paleobotanical circles.

Plants with dicotyledon seeds have flowers (roses, apple, almond, and cherry blossoms, for example) with multiple parts (petals, stamens, and so on) whose number is usually divisible by four or five. Monocots (lilies and irises, for example) have flower parts that are always divisible by three. Dicots have one to three furrows in their pollen grains and net-like veins in their leaves (such as oak and maple leaves), whereas monocots have only one furrow; and most have leaves with parallel veins along their length (such as tulip and daffodil leaves). The vascular bundles in a dicot's stem are in a ring-like form and similar to those of the primitive vascular plants, while vascular bundles in monocot stems are generally scattered at random. The early descendants of the primitive dicots had a variable number of petals, from three to sixteen, while the earliest monocots had fewer petals, always in multiples of three. Dicots produce wood, in the timber sense, just as gymnosperms do. Monocots do not.

The oldest angiosperm pollen is dicot and appears in the fossil record 134–136 MYA. The oldest monocot fossils yet found are dated about 120 MYA. However, no plausible intermediate evidence has yet been unearthed to account satisfactorily for the evolution of angiosperms from gymnosperms.

DARWIN CALLED THE ORIGIN of angiosperms "an abominable mystery," and to a limited degree their origin is still a mystery. Wallace drew attention to the odd distribution of plants that appeared on either side of his (Gondwanan-Laurasian) floral-faunal line. And Wegener proffered a possible solution when he wrote in *The Origin of Continents and Oceans*

> *... the displacement theory offers a simpler solution from the purely biological standpoint as well, because it adduces, in explanation of the distribution of plants and animals, not only a land connection, but also variations in distance of the continents concerned.*

So it seems that Darwin in the realm of plant evolution, Wallace in the realm of biogeography of plants and animals, and Wegener with his theories of continental displacement, bequeathed an intellectual challenge to modern paleobotanists, biogeographers, and paleogeographers—a challenge that has been met with vigor and enthusiasm, but not yet completely resolved.

7 First to Flower

Sanmiguelia lewisii

Angiosperms, the flowering plants, dominate the plant kingdom today and are crucial to the present food chain. Their 240,000 or more varieties outnumber all other plant species combined. They include trees, shrubs, all grasses and modern cereals, and other familiar nonwoody garden plants. But they do not include conifers and cycads, members of an ancient lineage of gymnosperms (Gr. *gymnos sperma*, naked seed) from which angiosperms derived (Gr. *angeion sperma*, capsule).

The term *Magnoliopsida* is the botanical name for all dicots—angiosperms that sprout two leaves in their early stage of growth. The reason for this nomenclature is that the magnolia family was long believed to be ancestral to all flowering plants. But the earliest magnolia fossil known is dated 100 MYA. The first undisputed angiosperm pollen is dated 135 MYA and the earliest flower 117 MYA. Both these fossils belong to the Order Piperales, which includes pepper plants. Some paleobotanists think that the ancestry of flowering plants goes back even further in time.

In 1974 fossilized angiosperm-like pollen grains were discovered in sedimentary deposits of the Newark Rift system. The sediments are dated 230–206 MYA [Essay 1]. Six years later the same paleo-botanist, Bruce Cornet, made a second discovery, this time in West Texas—an almost perfect fossil of *Sanmiguelia lewisii* dated around 206 MYA, which he and others thought to be a primitive angiosperm. That fossil *Sanmiguelia*, redrawn here to the left, is modest in size and complete with roots, stems, leaves, male and female flowers, pollen, and seed. However, other paleobotanists interpret this plant as perhaps a transitional gymnosperm that under extreme conditions of aridity evolved many angiosperm characteristics. Indeed, the anatomical parts of the *Sanmiguelia* fossil demonstrate an affinity with a group of very rare modern gymnosperms. These are represented by *Welwitschia mirabilis*, a giant West African coastal plant that is about 2 m (6 ft) in overall height and perhaps 4 m (12 ft) in diameter.

The brigantine *Romance* off Antigua, Leeward Islands, North Atlantic Ocean

THE BRIG HMS *BEAGLE* SET SAIL FROM DEVONPORT INTO THE ENGLISH CHANNEL ON DECEMBER 27, 1831, OUTWARD BOUND FOR THE CANARY ISLANDS AND THE CAPE VERDE ISLANDS OFF THE COAST OF WEST AFRICA, AT THE START OF A VOYAGE OF CIRCUMNAVIGATION THAT WOULD LAST ALMOST FIVE YEARS. THE SHIP'S NATURALIST WAS 22-YEAR-OLD CHARLES DARWIN. HIS APPOINTMENT WAS FORTUITOUS, THE RESULT OF SEVERAL TWISTS OF TENUOUS CIRCUMSTANCE THAT WERE TO TRANSFORM THE GENERAL PERCEPTION OF THE NATURAL WORLD. THE BRIGANTINE *ROMANCE*, PICTURED HERE OFF ANTIGUA, CAPTURES THE SPIRIT OF SUCH A VOYAGE.

ATLANTIC REALM
144–99 MYA

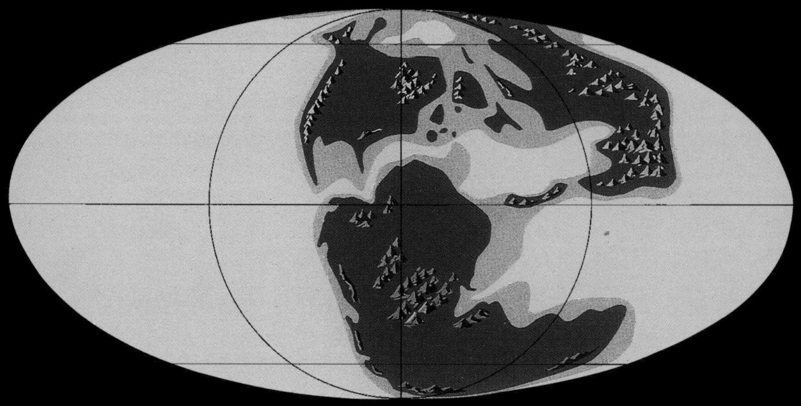

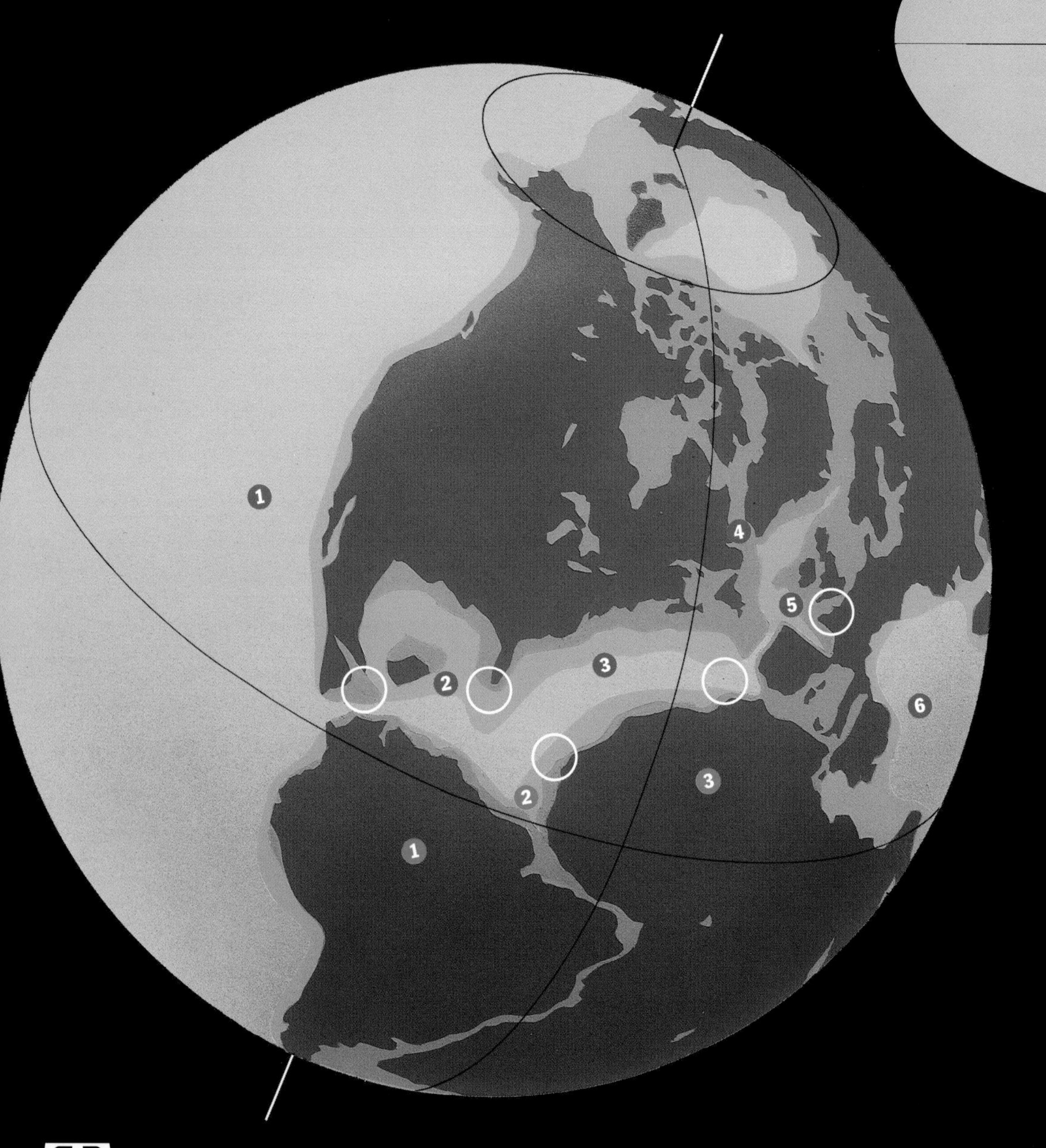

POST-PANGEAN LAURASIA

1. Pacific Ocean
2. Caribbean Sea and Gulf of Mexico
3. Central Atlantic Ocean
4. Labrador Sea
5. Bay of Biscay
6. Tethys Ocean

○ BASINS, ISLANDS (left to right)

Greater Antilles
Florida-Bahama Platform
Cape Verde Islands
Canary Islands and Azores
Anglo-Paris Basin

POST-PANGEAN GONDWANA

1. South American Gondwana
2. Proto-South Atlantic Ocean (north)
3. African Gondwana

The paleoglobe at 144 MYA shows that the tenuous seaway between the Tethys and Pacific Oceans had by now become an established ocean itself. Central Atlantic channels and branches reached between Iberia and France and between the British Isles and Greenland. Near the North Pole a small ocean had developed, and as it continued to widen an archipelago, the North Slope of Alaska, began to rotate and to collide with North America. But again the modern outlines on the paleoglobe are misleading: it is the oval projection that gives the more accurate view of a world of vast continental shelves and shallow continental seas basking in an increasingly balmy global climate.

In this chapter we will look at the *marine* world of 144–99 MYA—the Early Cretaceous. Later the South Atlantic Ocean will begin to open, and coincidentally the Bay of Biscay and the Labrador Sea. In the following chapter we will continue with the *terrestrial* world of 99–65 MYA, the Late Cretaceous, with a generous overlap between the two time frames.

In late 1831, the 22-year-old Charles Darwin arrived at Devonport in southwest England, assigned to join the naval survey ship HMS *Beagle* as the ship's naturalist. The *Beagle* was a refitted brig with a displacement of 500 tons when fully loaded, a crew of 65 naval officers and men, and many of its guns removed to make way for stores and "supernumeraries." The supernumeraries included three homeward-bound natives of Tierra del Fuego who had been brought to England on the *Beagle*'s previous voyage. After several setbacks due to violent early winter gales in the English Channel, the overcrowded ship successfully set sail on December 27. The vessel turned Lizard Point in rough seas and headed into the North Atlantic, bound for the Straits of Magellan near the southern tip of South America, the Pacific Ocean beyond, and a return voyage by way of the South Seas: a circumnavigation of the globe.

Having read Darwin's marvelous account in his *Voyage of the Beagle* (first published in its present form in 1845), one feels that when the ship returned on October 2, 1836, Darwin had grown from a youth, indulging his love of pastimes such as shooting and hunting, into a man far older than his years. In those four years and nine months Darwin accumulated a store of knowledge and experience that exercised his extraordinary intellect for the rest of his life. Curiously, the whole *Beagle* episode was actually the result of the most unpromising set of circumstances, even less promising than Alfred Wegener's presentation of a "hurried" thesis to a fledgling geological society on January 6, 1912.

After his graduation from Cambridge, Darwin was introduced to Adam Sedgwick, professor of geology at the University, by the Reverend John Stevens Henslow, a professor of botany who had popularized biology at Cambridge. Following a brief geological field trip to Wales with Sedgwick, Darwin was asked by Henslow if he would be interested in joining the *Beagle* as the ship's naturalist. He would be unpaid and would have to provide his own equipment and valet-assistant. Dr Robert Darwin, his father, adamantly opposed the proposition, but it was enthusiastically endorsed by his uncle Josiah Wedgwood (son of the founder of the famous Staffordshire pottery). Having challenged his son to find "any man of common sense" to support the idea, Dr. Darwin had little alternative but to capitulate and agree that Charles should participate in the *Beagle* voyage. However, there was a practical difficulty.

According to Leonard Engel, one of Darwin's many biographers, Robert FitzRoy the captain of the *Beagle*, had decided to appoint a friend as honorary naturalist. Having heard that his preferred candidate could not get the necessary leave of absence from his employer, FitzRoy interviewed Darwin, but was deterred by the shape of his nose, believing that it denoted idleness. However, Darwin put an end to his hesitation by producing a persuasive letter of support from a nephew of Lord Londonderry, a distant relative of FitzRoy. So it was by the slenderest of margins and by whim and circumstance that FitzRoy offered the naturalist's berth to Darwin.

The parallel between Darwin's entry into the field of natural science and Wegener's approach to the disposition of continents is irresistible. Both men were catalysts for revolutionary change. Both became central figures in a raging debate—and both were correct in their hypotheses. But at that point the comparison ends. Wegener's account of his ideas in its original form was stilted, perhaps arrogant, and often flawed. In contrast, from the outset of Darwin's account in *The Voyage of the Beagle* one is startled by the freshness and pertinence of his observations even though they were made over 150 years ago. For instance, as the *Beagle* was approaching the Cape Verde archipelago off Dakar on the coast of West Africa (**1**), Darwin commented on the general subject of microscopic organisms—organisms that are now recognized to be of fundamental importance to the very fabric of the natural world.

As the brig neared harbor on the island of San Jago on January 16, 1832, Darwin collected a small quantity of brown-colored dust that had been "filtered from the wind by the gauze at the masthead." The dust had been blown high into the atmosphere over the Sahara Desert and had then been carried far out to sea by the North Atlantic easterlies. The sample was later examined by Christian Gottfried Ehrenberg (1795–1876), of the University of Berlin, the most distinguished biological microscopist of his time. Ehrenberg was a founder of the science of micropaleontology—now a key area in the fields of geology, biology, botany, and climatology. Darwin's sample, together with others collected from vessels up to 2,560 km (1,600 miles) from African shores, was found to contain the fossil remains of 65 species of microscopic "animals and plants." All of these species were previously unknown to Ehrenberg, although he was very knowledgeable about African *infusoria* (as such fossils were then called).

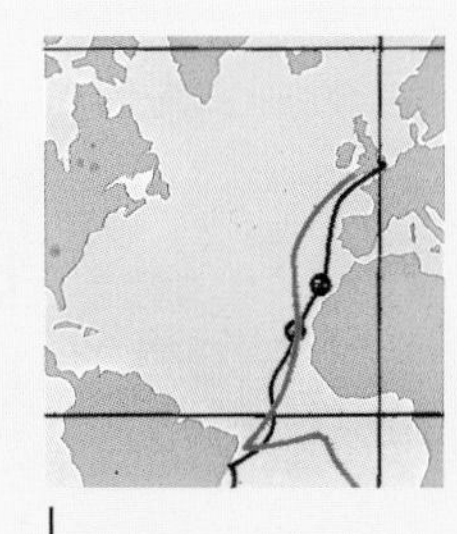

1

In his first chapter Darwin also describes a band of reddish-colored seawater that he had observed as the *Beagle* approached the Cape Verdes. Later he saw similar phenomena in a number of widely separated regions of the world ocean. Darwin attributed these discolorations to the presence of live *infusoria*. He describes how great and uncountable a number lived in the smallest drop of water that he could put under a microscope, and how the organisms burst and multiplied in numbers at an astonishing rate. But it was not just the fact that the sea surfaces off continental coastal regions were streaked for miles with reddish or yellowish or muddy *infusoria* that puzzled him. It was that the demarcation of these colored bands was so clearly defined. What kept the edges so sharp? What made the bands so long and narrow?

Darwin deduced that the blooms must be produced by particular conditions in the ocean, but could not guess their cause. He was in fact observing the limit of upwelling of nutrients from the deep seafloor (**2**). These are created by decomposing *infusoria*, the nitrogenous compounds that fuel both the colorful algal blooms and subsequently the frenzy of feeding upon them in which other inhabitants of the sea engage.

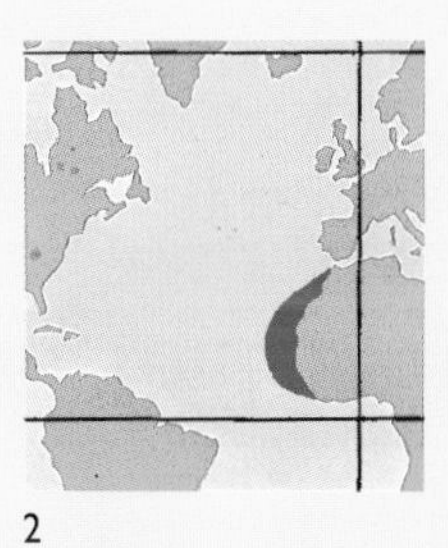

2

Finally, as the ship was entering the harbor in the Cape Verdes, Darwin noted that "a perfectly horizontal white band in the face of the sea cliff may be seen running for some miles along the coast, and at the height of about forty-five feet above the water." He recorded that the rock was formed of "calcareous matter, with numerous shells embedded," that it rested on ancient volcanic rock, and that therefore it must have been formed on the seafloor. In fact what Darwin was seeing was a relatively young terrace of "white" shallow-water carbonates that had formed on the surface of the "ancient volcanic rocks," after the latter had been raised from the seafloor. The volcanic rocks were parts of the Central Atlantic seafloor of 138–131 MYA in which Darwin may have noticed lenses of limestone—carbonates formed from the minute skeletons of *infusoria* in the proto-Central Atlantic Seaway around 165 MYA [Essay 1]. Only in 1840 did Ehrenberg establish that *infusoria* formed such rocks—Darwin did not know that the masthead dust, the colored bands in the sea, and the rocks he now described were linked. Similar deep-sea carbonate lenses, embedded in volcanic rocks or permeated by volcanic dikes, are found on other Cape Verde islands and on Fuerteventura in the Canary Islands. In fact, some carbonate lenses in the Cape Verdes contain deep-sea pillow lavas and shallow-water "hyaloclastites"—grains of basalt that formed during violent volcanic explosions.

These modern discoveries would have fascinated both Darwin and Wegener. This chapter reviews present scientific understanding of those *infusoria* and other denizens of the deep in Cretaceous times. The next chapter will emerge from the sea, so to speak, and review the terrestrial scene.

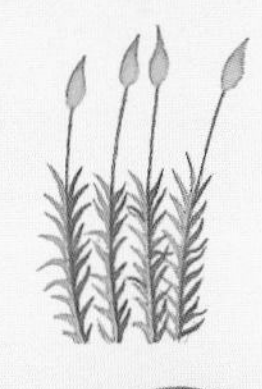

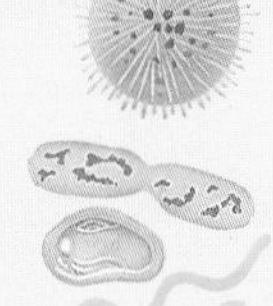

3

The five kingdoms of life on Earth—the Animal, Plant, Fungi, Protoctista (**3**), and Monera (prokaryote bacteria) Kingdoms are

[*continued on page 172*]

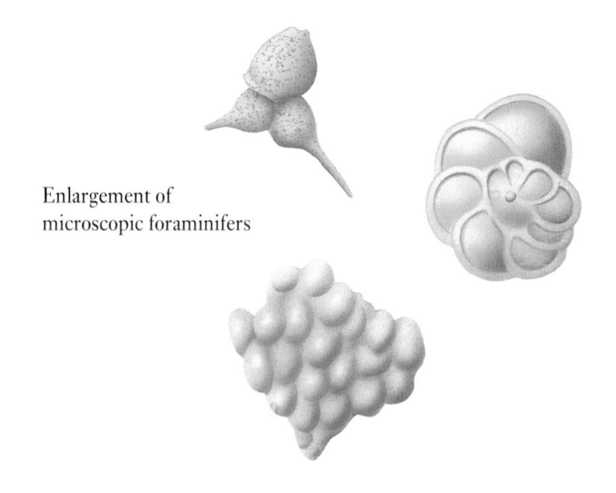

Enlargement of
microscopic foraminifers

1 Central Atlantic

This terrace of carbonate rocks is perched on a cliff top on the western shores of Fuerteventura, in the Canary Islands off the coast of Morocco. The terrace is a petrified seashore now raised considerably above sea level. However, the carbonate rocks in the cliff beneath the terrace and in the peninsula on the horizon were formed on the Central Atlantic seafloor 138–131 MYA. The cliffs are cut by vertical volcanic dikes and, together with the overlying terrace, form a band similar to that first observed by Charles Darwin on the Cape Verde Islands off Dakar, southwest of the Canaries, during HMS *Beagle*'s visit in January 1832.

The terrace and petrified seashore, with its pebbles of volcanic rock in the foreground, are comparatively young. They are no more than a few million years or so old. But the carbonates in the cliffs beneath the terrace are made up of debris that accumulated at the foot of the West African continental slope on the newly formed Central Atlantic seafloor. The debris was eroded from reefs and

carbonate platforms that began to form when the proto-Central Atlantic first separated Laurasia from Gondwana. Submarine volcanic activity is thought to have raised parts of this seafloor to the surface about 20 MYA.

The carbonates were primarily composed of the skeletal remains of minute life forms called "protoctista"—such as the foraminifer shells illustrated. In Darwin's day these organisms were thought to be microscopic animals: they are indeed animal-like and other fossil protoctista are plant-like, but both types of fossil are now considered by most biologists to be forms of life much more primitive than either animals or plants [see page 172].

Protoctista are truly the artisans of the geological world, for they make rocks—over a billion tons of carbonate rocks a year. They also make oil—in Cretaceous times, over a period of 30 MY, they produced perhaps 70 percent of the presently known global reserves of oil.

Nodules of flint, overlooking Beachy Head, Sussex, England

2 Anglo-Paris Basin

The Seven Sisters and the other chalk cliffs that form the coastlines of southeast England and northern France are spectacular testimony to the productivity of countless trillion trillions of coccolith protoctists. Their shells, each no more than a few thousandths of an inch across formed "benthic oozes" on the floors of the Late Cretaceous seas. Under their own accumulating weight, the oozes consolidated into chalk, particularly in an extensive Laurasian depression called the "Anglo-Paris Basin." The white cliffs gave the Cretaceous Period its name (L. *creta*, chalk); this geological period was a warm, balmy "Carbonate Age," the opposite of an "ice age." Similar chalk formations were deposited in the North Sea Basin in Late Cretaceous times, capping oil reservoirs forming in the rocks beneath them, and in some important cases becoming the reservoirs themselves.

There are two basic types of protoctist shells: those composed of carbonates such as chalk, like the coccoliths, and those formed from glass-like silica compounds, like the radiolarians illustrated. The latter were less numerous in the Cretaceous; they either combined with volcanic sediments formed from wind-blown volcanic ash, or with terrestrial dust, such as that blown out to sea from the Sahara Desert, to form red-colored abyssal clays. Or they formed "flint"—an extremely hard, microcrystalline rock of the kind that was flaked and fashioned into arrowheads and tools by our forebears. The flint nodules are formed from silica solutions trapped in cavities in the chalk: such potato-shaped nodules are protruding from the cliff in the left-hand picture. Chalk debris and flint pebbles litter the seashore at the foot of the cliff.

Coccoliths

Radiolarians

Enlargements of microscopic protoctista

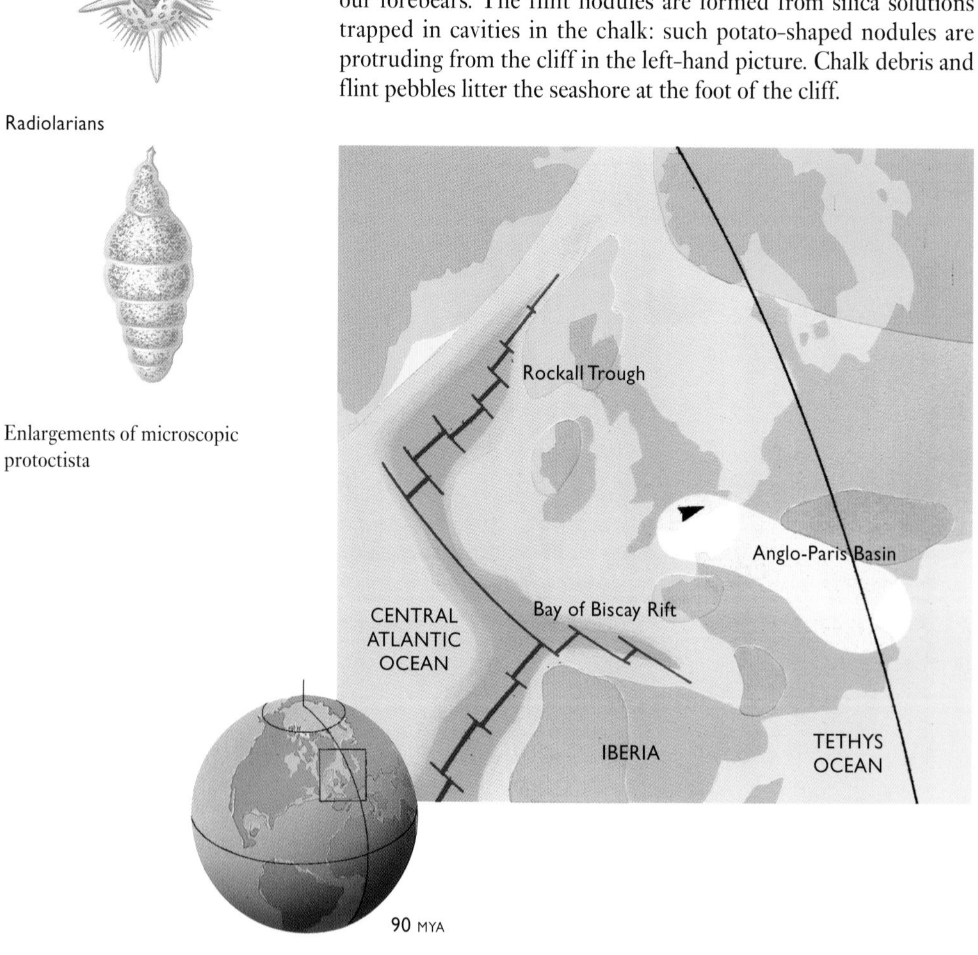

[*continued from page 167*]

illustrated on pages 50–51. Today, Darwin's *infusoria*, and the organisms that formed the Cape Verde and Fuerteventura carbonates, are considered members of the "Kingdom Protoctista."

The name protoctist was coined by John Hogg in 1861 to describe "all the lower creatures, or the primary organic beings: *Photophyta* ... having more the nature of plants, and *Protozoa* ... having rather the nature of animals." The five-kingdom classification, including Protoctista, was suggested in 1959 by R.H. Whittaker; it was then adopted by L. Margulis and K.V. Schwartz in 1982, as a step towards reducing the general confusion that had reigned for years in the classification of tens of thousands of species of unicellular and multicellular aquatic micro-organisms.

Protoctists are the most primitive of all eukaryotes, and like all eukaryotes they are aerobic; they require a supply of oxygen to live, but they can only exist in a watery environment. Many protoctists depend on photosynthesis to sustain life and cannot move about of their own accord, which is why some taxonomists classify them as plants. But without photosynthesis some protoctists feed on nutrients or on other organisms; this requires that they be "motile"—that they are able to propel themselves. They do this in a variety of ways, which is why some taxonomists classify these protoctists as animals.

Some protoctists are unicellular micro-organisms (**4**), such as radiolarians and foraminifers, while others form colonies. Some colonies look like plants but lack vascular systems—such as the red algae and green algae. Other protoctists may be more complex multicellular organisms, but also lack both vascular systems and the tissues that bind and characterize land plants. These are exemplified by the brown algae: kelp and common tide-line seaweeds, for example.

Protoctist organisms are the prime contributors to the food cycle that sustains life in the ocean and other watery environments. Some of them are the prime contributors to marine sediments in seas and oceans, in the form of both skeletal and organic remains. Skeletal remains make sedimentary rock: in special circumstances the organic remains form the base material that produces petroleum. These remains have accumulated at a rate of more than a billion tons each year in modern times, and far faster than that in Cretaceous times (144–65 MYA), the period that we are exploring in this chapter. "Cretaceous" means "chalky" in Latin; chalk was, and is still, produced from the skeletal remains of microscopic, golden, self-propelling, single-celled algal protoctists called "coccoliths."

THE WEB OF LIVING ORGANISMS in the *pelagic* environment (Gr. *pelagos*, the sea) is three-dimensional. It radiates in all directions from the surface and near-surface to the bottom of the sea. There are three distinct life zones in the pelagic environment, and these are categorized according to the penetration of sunlight. The well-lit region of the sea is termed the *euphotic* zone (Gr. *eu phos*, good light: **5**). In ideal conditions of clarity it reaches down to 150 m (500 ft), and this is therefore the greatest depth at which photosynthesis can take place. Organisms that live near the surface of the sea (the upper 200 m or 675 ft) are termed *epipelagic* organisms (Gr. *epi pelagos*, upon the sea). The region beneath that is the *mesopelagic* or middle zone, where sunlight is still sufficient for the human eye to see at around the 330 m (1,000 ft) level, but can be detected below that point only by delicate instruments. The third region, the largest by far, is the deep sea, where sunlight does not penetrate at all—the *bathypelagic* zone (Gr. *bathys*, deep), which starts at 1,200 m (4,000 ft) below the surface and continues into the abyss. The animals and bacteria that live on the seafloor from deep abyss to shallow seashore are *benthic*, and are collectively termed the *benthos* (Gr., the depths).

Plankton (from the Greek *planktos*, wandering) comprise the hordes of organisms (both minute and relatively large) that populate the epipelagic zone and to a lesser extent the mesopelagic zone; their common characteristic is that they have either seriously limited powers of self-propulsion or none at all. Plankton are simply swept by the waves, drift with the wind, or are carried along by ocean

4

5

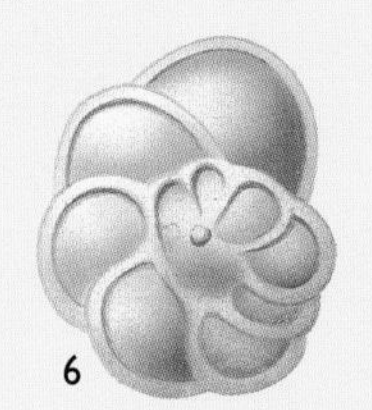

6

currents. They can be divided into two broad categories: protoctist plankton and zooplankton.

As a group, protoctist plankton make up a large proportion of the overall epipelagic population and contribute hugely to the food cycle in the shallows and the open ocean: they are the basic food supply for zooplankton—which, as the name suggests, are classified as animals. Zooplankton include the minute larvae of mollusks, such as clams and snails, crustaceans—such as shrimp-like krill—and other animals that are crafted by evolution to drift with the crowd. Zooplankton also include a multitude of infant and adult transparent animals like comb jellies, jellyfish, and others of both individual and colonial lifestyle. Some colonies have populations of thousands of individuals and form circular or snake-like "galaxies," some 10 m (30 ft) or more in diameter or overall length.

All remaining creatures that swim in the sea and are not hostage to the winds and currents are called "nekton" (Gr. *nektos*, swimming). Nekton, which include all fish and invertebrates like squid and nautilus, feed upon each other and upon plankton.

OF THIRTY DIFFERENT protoctist phyla, just seven, perhaps containing tens of thousands of individual species, contributed to the production of the countless tons of calcium and silica compounds that formed sedimentary rock in Cretaceous times and since. Three of these phyla had silica shells—these included diatoms and radiolarians, and usually frequented colder water. Two had carbonate shells; they usually frequented warmer water and included coccoliths and foraminifers.

These silica and carbonate shells make excellent index fossils (**6**). They do so because their shells were not digested by the animals that ate them. Many of the species evolved quickly, and their variations are readily identified and classified under a microscope. They also had a worldwide distribution. Such fossils are used by paleontologists to deduce the relative time at which the living organisms existed. Other forms were particularly sensitive to environmental conditions and their presence indicates the original climate, salinity, depth, and so on, of a fossil environment.

The remaining two phyla of the seven that "manufactured" Cretaceous rocks were "seaweeds," the familiar green and red algae of today. They had carbonate skeletons, but were anchored to the seafloor, not planktonic. Green and, to a lesser extent, red algae make by far the greatest contribution to the formation of carbonate rocks in shallow benthic environments [Essay 4].

In Cretaceous times as now, dead organisms, their skeletal remains, and fecal pellets from living organisms were consumed by zooplankton or nekton as the debris fell towards the seafloor. The detritus was ingested, redigested, reduced in bulk, and re-excreted several times, depending on the depth through which it fell and the population density of organisms down the water column. Meanwhile, bacteria were at work reducing organic content. Most phosphate nutrients were removed in the epipelagic zone, where they quickly intermixed with seawater and provided immediate nourishment. The even more nutritious nitrogenous compounds were removed from fecal pellets as they passed through the mesopelagic zone. Nitrates were only returned to the epipelagic zone by way of upwelling currents off Central Atlantic continental shelves.

Once the pellets reached the seafloor—which took only a few minutes in the shallows but several *weeks* in the deep ocean—their content was again reworked. This time it was by benthic fauna, such as marine worms and bacteria, which removed further organic matter. This final process left an ooze on the seafloor consisting of minute indigestible protoctist skeletons.

CONSTANT SUBMARINE "heavy weather," in the form of vigorous upwellings such as Darwin had witnessed, disrupted the benthic environment in the Central Atlantic. This turbulence was caused by several factors, including bottom currents, storms and eddies in the deep ocean, and the interaction of these currents with the mountainous topography of the ocean floor, as well as changes in the density of the sea produced by varying temperature, salinity, and turbidity. Prime examples of the latter include the Florida-Bahama carbonate platform and the proto-Mid-Atlantic Ridge.

Whatever its cause, this upwelling resulted in the release of nitrogenous nutrients from the seafloor and their transport to the surface. This produced a feeding frenzy, followed by a reproductive orgy. First the protoctists would feed and bloom in such prolific numbers that they discolored the sea—as Darwin had observed off the Cape Verde Islands and elsewhere. Then hordes of zooplankton would take their fill of the excess of protoctists, while schools of nekton assembled greedily to reap the combined harvest. But the general overpopulation that resulted from this excess frequently resulted in catastrophe for the participants. Too high a population depleted the oxygen in the epipelagic zone, and this was exacerbated by an increase in the release of organic toxins, such as urea. The consequence was a collapse of the ecosystem—either by suffocation or by poisoning of the vast numbers of protoctists, zooplankton, and nekton participating in the frenzy. They died en masse, and their remains then drifted down towards the seafloor in a form of detritus known as "marine snow."

Just as snow lines reflect the overall height and geographical latitude of a mountain range, marine snow forms its own "snow line"—a line of carbonate precipitate on the flanks of marine mountains. The line varies according to ocean depth and the altitude of the marine mountain ranges above the sea floor. Ocean snowfall often behaves in a similar way to terrestrial snowfall. The difference is that marine "snow clouds" are oversaturated with carbonate crystal flakes in place of ice-crystal flakes. As marine snow falls from epipelagic clouds, it settles and accumulates on ocean mountaintops and valley walls in the form of calcareous ooze. But at a particular altitude above the seafloor, the reduced temperature of the ocean water allows the carbonate crystals in the marine snow to dissolve and disappear: carbonates dissolve more readily in cold water than in warm, while terrestrial snowflakes "dissolve" in warm air and form in cold air. In the ocean there is thus a transition from "carbonate-covered" to "carbonate-free" conditions on benthic slopes—a carbonate "snow line" at the depth where seawater completely dissolves calcium compounds in the open ocean.

This snowline depth is called the "carbonate compensation depth." It adjusts to reflect ocean temperature and the amount of carbonate being produced in the epipelagic zone by protoctists. So during times of increasingly warm climates and warm oceans, the Earth will move towards a carbonate-deposition age, and during times of increasingly cold climates and cold oceans, towards an ice age. The Cretaceous Period was a time of high global temperatures, on average perhaps 5–11°C (10–20°F) higher than today's. It was therefore a carbonate-deposition age with an appropriately low carbonate snow line in the Central Atlantic.

The vast depth of carbonate deposits is evident from the amount of chalk in the English Channel cliffs [Essay 2] and in formations beneath the North Sea (**7**). An even more spectacular example of the protoctist carbonate industry in the Cretaceous lies beneath the surface of the modern Central Atlantic Ocean.

In the early Cretaceous (144–99 MYA) the base of the present Florida-Bahama carbonate platform covered an area equivalent to almost the whole of modern France—roughly twice the size of Britain. It was about 1.6 km (1 mile) thick. The familiar outline of the present Florida peninsula did not exist. From the air the whole area, including the Bahamian region, would have looked very much as the Bahama Platform looks today, but far greater in extent. The platform is now well over 5 km (3 miles) thick, and the atolls at its base, formed as Darwin so elegantly described [see page 162], are now deeply indented into the lithosphere. Global sea level during the Early Cretaceous was on average about 45 m (150 ft) higher than today. For this reason the northern Florida section (**8**) of the platform was separated from the Laurasian mainland by a shallow sea-filled depression called the Suwannee Channel. This depression marked the approximate line of the Pangean suture between Florida and North America.

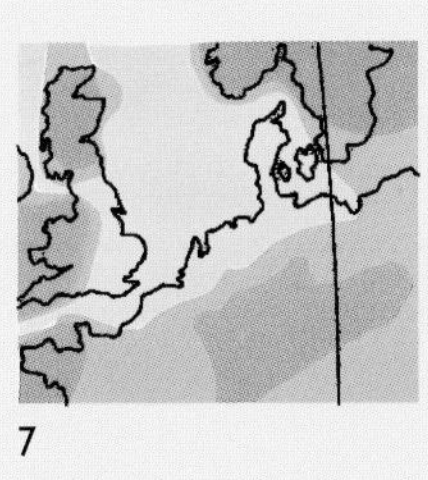

7

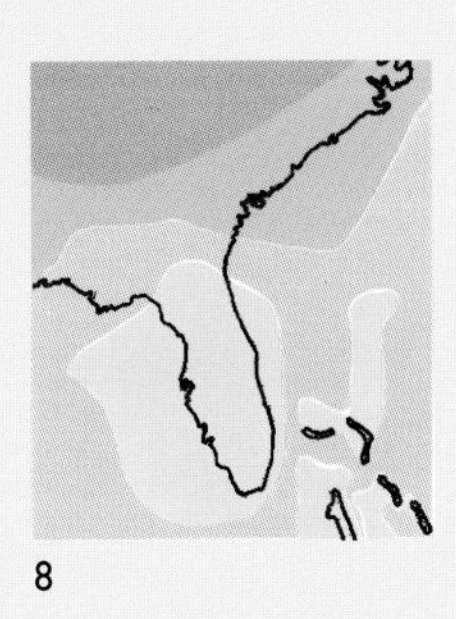

8

9

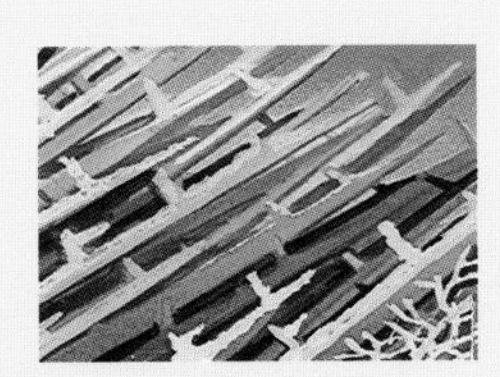

10

The sheer color and extent of the Florida-Bahama carbonate platform during the Early Cretaceous would have been overwhelming [Essay 3]. There were seemingly endless rudist and algal reef-rimmed islands set on the edges of the platform next to the most active regions of the sea (rudists were Cretaceous mollusks of coral-like habit: **9**). These fringing reefs separated the deep channels and open ocean from the shallow interior platform, where the seafloor was marbled by the belts and lobes of tidal sandbars. The overlying shallow sea, animated by light reflected from the shallows, would have been tinted by subtle shades of green, turquoise, and cyan, shot with occasional patches of vivid electric blue. But on the ocean side of the fringing reefs the color would have changed abruptly to a deep, rich, prussian blue flecked with white-capped waves.

In modern reefs competition for sunlight dictates the form of coral growth. In shallow well-lit ocean water, where the force from waves is high, branching corals grow in close-knit and short-cropped communities (**10**). Deeper colonies can be room-sized. The deepest corals need the greatest numbers of zooxanthellae [a form of algae—see Essay 3] to sustain photosynthesis from the fading light. They grow sturdy, horizonal, plate-like branches that reach out for the light—a quite different shape from their shallow-depth staghorn-shaped brethren.

The labyrinthine structure of the fringing reef attracts many encrusting organisms, such as green and red calcareous algae, and the more delicate corals that grow in sheltered places, as well as fish and invertebrates that eat the polyps and excrete feces rich in carbonate compounds. The excretions help to bind and cement the reef.

The branching corals form the reef structure, while the steady flow of detritus from dead organisms and their skeletal remains creates a waste chute on the reef wall beneath the branches. The chutes shape the contour of the reef into a pattern of channels and buttresses while disposing of waste materials that would otherwise smother the sensitive polyps and kill the reef. Detritus accumulates as a terrace beneath the widespread branches of the lower reef, where it is reworked by organisms such as sponges and bacteria. Beyond the terrace a near-vertical wall plunges into the abyssal darkness of the North Atlantic Ocean.

The Florida-Bahama bank, with its protective reefs and interior carbonate platform [Essay 4], is a supreme example of limestone production in the epipelagic zone of oceans and shallow seas during a carbonate age. In the last few fleeting seconds of geological time, this is the rock that has been quarried to build cities and skyscrapers, marble halls and national monuments. It is also the rock that has been pulverized to produce concrete, which makes possible the construction of such buildings and the highways to serve them.

But protoctist carbonates are of fundamental importance in yet another way. During the initial breakup of Pangea, protoctists were primarily responsible for the inconceivable quantities of organic waste that were preserved and later matured into oil and gas—the main bulk of the world's fossil fuel reserves today. These reserves were mostly created in the special conditions of the Central Atlantic Ocean and the connecting Tethys Ocean during a well-defined period of time in the Cretaceous.

In Early Cretaceous times (144–99 MYA) the Central Atlantic separated post-Pangean Laurasia, the northern continents, from post-Pangean Gondwana, the southern continents; the South Atlantic Ocean did not exist. The Central Atlantic stretched from

[continued on page 176]

95 MYA
Suwannee Channel
Equatorial currents
Florida Carbonate Platform
Bahama Platform

Eleuthera Cay, Bahama Islands

3 Fringing Reefs

The Florida-Bahama carbonate platform of Early Cretaceous times (144–99 MYA) was roughly 640 km (400 miles) wide and 1,440 km (900 miles) from north to south. In its Bahama region, it was cleft by deep channels scoured by equatorial currents. This immense peninsular structure was made possible by marine organisms that built protective fringing reefs and back reefs along its wave-washed margins—similar to those pictured here. Cretaceous reefs were largely built by rudists, specialized mollusks that could withstand high-energy wave action.

Modern coral reefs are built by polyps, soft-bodied animals that secrete a particularly tough and fibrous carbonate mineral called aragonite, which they deposit in the form of radial platelets. The platelets accumulate as cylindrical branches of a coral skeleton, with openings through which the animals fish for food with their tentacles. Fringing reef corals have many branches, and a multiplicity of such corals combine to form the fabric of the reef. The structure of this fabric is dictated by the position of individual branched corals relative to penetrating sunlight—they vary in pattern like the interwoven branches of trees in a forest.

Polyps require clear, sunlit seawater at not less then 21°C (70°F). Each polyp is symbiotically dependent upon the presence of zooxanthellae, protoctist algae that live and reproduce in the cell walls of the polyp gut. Zooxanthellae are fundamental to the existence of the reef. They photosynthesize sunlight and so produce carbohydrates that supplement the polyps' diet, and serve to enhance their rate of calcification.

The back-reef community is dominated by brain corals resistant to turbid tidal water. They form patches of coral growth on the back-reef floor. Debris from these "patch-reefs" accumulates to form a supporting buttress behind the fringing reef, and together they protect the vast interior platform from the onslaught of ocean waves. Rudists, as well as corals, performed the same function on Cretaceous carbonate platforms.

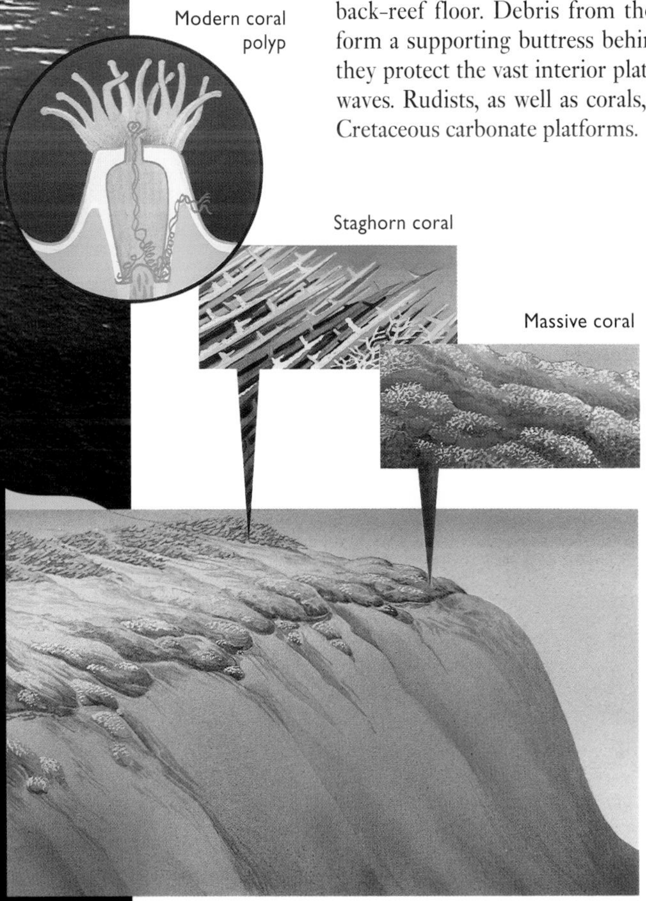

Modern coral polyp

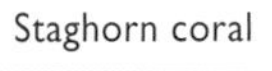

Staghorn coral

Massive coral

Rudists

Fringing reef

[*continued from page 173*]

the old Tethys Ocean in the east via the Strait of Gibraltar to the Pacific Ocean in the west. As a consequence of this geography, and with no landmasses to divert them, equatorial currents flowed around the globe from east to west (**11**). Although a surface connection, and therefore an oxygenated epipelagic zone, was kept open between the Tethys Ocean and the Pacific Ocean via the Central Atlantic *throughout* Cretaceous time, bottom circulation was inhibited by constrictions in the region of the Strait of Gibraltar and shallows adjacent to the Bahama Banks for a period of about 35 MY [see Essays, pages 154–55 and 156–57]. With this vital source of oxygen cut off, much of the Central Atlantic bathypelagic zone and the by now fast-closing Tethys Ocean became "anoxic" during this period. They were frequently devoid of oxygen (some think in a cyclic pattern) until the opening of the South Atlantic Ocean.

There were two deep basins in the Central Atlantic Ocean throughout this period—one west of the Florida-Bahama Platform (the Americas Seaway basin, now the Gulf of Mexico and the Caribbean regions), the other east (the proto-Central Atlantic basin, now the Central Atlantic Ocean). More than 3,600 m (12,000 ft) deep in parts, each was connected by a submarine spreading-center (**12**), a deep oceanic rift at the foot of the dog-legged Florida-Bahama Platform, itself immersed in relatively shallow sea [Essay 5].

During the early part of the Cretaceous the equatorial current that refreshed the Central Atlantic flowed through a deep but comparatively narrow trough that cut through the Strait of Gibraltar. About 125 MYA the trough became constricted to a point where *bottom* currents flowing from the Tethys Ocean to the east were first reduced and ultimately blocked. Meanwhile, the hot Cretaceous climate had caused excessive evaporation of shallow seas on the continental margins, creating excessively salty "supersaline basins" on continental shelves. These supersaline shallows were either confined to nearshore or offshore carbonate basins, or they poured off the shelf down continental slopes to the ocean floor, where the heavier saturated salt solutions accumulated. With no bottom currents to disturb stratification or cause intermixing with oxygenated bottom water, the bathypelagic zone in much of the Central Atlantic became devoid of oxygen—and thus lifeless. Similar conditions existed in the Gulf of Mexico region of the Americas Seaway basin to the west of the Florida-Bahama Platform; it is not known whether the proto-Caribbean seafloor was similarly affected or not as it has since been destroyed.

As WE SAW EARLIER [see Essay, pages 122–23], under anoxic conditions the organic content of marine snow cannot decay when it accumulates on the seafloor, because the oxygen-dependent organisms that normally consume or reduce such materials cannot live in anaerobic conditions. As a consequence, from 125 to 90 MYA the prolific marine snowfall in the Central Atlantic was preserved on epicontinental seafloors, in depressions on continental margins, and on the deep ocean floor, in unprecedented quantities. It permeated carbonate rocks that formed there, and found its way into the anoxic shales (consolidated silt or mud or clay) of marine environments.

As the preserved organic remains of Cretaceous protoctists accumulated to considerable thickness they were converted by heat and pressure into "kerogen" (Gr. *keros gen*, wax-producing). Generally during deep burial, kerogen was converted into a greasy, black, solid form of hydrocarbon, that subsequently converts to oil and gas—hence "fossil fuels." Kerogen forms in "source rocks," intensely compressed limestone or shale that contains organic remains. As kerogen matures under continued heat and pressure, it tends to liquefy, beyond a certain point degrading completely into methane—"natural gas." The degrees of maturation, together with the original composition of the kerogen base, largely determine the grade or quality of the end product.

As kerogen liquefies into oil, or the oil gasifies, oil, gas, or a mixture of the two begins to seep into neighboring porous rocks. In fact

[*continued on page 178*]

11

12

Oolite banks off Exuma Sound, Bahama Islands

4 Interior Platform

As can be seen here, tidal washes surge onto the surface of the Bahama Platform where it is not protected by reefs. Within these washes, seawater warms up, and its capacity for holding carbonates in solution correspondingly falls—so minute crystals of carbonate will form around any available nucleus, such as a fragment of shell. The crystals of carbonate continue to grow and aggregate together in a "mixing zone," until they eventually form spherical grains called "oolites" (Gr. *oion lithos*, egg stone), which accumulate on the platform surface.

Although oolitic sands were an important constituent of the Cretaceous Florida-Bahama Platform, the prime contributors to its structure were a group of protoctists—green colonial algae called Chlorophyta that grew profusely in the warm shallows and relative quiet of the interior platform, as indeed they still do. Chlorophyta and other forms of marine algae are commonly called "seaweed," but in this case each weed is a multicellular colony of individual organisms, and each colony has an extraordinary reproductive drive. Chlorophyta photosynthesize sunlight and thus generate both an inexhaustible supply of carbohydrates for the platform's food web, and an inexhaustible supply of oxygen that refreshes very warm shallows that might otherwise become stagnant.

The walls of individual Chlorophyta cells are encrusted with calcium carbonate and silica, which are indigestible and are therefore excreted by the marine invertebrates that eat the Chlorophyta. They are also left as indigestible fragments by bacteria and other micro-organisms that decompose the algae as they die. Excreted pellets sometimes form cemented sand grains called "grapestones," but mostly they are reworked by bacteria that reduce them to "pelletoidal" mud and ooze.

Under the pressure of accumulation, the muds, oozes, grapestones, and oolitic sands, plus fragments of shell, coral, and other animal debris, consolidate into "limestone." By 118 MYA the Florida-Bahama Platform was the most formidable structure in the Central Atlantic realm: its size and bulk largely determined the shape of the present Caribbean basin.

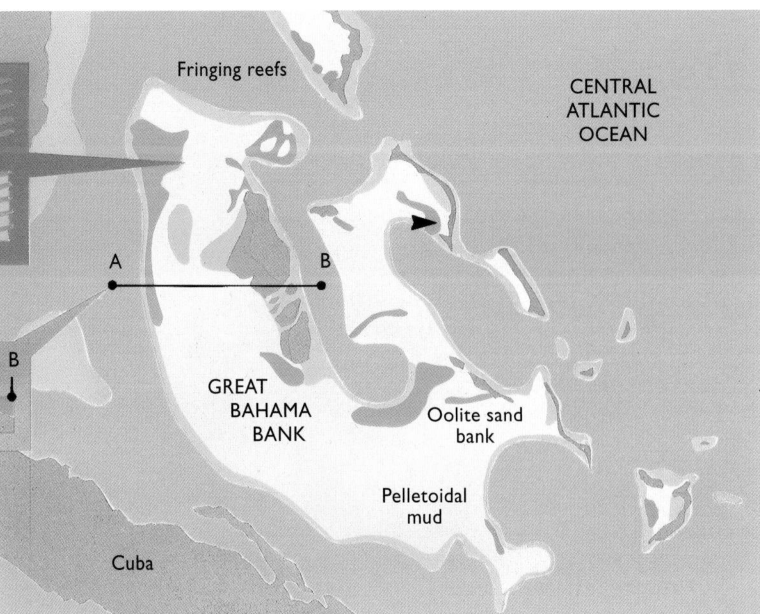

[*continued from page 176*]

the kerogen expands as it matures and is thus expelled from the source rock under its own pressure. The maturing hydrocarbons can only migrate into relatively porous neighboring sandstones or carbonates, such as buried fringing reef structures. Such porous formations are called "reservoir" rocks (**13**). They can be immediately above or beneath the source rock, or on the same plane, or at an angle to it. Permeability, not gravity, is the key to migration.

It follows that reservoir rocks can be considerably younger, older or of the same age as the source rocks. But once contained in porous and permeable rock the oil or gas, still under enormous pressure, will migrate, perhaps via cracks and faults, along the line of least resistance, generally towards the surface—until it escapes to the surface or is trapped beneath it.

After migration from their source rocks, hydrocarbons are said to be "trapped" in a reservoir rock when the route to the surface is blocked by impermeable rocks or minerals—when the reservoir is "sealed" by a "cap rock." The most productive reservoirs are capped by impermeable mudstone, or by shale or chalk, or by salt and other evaporites sometimes 1,500 m (5,000 ft) or more thick. Both source rocks and reservoir rocks are still subject to tectonic processes, and so can be folded, fractured, faulted, overthrust or otherwise disrupted while the kerogen is maturing into hydrocarbons, often contributing to its migration and trapping.

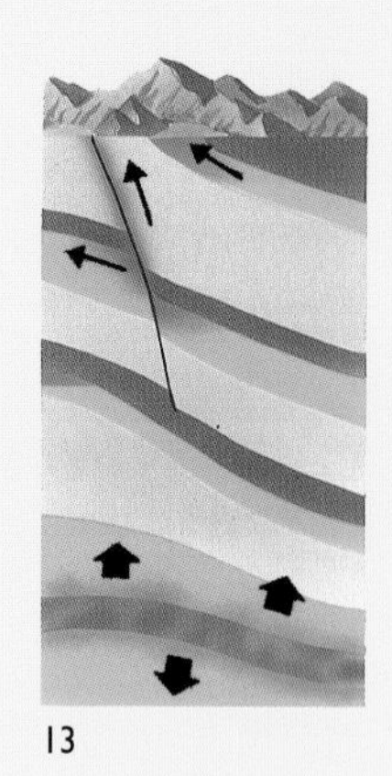

13

14

KEROGEN ACCUMULATED on an incredible scale from the start of the breakup of Pangea and the formation of the Central Atlantic Ocean. Worldwide, the accumulation could account for about 70 percent of the present petroleum reserves, according to estimates by G. J. Demaison and G.T. Moore. It gathered in offshore basins like those in the Gulfs of Mexico and Arabia (**14**). It accumulated in the general region of West and North African Gondwana on the shores of the now shrinking and confined Tethys Ocean, and in the Cretaceous muds of the failed rifts and grabens of the North Sea Basin. Some of this oil migrated into Jurassic river delta sandstones and was capped and sealed by Permian salt from the Zechstein Sea or by compacted coccolithic remains—chalk like that which formed in the Anglo-Paris Basin and grows the best champagne grapes. The hydrocarbons produced by Darwin's *infusoria* during this period fuel the progress of the present Western World.

While the South Atlantic Ocean did not exist during much of Early Cretaceous time [Essay 6], the general breakup of post-Pangean Gondwana resulted in a shallow seaway being formed between Africa and South America by 118 MYA; by 84 MYA the young South Atlantic was an established ocean, and about a third of its present width. This evolution modified the east-west circulation pattern of the Central Atlantic Ocean, which was now refreshed by a flow of oxygenated bottom water drawn via the South Atlantic from South Polar regions. So it was that after 35 MY of cyclic deprivation the Central Atlantic benthic community was able to regularly breathe oxygenated water once more and gradually to re-establish itself. Meanwhile, the epipelagic nekton community had thrived in the Central Atlantic realm and so, in this overview of life in Cretaceous seas, we now turn to larger fry.

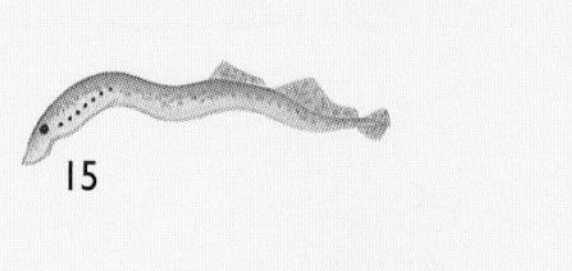

15

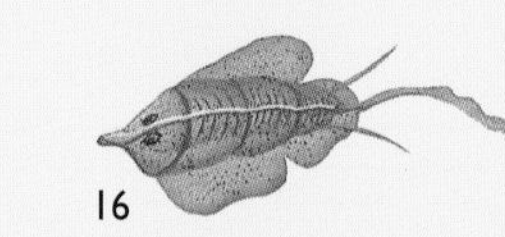

16

FISH ARE EXTRAORDINARILY DIVERSE ANIMALS, yet only four basic types of fish have evolved in the almost 500 MY of their existence: the jawless, the cartilaginous, the lobefin, and the rayfin. This limitation is a result of "convergent" evolution—the style of evolution that results when the conditions for life are similar for all participants. Fish are essentially similar in appearance and structure because they evolved in a stable environment. They evolved in a three-dimensional world in which hydrodynamics and swift maneuverability, in near-weightless conditions, had become paramount factors in their evolution. In contrast to fish, terrestrial animals were bound by gravity during their "divergent" evolution. They suffered extremes of environmental stress in a great variety of circumstances that were forced upon them. This resulted in an evolutionary process that produced great differences in both appearance and structure (as we shall see in the next chapter).

Perhaps it is not surprising, therefore, to learn that jawless fish are very much in evidence today—over 500 MY after their first appearance (traces of early jawless fish have been found in Cambrian rocks aged 510 MYA). Sometimes called "slimefish," these animals (**15**) are the sucking, scavenging, eel-like hagfish and lampreys (of loathsome feeding habit), which are the sole survivors of the original toothless, soft-bodied, fish-like creatures that appeared in Cambrian seas. But with the evolution of jaws in Early Silurian times (438–428 MYA), the life of the fish changed. They no longer had to rely on the consumption of protoctists in the euphotic zone, or food particles on the seafloor, or soft tissue of other animals. They could now pursue prey, seize it in their jaws, and manipulate the prey into the best position for swallowing.

The early jawed fish were armored with plates, which later developed into the now-familiar scales. Their skeletons were cartilaginous, not bony, and because of this, and the general design of their fins and tails, the more advanced varieties are rather confusingly termed "spiny sharks." Despite their name, most paleontologists believe that these fish actually evolved into the first bony fish, and not into the commonplace cartilaginous fish of today—the sharks, dogfish, and rays.

Grand Banks, Bahama Islands

CARTILAGE IS A TOUGH flexible material similar to bone, but lacking the mineral apatite that causes it to harden into bone; the only mineralized parts of cartilaginous animals were (and are) their teeth and skin. The skeletons of all other vertebrates convert from cartilage to bone as their embryos develop. It is thus reasoned that cartilaginous fish (**16**), which first appeared about 410 MYA, evolved separately from bony fish and were not, as long thought, ancestral to them.

In addition to being cartilaginous and having a tendency towards being warm-blooded, modern sharks give birth to live young and deliver them tail first from a placenta-like sac. Some modern sharks are little changed from their Cretaceous counterparts. It is quite likely that shark pups were born in the quiet shallows of the interior Bahama Platform in Cretaceous times, just as they are delivered today. But like terrestrial animals, sharks and rays are not naturally buoyant (**17**); they largely depend on movement to keep themselves afloat in water, and must either swim or slowly sink to the bottom of the sea. Also, because they only have gill slits, not "flapping" gills like bony fish, they must swim continuously to respire—to "breathe."

17

The primitive scale-covered bony fishes, the lobefins, which first appear in the fossil record about 405 MYA, appear to have been buoyant. They had internal air sacs on the underside of their bodies that may have enabled them to move to different water levels. Today's lobefins are a few species of lungfish and the coelacanth—only survivors of a once-diverse group (**18**) that led to the evolution of amphibians, and through them to reptiles, mammals, and humans.

18

The most diverse group of modern marine and freshwater bony fish is descended from the rayfins, which evolved from the so-called "spiny sharks" in the Late Silurian (421–408 MYA), and had an explosive evolutionary surge in Cretaceous times. Rayfin predominance arose from several evolutionary advances. These included the development of lighter scales, improved hydrodynamic shape including more symmetrical tail fins (the earliest known rayfins were herring-like in appearance), and tails that allowed the fish to steer in a straight line. Paired fins at the front and rear of the body stabilized the fish, propelled it through the water, and allowed it to change direction as it moved. But the rayfin's outstanding achievement was the evolution of a primitive air sac into a "swim bladder"—an automatic buoyancy control device.

The swim bladder is a small inflatable sac located between the rayfin's gut and backbone, and is filled with gas that changes volume according to water pressure. Because the sac wall is covered with thin-walled capillary blood vessels, blood passing through these walls can remove or replace gas in the bladder and so adjust the fish's buoyancy as it undergoes changes of pressure at varying depths. The earliest surviving representatives of the rayfins that developed this mechanism are the sturgeons (**19**), of caviar fame, and the paddlefish; both species have changed little in their general appearance since Cretaceous times.

19

But these ancient rayfins are now vastly outnumbered by the "teleosts"—the salmon and trout, the cod and flounders, the herring and mackerel, and a multitude of other familiar rayfins whose ancestors originated in Cretaceous times. They account for about 28,000 marine and freshwater species. These rayfins are called teleosts (Gr. *teleos*, perfect, the ultimate: **20**), because to taxonomists they represent a plateau of perfection, a supreme example of the adaptation of one class of animals into every conceivable habitat and niche available to them.

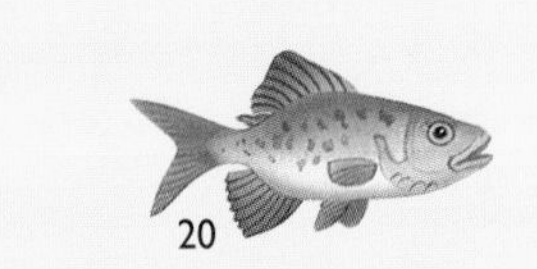
20

The main anatomical features that distinguished the early teleosts from more primitive rayfins were their concertina-like gills. With expandable gills behind their jaws, increased volumes of oxygenated water could be made to flow through teleost mouths and over the much-enlarged surface areas of their gills. This process replenished the fish's blood-oxygen supply much more efficiently. The result was a better means of boosting muscular activity, and therefore of increasing the potential speed of the fish through the water, which in turn resulted in further improvement to their hydrodynamic design.

The teleosts could also draw their prey towards them by pursing their tube-shaped mouths and flexing their enlarged gills. Good camouflage, maximum energy efficiency, and the ability to be highly selective made the teleosts extraordinarily effective predators. Highly carnivorous teleosts, pike for example, developed powerful jaws to grab at their prey. If the predator itself was threatened, the fish's acceleration was remarkable—as any angler knows.

A hundred million years after they surged to dominance in Cretaceous seas, the teleosts would lure the maritime fishermen of Western Europe to venture across the Atlantic Ocean. The fishermen sailed from the Bay of Biscay and the eastern Atlantic to the Labrador Sea and the western Atlantic—and eventually colonized the shores of Newfoundland and Labrador early in the 16th century. But in Cretaceous times teleosts had other and perhaps less devastating predators with which to contend, among them the ammonites and the marine reptiles.

Ammonites are crucial to paleontologists as index fossils; they were marine invertebrates, mollusks of very diverse habit that swam and lived on the seafloor, in fore reefs (**21**) and probably in the open ocean. They built and lived in often decorative coiled shells which, according to species, varied in size from minute to huge—a fraction of an inch in diameter, to several feet. Ammonites were cephalopod mollusks (Gr. *kephale*, head; *pod*, foot: thus a head surrounded by tentacles). They first appear in the fossil record about 400 MYA, and were completely extinct by the end of Cretaceous time (65 MYA). In all, thirteen orders of cephalopods evolved in the period from Late Cambrian (from 505 MYA) to the present. Each order had its subdivisions of families, genera, and species. The whole cephalopod class embraces a *vast* number and variety of animals whose fossils contribute mightily to establishing the relative time scale in Paleozoic and Mesozoic times.

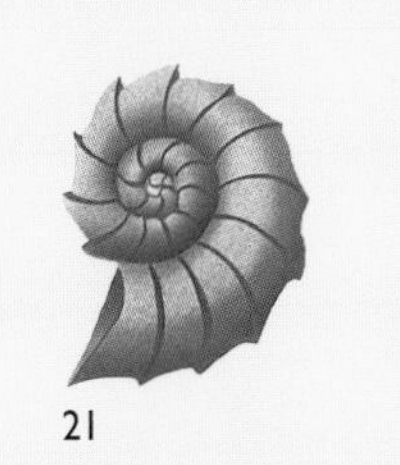
21

Today's cephalopods include all species of squid and octopus. The last remaining cephalopod living in a shell is the beautifully designed nautilus (**22**), a distant relative of the ammonites. About a dozen species of nautilus live in the modern ocean, principally in Malaysian, Australian, Japanese, and Pacific island waters.

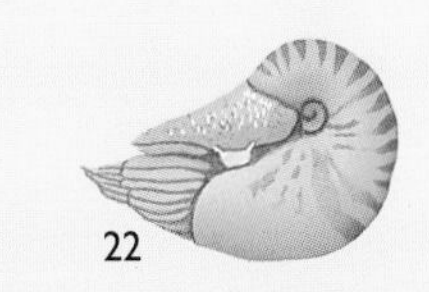
22

Fossil nautiloids going back to Late Cambrian times around 505 MYA are distinguishable from ammonites by two main characteristics: ammonites had convoluted internal chamber walls (called "septa"), while the modern nautilus has smooth-walled septa. The siphon conduit (the "siphuncle") that connects the chambers within the coiled shell allows the living animal to control its buoyancy by transferring fluid from one chamber to another. It follows the external line of the coil in ammonites but passes through the center of the septa in the nautilus. The modern nautilus is restricted to a depth of about 540 m (1,800 ft): their shells collapse (implode) under pressure at depths around 690 m (2,300 ft). They also cannot tolerate water temperatures above 27°C (80°F), two factors believed to control their present distribution. Similar limitations may have controlled the distribution of ammonites in Cretaceous times.

In pre-Pangean times some nautiloid cephalopods grew elongated conical shells and a few rare specimens of such fossilized shells are 13.6 m (2 feet) or more in length. Most of the Paleozoic species were extinguished during the great Permian extinction. Some ammonites, which had evolved around 400 MYA, survived this extinction, and they reached their zenith in both numbers and variety in post-Pangean times—especially during the formation of the Central Atlantic Ocean and during the existence of the epicontinental seaways which followed.

Fossil ammonite shells, in all their infinite variety, are therefore catalogued and used by paleontologists as index fossils (**23**). Ammonite species are set in their correct order of appearance and disappearance in the fossil record. In this way a chronology is produced that makes it possible to judge the relative age of other marine fossils in the same or in allied rock formations. The ammonite catalog (and similar indices of the protoctist foraminifers) also provides evidence of the age, depth, extent, and temperature of epicontinental seas through Cretaceous time, and so plays an important role in the reconstruction of Cretaceous paleogeography.

23

Marine reptiles were the Cretaceous equivalent of today's marine mammals. Marine reptiles were air-breathing; they were basically land animals that had adapted to life in the sea. They are divided into two main groups: those that lived their entire lives in

[*continued on page 184*]

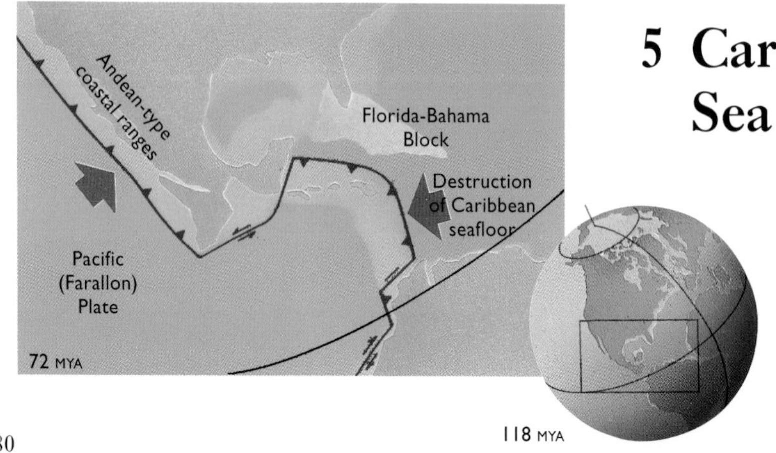

5 Caribbean Sea

The proto-Caribbean, the Americas Seaway featured in earlier essays, reached its maximum width (2,800 km or 1,800 miles) and depth (4,000 m or 12,000 ft) as the South Atlantic Ocean began to open. As South America and North American Laurasia swung away from Africa in unison, seafloor spreading in the proto-Caribbean came to an end.

Up to this point, part of the Pacific Ocean floor called the Farallon Plate had been subducting beneath the proto-Caribbean seafloor as well as beneath the west coasts of South America and North America (Figure 1: 118 MYA). Subduction caused volcanoes to erupt along the continental margins. This resulted in the formation of the Andean, Mexican, and Californian coast ranges,

and caused a volcanic island arc to form on the proto-Caribbean seafloor. The island arc included parts of the future Greater Antilles—Cuba, Hispaniola, Jamaica, and Puerto Rico.

Later, an oceanic plateau similar to modern Iceland was conveyed towards the area by the Pacific Ocean floor. The plateau collided with the island arc, and it is thought that as a consequence the direction of subduction switched 180°. In effect a shallow part of the Farallon Plate now began to invade the region and to override the Caribbean seafloor, conveying the Greater Antilles island arc with it (Figure 2: 84 MYA).

In this panorama we are facing due west. The scene is analogous to the spectacle about 72 MYA (Figure 3), when the Greater Antilles were being carried by the Farallon Plate towards an "immovable object"—the Florida-Bahama Platform. Parts of the Greater Antilles island-arc system overthrust the leading edge of the carbonate platform's basement rocks to form the upper sections of modern Cuba, much of Hispaniola, and all of Puerto Rico. During this collision, the platform was raised above sea level, resulting in a regional hiatus in the laying down of sediment (and causing an unconformity that dates from 60–50 MYA).

6 South Atlantic

This panorama is of a "seafloor" spreading-center that crosses São Miguel, in the Azores archipelago of the Central Atlantic Ocean. We are looking west across a volcanic zone that stretches from left to right in the picture. Although far from obvious to the eye this is an anomalous spreading-center. It is anomalous because it considerably offsets the Mid-Atlantic Ridge (MAR) on the ocean floor to the west—yet the region we see here mimics the spreading action of the MAR. The two halves of São Miguel are moving apart at the same rate that Africa and North America are separating—about an inch a year. Other islands in the Azores group to the east of the MAR are similarly affected.

The volcanic rocks of São Miguel in the map of the island which appears on page 183 are shown in two colors. Those at the extremities were formed in the anomalous spreading zone at a time when the magnetic pole was reversed. They therefore bear a magnetic imprint aligned to the south. The Earth's magnetic field switched from south to north about 700,000 years ago, so the volcanic rocks that have formed in the center since that time have a magnetic imprint aligned to the north, and are differently colored on the map. The Central and South Atlantic Ocean floors (and all others) are similarly imprinted with roughly parallel bands of either south (reversed) or north (normal) magnetized basalt that originated in a central spreading zone.

The zebra-striped portion of the diagram to the right relates magnetic reversals on São Miguel to a chart of such reversals dating back 150 MY to the period when South American and West African Gondwana were still attached. The paleoglobes show three stages in the opening of the South Atlantic Ocean, from 144 MYA when rifting began to 84 MYA, by which time the ocean was well established. The timing of the opening stage has been established by dating the oldest magnetic striping immediately adjacent to the modern South Atlantic seafloor spreading-center.

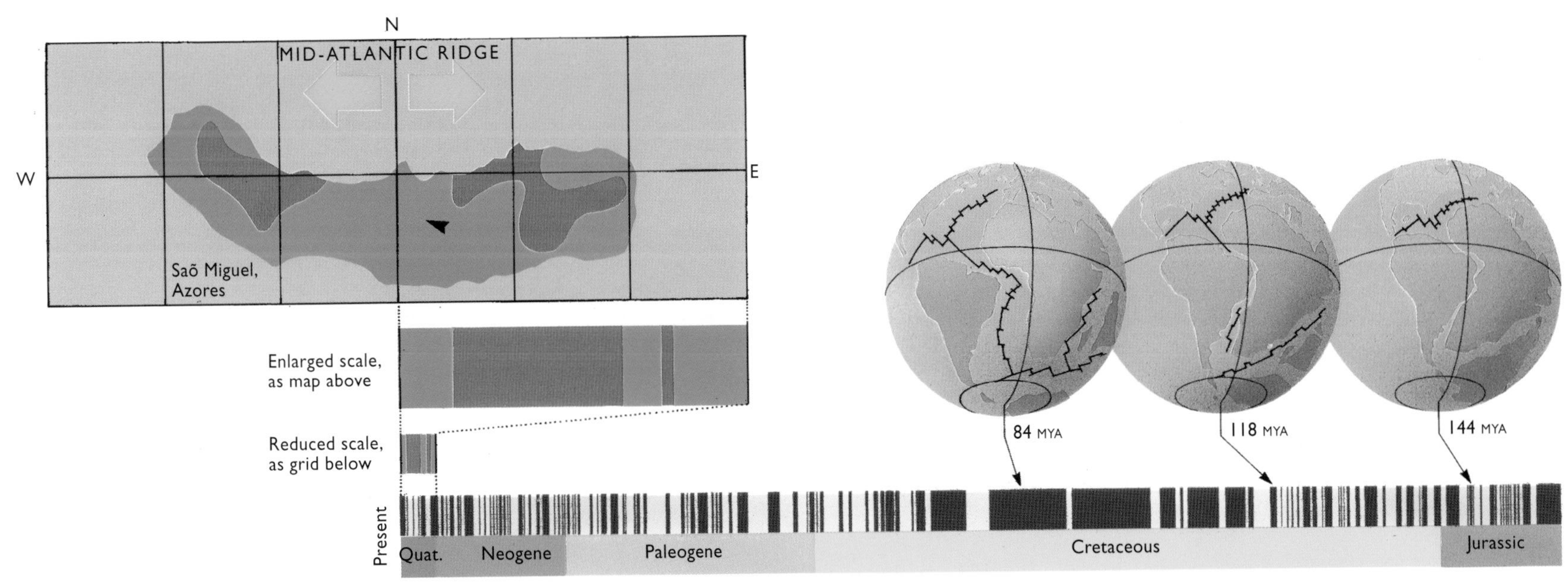

N
MID-ATLANTIC RIDGE
W
E
Saõ Miguel, Azores
Enlarged scale, as map above
Reduced scale, as grid below
84 MYA
118 MYA
144 MYA
Present
Quat.
Neogene
Paleogene
Cretaceous
Jurassic

[*continued from page 179*]

the sea and those that are thought to have laid their eggs on land. The ichthyosaurs (Gr. *ichthys sauros*, fish lizard) were the former, and the plesiosaurs (Gr. *plesios sauros*, near-lizard) the latter, but their ancestral lineage is unknown. However, their reason for adapting to a marine life was the simple attraction of a rich source of food.

The ichthyosaurs (**24**) first appeared in late pre-Pangean times and, like modern sharks, gave birth to live young: fossils of ichthyosaurs have been discovered with a fetus in place at the moment of its birth. Whatever their origin, at some point in their evolution ichthyosaurs abandoned the use of paddle-like flippers as a prime means of propulsion in favor of a fish-like tail that was shaped like that of modern sharks and tuna, but anatomically distinct.

Ichthyosaur tails were capable of lateral movement that may have propelled the animal through the water at speeds up to 25 mph (or so it is believed). Researchers have discovered that some sharks and all tuna (both descend to great depths) are several degrees warmer in body temperature than the water in which they swim. This is necessary to allow the animal to change from warm to cold water to chase prey or to avoid predators, while sustaining a comparable speed. Some sharks and all tuna are "regionally" heterothermic: they possess some degree of warm-bloodedness (according to researchers A.R. Cousins and R. Bowler). The apparently similar speed and tail shape of the ichthyosaurs raises the possibility that they too were heterothermic—an intriguing but perhaps unanswerable question.

The ichthyosaurs reached peak diversity during the breakup of Pangea, but were almost extinct by the time the South Atlantic Ocean became established about 90 MYA. It is thought likely that they succumbed to the growing number of avaricious "modern" sharks that infested the epicontinental and deep seas of that time. However, there could have been a more subtle cause that is not evident at this distance in time.

The plesiosaurs (**25**) were an entirely different type of animal, with savage teeth (they fed on fish and cephalopods), and deep sturdy bodies up to 12 m (36 ft) or longer. They had either short tails, long necks and flippers (of the character attributed to the Loch Ness Monster), or short necks and flippers. In either case, the flippers, which had evolved from reptilian limbs into huge paddles, are believed to have been used to drive the animals through the water beneath the surface in a form of submarine flight. This may have resembled the "flight" of a modern penguin swimming beneath the surface. But it is thought that plesiosaurs had to return to land to lay eggs, and if they did so, they probably followed the seagoing reptilian habit still practiced by turtles. They had to engage in an ungainly struggle up a beach on a heavily reinforced rib cage to deposit eggs in the sand above the high-tide mark. Here eggs were left to hatch and later young hatchlings had to fend for themselves. Plesiosaurs first appeared in Early Jurassic times and survived throughout the Cretaceous (208–65 MYA).

MOST OF THESE EVOLUTIONARY EVENTS took place in Cretaceous times when all the Earth's continents were flooded to a considerable depth by epicontinental seas. The flooding was caused by a combination of factors. There was no ice at the poles and a high degree of activity at swollen spreading ridges. The extent of inundation and fluctuation in specific continental regions varied, but the general inundation resulted in the flooding of up to 60 percent of the Earth's land surface as we know it today. This had as profound an effect on terrestrial evolution in Late Cretaceous times as such a degree of flooding would have on *our* world today. Late Cretaceous seas confined animals such as the dinosaurs, and plants such as angiosperms, to strictly limited regions. These limitations caused environmental stress and enforced competition. In contrast to the convergent evolution in spacious Cretaceous seas, the stress of confinement on land induced divergent and environmentally specialized styles of evolution—as we shall see in the next chapter.

Aiguilles Rocheuses, Belle Isle off southern Brittany

7 Bay of Biscay

This is Aiguilles Rocheuses on Belle Isle, off the Brittany coast—a remnant of Armorican basement rocks that once flanked the Bay of Biscay rift. The modern bay is bounded by the coasts of Brittany and northern Spain; they meet at the apex of an angle, from which the Pyrenees extend to the shores of the Mediterranean Sea. The bay itself is a remnant of an ocean that failed to develop. Nevertheless, its formation about 118 MYA was a key step in the shaping of Western Europe.

As North American Laurasia and South America began to rotate away from Africa in unison, the northern end of the Atlantic spreading-center propagated towards the British Isles. In Early Cretaceous times the *ancestral* Azores straddled both this center at the head of the ocean, and the Azores Fracture Zone, which swept in a huge semicircle, from the edge of the Newfoundland Grand Banks into the Tethys (now the Mediterranean). Within the fracture zone there lay a crowded amalgam of jostling continental fragments etched by rifting in the regions that are now the English Channel, the Bristol Channel, and the Irish Sea—rifts that failed to develop.

Seafloor spreading, and therefore crustal separation, first started between the Newfoundland Grand Banks and Iberia, and by 118 MYA several additional spreading-centers had developed, first between Brittany and Iberia—the Bay of Biscay Rift—and second between Greenland and Labrador. Iberia was now an independent microcontinent with a life of its own: it drifted to the southeast in accord with the rate of seafloor spreading in the Atlantic. But neighboring continents rotated, and Iberia was caught between southern France and Morocco. The resulting wedging inhibited seafloor spreading in the Bay of Biscay, and the colliding continents caused mountains to form between northern Spain and southern France—the proto-Pyrenees.

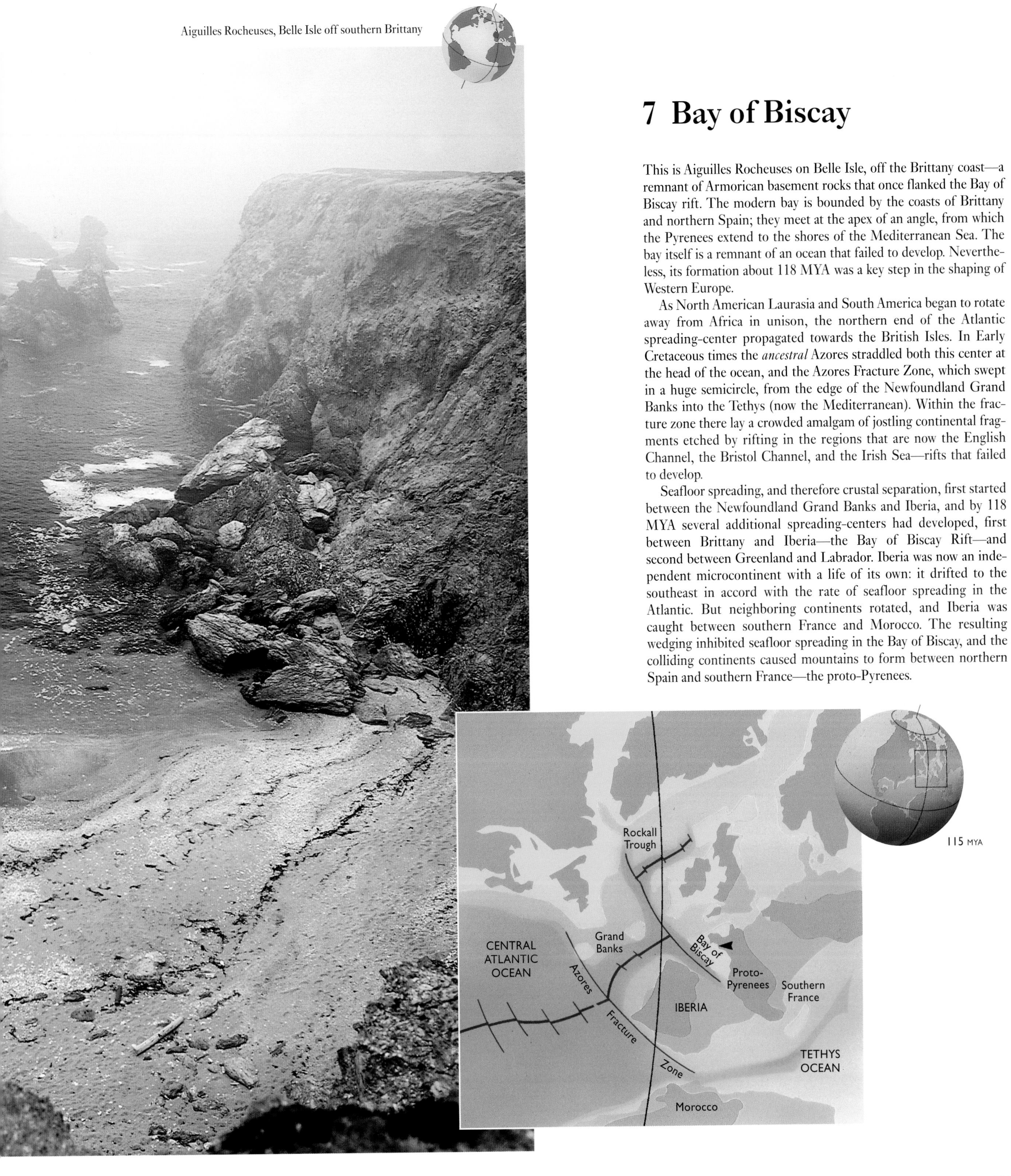

8 Labrador Sea

Although the notion is controversial, the hot-spot volcanic island of Madeira pictured here is thought by some scientists to be the surface manifestation of a plume in the Earth's mantle that contributed to the formation of the Labrador Sea now 4,800 km (3,000 miles) from Madeira.

The global figures to the far left show the trace of the Labrador-Madeira mantle plume as it may have appeared 100 MYA, and as it appears today. A second plume shown in the diagrams (there are many others not illustrated) formed a contiguous chain, the New England Seamounts and the White Mountains of New Hampshire. The two traces are roughly parallel on both globes, with a dogleg shape that corresponds to a change in direction of the lithosphere's movement. The mantle plume that formed the volcanic island of Madeira may have first penetrated the rotating lithosphere when the Labrador-Greenland region was above it in Pangean times.

In an analogous sense only, the Earth's fragmented outer shell rotates over the fixed surface of an inner sphere, rather like a soft-metal housing that rotates over a ball bearing. In this case, the surface of the gigantic "ball bearing" (the Earth's mantle) is spiked with a hundred or more intensely hot mantle plumes. The plumes remain comparatively fixed, while the outer casing (the lithosphere) rotates over the "ball bearing." Each plume leaves a volcanic imprint on the surface of the "casing," the moving lithosphere.

According to current theory, when Pangea fragmented, the plume continued to puncture the lithosphere, leaving a deep elongated scar that filled with solidified magma and sedimentary rocks. The scar was a structural weakness and later became the site of renewed volcanism when the Mid-Atlantic spreading-center began to propagate towards the British Isles [Essay 7]. Thus, when the spreading-center failed to penetrate the Rockall Bank, between Greenland and the British Isles, it found the next line of least resistance—the ancient trace between Greenland and Labrador. Seafloor spreading began in what is now the Labrador Sea as spreading in the Bay of Biscay ended, about 84 MYA.

IX MARITIME WEST

A GREAT DEBATE IN NATURAL SCIENCE TODAY CONCERNS THE POSSIBLE CAUSE OF THE END OF THE MIDDLE AGES OF LIFE'S EVOLUTION—THE TERMINATION OF THE MESOZOIC ERA 65 MYA. WAS THERE A SUDDEN MASS EXTINCTION OF DINOSAURS AND OTHER LIFE FORMS AT THIS TIME? AND IF SO, WAS THE CATASTROPHE CAUSED BY THE IMPACT OF A BOLIDE FROM OUTER SPACE? OR WAS THERE A LESS SPECTACULAR BUT EQUALLY INTRIGUING REASON? THIS PANORAMA WAS TAKEN ABOVE THE POINT OF IMPACT OF A 1.6-KM (1-MILE) BOLIDE WHICH STRUCK DEVON ISLAND IN THE CANADIAN HIGH ARCTIC 15 MYA. ONLY A MISSILE WITH SEVERAL HUNDRED TIMES THE DESTRUCTIVE POWER OF THIS "HAUGHTON BOLIDE" COULD HAVE CAUSED THE MESOZOIC CATASTROPHE.

MARITIME WEST
99–66.4 MYA

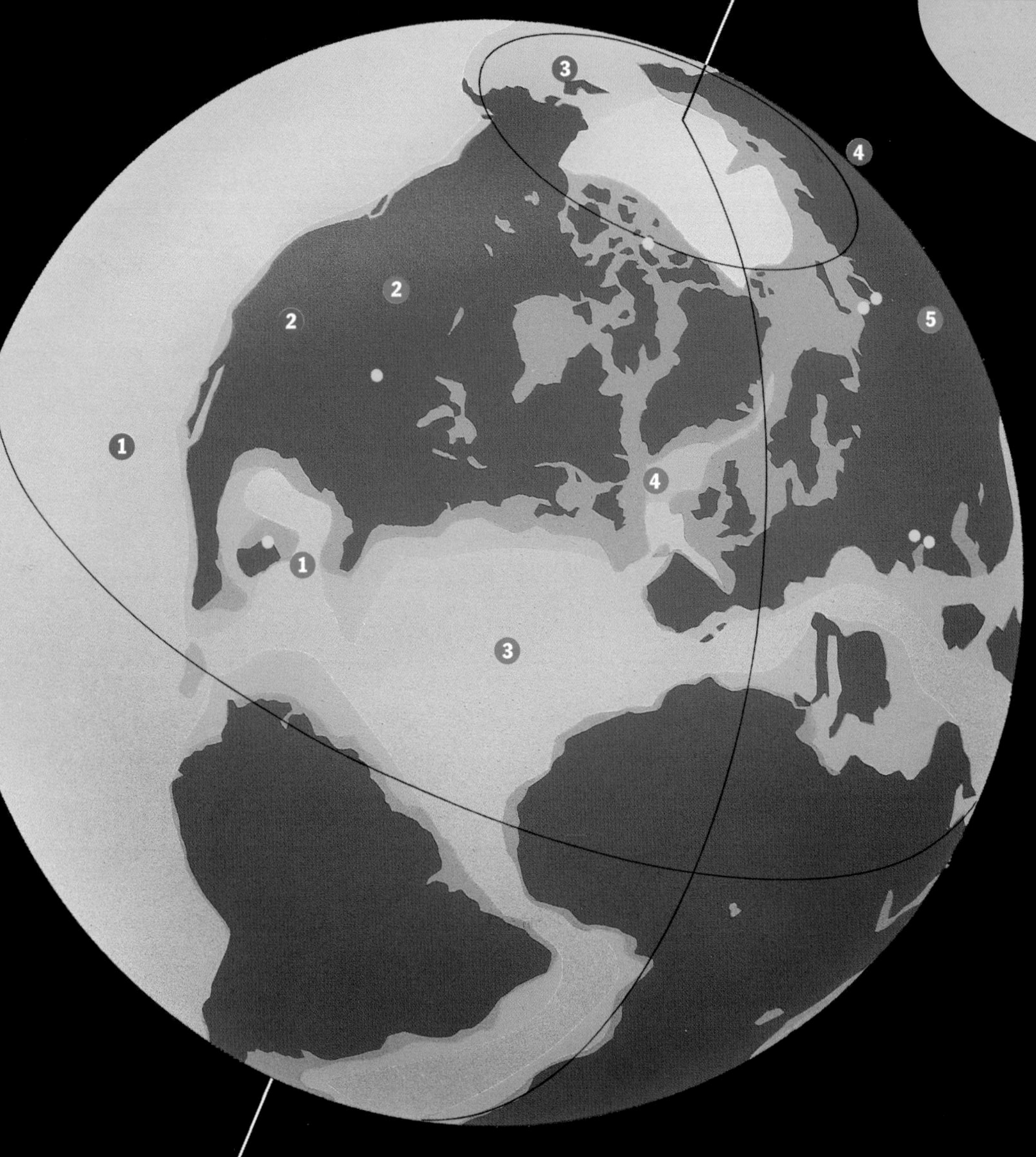

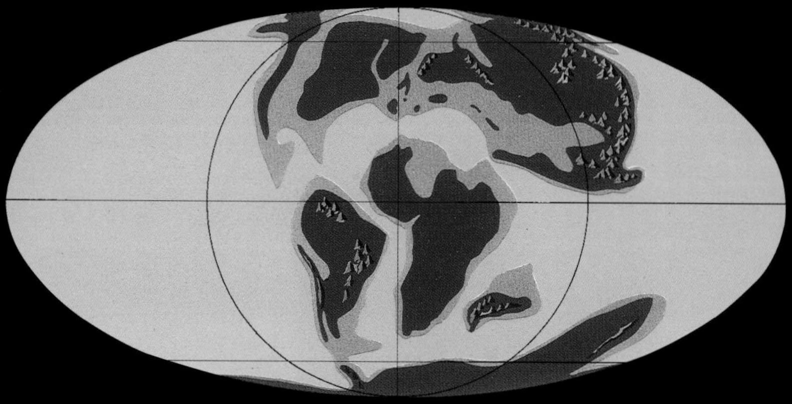

POST-PANGEAN LAURASIA

1. Farallon plate (Pacific Ocean)
2. Laramide Sector
3. Chukchi Peninsula and North Slope of Alaska
4. Mongolia

IMPACT SITES (Read left to right)

- Yucatán (c.65 MYA)
- Manson: Iowa (c.66 MYA)
- Devon Island (c.27 MYA)

- Kara and Ust-Kara (c.75 MYA)
- Gusev and Kamensk (c.65 MYA)

SEAWAYS AND OCEANS

1. Gulf of Mexico
2. Western Interior Seaway
3. Central Atlantic
4. Labrador Seaway and Nordic Sea
5. Turgai Seaway

The Central Atlantic was still connected to the shrinking Tethys Ocean 95 MYA although its circulation was restricted. The newly formed South Atlantic connected the central ocean to a supply of oxygenated polar waters from the Southern Ocean. This ended a prolonged period of anoxic conditions and massive formation of Cretaceous hydrocarbons in the basins of the Central Atlantic.

Ocean floors were swollen by the heat released at spreading centers during these and other continental movements; the oval projection reveals the consequences—displacement of the global ocean, causing epicontinental seas to flood low-lying regions of continents everywhere on Earth.

Although several interior seaways flooded much of Laurasia, we will concentrate on the North American West—the *Maritime West* of the chapter title—where an interior seaway frequently separated Western Laurasia from the rest of the megacontinent. The uplifted maritime province was a haven for dinosaurs at a time when low-lying parts of Europe were inundated to depths over 300 m (1,000 ft). From 85 MYA the North Slope of Alaska, having sutured to North America, was on the migration route for dinosaurs between Mongolia and the Maritime West.

Charles Darwin's failure as a teenage medical student at Edinburgh University (1825–27) is generally attributed to his exposure to the grim reality of brutal 19th-century surgical practice. The record also suggests that at Edinburgh Darwin was more attracted to the performance of seals in the nearby firth than to that of surgeons at the medical school. Yet the medical profession seemed always to dog Darwin's life in one sense or another. After his return from the voyage of the *Beagle* he was afflicted by the then unknown Chagas' disease (or so it is now thought), which he may have contracted during the voyage. It is a seriously debilitating and often fatal disorder caused by a parasitic protoctist (*Trypanosoma cruzi*), transmitted to the victim by bloodsucking insects that are endemic in South America. Fevers, lassitude, and particularly heart problems are all symptomatic of the disease, and Darwin had a prolonged history of such symptoms; he suffered a severe heart attack in 1873 and another, this time mortal, in 1882.

Two medical associates among many had a profound influence upon Darwin: Joseph Dalton Hooker (1817–1911), later Sir Joseph, one of Darwin's closest lifelong confidants, and Richard Owen (1804–92), later Sir Richard, who became an embittered and ruthless adversary. Remarkably, none of them earned their fame (Darwin as a biologist, Hooker as a botanist, or Owen as a paleontologist) in pursuit of the medical careers on which they at first embarked.

Joseph Hooker's first professional appointment was that of assistant surgeon and botanist aboard HMS *Erebus*, during Sir James Clark-Ross's expedition to the Antarctic in 1839–43. His qualification as a botanist was derived from familiarity with the work of his father, Sir William Jackson Hooker. Sir William was a preeminent British botanist who specialized in ferns, algae, lichens, and fungi, as well as being an expert on the higher plants. Joseph Hooker had read the proofs of his friend's *Voyage of the Beagle* before departing on the Clark-Ross expedition and was aware of Darwin's microscopic examination of seawater discolorations. Young Hooker observed similar discolorations during his Antarctic voyage and decided that they were caused by a "herbage," the classification that stood for a century or more (**1**). After many years of travel and recording his observations on plants and their botanical and geographical relationships, in 1855, Hooker was appointed assistant director of the Royal Botanic Gardens at Kew, near London. He became director in 1865, after his father's retirement. It was Joseph Hooker, along with the geologist Charles Lyell, who had explained to the Linnean Society in July 1858 how Wallace and Darwin had independently "conceived the same very ingenious theory" of natural selection. The younger Hooker's intellectual friendship provided consolation to Darwin later in life, while the uproar over Darwin's books, *The Origin of Species by Means of Natural Selection* (1859) and *The Descent of Man and Selection in Relation to Sex* (1871), continued in the background.

Indeed, after their publication (and that of two related books), and with Hooker's prolific knowledge of plants to draw upon, Darwin focused his attention almost exclusively upon natural selection in the context of plants. He made many fascinating discoveries: that plants are light-sensitive, why some creepers twine, that plant roots respond to gravity, and that the key role of earthworms is in both "plowing" soil and forming vegetable mold. Most importantly, although Darwin did not coin the modern expression "coevolution," he was able to show that flowers and insects had indeed coevolved, and that such coevolution was the probable cause of the extraordinary advance of both angiosperms and insects in Cretaceous times.

Richard Owen, who had been a medical student at Edinburgh in 1824 before transferring in 1825 to St. Bartholemew's, a London teaching hospital, became both a distinguished physician and a teacher of physiology and anatomy. However, in 1830 he met Georges Cuvier, the great French paleontologist, who turned his interests in the direction of paleontology, particularly towards the comparative anatomy and physiology of living and fossil animals. It was in his paleontological role that Owen had identified and named a new order of reptiles (**2**): he coined the word *Dinosauria* in 1842 and later described *Archaeopteryx*, the first-known fossil bird (**3**). By 1856, when he was appointed superintendent of the natural history departments at the British Museum, he too had been completely seduced by natural science. In addition to his scientific work, Owen's main contribution to posterity was the development of the British Museum of Natural History at South Kensington in London and the fostering of that museum's great paleontological tradition.

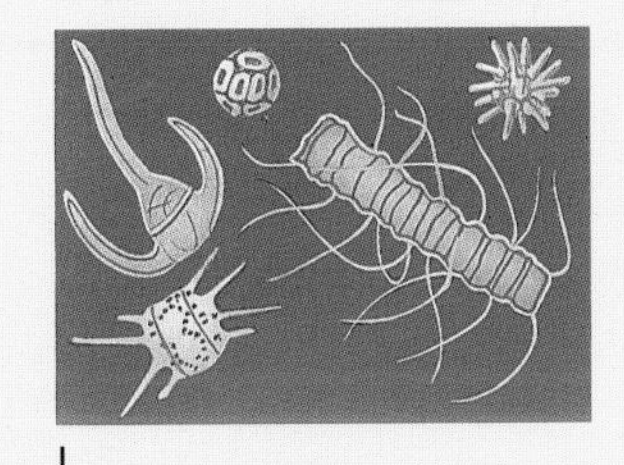
1

2

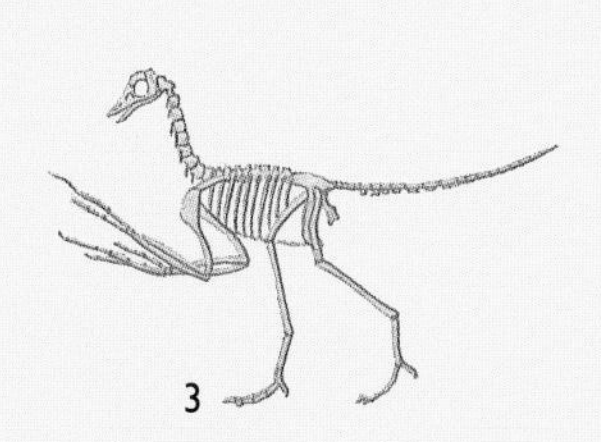
3

Darwin's problem with Owen was precipitated in 1858 with the publication of *The Origin of Species*. Until then Owen was considered to be the leading biologist in England, holding his own theories of evolution. These were based on the concept of similarity of anatomical structure (analogy), and similarity of origin (homology). This may be best exemplified by the fallacious idea that vertebrate skulls are a modified form of vertebrae. Darwin was convinced by the evidence for similarity of origin—indeed this belief was the substance of his thesis, but rejected similarity of anatomy as Owen proposed. When Owen realized that he was about to be eclipsed as an evolutionist, he became, in the words of one writer, "mad with jealousy ... devoid of all scruple, seeking to discredit Darwin at all costs."

One result of Owen's scheming to this end was a debate at a meeting of the British Association for the Advancement of Science at Oxford University in 1860. Bishop Horace Wilberforce, coached in advance by Owen, sought to destroy both Darwin and his dangerous new theory of evolution. But he was up against Thomas Henry Huxley (1825–95) who was to reply at the meeting. Huxley, a one-time medical student and now a well-known biologist, was (along with Hooker and Lyell) an avid supporter of Darwin's theory of evolution. Huxley reported later that the bishop spoke with "inimitable spirit, emptiness, and unfairness," and made two cardinal errors—by confusing philosophy with science, and by being offensive about Huxley's antecedents. Huxley, however, utterly demolished the bishop's case, and by implication that of Owen, his mentor.

We turn now to an ongoing debate in natural science today, in the sense that the opposing sides are equally passionate, and that much publicity attends the more sensational claims—which often do not acknowledge that there are strong counterclaims. This debate is about why the dinosaurs became extinct, and how and why the Mesozoic Era ended and the Cenozoic Era began. Did a bolide (an impacting object from outer space) cause geologically instant annihilation of many terrestrial and marine species? Or was a bolide impact just a *coup de grâce* after a different primary cause of mass extinction? To judge these issues we have to consider a time when life on land was facing a prolonged period of stress, a period not unlike that which faces life on Earth today: the gradual limitation of resources for life's continued existence in its established form.

In Cretaceous times stress on terrestrial populations of plants and dinosaurs, the leading subjects of the present Cretaceous boundary extinction debate, was caused by the advance of shallow epicontinental seas. These inland seas were up to 300 m (1,000 ft) or more in depth, and at times they reduced the terrestrial domain to less than half its present extent [Essay 1]. One might expect such enormous changes in global sea level to have affected all regions of the megacontinents equally. In fact, in Laurasia the ebb and flow of such inland seas varied, affecting the distribution, confinement, and perhaps the general evolution of plants and animals in Laurasia and elsewhere. But why were there such excessive epicontinental seas? Why did their depth and extent vary in different regions?

After Pangea began to fragment, and the Central Atlantic opened, seafloor spreading started in the Bay of Biscay. Later it commenced in the Labrador Sea [see Essays 7 & 8 previous chapter—pages 184–85 and 186–87], while the South Atlantic Ocean had also begun

[*continued on page 194*]

1 Great Inundation

During Cretaceous times the global ocean was gradually being displaced. It was as if bricks, in the form of swelling spreading-centers and massive sedimentary structures, were being added to an already full pail of water. Sea level was already high because by mid-Cretaceous times the global mean temperature was an additional 11–15°C (18–27°F) warmer than today and there was little if any permanent ice at the poles. From 113–63.6 MYA low-lying continental platforms the world over were inundated, and often the sea rose 300 m (1,000 ft) or more above today's global level. The resulting epicontinental seas were often anoxic, and because of their adverse chemistry sometimes barren of any kind of life. They were almost tideless, with low-energy waves, and may at times have resembled this remote uninhabited island in the Bahamas.

The North American epicontinental sea is termed the "Western Interior Seaway." It linked the Gulf of Mexico to the Arctic Ocean in the region of the Mackenzie Delta for about 50 MY. During this time there were five peak inundations, each coinciding with maximum global sea levels, each with a different geographical configuration, and each with a different name (shown in the graph below). The western edge of North America became an isolated maritime province, bounded by the Pacific Ocean to the west and the Interior Seaway to the east. This island microcontinent (which we will call the Western Maritime Province) was dominated by the remnant Ancestral Rockies. Throughout this period, these mighty ancient mountains were assaulted by folding and overthrusting, torn apart by batholiths, and injected with mineral belts.

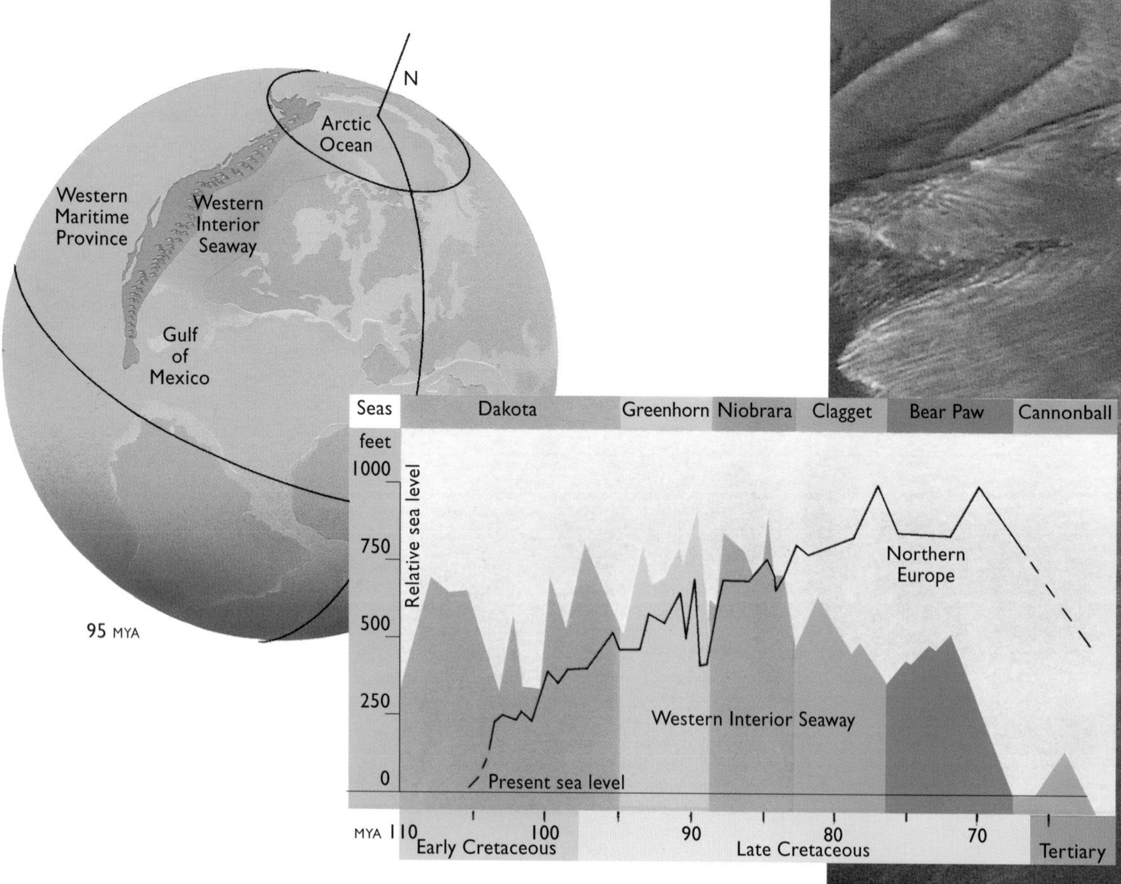

Little Wax Cay,
Exuma Cays,
Bahama Islands

[*continued from page 191*]

to open. While this tectonic activity was in progress in Early Cretaceous times, the Florida-Bahama carbonate platform had been forming, and the Anglo-Paris Basin filling with chalk. The North Sea Basin filled with even greater quantities of chalk in Late Cretaceous times. Simultaneously, seafloor spreading-centers had been developing in the proto-Indian and proto-Antarctic Oceans during the general breakup of post-Pangean Gondwana. Meanwhile innumerable other reefs, carbonate banks, carbonate shelves and basins were being formed and filled by industrious protoctists and shelled invertebrates the world over. All these activities contributed to the gradual displacement of the global ocean.

It was as if bricks (in the form of swollen spreading-centers and massive sedimentary structures) were being added to an already full pail of water. The "bucket" was full because there was little if any ice at the poles by mid-Cretaceous time, and none at all thereafter until perhaps near the end of the period, 65 MYA. The net result of all these swollen seafloors and carbonate-burdened continental shelves and basins, in an already brimming global ocean, was that the ocean spilled over continental margins, inundating low-lying interior platforms (**4**) to depths of 300 m (1,000 ft) or more. Spread over an interval of 50 MY (113– 63.6 MYA), there were five peak periods of exceptionally high global sea level.

According to J.M. Hancock, E.G. Kauffman, and W.G.E. Caldwell, the first three peaks occurred between 113 and 84 MYA. Epicontinental seas flooded much of central North American and European Laurasia to depths up to about 300 m (1,000 ft) above today's sea level. During the last two global peaks, between 84 and 63.6 MYA, different rules seemed to apply. Epicontinental seas reached a maximum of only 150 m (500 ft) above today's level in North America, while in Europe the maximum reached 350 m (1,150 ft) above today's level. Areas of flooding in North America thus decreased, while those in Europe tended to increase, yet global sea level generally rose through most of this period.

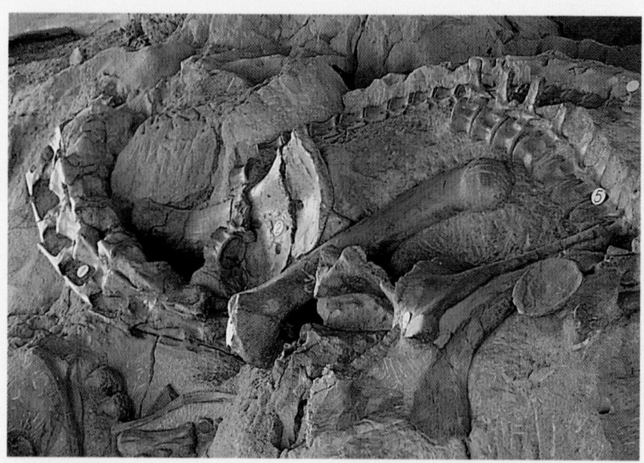

Dinosaur quarry face, and Sauropod fossils, Dinosaur National Park, Utah-Colorado

There was a reason for this apparent paradox. It was caused by uplift of the western edge of North America during two overlapping mountain-building periods. The first was the Sevier orogeny (named after a desert in Utah) and the second, the Laramide orogeny (named after Laramie, a town in Wyoming). Both were the result of tectonic events in the Atlantic Ocean. Lower epicontinental sea levels in North America during the later Cretaceous drastically increased the extent of land available for the growth of land plants and the movement of dinosaurs.

From about 113 MYA both the Central Atlantic and the young South Atlantic Oceans opened simultaneously while, in relative terms, Africa remained in place. This interaction caused North America and South America to rotate away from Africa in unison at twice their normal speed—roughly four inches a year instead of two.

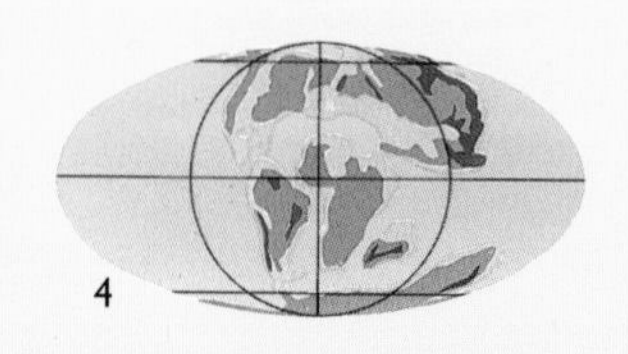

4

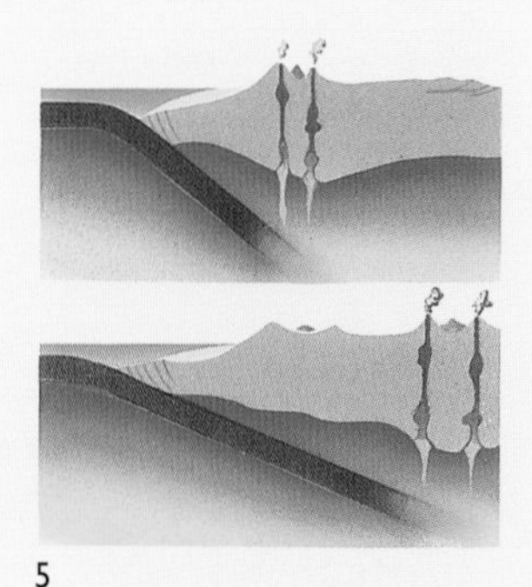

5

Since Pangean times the Farallon Plate, part of the Pacific seafloor off the West Coast of North America, had been descending beneath the continent edge. Now, with the increased rate of westward movement, the Farallon was subducted beneath the West Coast at a shallower angle than before (**5**), causing the subducting ocean floor to behave like a burrowing mole in a field and create disruption and uplift of the already mountainous continental surface above.

The burrowing of the Farallon Plate was on a gargantuan scale; it enhanced the effect of an orogeny already in progress in Idaho and Utah, the Sevier orogeny (152–58 MYA). Progressing eastwards, it began to push up the Laramide Rocky Mountains in place of the remnant Ancestral Rocky Mountains—a period called the Laramide orogeny (80–40 MYA). The progress of both orogenies effectively changed the rules for central North American Laurasia, reducing the extent of epicontinental flooding. Meanwhile, because the same huge amount of sea water was still being displaced, European Laurasia and other parts of the world were subjected to an *increased* depth of inundation [Essay 1].

During the three global sea-level peaks that occurred *before* 84 MYA, a sea called the Western Interior Seaway stretched continuously from the Gulf of Mexico to the Beaufort Sea, in the region of the Mackenzie Delta in the Arctic. The remnant Ancestral Rocky Mountains lay to the west. In effect, the western third of North America became a maritime province, bounded by the Pacific Ocean and the variable Western Interior Seaway (**6**). Since this mountainous maritime region has no geological name, we will call it the "Western Maritime Province" [Essay 2].

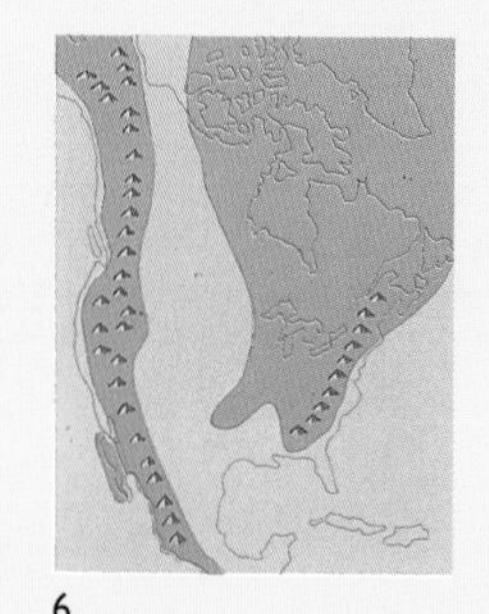

6

As the Laramide uplift began to take effect, *after* 80 MYA, the province slowly assumed a higher profile. The Interior Seaway was not quite as extensive as before, although during the two remaining sea-level peaks the Gulf of Mexico and the Arctic Ocean were linked. After 75 MYA the seas gradually withdrew; before the end of Cretaceous time the western region was no longer a maritime province.

During its existence the eastern flanks and shores of the Province stretched from present New Mexico in the south up to Alberta and beyond. This eastern coastal area, which we will call the "Laramide sector," became the focal point of angiosperm and dinosaur evolution in North American Laurasia in Late Cretaceous times.

Until 90 MYA ferns, conifers, and cycads were the dominant plants. Although spore-bearing plants continued to flourish in Late Cretaceous times, these gymnosperms were gradually superseded by flowering plants, the angiosperms [Essay 3]. There are two possible reasons for this change in dominance: the coevolution of insects and angiosperms, and the proliferation in the number and diversity of herbivorous dinosaurs.

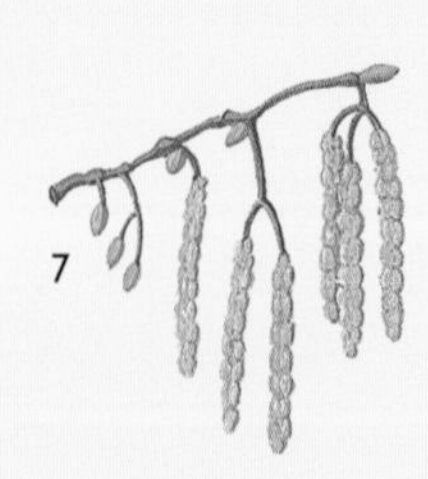

7

During coevolution, insects and plants evolve interactively. Plants naturally develop structural and chemical barriers to discourage some insects, and evolve other strategies to lure beneficial insects. The insects in turn adapt to changes in the plant by developing physical and other stratagems of their own. Thus while gymnosperms cross-fertilize by the random process of shedding wind-borne pollen, angiosperms have specific flowers colored to attract flying insects and shaped to receive them; the insect is fed by the flower's nectar and dusted by its pollen. The pollen is then transferred to other flowers, which are cross-fertilized by the pollen carried by the insect. For example, magnolia pollen is transported by beetles, which feed upon parts of the flower and cross-fertilize when moving from one flower to the next. Thought to be among the more primitive flowering plants, the magnolia's dependence upon beetles is considered by some botanists to confirm the notion of coevolution. The change to insects from wind pollination was occasionally reversed following the scarcity or abundance of pollinating insects (hence the evolution of angiosperms with catkins: **7**).

However this may have been, there is general agreement that the earliest angiosperm flowers developed from ordinary leaf-bearing shoots, and flower parts evolved from leaves. During their evolution, the uppermost leaves of future angiosperm plants were transformed into the female parts of the flower (the ovule). These bear the female reproductive "gamete" cells, that later become seeds. Ovules are wrapped by a leaf-like organ (the carpel), which with the ovule ultimately form the fruiting body of the plant.

The next set of leaves adapted to form male organs, the stamens (which produce pollen containing male gametes), while a neighboring set of leaves developed glands (the nectary) that secrete sugar-based solutions and perfume—and sometimes strong poisons to repel predatory insects.

The leaves that form the colored petals of the plant lost all trace of their chlorophyll and evolved shapes to attract (**8**) and direct the right insects towards the nectary and therefore to the pollen repository—while discouraging others. The flower itself is contained and protected by the sepals—a set of specialized leaves that have often kept their chlorophyll. So the very anatomy of flowers evolved to attract and to accommodate the flying insects that were to be the plant's means of perpetuation. The fact of coevolution is confirmed by the increasingly high degree of coordination between the features of the insects' bodies and the features of the angiosperm flowers. For example, the development of insects' elongated sucking mouth parts for trumpet-shaped flowers, and the evolution of orchids that mimic female sex organs to attract male insects.

Beetles, which account for about a third of the million or so modern species of insects, have the longest association with plants of all kinds (**9**). Beetle-plant coevolution probably began over 300 MYA in Carboniferous forests, yet beetles have not become completely dependent on plants, as other flying insects have done. Beetles get nourishment from sap, dung, and carrion, which explains why some flowers have evolved a strong and (to us) objectionable odor. Bees are believed to have evolved about 100 MYA, and to have participated in the gradual evolution of angiosperms. The earliest known fossil bees are dated at only 40 MYA, but they are of an advanced and highly specialized variety, which is why specialists assume that bees evolved from wasps in Late Cretaceous times.

Beetles, bees, wasps, and most other flying insects are "endopterygotes" (Gr. *endo pteron*, inside wing). They lay eggs that produce grubs which live their lives inside fruit or seeds, or in soil or wood, and thus survive when conditions of climate are fatal to "fair-weather" adults. The grubs go through a pupal stage involving a complete transformation of their body structure and the development of wings (from chrysalis to butterfly, in effect). Adult endopterygotes thrive on angiosperm nectar and pollen, unsuitable foods for their offspring. Winged adults are independent, and can fly off to forage, mate, and lay eggs—cross-fertilizing the angiosperms that attract them in a never-ending cycle.

About 12 percent of flying insects—the more primitive dragonflies, mayflies, termites, cockroaches, and their kind—have no coevolutionary bond with angiosperms. These insects are "exopterygotes" with no pupal stage; the young grow wings as they develop.

As THE SEED of a flowering plant matures, the leaf-like carpel develops to produce a "fruit"—a pod, a nut, an acorn, a burr, or a soft-fleshed fruit. All angiosperm fruits are adapted for wide dispersal, whether by the drifting parachutes of dandelions, the flight of sycamore wings, or by explosive force, flowing water, or wandering animals.

On the other hand, soft-fleshed fruits such as dates, figs, and grapes *depend* on being eaten by animals. Seeds are suitably protected, so they can pass through the animal and be deposited on the ground unharmed. Some even *need* to pass through the digestive tract to trigger germination. Most fruits have a laxative effect, speeding up the passage of the seed through the animal to minimize the

8

9

10

11

damage by the carrier's digestive system. In the Laramide sector of the Western Maritime Province in Late Cretaceous times, "pulpy" and "meaty" fruits had not yet developed; dinosaurs were probably not significant agents of seed dispersal, but after their extinction, many small mammals evolved to fill this niche.

DINOSAURS FIRST APPEARED in the fossil record in late Pangean times: they diversified rapidly because their superior bipedal posture gave them a competitive edge over the dominant but less agile quadrupedal mammal-like reptiles. Except for carnivorous dinosaurs, which remained bipedal throughout their evolution, the later dinosaurs were either bipedal or quadrupedal. By the end of the Triassic Period (208 MYA), the "early" dinosaurs had evolved from orders, suborders, and infraorders, into many families and thirty or more genera (**10**). Two other quite distinct and overlapping groups of dinosaurs evolved later. In the 150 MY of their existence (counting the early dinosaurs as one regime), three distinct regimes of dinosaur ruled [Essay 4]. To avoid a plethora of dates and categories we will use the term "regime" as a broad and arbitrary division.

The dinosaur regimes can be likened to political parties that achieve power, rule, and ultimately give way to a different political persuasion. Each of the three regimes also had a hierarchy of family names, including any number of genera and species. When one regime was replaced, perhaps after a period of coexistence, the original regime continued in opposition but did not necessarily expire. Each genus may imply one, several, or many species, for dinosaurs are not generally discussed at species level.

AS PANGEA began to break up (208–163 MYA), the first regime of dinosaurs declined in numbers and diversity, and much larger animals took their place. This second regime of about 38 known genera included some of the largest dinosaurs of all time: the Sauropod infraorder (Gr. *sauros pod*, reptile foot), herbivores that included

Cretaceous scene: salt evaporating pan, Great Inagua, Bahama Islands

Apatosaurus (once called *Brontosaurus*: **11**) and some of those with the most familiar shape—for example, the Stegosaur suborder of herbivores distinguished by the triangular plates on their backs. All were hunted by attendant carnivores like the Allosaurs.

Apatosaurus is estimated to have weighed about 30,000 kg (33 tons), and to have been 21 m (70 ft) in length. With only small teeth in the front of the jaws and no cheek teeth, it had to swallow cobble-sized rocks to reduce food to digestible size in its gizzard.

If Sauropods were "warm-blooded," as some experts conclude, *Apatosaurus* would have needed to consume up to 700 kg (1,500 lb)

[*continued on page 200*]

1 Ancestral Bering Landbridge
2 Idaho Batholith
3 Colorado Mineral Belt
4 Bear Paw Seaway
5 South Atlantic Ocean
6 Farallon Plate
7 Mid-Atlantic Ridge

2 Western Uplift

By the end of Cretaceous time the Western Interior Seaway had withdrawn, although global sea level was still at peak height. Thus while central North America was high and dry, much of European Laurasia was flooded up to 350 m (1,150 ft) deep. This extraordinary difference was the result of the whole of the Western Maritime Province being lifted by tectonic forces. This first caused the seaway to migrate eastwards, and then to drain off the continent altogether. The prime cause of this uplift was far away.

As the young South Atlantic Ocean opened (about 113 MYA), both North and South America rotated away from Africa in unison, while Africa remained fixed in place. The net result was that North America separated from Africa at effectively double the seafloor-spreading rate. This increased rate of drift caused a section of the Pacific Ocean floor called the Farallon Plate to subduct beneath the western edge of North America at a shallower angle than before.

The subducting plate, possibly with a huge seafloor plateau on its surface, had already been interacting with the edge of the

Maroon Bells region, Aspen, Colorado

continent. The Sierra Nevada batholith was formed as a consequence [see Essay, pages 134–35]. The Farallon plate now behaved like a gargantuan burrowing mole disrupting the surface of a field. As it pushed eastwards, it caused violent disturbance of the surface for about 1,600 km (1,000 miles) inland. The Idaho batholith and its associated mountains were the first product of this disruption, forming during the Sevier orogeny 152–58 MYA. The Laramide Rocky Mountains and the Colorado Mineral Belt were the second result, during the Laramide orogeny 80–40 MYA.

One locality uplifted during the Laramide orogeny is pictured here: Maroon Bells near Aspen, a Colorado Mineral Belt silver-mining town—now a famous ski-resort. The inset picture is an aerial view of Central City, one of the principal gold-mining areas of the Colorado Mineral Belt in the 1860s, once known as the "richest square mile on Earth."

Central City, Colorado (2,568 m, 8,560 ft)

3 Angiosperm Advance

The key to increasing angiosperm (flowering plant) success in Late Cretaceous time from 85 MYA to 65 MYA was their ability to produce an encapsulated seed that incorporated the nutrients to sustain the plant's embryo during germination. This facility enabled flowering plants to grow and reproduce themselves in only weeks or months. In contrast, gymnosperms (coniferous plants) employed a much more elaborate system of seed production and required several years to reproduce; they were gradually superseded by angiosperms.

Gymnosperms depended on haphazard wind-borne distribution, whereas angiosperms attracted insects that provided a systematic method of distributing pollen. Because particular insects tended to specialize in feeding from particular flowering-plant nectaries, new kinds of insects presented opportunities for "coevolution" with new species of angiosperms. This tendency accelerated the rate of speciation in both the flowering plants and the insects. Contrary to the popular view, until the extinction of third-regime dinosaurs and the appearance of vast numbers of small mammals, large-scale dinosaur browsing inhibited angiosperm diversity and allowed many archaic forms to survive.

Paleobotanists began to reexamine the fossil record of Cretaceous plants in the late 1970s; they have since discovered that for a century or more many angiosperms had been incorrectly classified. It seems that although superficially similar in appearance, fossil leaves found in association with dinosaur remains are often relics of extinct dead-end species. These include fossil leaves previously identified as viburnum, grapevine, birch, beech, and oak, and water-loving plants of the poplar family, such as willow and cottonwood.

During the time of third-regime dinosaurs, besides conifer-clad mountainsides and fern-clad glens, one would have seen mixed forests of flowering plants. These would have been found at appropriate elevations and latitudes in the often Mediterranean-like maritime climate of the Western Province. The scene would have included groves of sycamore, ancestral to the modern plane tree, forests of ancestral laurel, and stands of palm (inset), alongside many long-extinct flowering plants. Some areas may have looked very much like this "fossil" laurel forest (*Persea indica*) in Madeira, which is interspersed with stands of cypress-like gymnosperms and fringed with fern.

Balcões Forest, Madeira

Royal Palms, Barbados,
Lesser Antilles

[*continued from page 195*]

of vegetation a day. If they were "cold-blooded," they would have needed only half that amount. The reason for this difference is the length of time taken to metabolize such great quantities of food. With such primitive means, could Sauropods have digested enough low-calorie vegetation in a day to sustain themselves? This is one of the factors used to argue that these animals could not have been endothermic: that they would have *had* to absorb solar heat to supplement their food-calorie intake, and were therefore ectothermic. Sauropods are now generally thought to have been land-dwelling animals, and not the semiaquatic creatures once portrayed.

At the start of their period of dominance, second-regime dinosaurs were free to roam anywhere in Pangea without a natural barrier. Although they lost their ascendancy after the breakup of Pangea, second-regime animals lived on in diminishing numbers almost to the end of Cretaceous times. For example, the last-known Stegosaur is dated 113 MYA, but the last known Sauropod in the Laramide sector, *Alamosaurus* (**12**), was found in New Mexico, and has been dated 67 MYA. There is a gap of over 30 MY between this enormous creature and its last known regional predecessor.

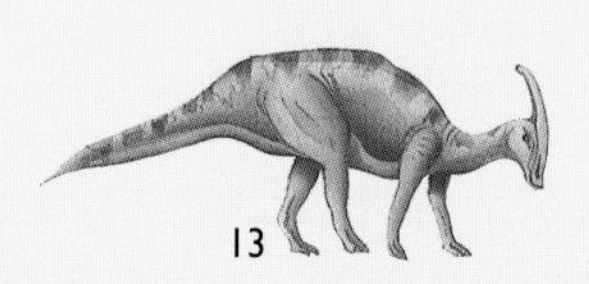

THE THIRD REGIME of dinosaurs, both the herbivores and carnivores such as the Tyrannosaurs, appeared in the Mid-Cretaceous, but only became dominant very late in Cretaceous times. Although generally smaller than the largest second-regime dinosaurs, they were still very big. They looked more bizarre than their predecessors, and were often more physiologically advanced. With a range of movement restricted by epicontinental seas, their evolution was divergent, with unparalleled diversification.

Two families dominated the third and last dinosaur regime in the Laramide sector, numbering 90–95 percent of all genera of herbivores: the Hadrosaurs (Gr. *hadros*, bulky: **13**), which appeared in the fossil record about 85 MYA, and the Ceratopsians (Gr. *keras opsis*, horn appearance), which arrived 80 MYA. The origin of the Hadrosaurs is unknown, although a fossil of an early Hadrosaur has been found in the Laramide sector. The *Protoceratops* (**14**), the ancestral Ceratopsians, seem to have originated in the Gobi Desert region of Outer Mongolian Laurasia, and were confined there until a landbridge developed between Asia and the Western Maritime Province about 80 MYA, after the Laramide orogeny had commenced. Both ancestral Ceratopsians and Hadrosaurs could now migrate between Outer Mongolia and the Laramide sector [Essay 5]. As a result, the diversity of third-regime animals in the Laramide sector increased exponentially, with a peak diversity of more than 30 genera at 75 MYA [Essay 4]. Thereafter, they declined to extinction. Meanwhile, members of the second regime contineud to maintain a plateau of low diversity.

OF THE MANY THOUSANDS of animals in the Laramide herbivore population, about half were quadrupedal and half bipedal. The quadrupeds were mainly members of the Ceratopsid family of "horned" dinosaurs, and the bipeds were mostly members of the Hadrosaur family (the "duckbilled" dinosaurs). Collectively, Ceratopsians, Hadrosaurs and their kind are called "beaked" dinosaurs because they had developed ways to prepare food before digestion: they used their beaks to gather food and their cheek teeth to crush and grind it. The balance of the Laramide-sector herbivores, between 5 and 10 percent of the population, was made up of a variety of specialized animals, including other beaked herbivore browsers and the remaining Sauropods, comparatively rare descendants of gigantic second-regime animals.

The horned dinosaurs resembled the modern rhinoceros in appearance, but there the comparison stops: the formidable modern animal weighs 2,700–4,500 kg (3–5 tons) and is 2.5–4 m (8–13 ft) long. Some Ceratopsians weighed as much as 9,900 kg (11 tons) and were up to 9 m (30 ft) long. They had massive jaws and skulls engineered to accommodate enormous jaw muscles. Ceratopsians ate tough and fibrous cycad fronds and ferns, as well as angiosperms. To accommodate the heavy wear upon them, rows of worn-down teeth were constantly being replaced. The consequence of this replacement was that Ceratopsians did not need to deliberately swallow stones to grind ingested food in their gizzards . The modified pelvic structures of Ceratopsians suggest that they were pot-bellied, with more room for intestinal gut. This indicates that they made use of microbial fermentation to aid digestion much more than their distant relatives the Sauropods.

The last of the twelve genera of Ceratopsians to appear in the fossil record is called *Triceratops* (**15**), and was probably the most formidable. These three-horned animals had a huge frill protruding from their skulls and extending over the neck region. The frill may have served both as a heat-exchanger and as a protective shield. When they were attacked, these ten-ton-plus herbivores fought back. They were very impressive animals.

The bipedal duckbilled Hadrosaurs (**16**) are thought to have been among the dinosaurs best adapted to their environment. They weighed up to 4,500 kg (5 tons) and were up to 13 m (43 ft) in length. In contrast to Ceratopsians, the duckbilled Hadrosaurs' beaks were lined with a sharp, horny edge for cutting tough vegetation. They grazed on low-lying plants and probably browsed on anything edible within their reach. They too had an arrangement that constantly replaced worn teeth with new teeth; and even more importantly they could move the upper jaw across the line of the lower jaw. This movement produced a grinding and shearing action that aided grinding up of food in the mouth, behind their beaks. It is probable that they also had cheek pouches in which they could hold food before grinding it and then despatching it down a long gastric tract. Like the Ceratopsians, their pelvic bones were also modified to accommodate a large stomach.

Hadrosaurs may have grazed on all fours and browsed by rearing up on their hind legs, but they also inhabited swampy environments where leafy angiosperms were plentiful. Some paleontologists think that for mutual protection against predators they wandered in herds like elephants; they might also have used their bipedal stance and counterbalancing tails for speed when attacked. Also, they may have avoided predators by wading and perhaps swimming. However, some experts discount their speed and suggest that the Hadrosaurs' herding instinct was their main defense—they sought safety in numbers.

FIVE TO TEN PERCENT of the balance of the dinosaur population in the Laramide sector at this time (85–68 MYA) consisted of two suborders, the Ankylosaurs and Pachycephalosaurs. The rather clumsy but well-defended "armored" Ankylosaurs (Gr. *ankylos*, fused: **17**) grew up to 10 m (33 ft) and possibly weighed over 3,600 kg (4 tons). They had small, weak teeth, and cropped vegetation with their beaks, which were strong and sharp. They too had cheek pouches for

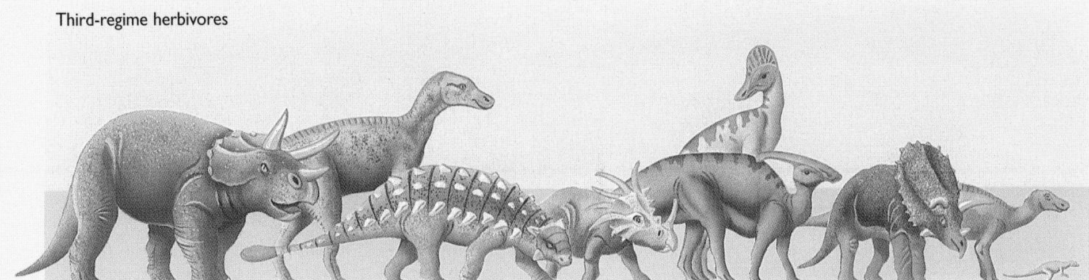

storing food before swallowing, and it is thought that they had a primitive fermenting chamber in the rear of their bodies for digesting food. The bipedal Pachycephalosaurs (Gr. *pachys kephale*, thick head) were "helmet-headed" animals. They were either small, from 2 to 4.5 m (6–15 ft) in complete length, and weighing from 45 to several hundred kg (100 lb upwards), or as large as Hadrosaurs, and weighed thousands of kilos (several tons). The smaller animals depended on their speed to outpace predators—like modern gazelles. The larger ones (**18**) may have depended on the helmeted skulls to ward off attackers by butting—like modern bovine animals.

18

LARAMIDE-SECTOR PREDATORS were terrifying brutes. They include the Tyrannosaurs (**19**), which weighed about 6,300 kg (7 tons), grew up to 6 m (20 ft) high and 15 m (50 ft) long, and are thought also to have originated in Mongolia. Tyrannosaurs may not have been able to run as far or as fast as originally thought; rather than chase duckbilled Hadrosaurs and other prey over a distance, they may have lain in wait to pounce on them. Although the "kings" of the third-regime dinosaurs, they were also scavengers. These huge and fearsome predators would have chased off smaller carnivores feeding on their successful kill as a regular means of gaining a meal.

19

Judging by the relative size of their brain cases, the carnivorous dinosaur equivalents of "lions and jackals" were smaller, faster, and more intelligent than Tyrannosaurs, and may have hunted in packs. These smaller carnivores were 2–3 m (6–10 ft) in length, and some of them possessed extremely savage talon-like claws—they are sometimes called the "sickle claws" for this reason. They are thought to have been swift-footed, and some had large bird-like eyes, which suggests that these animals may have hunted in the dark. An example from the Montana fossil beds of the Laramide sector is *Deinonychus* (Gr. *deinos onyx*, terrible claw: **20**), perhaps the most widely illustrated animal of its type; this genus was larger than most, over 3 m (10 ft) in length. One fossil find in Montana includes a pack of five of these animals near an 8-m (24-ft) herbivore that shows evidence of having been attacked. While Tyrannosaurs suffered extinction before the end of Cretaceous time, it is claimed that the "sickle claws" survived for tens of thousands of years after the mass extinction that marked the Cretaceous boundary; fossil parts have certainly been found above the boundary layer that marks the extinction.

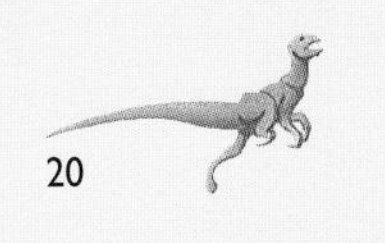
20

MOST, AND POSSIBLY ALL, dinosaurs laid eggs, so the question of where and when they laid their eggs is an important consideration for paleontologists trying to reconstruct their way of life. There are still many intriguing and yet unanswered questions about nesting and breeding habits.

Pregnant females in a modern herd of mammals carry their unborn young with them. But most dinosaurs were more like birds, in the sense that they were bound to a particular site for laying and hatching their eggs, and for caring for their young. It is thought that Ceratopsians, Hadrosaurs, and other large dinosaurs did not reach maturity until 5–12 years after their hatchling stage. These issues raise questions about food supply: how did so many large animals in an egg-laying community sustain themselves and their immature young in the Laramide sector during the nesting season?

Where dinosaur nests have been found, they are often clustered together in rookeries, a means of reducing predation. How female Ceratopsians and Hadrosaurs managed to deposit eggs in their nests is a matter for interesting conjecture: the nests have been either scooped out of the ground or raised above ground level (**21**). They are usually about 2 m (6 ft) across round to oval in shape, and have sometimes contained clutches of up to 25 eggs. Eggs have also been found in lines two abreast— these are attributed to the activity of a small predatory dinosaur called *Troödon*, meaning "wound-tooth." From the great number of eggs in a nest, and the relative proximity of nests in the rookery, paleontologists believe that predation on dinosaur eggs around the edges of a rookery must have been an industry in itself. It is also probable that breakage of eggs was commonplace and that infant mortality rates generally were extremely high.

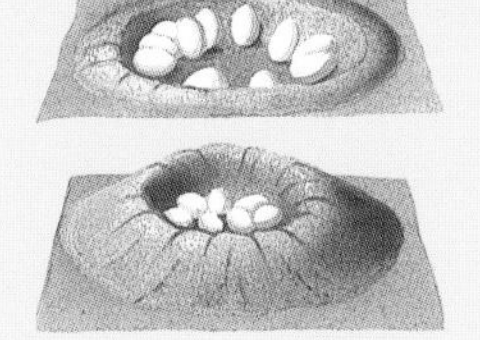
21

Some nests also contain evidence of partially decayed, possibly fermented, vegetation; this decayed material is presumed to have been a source of heat for incubating the clutch. Nest temperature is crucial to modern reptiles such as alligators and crocodiles, because it determines the sex ratio of the young nestlings. In many modern reptiles, temperatures between 31–33°C (88–91.4°F) produce either all-male or a mix of male and female hatchlings. Outside this temperature range, either all females or a markedly reduced number of males are hatched. Nest temperature might also have determined dinosaur sex ratios.

The climate of the Laramide sector may have determined the time of year when eggs were laid. Generally cooler air temperatures could have delayed egg-laying. This would have led to hatching times that were too late in the year to allow the juvenile dinosaurs to mature before the winter season, when the herd might have migrated southwards to feed.

The number of fossil adults and hatchlings that have been found near nesting sites suggests that both horned and duckbilled dinosaurs practised parental care for prolonged periods. However, some predatory dinosaurs have been found with juvenile skeletal remains of the same species in their stomachs. This suggests that like many modern animals, given the opportunity, carnivorous dinosaurs may also have eaten their own wayward young.

THE HEAVY FORAGING of so many large dinosaurs undoubtedly put a major strain on the Laramide-sector food chain. "Dinosaur stress" may have led to the rapid evolution of angiosperms—or at least this was once the established view [Essay 3]. The reasoning that had led to it was that gymnosperms are slow-growing plants, whereas the angiosperms are fast-growing. Since flowering plants produce seeds that germinate quickly, they could have produced several crops a year in the prevalently warm and moist conditions of Late Cretaceous times, particularly when branches were being regularly pruned by browsing dinosaurs. The difference in rates of reproduction and growth in gymnosperms and angiosperms, the argument went, resulted in a special "synergistic" association between the dinosaurs and the flowering plants. The best interests of both the animals and the plants were served: more food for browsing dinosaurs and greater distribution and diversity for angiosperms.

During the 1970s and 1980s the views of leading paleobotanists about the rapidity of angiosperm evolution underwent a major change. Their view today is that little evidence supports the dinosaur-dispersal model: judging from the results of vastly improved dating and identification techniques now available, angiosperm evolution was not explosive but steady. It is claimed that

[*continued on page 212*]

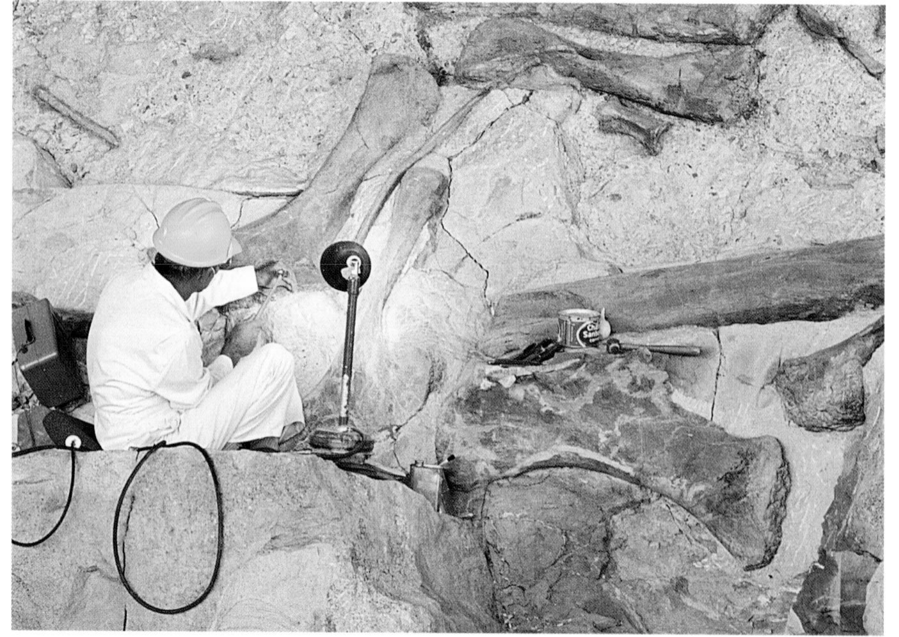

Second-regime dinosaurs, Dinosaur National Park, Utah-Colorado

4 Dinosaur Regimes

This grove of aspen (poplar family), sagebrush (woody shrub), and ponderosa pine (gymnosperm) stands in the Cedar Breaks region of the Colorado Plateau. About 80 MYA this region was near sea level and part of the Western Maritime Province. Grasses had not yet evolved, but woody shrubs and trees resembling, but unrelated to, those pictured here were beginning to colonize some regions. The gradual replacement of slow-growing gymnosperms with fast-growing angiosperms was aided by browsing herds of third-regime dinosaurs. Heavy browsing promoted growth but inhibited speciation of the flowering plants.

There were three overlapping dinosaur regimes in the 150 MY of their worldwide rule, and roughly 95 percent of the animals were herbivores [see pages 195, 200–201]. As shown in the diagram to the right, the first Western Province regime evolved late in Pangean times and became extinct as Pangea began to disintegrate. They reigned in the maritime west from about 235 to 187 MYA.

Because of mountain-building events and subsequent erosion of sedimentary rocks, there is no record of second-regime animals in

Cedar Breaks, Utah (2,700 m, 9,000 ft)

the province until about 163 MYA. From the fossil record it seems that after that time these huge animals (inset) predominated until 100 MYA, then began a general decline in diversity from 91 to 86 MYA. Near the end of this period, animals of the third and final dinosaur regime begin to appear in the provincial record.

In the following 6 MY third-regime animals achieved the most astonishing diversity. Indeed, in the period 80–75 MYA, they appear to have become the most diverse and populous dinosaurs on Earth. Judging from fossil remains in particular localities they were about as numerous and diverse as herding animals and their predators in Africa in recent times. But after they reached a peak of diversity, numbers of genera began a rapid decline—from 32 to 12 genera in the last ten million years of the Cretaceous Period. This precipitous extinction is shown in the block diagram to the right. The essays that follow will describe some circumstances that may have contributed to this situation.

5 Mongolian Connection

During the progress of the Sevier orogeny around 100 MYA, violent volcanic eruptions produced smothering ash falls in the Western Maritime Province. At this time bolides struck Texas and Alberta, forming craters about 12.9 km (8 miles) and 25 km (15.5 miles) in diameter. Later, North America collided with Asia (99–84 MYA), completing the encirclement of the Cretaceous north polar sea by forming the proto-Bering Landbridge.

During these events, second-regime dinosaurs reached the height of their dominance. The animals seemed unaffected by either excessive volcanism or the effect of bolide impacts, and did not begin to decline in numbers and diversity until after 90 MYA. Meanwhile, the ability of angiosperms and ferns to rapidly colonize barren landscapes was probably influencing the numbers of third-regime herbivores that could be supported by the ecosystem.

These new herbivores had teeth that could grind, jaws that could chew, and more efficient digestive systems. Hadrosaurs and Ceratopsians, the duckbilled and horned dinosaurs, accounted for 90–95 percent of provincial third-regime herbivores from 85 to 65 MYA. Hadrosaurs were established by 85 MYA, but Ceratopsians did not appear in the province until 80 MYA. They are thought to have originated in Mongolia, where their ancestors date from about 100 MYA. They migrated to the Western Province via the Ancestral Bering Landbridge, which formed between Asia and the Laramide sector of the province about 80 MYA.

Fossil leaves, the pollen of cool-climate forest plants, fossil Hadrosaurs (interpreted by some to have been herds complete with young), and Tyrannosaur predators have been found on the North Slope of Alaska. This evidence dates from 76 MYA and suggests a dinosaur migration route 1,120 km (700 miles) from the North Pole. The fossil plants show that the arctic climate was now 6–8°C (10–15°F) less than before the Arctic enclosure. The annual mean temperatures on the slope fell to 2–6°C (36–42°F).

The panorama shows a similar cool-climate rainforest in Alaska today: the canopy trees are gymnosperms, and lush flowering plants are growing shoulder high on the dank and dripping forest floor. How dinosaurs survived in cool summer conditions such as this—and possibly through the arctic winter night—is a matter for conjecture [see page 212].

EQUIVALENT WIND CHILL TEMPERATURE °F

THERMOMETER READING °F	0	5	10	15	20	25	30	35	40
50°	50°	48°	40°	36°	32°	30°	28°	27°	26°
40°	40°	37°	28°	22°	18°	16°	13°	11°	10°
32° NORMAL FREEZING POINT									
30°	30°	27°	16°	9°	4°	0°	−2°	−4°	−6°
20°	20°	16°	4°	−5°	−10°	−15°	−18°	−20°	−21°
10°	10°	6°	−9°	−18°	−25°	−29°	−33°	−35°	−37°
0°	0°	−5°	−21°	−36°	−39°	−44°	−48°	−49°	−53°
−10°	−10°	−15°	−33°	−45°	−53°	−59°	−63°	−67°	−69°
−20°	−20°	−26°	−46°	−58°	−67°	−74°	−79°	−82°	−85°
−30°	−30°	−36°	−58°	−72°	−82°	−88°	−94°	−98°	−100°

ESTIMATED WIND SPEED MPH

Data from National Science Foundation, Washington, D.C.

6 Cooling Climate

The summer climate in this Canadian High Arctic locality is close to that of the North Slope of Alaska 85–65 MYA, and is about the same distance from the North Pole. The panorama was taken in the calm that followed a storm in this sheltered fiord. During an Arctic gale the chilling effect of a 65-kph (40-mph) wind is remarkable, as the chart to the left shows.

Because of wind chill, the temperature in the top fraction of an inch of the sea's surface can be reduced by 17°C (30°F) or more below that of the air above it. In a continuous process, the cooled skin of surface water sinks and mixes with the warmer water beneath, gradually cooling the sea as a whole. Given that conditions of climate fluctuated from time to time, the fossil evidence of flora from the North Slope suggests that cool summers and cooler winters often prevailed from about 85 MYA, and their effect extended as far south as Alberta. In such conditions, the enclosed Arctic basin may gradually have become a vast reservoir of ice-cold

Alexander Fiord, Ellesmere Island, Canadian High Arctic

seawater well before the end of Cretaceous time 65 MYA. However, in the same period warmer and even excessively hot climate prevailed to the south—right up to 70 MYA when the climate cooled rapidly before the end of Cretaceous time.

North America was sutured to Asia during the Late Cretaceous, and was already joined to Greenland and Norway. Consequently there were no outlets for Arctic bottom-water to cool adjacent oceans as it does now. However, if one could have looked down at the Earth from a satellite above the North Pole, one would have seen continental seaways radiating from the polar region like spokes from the hub of a wheel. The seaways would have contributed to fluctuations and general changes in the Cretaceous climate, and the evident decline of the third-regime dinosaurs in the Laramide sector could have been one consequence. They reduced from 30-plus genera at their peak diversity 75 MYA to just 12 genera at 67 MYA.

75 MYA

1 Bear Paw Seaway
2 Labrador Seaway
3 Nordic Seaway
4 Chukchi-North Slope Connection
5 Enclosed Arctic Ocean
6 Outer Mongolia
7 Turgai Seaway

7 The End of an Era

The Middle Ages of life's evolution came to a geologically abrupt end 65 MYA. This is the point in time at which the Mesozoic Era concluded and the present Cenozoic Era commenced. But its significance is not simply that some marine animals became extinct, or that some plants vanished from the fossil record, or even that the remaining dinosaurs disappeared. The importance of this boundary is that the whole way of life on Earth changed from its past mode to its present style. From this point in time, mammals and flowering plants began to come to the fore and ultimately to dominate terrestrial life.

In preceding essays we have seen how global change, most often triggered by tectonic events, can be the driving force for radical and stressful change in the environmental conditions for life. We will meet other factors in following chapters, but the end of Cretaceous time may have been caused by an exceptional event—the impact on Earth of an unidentified object, a bolide from outer space.

The bolide impact is termed the "K/T event" (an abbreviation of Cretaceous [K] and Tertiary [T] Periods). The K/T boundary is clearly registered in rock formations in many parts of the world—it is marked by the presence of a layer of clay impregnated with iridium. Iridium is a rare element in the platinum group that occurs in the Earth's mantle (though it is sometimes concentrated by the action of bacteria if it reaches the surface). It also occurs in comparatively high concentrations in some extraterrestrial objects that impact the Earth's surface. In an impact event, fine particles of iridium form an aerosol that ultimately falls back on the Earth's surface with other ejected material. Furthermore, the iridium is only one component of the K/T clay layer. It also contains pressure-shocked minerals and tektites (glass-like spherules), substances that testify to a violent origin.

Most scientists now accept that a massive bolide did strike the Earth at the end of Cretaceous time. Geophysicists are generally of the view that the impact caused a geologically instantaneous mass extinction that included the dinosaurs. However, most paleontologists do not accept this point of view. They believe that the bolide impact delivered the *coup de grâce* to declining populations, a theory the diagram on page 203 tends to support.

Unless given scales by which to make a judgment, each of us interprets words like "sudden" or "massive" according to our own experience of time and size. The diagrams on this page apply a scale to both the time taken for the "sudden" extinction of the dinosaurs and to the "massiveness" of the K/T bolide impact. The square marked on the smaller graph (the same graph used in Essay 1) has been enlarged to the right. Consequently, details of epicontinental sea level, the decline of third-regime dinosaurs, and the timing of known bolide impacts, can be contrasted with the date of the iridium trace at the K/T boundary (the vertical red line). There were two bolide impacts in Arctic Russia at 74 MYA, as well as the Manson, Iowa, impact at or near the K/T boundary. Most importantly a major impact event occurred at Chicxulub on the northern tip of Yucatán in Mexico, also at the boundary: traces of giant tsunami waves of the right date have been found in Texas near the Gulf Coast. Some astrophysicists think that the Manson bolide may have been a detached fragment of a larger object, possibly the Chicxulub bolide.

Before the discovery of Chicxulub, geophysicists had calculated that to have caused a sudden mass extinction, an object at least 16 km (10 miles) across, traveling at several times the speed of a bullet, must have struck the Earth's surface at the K/T boundary. They calculated that a bolide of this size would form a crater about 160 km (100 miles) in diameter. The diagram of the Northern Hemisphere brings perspective to this calculation by relating correctly scaled "160 km" circles to well-known geographic features. The circled locality of the "small" 20-km-wide (12-mile) Haughton Astrobleme on Devon Island in the Canadian High Arctic (panorama overleaf) is also to scale. The Chicxulub crater is deeply buried beneath limestone formations and difficult to measure. But

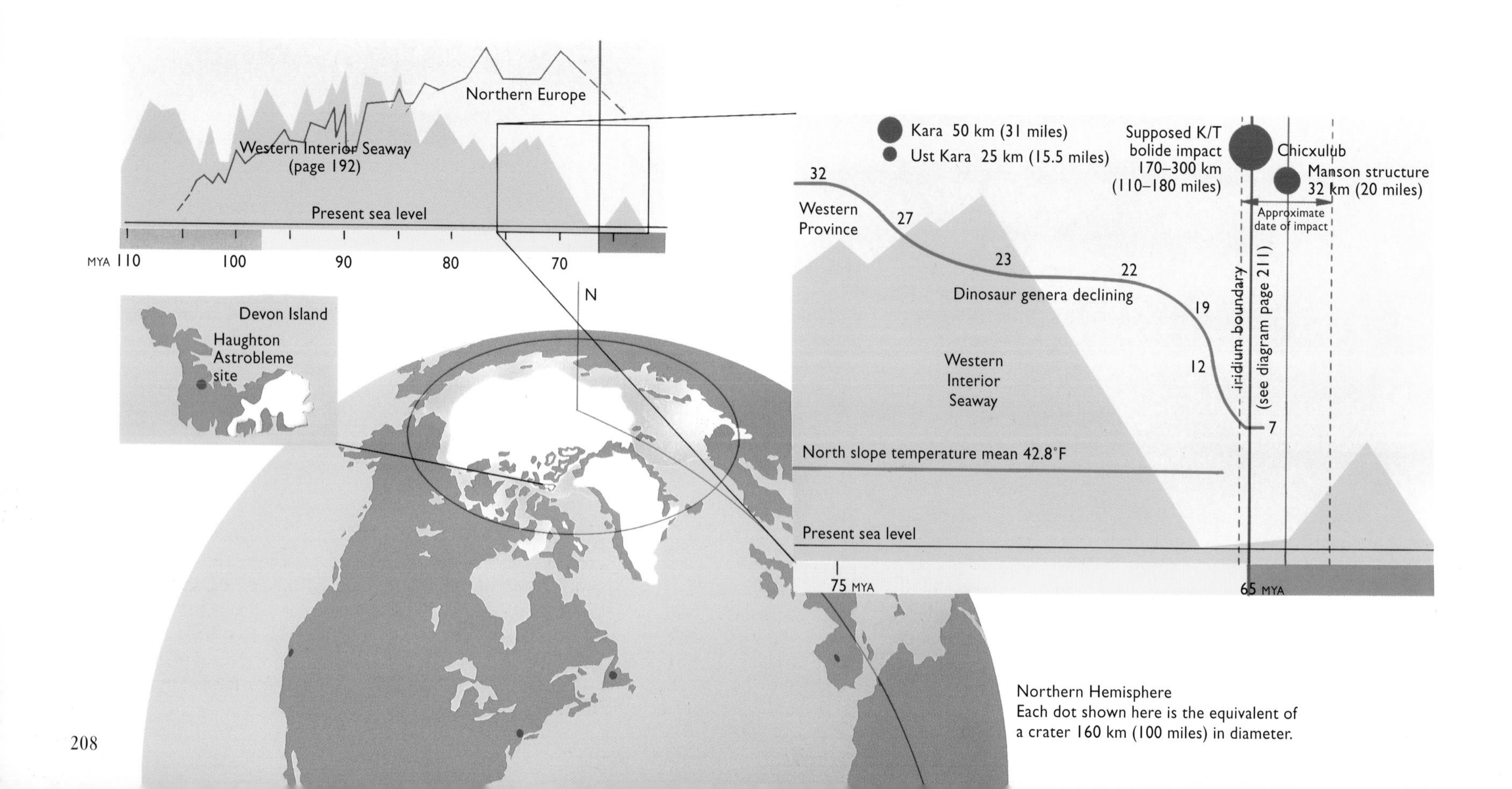

Northern Hemisphere
Each dot shown here is the equivalent of a crater 160 km (100 miles) in diameter.

Haughton Astrobleme
The bolide that caused the impact structure pictured here is considered to have been far too small an object to have produced a global mass extinction, although its destructive force was the equivalent of more than 25,000 Hiroshima-sized atom bombs. The Haughton Astrobleme is 20 km (12 miles) in diameter and was formed by a missile perhaps 1.6 km (1 mile) wide, that hit the Earth 15 MYA. In this aerial panorama we are approaching the second of the three eroded but once clearly concentric rings that formed a series of crater walls after the impact. The point of impact, the so-called target zone, is at the upper left-center of the picture.

best estimates suggest a structure measuring from 170 to 300 km (110 to 180 miles) in diameter.

The size of an impact crater depends on the bolide's size, composition, speed, and angle of impact. Apart from an estimate of size, these data are unknown. However, the object that formed Chicxulub could have been either a nickel/iron asteroid measuring at least 6 km (3.6 miles), a stony asteroid measuring at least 10 km (6 miles), or a comet nucleus measuring at least 16 km (10 miles). Calculations suggest that the range of sizes estimated for the Chicxulub crater "bracket the impossible to the possible conditions for mass-extinction" on Earth (J. Morgan et al., Imperial College London, 1996). In other words, Chicxulub could not be the prime cause of the K/T mass extinction if it turns out to be closer to 170 km (110 miles) in diameter, but it could have been responsible if it proves to measure 300 km (180 miles).

Looking now at dinosaur evidence dating from the end of Cretaceous time (see diagram overleaf), the lines that dissect the curve in the diagram indicate the number of genera of dinosaurs that lived at that particular time in the Western Maritime Province. According to these data, compiled by R.E. Sloan of the University of Minnesota and others, a dinosaur decline began about 75 MYA and leveled off for a few million years before going into "rapid" decline after 68 MYA. The number of genera of dinosaur fossils that have been found immediately below and above the iridium trace is approximately the same. According to Sloan and his co-workers, complete dinosaur extinction occurred about 40,000 years *after* the iridium trace was deposited.

Bolide Extinction

Recent calculations show that a bolide with several hundred times the destructive energy of the Haughton missile would have been necessary to have caused a global mass extinction. Such an object (or objects) could have struck either the land or the sea (or both). In any event the impact would have caused a partial loss of the atmosphere to space, the destruction of the ozone layer, and the mass production of acid rain. In addition, an impact on land would have produced sufficient aerosols (microscopic particles) to have temporarily blocked sunlight. This would have halted photosynthesis, and cooled the global climate for sufficient time to have reduced plants and starved animals, perhaps to the point of their extinction. An impact on the deep ocean, which is six times more likely, may have had a different effect because of the tremendous volume of water that would have been vaporized.

The Chicxulub crater, discussed in more detail on the previous page, is a structure from 170 to 300 km (110–180 miles) in diameter. Although Chicxulub is evidence for a K/T impact, recent calculations suggest that in spite of its size it might not have been big enough to cause global mass extinction. However, another bolide struck North America, near Manson, Iowa, in the same period as the Chicxulub bolide. This raises the intriguing possibility that the two are related, although this theory has not been thoroughly investigated yet. The Manson structure, which is buried beneath glacial till, is 32 km (20 miles) wide, and was formed by an object

[*continued overleaf*]

Bolide Extinction [*continued*]

that released about four times the energy of the Haughton bolide—well over 100,000 megatons. It is approximately 4,000 km (2,500 miles) north of Yucatán, near Manson, Iowa, and is dated 65.7 MYA (±1.0 MY).

If one assumes a shallow marine impact on a limestone base, as would have been the case for the Chicxulub bolide, then instant vaporization of a colossal volume of ocean would have produced deadly sulphuric and carbonic acid rain. Depending on the volume of material ejected, this could have had a profound adverse effect on the plankton at the base of the regional or global oceanic food-chain. A cataclysmic "tsunami" (Japanese for "harborwave") would also have swept across the Gulf of Mexico and far inland, ripping up and displacing carbonate reefs and other structures. Near the Brazos River in Texas, there is indeed evidence for such a giant tsunami in the Gulf of Mexico, on the opposite shore to Yucatán, at the K/T boundary.

Also at this site is an otherwise anomalous layer of sandstone capped with a thin layer of iridium-bearing clay. The formation is consistent with deposits of the kind that would be left inshore after being swept by a tsunami 50–150 m (150–300 ft) high. Such a wave would have been generated by the impact of a 10-km-wide (six-mile) bolide hitting the deep ocean up to 5,000 km (3,000 miles) away, or of a smaller bolide hitting shallow water a few hundred miles away. The size of the tsunami estimated from the Brazos River deposits actually suggests a mid-sized Chicxulub bolide, not the potential maximum size.

The K/T Boundary

The Cretaceous extinction, although considerable, was the smallest of the six major mass extinctions that have occurred in the past 600 MY. The block diagram provides a *global* picture of the period on 5 MY either side of the K/T boundary. As one can see, certain marine animals (index fossils) were in decline before the boundary, some expired at the boundary, and some survived the event. The dinosaurs were also in decline, and their remaining members were extinguished either at the boundary or shortly afterwards.

Although plant species lost about a third of their number, plant survivors proliferated. Existing mammals were little affected by the event—in fact the oldest known *primate* fossil was discovered at Hell Creek, Montana ... just one solitary tooth in the same formation as some dinosaur teeth.

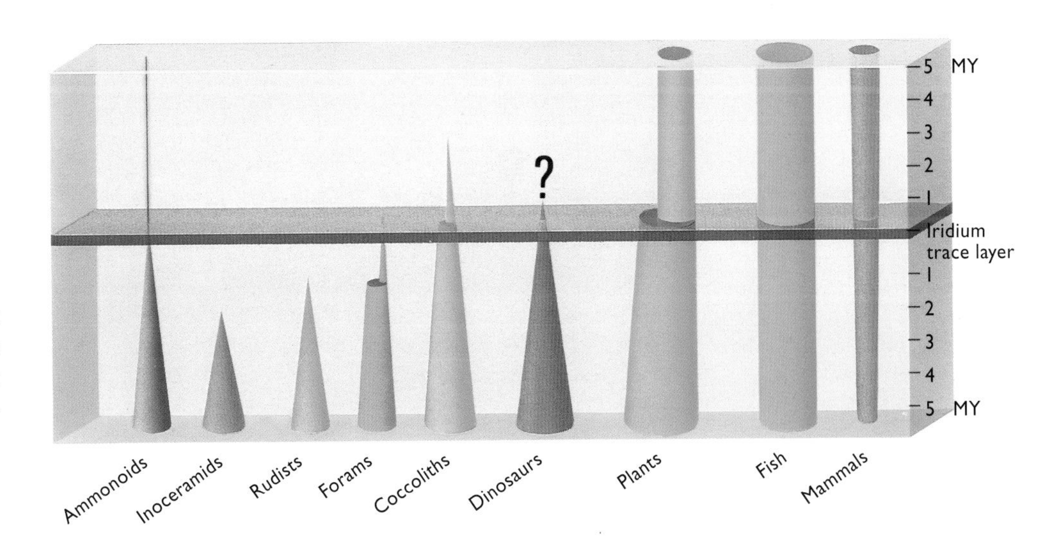

[*continued from page 201*]

wrong conclusions were drawn in the past, due to early misidentification of fossil plants by leading 19th-century paleobotanists (interpretations that were not critically reexamined until the early 1990s).

There is little doubt that dinosaurs ate both types of plants and that seeds were churned and trodden into the topsoil on the open plain, in the pine forests, or in laurel groves where herbivores sought shade. In fact heavy browsing inhibited the speciation of angiosperms, and archaic forms maintained over long periods.

Nor can there be much doubt that seeds were deposited on what is now the North Slope of Alaska. Here, 76 MYA, dinosaurs roamed near the edge of the polar sea on the migration route that linked the Western Maritime Province via Alaska to the Gobi Desert region of Outer Mongolia.

Today, the Chukchi Peninsula of Siberia marks the most eastern extremity of Asia. It is separated from the western extremity of North America by the Bering Strait, just south of the Arctic Circle and 2,800 km (1,750 miles) from the North Pole. In Late Cretaceous times this region was only 1,120 km (700 miles) from the pole. Although the plate boundary is diffuse, in tectonic terms the Chukchi Peninsula of Soviet Russia is thought to be part of the North American plate and the Bering Strait itself a recent feature. Around 99 MYA the Chukchi region began to collide with Siberian Laurasia, completing the enclosure of the North Polar region by 84 MYA. As the landmasses collided, they completely trapped the polar sea, although the contained Arctic Ocean might still have spilled over the landbridges from time to time.

The uplifted area of collision between Chukchi and Siberia was a physical extension of the Brooks Range region of Alaska and the Laramide Rocky Mountains. It provided the only means by which Hadrosaurs and Ceratopsians, could have migrated between Mongolia and the Laramides, assuming a transfer from Asia to North America (**22**). Between 84 and 75 MYA, it would have been possible to drive a metaphorical jeep from the Gulf Coast of Texas via the Chukchi Peninsula to Mongolia in Eastern Laurasia—without a map. Geographic constraints of inland seaways and fringing mountain ranges alone would have determined the route, but crossing innumerable rivers, alluvial fans, and deltas would undoubtedly have been daunting tasks.

Following the Laramide sector shoreline, the jeep would eventually have reached the present Colville River on the North Slope of Alaska. The river now drains the flanks of the Brooks Range, but the region was once washed by the Cretaceous Arctic Ocean. Where the Colville River flows on the North Slope (**23**), one would certainly have seen groups of Hadrosaurs. In Late Cretaceous times, that vicinity was also within 1,120 km (700 miles) of the North Pole.

The dinosaur-bearing sedimentary rocks of the Colville River were first discovered in 1961 by R.L. Lipscombe, but their significance was not fully realized until 1984. Since then the area has been intensively studied by paleontologists and paleobotanists. Fossils of duck-billed Hadrosaurs (**24**), the principal animals in that locality from 76 to 68 MYA, are so numerous in the Colville beds that they may represent herds of adult animals. They are complete with juveniles and with attendant predators like Tyrannosaurs and Troödonts.

The plants found in the fossil assemblages are mainly low-lying horsetails, ferns, and herbaceous shrubs. However, stands of canopy-forming *deciduous* conifer trees, cedar-like gymnosperms, grew on higher ground, and these trees were intermixed with broad-leaved angiosperms [Essay 5]. This evidence suggests to paleobotanists that parts of the Colville region at this time were clothed in cool-to-temperate forests. Such forests became increasingly dominant and stretched as far south as the Montana region of the Laramide sector.

More questions are posed by these intriguing fossil data than are answered by them, and conjecture is rife. For instance, did the Colville Hadrosaur herds migrate seasonally like the modern caribou? If so, bearing in mind that the dinosaurs were supposedly tied to their nests for prolonged periods, how could infant Hadrosaurs have survived the trek? Or perhaps the Colville Hadrosaurs genus were ovoviviparous (L. *ovum vivus parere*, literally 'egg alive to give birth'). This tongue-twister implies that the Colville Hadrosaurs, instead of just laying eggs in a nest, might have given birth to live young without placental attachments, from internal eggs hatched within the body. Species of modern lizards that are exposed to cold temperatures either at high latitudes or at high altitudes are often ovoviviparous; for example, all New Zealand gekkos bear live young, but all other gekkos lay eggs. If Hadrosaurs were ovoviviparian. this could have overcome both the difficulty of local movement and migration. It would also have solved the problem of egg temperatures and sex ratios—assuming egg temperature was as crucial to dinosaurs as many paleobiologists believe. Even so, if they lived in the Arctic, how did Hadrosaurs survive the winter nights?

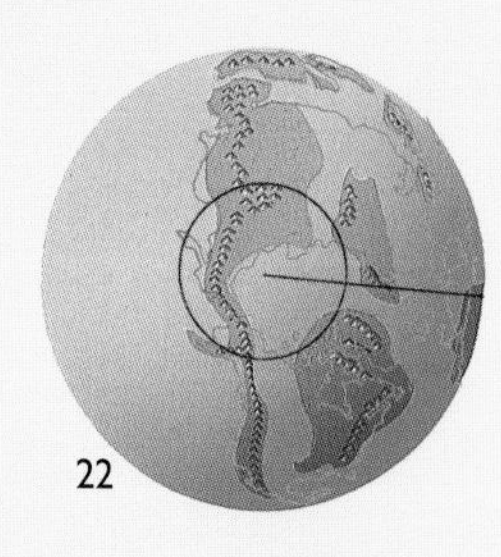

22

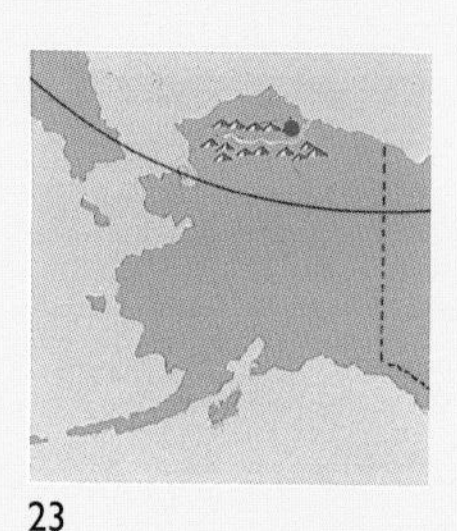

23

24

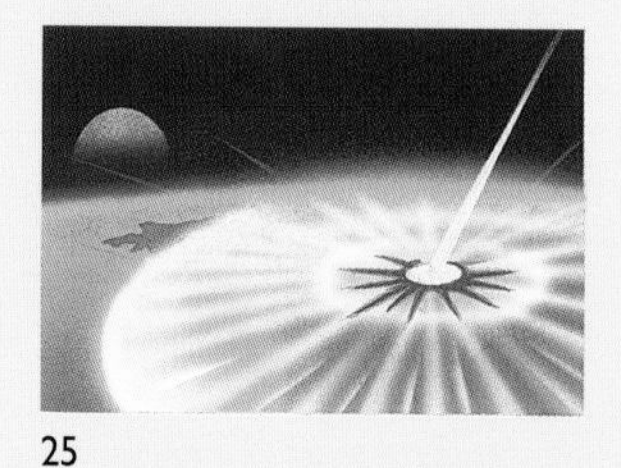

25

The absence of sunlight in high latitudes in winter meant that plant photosynthesis just did not occur for a prolonged period each year. This accounts for the deciduous character of normally evergreen Cretaceous plants, but also raises a question about how dinosaurs survived for at least the four months of darkness each year. Would this prospect have favored "warm-blooded" dinosaurs? Is it an argument for a hibernating genus of Hadrosaur? Or did the animals keep warm by fermenting woody food instead of leafy food in their primitive stomachs in wintertime, as some suggest? To add to the unsolved mystery, although the average annual temperature for the North Slope of Alaska was an attractive 10–13°C (50–55°F) until about 88 MYA, four million years later—when the proto-Bering Landbridge was in being—North Slope temperatures had fallen on average to an unwelcoming 2–6°C (36–42°F) [Essays 5 and 6].

With the rearrangement of continents and deterioration of climate, the great Cretaceous carbonate age was coming naturally to its end. Then, like a bolt from the blue, 65 MYA, the Earth was struck by a single large object, or perhaps by several objects from outer space (**25**). This bolide (or bolides) could have hit either the land or the ocean, or both land and ocean [Essay 7]. It could have been a stray asteroid from the asteroid belt that orbits the Sun between Mars and Jupiter, where asteroids measure from less than 1–770 km (0.6–480 miles) in diameter, or a comet from the Oort Cloud of icy objects about a light-year from the Sun.

Evidence that a violent impact took place at the end of Cretaceous time has accumulated since a hypothesis to this effect was first proposed by L.W. Alvarez and his father W. Alvarez and others in 1980. Evidence is now so formidable that most scientists today accept that a major impact or several impacts took place at this time. The prime proof lies in a thin layer of clay that sometimes overlies the youngest strata of Cretaceous rock in widely separated parts of the world.

The term "clay" is a definition in geology: it implies a mix of particles of specific microscopic size. The actual mineral content of lithified Cretaceous clay varies from one locality to another. However, the clay deposit at the Cretaceous boundary frequently contains a much higher concentration of iridium than that which naturally occurs on the surface of the Earth. Iridium is a rare metal in the platinum group of metals, found in high concentration in some bolides.

Other evidence of an impact at or near the end of the Cretaceous includes the presence of "shocked quartz" (crystalline silica), bead-like spherules of calcium silicate ("tektites"), and soot. All or some of these substances are occasionally found either in the clay layer or at the boundary if clay is not present. Shocked quartz and tektites are produced when rock is subjected to the pressure and heat of a powerful explosion. The presence of soot at the boundary has been attributed to forest fires, probably ignited by the blast.

Obviously the best direct evidence of impact takes the form of a crater—called an astrobleme (Gr. *astron blema*, meaning star wound).

For example, what appears to have been a cluster of bolides formed two sets of twin craters in Russia about 75 MYA, but these are not yet reliably dated. They fell at Kara (48-km-diameter crater: 30 miles), and Ust-Kara (24–48 km; 15–30 miles) within the Arctic Circle, and at Gusev (3.2 km; 2 miles) and Kamensk (24 km: 15 miles) near the Black Sea, then part of the closing Tethys Ocean. In addition, and more significantly, a subterranean astrobleme has been identified at Manson, in Iowa, 960–1120 km (600–700 miles) east of the Laramide sector. Originally, this was dated *c.*70 MYA, subsequently revised to be 65.7 MYA (plus or minus 0.5 MY), but the structure is only 32 km (20 miles) across. The bolide that caused this crater was far too small to have been the prime cause of a global mass extinction. However, the discovery in 1991 of another much larger subterranean impact structure at Chicxulub, on the northern end of the Yucatán peninsula, is the most promising discovery so far. The Yucatán impact at Chicxulub is believed by some to have coincided with the Manson event, and the Manson bolide may have been a fragment of the larger Chicxulub object (**26**).

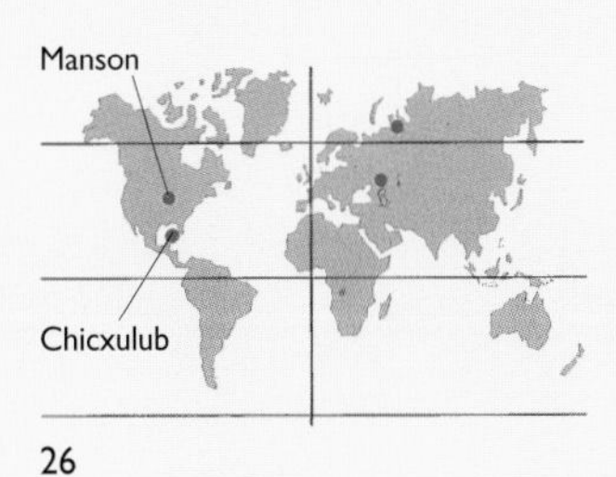

26

THE CHICXULUB BOLIDE formed a crater estimated to be from 170 to 300 km (100–180 miles) in diameter. It is buried beneath a kilometre (0.6 mile) thickness of limestone rocks now deposited on top of it, and although it is difficult to measure accurately, it has been modeled by 3D imaging of seismic waves passing through it. When Chicxulub was discovered, it was promptly dubbed "the smoking cannon" by American geophysicists, for the crater seemed to be the right size to qualify as the long-sought K/T astrobleme [Essay 7].

A major survey of the Chicxulub astrobleme was carried out in 1996 by an international team headed by J. Morgan, L. Warner (Imperial College, London) and P. Maguire (Leicester University). According to its summary of findings, the team estimated that the Chicxulub crater could have been produced by the impact of a nickel-iron asteroid at least 6 km (3.6 miles) across, a stony asteroid 10 km (6 miles) in diameter, or an icy comet at least 16 km (9.6 miles) wide. According to the report, if the crater proves to be nearer to the lower estimate of size, whatever angle of impact and speed of approach, the ejecta would have been insufficient to have affected the global climate. If nearer to 300 km (180 miles), the ejecta could have affected climate, perhaps sufficiently to have caused a mass-extinction of species. The report states that "the end-effect is not linearly proportionate to the impact size. The 170–300 km range brackets the impossible to the possible conditions for mass extinction."

Additional factors must be allowed for in any judgment. Computer models based on the larger diameter suggest that global temperature would have been reduced by several degrees for a time after impact. The most important side effect would have been caused by sulphuric and carbonic acid falling as rain, the result of an impact on marine limestone. Fatal for plankton at the base of the oceanic food chain, but how widespread would the effect have been? And so the argument goes on ...

NOW THAT A POSSIBLE IMPACT SITE of the right size has been found, the continuing debate is whether a bolide impact, even of this considerable size, could have induced a *geologically* instant change in the direction of evolution on Earth. Most physicists and geochemists believe that this is possible. Most paleontologists believe that it is not—that the impact was a *coup de grâce* and not a prime cause.

In a countdown to the impact and beyond, we will look at the period 74.5–65 MYA through the eyes of R.E. Sloan of the University of Minnesota and his fellow paleontologists. They work at Hell Creek Canyon in Montana (**27**), a region that is extraordinarily rich in dinosaurs and other fossils of very Late Cretaceous time.

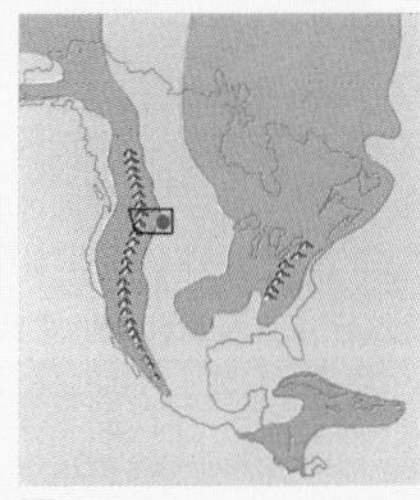
27

By measuring the thickness of a section of sedimentary rocks at Hell Creek accurately, estimating the time required for its deposition, and then carefully excavating that section for dinosaur fossils, Sloan and his associates have shown that there were 30 genera of third-regime dinosaurs in that region 75 MYA. This is the date of the peak of the dinosaurs' diversity in Hell Creek. An overlying and therefore younger 16-m (53-ft) section of fossil-bearing rocks revealed that in the period 74.5–69.5 MYA the local population of third-regime dinosaurs declined to 22 genera, and by 68 MYA to 18. Between 68–67 MYA there was a strong influx of rat-sized mammalian species, both marsupial and placental animals, and by 66.4 MYA—still *below* the iridium trace—Laramide-sector dinosaurs had declined to twelve genera. Furthermore, according to K.R. Johnson and others, in the quarter-million years *before* the impact shown by the iridium layer, one Laramide-sector forest consisting of 32 types of trees and shrubs was replaced by another of entirely different character. This pattern of decline included the extinction of at least one member of the laurel family. But K.R. Johnson and L.J. Hickey have observed that there was also a sudden, drastic, and short-term change which occured at the Cretaceous boundary itself. The first event is attributable to climatic change, they suggest, and the second to the bolide impact.

Subsequent to the impact—that is, *above* the iridium trace—the fossils of at least seven and possibly 11 genera of dinosaurs that existed before the impact have been found, according to Sloan and his colleagues. These include Ceratopsians, Hadrosaurs, and the smaller, "sickle-clawed," carnivores, which included one new genus. From the thickness of the post-impact sedimentary rocks in which these fossils are found, and the length of time it is estimated that the rocks took to form, it seems that dinosaurs might have continued to live on in the Laramide sector at least 40,000 years after the bolide impact. Genera of dinosaurs claimed to be an even younger age have also been found in Mongolia and the Pyrenees. In the Pyrenees, five successive post-impact strata, dated and cross-checked in many ways, contain nests and eggs of "*post*-Cretaceous dinosaurs"—a self-contradictory term since the end of Cretaceous time is by tradition marked by the end of the dinosaurs!

There are of course counter-arguments: the validity of paleontological evidence is challenged by supporters of bolide-impact extinction; the global destructive power of the presumed bolide is questioned by dissenting paleontologists. However, no matter how intense the debate, neither side has stooped to follow in the Owen-Wilberforce tradition by questioning the doubtful parentage of the other. The debate is still being conducted within the unwritten rules of the Darwinian scientific tradition.

THE FEW DAYS DARWIN SPENT with Adam Sedgwick, in the Vale of Clwyd and at Great Ormes Head in North Wales, appear to have been the extent of his induction into the realm of practical geology before his departure on the *Beagle*. The result of Sedgwick's work that season led to a radical change in the understanding of rock formations along the coast of North Wales from Denbigh to Anglesey; but perhaps the most important result was that Sedgwick had shown a young and reluctant classics graduate that science is a way of organizing and interpreting pragmatic observations. Darwin's latent talents, fired by his experiences during the voyage on the *Beagle*, and by Charles Lyell's *Principles of Geology*, had done the rest.

During the voyage of the *Beagle* Darwin developed his own scientific method: he formulated a hypothesis based on an initial observation, and deduced the consequences that should follow if the hypothesis was true. He (or others) then accepted, modified, or rejected the original thesis by making further and continuing observations. And it is in that empirical spirit, admittedly heated at times, that the present debate about the mysterious demise of the dinosaurs will ultimately be resolved.

X

Midland and Nordic Seas

The billowing landscapes that characterize much of southern England include gigantic steep-sided folds called "monoclines." Lulworth Cove and Stair Hole (foreground), pictured here, are the heavily eroded step-like remnants of one such monocline. The three rock formations that comprise this scene were deposited one above the other at widely separated times. The tightly folded limestones were laid down in horizontal layers about 200 MYA. The younger, softer, and therefore more easily eroded sands and clays, of the sort immediately to the left of Stair Hole and in Lulworth Cove beyond, were deposited on top of the limestone formation about 130 MYA. The hills that rise above and to the left of Lulworth are formed from chalk deposited on top of the sand and clay formation about 84 MYA. Finally, 20 MYA, the region was uplifted and folded by a powerful force generated by the collision of Africa with Eurasia, during the closure of the Midland Sea.

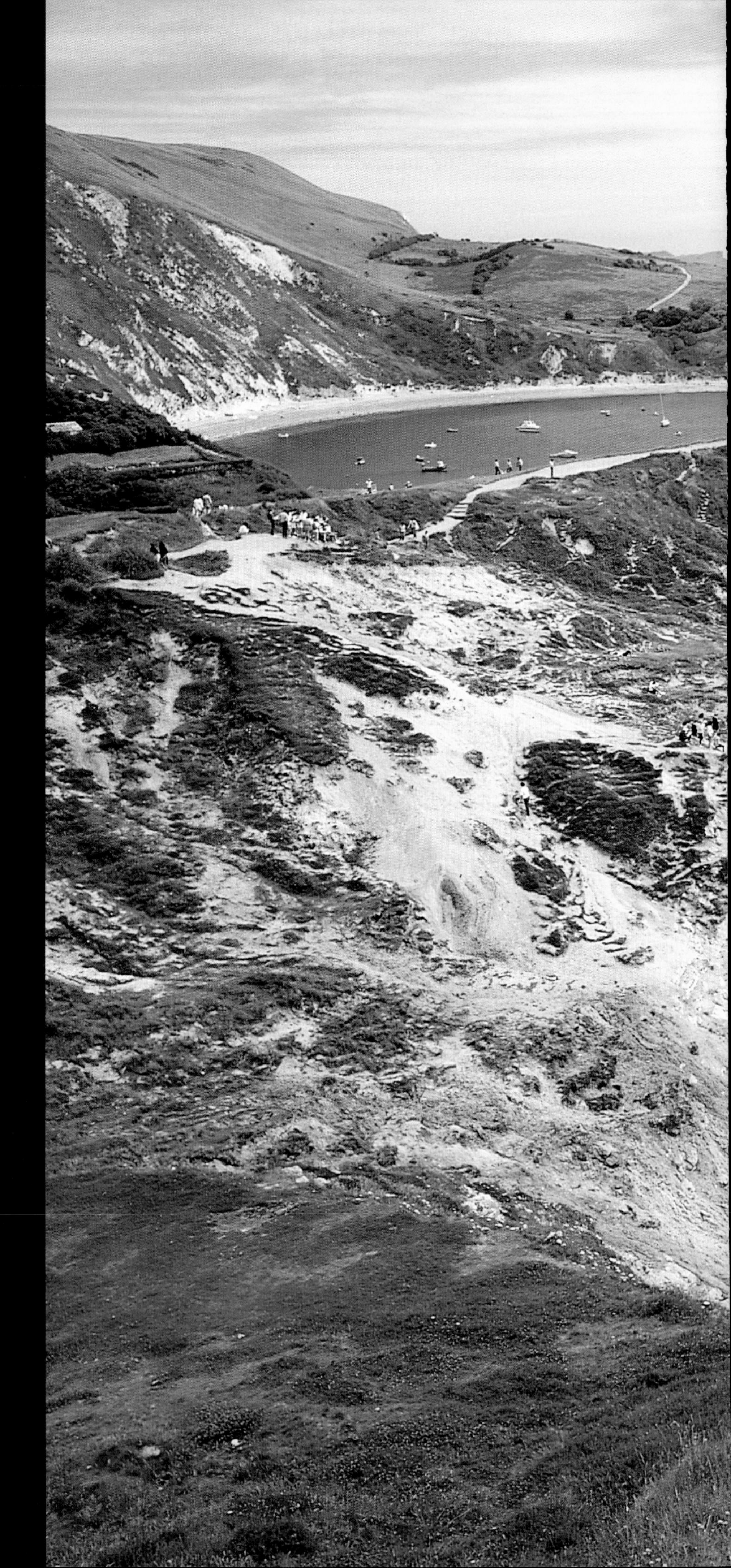

Lulworth Cove
and Stair Hole,
Dorset, England

MIDLAND AND NORDIC SEAS
65–1.6 MYA

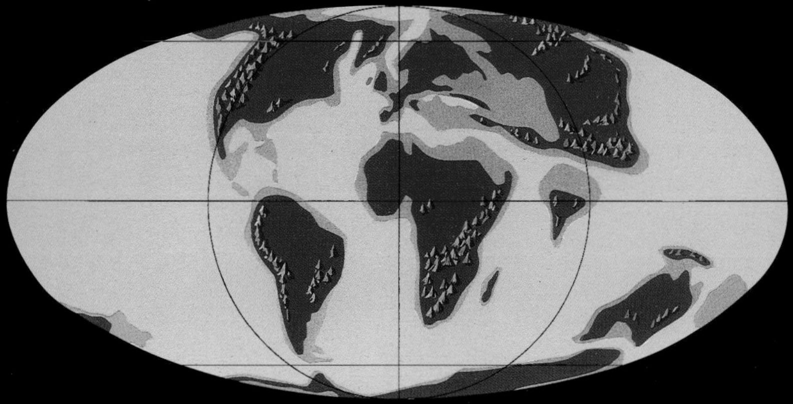

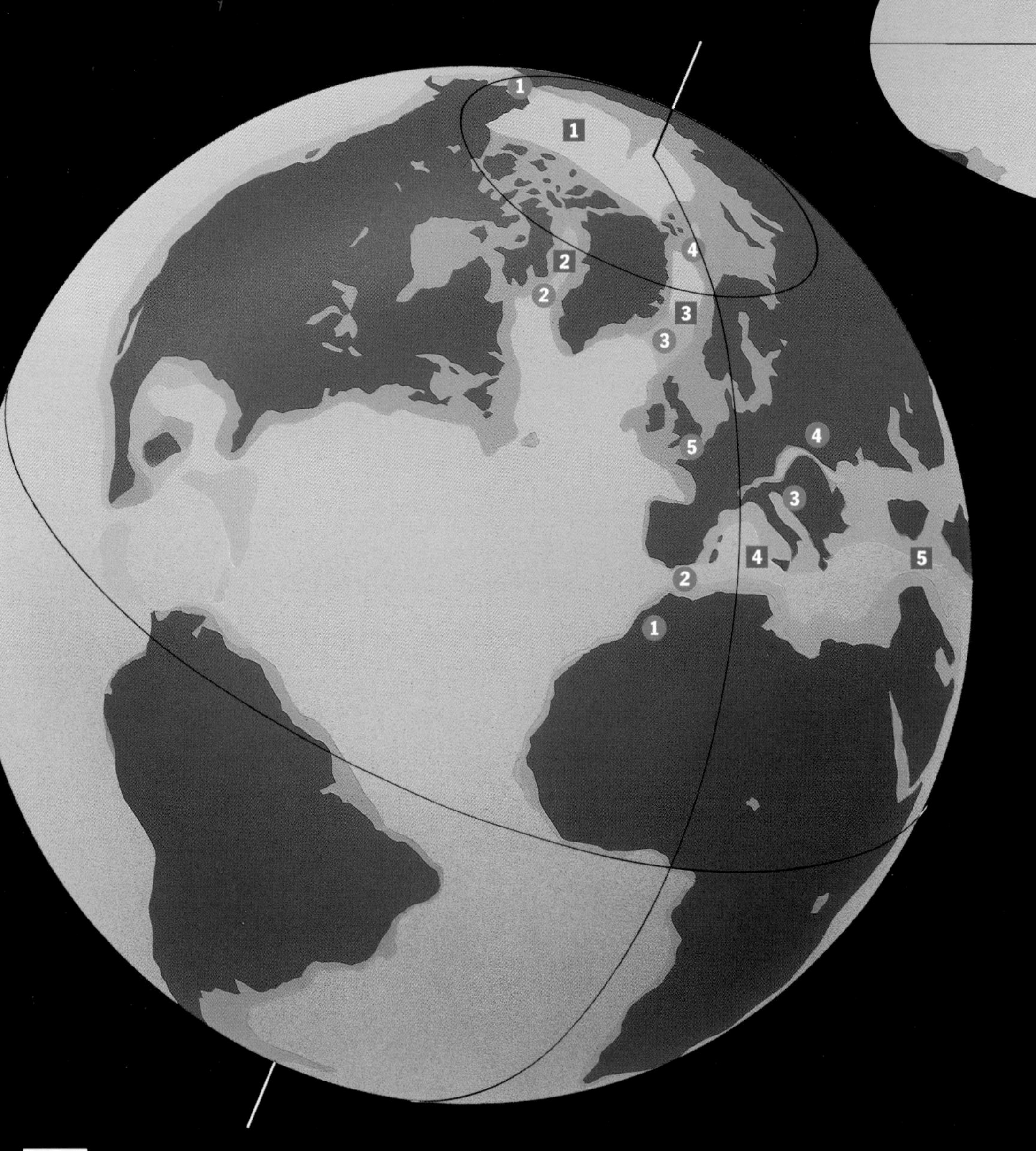

MIDLAND and NORDIC Marine Realm

1. Arctic Ocean (enclosed)
2. Baffin Bay-Davis Strait
3. Nordic Sea
4. West Tethys-Midland Sea
   Mediterranean
5. East Tethys
   Black, Caspian, and Aral Seas

LANDBRIDGES

1. Bering
2. Thulean
3. Greenland-Scotland Ridge
4. DeGeer
5. Dover-Calais

MIDLAND and NORDIC Terrestrial Realm

1. Atlas Mountains
2. Strait of Gibraltar
3. Apulian Prong
   Italy, Yugoslavia, Western Greece
4. Alps and Carpathians

These projections at 48 MYA show that transglobal ocean circulation continued despite the constriction of the Tethys Ocean and its reduction to the Midland Sea. Ultimately India's collision with Asia closed the Eastern Tethys, to form the Black, Caspian, and Aral Seas. Later, the collision of Africa-Arabia with Laurasia closed the Western Tethys, now the Mediterranean. Meanwhile a prong-shaped microplate, which included Italy, Yugoslavia, and western Greece, calved off North Africa and collided with Laurasia during the Alpine orogeny.

As these events progressed, the Nordic Sea was forming in the region between the Arctic Ocean and Greenland-Britain, the final stage in the formation of the North Atlantic Ocean as we know it. The paleoglobe shows an isolated Arctic Ocean before the separation of North America from Eurasia, and the location of the landbridges that joined the two continents.

The oval projection shows the island arcs that were being formed in the Caribbean region at that time. It also shows the movement of India relative to Australia, and the failing connection between South America and Antarctica. These distant tectonic events played a major role in the onset of the present Cenozoic Ice Age.

The global shift of continents during the present Cenozoic Era established today's familiar geography of the North Atlantic realm. Realignment also resulted in significant changes in the pattern of global ocean circulation and wrought geographical transformations that determined the distribution of life on northern and southern continents. It sparked a climatic revolution that determined the prehistory of the Northern Hemisphere.

The prime factor in this many-faceted set of changes was that ultimately Atlantic Ocean currents were forced to adopt a north-south "interpolar" flow. The previous equatorial flow of the global ocean had dominated the global climate since the breakup of Pangea, but it now gradually reduced and ultimately stalled. During this process the North and South Atlantic Ocean circulation patterns, called gyres, established an intricate pattern of interchange—a prototype for the system of global heat-exchange engine we see today. However, the current model did not develop until 3.5 MYA.

The realignment of global circulation was a result of several major tectonic transformations, culminating in a minor, but crucial, event: the formation of the Isthmus of Panama about 3.5 MYA. At that point, Northern Hemisphere climate began to deteriorate towards the present Ice Age [see Chapters XI: *Fountains of Youth* and XII: *Children of the Apple Tree*]. But by that time Antarctica was already sheathed in ice, and had been for at least 15 MY (and possibly for twice that time). Since the end of the Carboniferous Ice Age 260 MYA, the natural state of the global climate averaged about 16°C (30°F) higher than today's average temperature. Previous ice ages occurred at intervals several hundred million years apart, and, once established, most persisted for periods in excess of fifty million years. By this measure ice ages are both rare and prolonged events.

We are at present experiencing an interglacial interlude with the possible threat of a "greenhouse" stage. These tendencies are quite misleading indicators of the Earth's long-term climate. Warm interludes during ice ages are commonplace. The fossil record also shows that it is usual for prolonged ice ages, or even deep glacial stages within such ice ages, to be followed by an extinction event. The extinction that followed the last glacial advance, when most ice-age mammals died out between 10,000 and 8,000 years before the present (BP), is a good example of a minor extinction. That which followed the Carboniferous Ice Age in Permian times is a good example of a major mass-extinction event.

The pattern of global ocean circulation was a determining factor in establishing a particular style of global climate. An equatorial circulation, as in Cretaceous times, secures a prolonged greenhouse age. An interpolar circulation, as now, ensures a prolonged refrigerator age. Thus the tectonic events that determine changes in global-ocean circulation are key mechanisms that drive both mass extinction and the resurgence of life thereafter. We will now attempt to reconstruct the key tectonic and evolutionary events that preceded the closure of Panama, and the start of the present Cenozoic Ice Age. This story naturally divides into two sequences—those that occurred before 36 MYA, and those that occurred after 36 MYA.

Before 36 MYA, sweeping changes in paleogeography took place in the Southern Hemisphere. In the Northern Hemisphere, Europe remained attached to North America by landbridges. Meanwhile, equatorial ocean currents flowing via the shrinking Western Tethys (now the Mediterranean Sea—the "Midland Sea" of the chapter title) continued to influence the global climate (**1**). By 36 MYA, these warm currents were greatly reduced in volume. In the evolutionary field, mammals had evolved rapidly after the demise of the dinosaurs 65 MYA. Their evolution had developed along three almost distinct lines, two of which predominated: marsupials in the Southern Hemisphere and placentals in the Northern Hemisphere.

After 36 MYA, ocean circulation in the North Atlantic began slowly to change from equatorial to interpolar. This partial change was largely caused by the separation of Australia from Antarctica, and Antarctica from South America. These events were followed by the isolation of Antarctica over the South Pole. In the Northern Hemisphere at this time, North America separated from Europe in the region between Greenland and Svalbard: the Nordic Sea of the chapter title, that part of the North Atlantic beyond Iceland. With all landbridge connections severed between North America and Europe, the evolutionary picture changed dramatically.

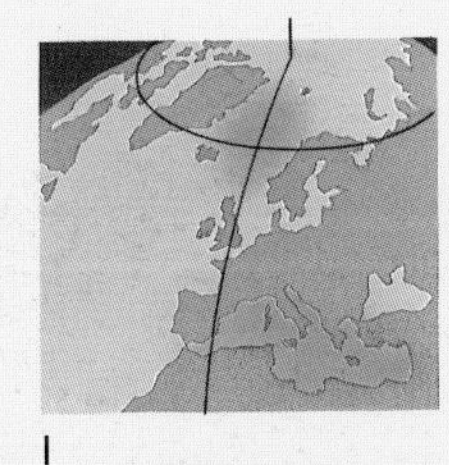

1

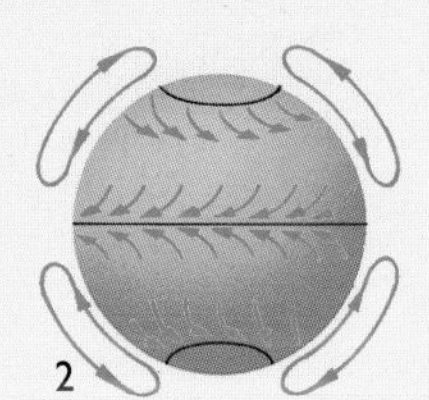

2

Since changes in the pattern of global ocean circulation play a major role in the story that follows, we must start by establishing the forces that drive it. These forces are easily seen in the phenomenon of draining bath water—why should it twist in opposite directions in opposing hemispheres of the planet, rotating clockwise down drains in the Northern Hemisphere, and counterclockwise down drains in the Southern Hemisphere?

The same force that governs the direction of the bath water also governs the direction of weather circulation. It determines the direction of rotation of the northern and southern ocean currents, and also accounts in part for the behavior of even more complex stratified ocean currents and eddies; the east-west path of hurricanes in tropical regions of the North Atlantic; and the clockwise rotation of tornadoes, waterspouts, and hurricanes in the Northern Hemisphere. South of the Equator corresponding phenomena rotate in the opposite direction. The discovery of the prime cause of this phenomenon took a century of investigation.

In an address to the Royal Society of London in 1735, George Hadley (1685–1768), an English physicist and meteorologist, first offered an explanation for the cause of the trade winds that blow westwards and away from the Equator in both hemispheres. According to Hadley, there are rotating cells of air above the Earth's surface (**2**) and forming from both Poles to the Equator. They are caused by currents air that rise as they are heated at the equator, and descend as they cool closer to the poles. But Hadley's idea did not take the rotation of the Earth into account—that factor did not arise until a French academic published a paper on another subject a century later.

Gustave Gaspard de Coriolis (1792–1843) was a mathematician and engineer who specialized in the study of relative motion and inertial forces in the field of mechanical engineering. In 1835 he showed that in addition to the ordinary effects of motion upon a wheel, there is an inertial force that acts upon the rotating surface, at right angles to its direction of motion. This component is now called the "Coriolis force," and the result of its action on free-moving bodies is called the "Coriolis effect." It is clearly seen in the apparent deflection of a free-moving object such as a spacecraft, relative to the rotation of the Earth's surface. Coriolis's abstruse engineering theorem explained many previously unexplained anomalies in unrelated fields of science—particularly in astronomy and military ballistics. The theorem also provided a basis for explaining the east-west direction of trade winds and surface-water currents at the Equator. The best way to understand this is by analogy.

Most table-top globes of the Earth are made to revolve on the globe's south-polar axis. They are connected to the north-polar axis by a semicircular bracket on which degrees of latitude are inscribed. Using this bracket as a guide for a felt marker, a straight line can easily be drawn on the surface of a stationary globe from its north or south pole to its equator. If the line is drawn as the globe is rotating in a counterclockwise direction (the direction of the Earth's rotation), the line produced by the marker will be curved. It will be deflected to the left—to the "west" on the table-top globe's surface. By holding the marker firmly against the cursor we caused an "inertial force" to act upon the marker, a force that acted at right angles to the moving surface of the globe—thus producing a Coriolis effect.

Suppose now that the table-top globe is motorized and that we

[*continued on page 220*]

DETAIL
TOP GLOBE

1 Central Atlantic Ocean
2 Rotation of Africa
3 Apulian Prong
4 Paratethys (Black, Caspian, and Aral Seas)
5 Indian Ocean
6 Restricted circumequatorial circulation
7 India

DETAIL
MIDDLE GLOBE

1 Rotation of Africa
2 Apulian Prong
3 Western Tethys
4 Arabia
5 Turkey
6 Iran
7 Eastern Tethys
8 India
9 Circumequatorial circulation

DETAIL
BOTTOM GLOBE

1 East Central Pangea
2 Apulian Prong
3 Arabia
4 Turkey
5 India
6 Iran
7 Tethys Ocean
8 North China
9 South China

1 Midland Seaway

About 250 MYA a great ocean, named after Tethys, the wife of Oceanus in Greek mythology, stretched from the entrance of the Mediterranean Sea (inset) to the China Seas. Today, the Black and Caspian Seas are all that remain of the Eastern Tethys Ocean—which was closed during the collision of India with Asia—and the Mediterranean Sea is all that remains of the Western Tethys—which was closed during the collision of North Africa and Arabia with Eurasia. Such collisions are continuing events. In the past they have had a profound effect upon both global climate and the geography of the Western World.

As Laurasia and Gondwana rifted apart during the breakup of Pangea, the newly formed Central Atlantic linked the Tethys and Pacific Oceans, allowing the global ocean to adopt a circumequatorial flow. Warm Indian Ocean currents then flowed via the Western Tethys through to the Central Atlantic, and on to the Pacific via the Panamanian gap between North and South America. As the paleoglobes show, this flow was pinched off after 66 MYA. The

Sierra Nevada, Cordillera Penibética (Betic Alps), near Granada, Spain

seaway between the Eastern and Western Tethys was first constricted by the collision of Arabia (including Syria and Iraq) with Eurasia (Turkey and Iran), and finally blocked after 37 MYA.

During the closure of the Western Tethys, Eurasia had remained almost motionless, even as North African Gondwana rotated towards it. As Africa advanced, a prong-shaped microplate off the North African coast was conveyed northwards. This assemblage included Italy, Yugoslavia, and western Greece and is called the "Apulian prong," after Apulia, a region of southeastern Italy. Around 66 MYA the Apulian prong collided with Europe, with the result that the Swiss-Italian Alps began to form, a mountain-building event that gave its name to the Alpine orogeny, one that affected the topography of Mediterranean Afro-Eurasia. During the course of the orogeny the Ancestral Pyrenees received an Alpine overprint, the Betic Alps in Southern Spain (panorama) were formed, and the Atlas Mountains in Morocco were elevated to their present stature.

Strait of Gibraltar and the Pillars of Hercules, from Mt. Hacho, Morocco

[*continued from page 217*]

could use two felt markers simultaneously—one from each pole, meeting at the Equator. If each marker represents free-moving masses of air or water on the Earth's surface, they will make marks on the globe that curve westward until they meet at the equator.

Similarly, being detached from the Earth's surface, either air or water flowing from the North Pole to the south, or from the South Pole to the north on the rotating surface of the Earth, must also deflect to the west, meeting at the Equator. As they are pushed along by following masses of air or water, these masses cannot do other than return polewards along the west coasts of their respective oceans. They form gyres rotating counter to each other. To make a viable comparison between the table-top model and the Earth, we have to consider many other factors, such as the relative sizes and thus speeds of rotation of the experimental globe vis-à-vis the Earth.

Although the Earth rotates very slowly at its Poles, (half the speed of the hour hand on a wristwatch), at the Earth's Equator the surface is moving at about 1,600 kph (1,000 mph). Instead of the seconds that it would take to draw a line with a felt marker on the surface of a table-top globe, air masses take days or weeks to circulate between the Earth's poles and the Equator. Seawater takes even longer to flow great distances—centuries for ultracold bottom water on the ocean floors. Moreover, the Earth's surface is not smooth like the globe. It has mountain ranges that obstruct and deflect the even flow of air between polar regions and the tropics and continental slopes, and ocean topography that constrains the movements of currents in ocean basins. Even so, we can visualize the "felt-marker" Coriolis effect: the tendency for both air and water masses to be deflected to the west —just as draining bath water always flows west, but when constrained by a pipe forms a mini-gyre that rotates clockwise in London and counterclockwise in Melbourne (**3**).

To this greatly oversimplified picture we must now add a series of complexities that confound climate modelers, even with highly sophisticated modern computers. These include the effect of the seasons, the Earth's varying around the Sun, and the slow wobble of the planet's axis. Allowance must also be made for the effect of oblique and therefore weak rays of the Sun at the Poles in polar summers, and the complete absence of sunlight in these regions in polar winters. The effect of the Sun's direct radiant heat at the Equator at midday has to be considered, and its absence at night. The reflection of sunlight from ice, snow, and clouds has another important effect on climate, as do the absorption of heat by land surfaces, the cooling effect of evaporation from the surface of the sea, and the release of greenhouse gases by volcanic activity. Present and future climate models also allow for greenhouse gases released into the atmosphere by rice paddy fields, animals' digestive tracts, and the burning of rainforests and fossil fuels.

Such parameters had an immediate effect upon the Earth's climate in the distant past, and some caused cyclic effects tens of thousands of years apart. But there is one inexorable parameter with which all the others interact—the effect of shifting continents upon the circulation of the global ocean. It was the change in the configuration of the continents that most probably triggered the Cenozoic Ice Age. During the countdown to that significant moment 36 MYA, when North America separated from Europe, the key change in world geography was the destruction of the Tethys Ocean [Essay 1].

THE DESTRUCTION OF THE EASTERN TETHYS (**4**) began with the separation of Madagascar and the Indian subcontinent (including Afghanistan, Pakistan, and Sri Lanka) from East Africa about 130 MYA. About 100 MYA the subcontinent separated from Madagascar and moved off like a charging rogue elephant towards an arc of island microcontinents off Eurasia. As the gap between the subcontinent and Eurasia narrowed (**5**), first the seafloor between the two was destroyed, and then the two continents met head-on. The net result of the collision, which commenced about 50 MYA, was the formation of the Himalayas and the uplift of the Tibetan Plateau.

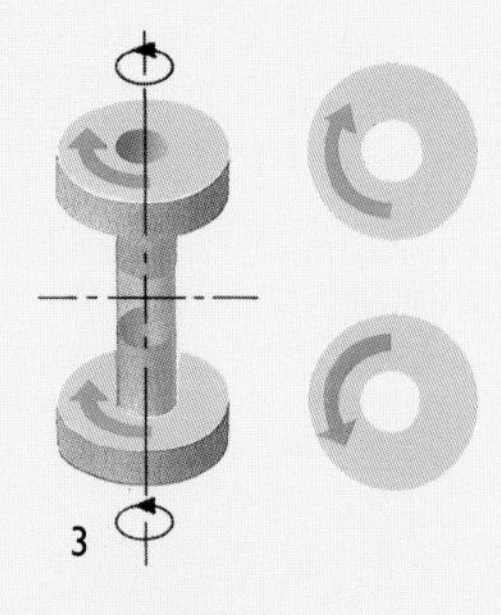

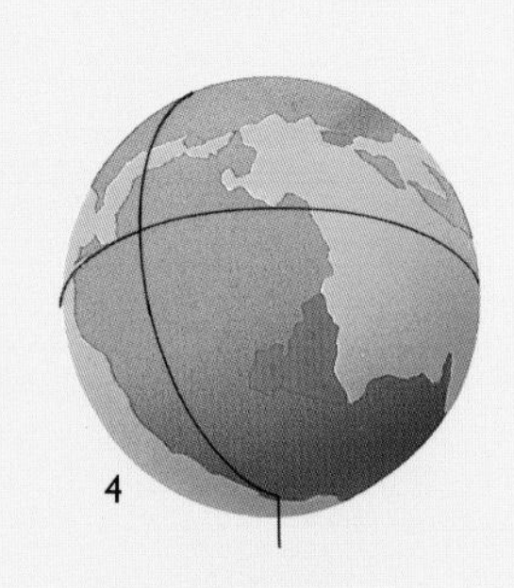

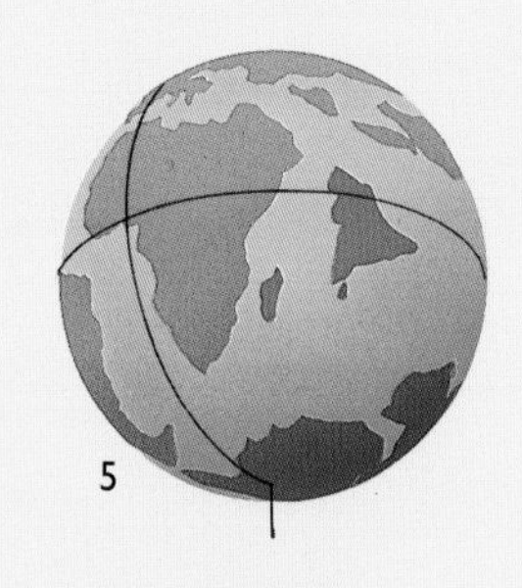

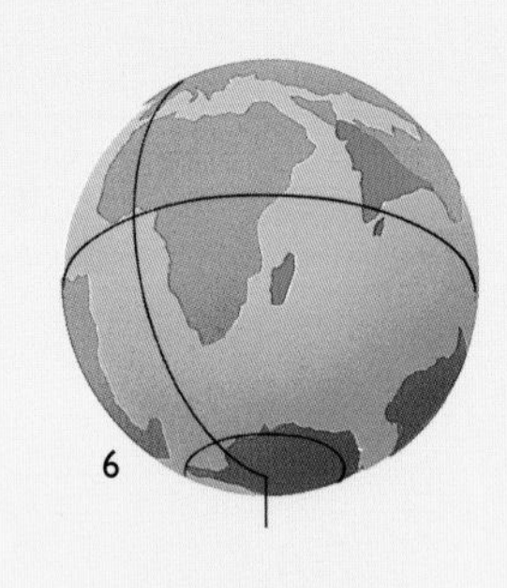

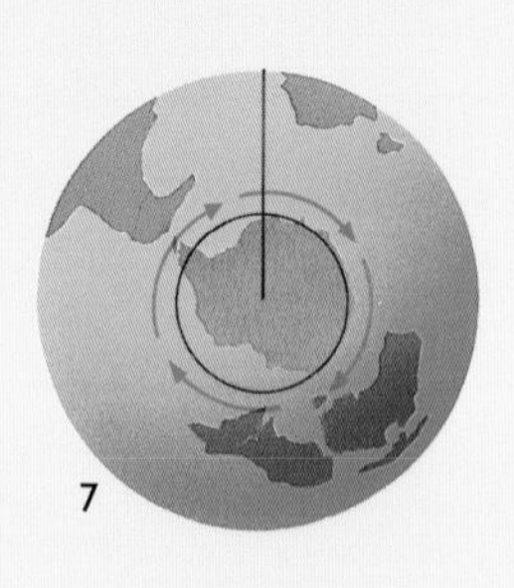

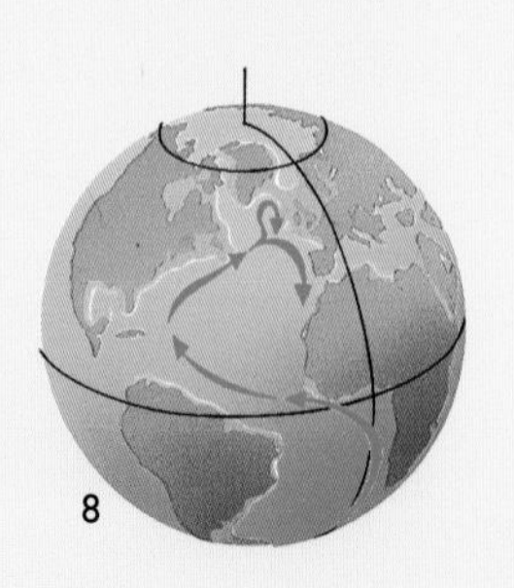

India, Antarctica, South America, and Australia had been as one when they were part of Gondwana. The breakaway of the Indian assemblage caused the South America-Antarctica-Australia block to detach itself as a separate plate, which rifted in two about 44 MYA. At this point South America and Antarctica began to drift towards the South Pole, where Antarctica appears to have stalled in quasi-isolation, while Australia moved off to its ultimate destination—a collision with Asia. Meanwhile, the closure of the Eastern Tethys had been completed (**6**), and about 36 MYA the Western Tethys was constricted to the point that the equatorial circulation of the global ocean was inhibited.

Although Australia and Antarctica were now separate continents, South America and Antarctica had retained their Gondwanan connection, and had been drifting northwards almost in unison. The region of Patagonia (the tip of South America) and its exceptionally wide continental shelf (the now largely submerged Falkland Island Plateau) were attached to the Antarctic Peninsula. As Antarctica stalled over the South Pole (**7**), South America continued to drift northwards. The link between the two continents became tenuous and was eventually severed completely, crustal separation leaving deep water between them by 36 MYA. This left Tierra del Fuego and the Falkland Islands on the north of the Drake Passage, and the South Orkney Islands and the Antarctic Peninsula on the south side of this seaway.

BECAUSE OF THIS BREAK, the continent of Antarctica was now isolated, surrounded by an ever-broadening, cool, circumpolar ocean with unrestricted circulation around the continent. Antarctica had lost the modifying influence of warm-surfaced neighboring continents and become thermally isolated. Glaciers now began to accumulate on the main transcontinental mountain range, and the first significant permanent sea ice had formed around the continent by 37 MYA (although some authorities, notably J.P. Kennett in 1982, doubt that an ice sheet of modern proportions developed in Antarctica until 15 MYA or later).

Extremely cold surface water, loaded with the brine released by seawater as it froze, now began to sink to the surrounding ocean floor, where it formed a layer of dense briny water that gradually accumulated around the continent. The newly formed Drake Passage allowed this layer of dense cold water to circulate around the continent, increasing in volume and starting to spill over from its low-lying ocean-floor basins. It spiralled out to form a global network of ice-cold bottom-water (**8**), called the "psychrosphere" (Gr. *psychros*, cold, frigid). The psychrosphere's river-like system [Essay 4] flowed through defiles in the South Atlantic and Central Atlantic abyss, and extended into the Indian and Pacific Oceans. In this way warm surface waters were displaced by equal volumes of cold bottom waters in tropical regions. Warm surface waters therefore flowed back towards the Southern Ocean and the Antarctic circulation system, where moisture released as they cooled fell as snow on high ground. As Antarctica's surrounding sea ice increased, it drove a general cooling trend in the global ocean, coinciding with a drop in global sea level. The latter was a consequence of the growing accumulation of terrestrial ice on Antarctica—by 23 MYA major deposits of rock debris on the seafloor, dropped by icebergs as they melted, suggest that the Antarctic ice sheet had reached the coast. By 20 MYA the diverse trees and shrubs that had once flourished on the continent seem to have been eliminated by the severe climate.

The development of the psychrosphere also coincided with the closure of the Western Tethys and the constriction of the previously dominant equatorial ocean circulation. At this time (36 MYA) the global climate was still warm and comparatively equable—although cooler than before the isolation of Antarctica. Also, although sea level was decreasing because of the accumulating ice on the southern continent, the general level was still higher than it is today. North America and Eurasia were still joined. The Nordic Sea had not yet developed spreading-centers, and Greenland was still attached to Eurasia by landbridges. Significantly, a generally balmy global environment had prevailed in the Northern Hemisphere in the 30 MY that had passed since the extinction of the dinosaurs, and this was to prove the critical period in which mammals rose to become the Earth's dominant animals.

An early evolutionary advance in the seas of this period was the appearance of ancestral whales in the equatorial Tethyan seas. The oldest known cetacean fossil (L. *cetus*, whale: **9**) was discovered in East Tethyan deposits in Pakistan (which was part of the Indian plate), and is dated 55–50 MYA. All cetaceans are mammals; today they include the *toothed* whales—dolphins, porpoises, and narwhals— and the *toothless* (baleen) whales—such as the gray whale and the blue whale, the largest creature on Earth. Ancestral cetaceans had teeth and skulls that resembled those of the carnivorous land mammals of the time; they also had similarities to the early ungulates—the hoofed animals (L. *unguis*, nail).

It is believed that cetaceans made a gradual transition from terrestrial to marine life. They spent progressively more time feeding on shallow-water fish in the highly productive seas and embayments that were characteristic of the closing Tethys Ocean. From their anatomy, and from the estuarine environment in which they lived, ancestral whales are judged to have been amphibious, for they lacked the auditory equipment necessary for directional hearing under water. Within another 10 MY or so (45–40 MYA) all the known whales had developed a hairless, streamlined, fish-like body, and evidently had become marine animals that quickly achieved wide global distribution. Their hind legs had lost their usefulness, their tail had developed into a fluke, and their forelegs had become flipper-like. In all other respects they retained the character of terrestrial mammals.

The earliest known terrestrial mammals, which evolved from mammal-like reptiles, date to about 210 MYA, and by 84 MYA a breed of squirrel-sized insectivores and omnivores was securely established in the shadow of the dinosaurs. Although by the end of Cretaceous time only a few herbivorous mammals had appeared, after the demise of the dinosaurs these few came to the fore. Their evolution was rapid, and by 62 MYA they were numerous and up to 1 m (3 ft) from shoulder to the ground. By 55–52 MYA the very early modern orders of mammals had evolved, and these can be divided broadly into three basic physiological types that persist to this day: monotreme, marsupial, and placental.

Monotremes (Gr. *monos trema*, single hole: **10**) have a single opening for their combined reproductive system, urinary tract, and anal canal. The ancient monotremes were probably oviparous, like the duckbilled platypus of today, and some had egg pouches (like the echidnas) in which to deposit and incubate their eggs, and then suckle their young. Fossil and living monotremes are found only in Australia, Tasmania, and New Guinea. They were evidently not able to compete with physiologically more advanced marsupials and placentals—which were far more adaptable animals.

Marsupials (Gr. *marsippos*, pouch: **11**) have a double opening, an anal canal plus a combined urinary tract and reproductive system. Marsupial placentas are inefficient in transferring oxygen from mother to fetus; consequently, marsupial young are born after about two weeks of development—the fetus would otherwise suffocate in the womb. After its precipitate birth, the young marsupial usually transfers to a pouch, where the infant suckles and completes its embryonic development.

Marsupials established themselves and diversified in South America, Antarctica and Australia. There is evidence that they originated either in North or South America and then migrated to Antarctica and Australia. They subsequently became isolated on the southern continents as they separated between 45 and 36 MYA. The oldest

9

10

11

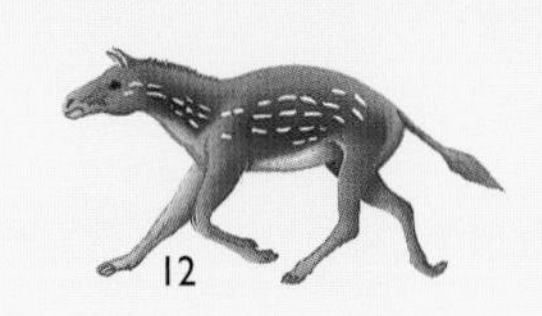

12

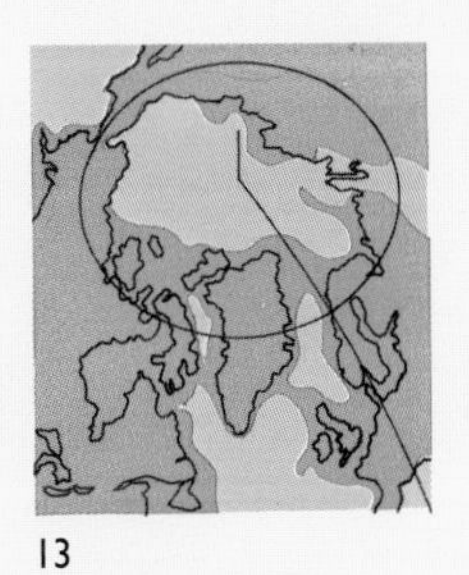

13

known fossil marsupials in Australia are dated about 23 MYA, but they are a very diverse group, which suggests that these animals had been evolving on that continent from about 50 MYA.

Placentals (Gr. *plakounta*, flat cake) have a triple opening: an anal canal, a urinary tract, and a separate reproductive system. The placenta permits oxygen to transfer from the mother's blood to the fetus's blood, allowing the processes of metabolism to continue throughout a longer gestation period. Additionally, the part of the placenta attached to the wall of the uterus secretes hormones that prevent the mother's immune system from rejecting the fetus; although it is enveloped in a watery environment, the fetus may literally be considered a "foreign body."

Some indigenous placentals, showing little diversity, were established in South America, but not in Australia or in Antarctica. Marsupials were dominant on all three of these continents. This evolutionary situation was reversed in the northern continents, where placentals became the dominant animals and marsupials failed to establish high-diversity populations. Yet the extraordinary fact is that the northern placentals and the southern marsupials developed in very similar ways: broadly, they looked alike and occupied similar ecological niches, but physiologically they were unalike—an almost perfect example of convergent (parallel) evolution.

The differences between southern and northern animals remain today: in the wild we see placentals like polar bears in the north but not in the south, and marsupials like kangaroos and others in the south but not the north. Such fundamental differences extend to innumerable other animals, birds, and plants. Fascinating though the subject is, we must not be drawn into comparisons and the reasons for the persistent separation of northern and southern flora-fauna: our prime focus is upon the "wholly" northern animals and plants (**12**), which are termed "holarctic" (Gr. *holos arktikos*, whole north).

Holarctic animals and plants evolved in warm and moist Malaysian-type climatic conditions on the still-intact post-Pangean megacontinent of Laurasia, which had crowded around the Arctic Ocean. Up to about 36 MYA North America was connected to Asia by the Bering Landbridge (**13**), and to Europe via two other transitory landbridges, the DeGeer and Thulean Landbridges. The latter connected Greenland to Europe across shallow sea-filled rifts where the Nordic Sea now exists [Essay 2].

It is thought that the Arctic Ocean and its surrounding continents remained warm because the psychrosphere's northward progress from the Central Atlantic was stymied by an obstruction. The vegetation in the middle to high latitudes of the Northern Hemisphere changed in character, but was far from being reduced and killed off as vegetation in Antarctica had been. It appears that *perennial* ice (as distinct from seasonal ice) did not form in the Arctic Ocean much before 3.5 MYA, and many scientists believe that it did not form until as recently as 1.6 MYA.

The Nordic Sea region had never fully recovered from the multicontinental collision that resulted in the formation of East Central Pangea. Laurentia, Baltica, and Siberia had collided first, then Avalonia had collided with Laurentia-Baltica, and finally Armorica-Iberia with the previous assemblage. And, as we saw in earlier chapters, each of these elements had rotated relative to each other as they collided. Although rifting in the zone between Greenland and Europe had gradually diminished, the weaknesses in this Nordic lithosphere had remained.

Apart from the DeGeer Landbridge, and the short-lived Thulean Landbridge, the rest of the Nordic intermediate zone was a no-man's land. The landscape was a mixture of flooded rift valleys, shallow and temporarily isolated inland seas, half-submerged microcontinents, and deep troughs (the latter now form the fully submerged Rockall-Hatton Basin that lies between Britain and Greenland). In addition, there were volcanic eruptions on a gargantuan scale, as flood basalts were released from a mantle plume—a decompressed

[*continued on page 228*]

Unnamed valley off Baillarge Bay, North Baffin Island

2 Northern Connections

During the breakup of Pangea, following the opening of the Central Atlantic 165 MYA, post-Pangean Gondwana fragmented, but Laurasia was to resist division until about 36.6 MYA. Laurasian continents clustered around the North Pole rather like defiant battalions under siege as Gondwanan armies (Africa and India) assaulted the Laurasian flanks.

The opening Labrador Sea (84–46.2 MYA) stalled and failed to develop into an ocean. Seafloor spreading commenced in Baffin Bay and Lancaster Sound (58–30 MYA), but these centers also failed. As a result, the entire Arctic archipelago was criss-crossed with failed rifts, and ultimately became a site of permanent geographic disarray. Meanwhile, although preliminary rifting between Greenland and Scotland-Norway had progressed, North America remained connected to Europe by two landbridges, the Thulean and DeGeer Landbridges. These are represented here by the inset picture and the panorama above respectively.

Both these now-barren regions are illustrative of the migratory routes between North America and Europe used by the early mammals—animals whose diversity had burgeoned after the demise of the dinosaurs at the end of Cretaceous time. Northern climates at

that time were still warm and moist, and food for browsing animals was plentiful. Because of the equable global climate and the existence of several landbridges, the early mammal populations throughout Laurasia were remarkably alike.

Asia and North America were connected by the Bering Landbridge, named after explorer Vitus Bering (1680–1741). They were also linked by the Thulean Landbridge, named from the Latin *ultimus thule*, the farthest north. This landbridge consisted of a chain of connections that were linked until about 50 MYA. The first connection was between Baffin Island and Greenland near the present Davis Strait. In turn, Greenland (then ice-free) was connected to Britain via the Greenland and Scotland Ridge, while Britain was connected to continental Europe by the Dover-Calais Landbridge (an uplifted region of the Anglo-Paris Basin).

The most enduring of the Euro-American landbridges is named after Gerhard De Geer (1858–1943) a Swedish geologist. The DeGeer was an intermittent means of animal migration from the Canadian archipelago via northern Greenland and Svalbard to Scandinavia and vice versa. This connection was only severed when Greenland separated from Svalbard 36.6 MYA [Essay 4].

Heimaey, Vestmannaeyjar, Iceland

3 Greenland Hot Spot

The division of continents may result from one of two basic types of rifting. One mechanism becomes effective when continents drift over stationary hot spots deep within the Earth's mantle, leaving a high-profile feature called a hot-spot track. The other occurs at rift valleys, and leaves low-profile features as evidence. The broadly accepted theory about the division of North America and Europe is based upon the former model—the main evidence for this is the existence of a hot-spot track called the Greenland and Scotland Ridge, illustrated below.

About 62 MYA a mantle plume beneath Greenland formed domed uplifts on the surface over 1.6 km (1 mile) high, causing volcanic disruption for up to 1,600 km (1,000 miles) around. This hot spot remained fixed as the Laurasian megacontinental assembly drifted northwest. Ultimately, the region of Greenland's present east coast moved over the center of the plume. By 50 MYA the circle of volcanism extended north and south along the Greenland margin, and across an intermediate zone between Greenland and Scotland. Old rifts in the intermediate zone were reactivated, new rifts opened, and huge volumes of flood basalts were both channelled under the surface and released onto it. The Faeroe Islands and many other familiar geological structures were formed as a result. These included the Scottish islands of Skye, Rhum, Canna, Eigg, and Muck, and the basalt columns that now form Fingal's Cave (Isle of Staffa) and the Giant's Causeway (northwest Ireland).

About 16 MYA, as North America and Greenland continued to rotate westwards, an island that was to become modern Iceland appeared above the surface of the Nordic Sea: it may have looked like the active volcanic island of Heimaey pictured here. Iceland marks the hot-spot track directly above the mantle plume, where the associated ridge, now the Iceland-Faeroe Ridge, is about 32 km (20 miles) thick. It stretches 1,120 km (700 miles) from the Denmark Strait off Greenland to the Faeroe-Shetland Channel.

36 MYA

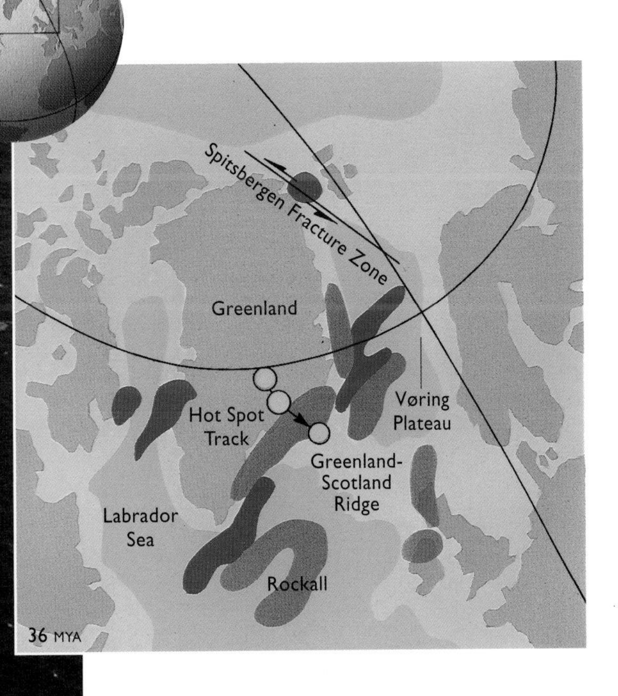

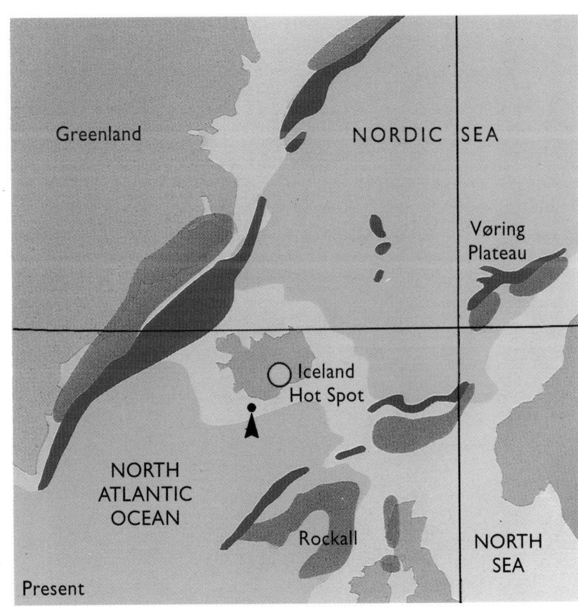

36 MYA
ARCTIC OCEAN
Svalbard
Archipelago
Spitsbergen
Fracture Zone
Baffin
Bay
Greenland-
Scotland
Ridge
Labrador
Sea
NORTH
ATLANTIC
OCEAN
Warm-surface current
Cold-bottom current

4 Nordic Seaway

We have now arrived at the nub of the geological origin of the Western World: the severing of the final link between Europe and North America around 36 MYA. It was at this point in geological time that global ocean circulation slowly began to change from equatorial to interpolar. The global climate now began to destabilize and trend towards an ice age, as the diversity of northern "holarctic" plants and animals began to increase, to segregate into more sharply defined climate zones, and to assume their present appearance.

Norway and Greenland began to move apart as rifts north of the Greenland and Scotland Ridge became an extension of the Mid-Atlantic spreading-center. To accommodate the widening Nordic Sea, a transform fault, the Spitsbergen Fracture Zone, developed off northern Greenland, severing the DeGeer Landbridge. As North America and Europe pulled apart, northern Greenland slid along the fault past the Svalbard archipelago. By 36.6 MYA the North Atlantic and Arctic Oceans had established a connection through the Fram Strait, along the strike of the fault.

Although Arctic and North Atlantic *surface* water could flow through Fram Strait, the Greenland and Scotland Ridge prevented Arctic Ocean bottom-water from flowing south and Central Atlantic bottom-water from flowing north. For this reason the Arctic Ocean and its surrounding continents remained relatively warm during a time of generally deteriorating global climate.

Meanwhile, the first significant perennial sea ice had formed around Antarctica, and this led to a general cooling trend in the global ocean. After 36 MYA, Antarctic circumpolar circulation permitted the refrigerated water that had accumulated in the deep Southern Ocean around that continent to feed a developing global bottom-water circulation system, the "psychrosphere" (Gr. *psychros*, cold). This system was fully formed in the South and Central Atlantic by the time Fram Strait opened, but the Greenland and Scotland Ridge prevented it connecting with Arctic bottom water. Most scientists think that the final link was not made until 15 MYA, while others accept a date nearer to 30 MYA.

[*continued from page 221*]

region of the asthenosphere beneath the eastern edge of Greenland. It was the flood basalt released from this mantle plume through such volcanic eruptions that formed a part of the Greenland and Scotland Ridge on which Iceland now stands [Essay 3].

BY 50 MYA, THE MID-ATLANTIC RIDGE had propagated from a point off the southern tip of Greenland, north towards the Greenland and Scotland Ridge. This was influenced by the activity of the Greenland mantle plume, which had further weakened the lithosphere. Rifting to the north of the ridge was also established by this time, and by 36.6 MYA seafloor spreading, and therefore crustal separation, had extended beyond the ridge to the northern tip of Greenland. As the two continents began to pull apart, Greenland remained attached to North America, while Svalbard remained attached to Europe [Essay 4]. The DeGeer Landbridge from Svalbard to Greenland was now severed, and the present Fram Strait was formed. Thus the Arctic Ocean became directly connected to the young and shallow Nordic Sea, while North America became a new continent on its own tectonic plate.

In the submarine world, the Greenland and Scotland Ridge remained an effective barrier between the Central Atlantic and the Nordic Sea. The psychrosphere was simply stymied at this point in its northern hemisphere circulation. Consequently, many specialists think that the modern psychrosphere may not have formed until about 15 MYA, but some argue for a much earlier time—30 MYA. Whichever school of thought is correct, the fossil record indicates that the northern continents remained warm during a time of generally deteriorating global climate.

During the six million years following the separation of North America from Europe, there was a steady and quite steep decline in the global climate. The generally warm and moist conditions that had prevailed on the northern continents gradually gave way to cool and dry conditions. By 30 MYA high-latitude northern climates were about equivalent to today's northern temperate zone. Even so, sea level at this time fell sharply to well below present levels. Such a sharp drop is explained by the decrease in global tectonic activity, and also coincided with the accumulation of substantial snow and ice on and around Antarctica. As the global climate continued to deteriorate, occasional land connections between Eurasia and Africa at times of exceptionally low sea level may have offered holarctic animals an opportunity of escape to warmer southern latitudes.

For the next 15 million years global climate fluctuated as if it had been thrown out of equilibrium. In this period (30–15 MYA) there were four warm-climate peaks followed by cold-climate troughs. Sea level remained generally below today's level, though it too fluctuated in tempo with the oscillations of global climate. With the formation of a fully fledged Antarctic ice sheet (certainly by 15 MYA, if not before) there was yet another marked downward slide in the global climate, and conditions became approximately similar to those of today. These changes seem to have had a stimulating effect upon the evolution of holarctic flora and fauna: from 15 MYA holarctic evolution began to accelerate at a frenetic pace, with an explosion of new forms.

JUST SO LONG AS the Thulean, DeGeer, and Bering Landbridges were in existence, holarctic animals had been free to migrate from one region of Laurasia to another, if the climate was hospitable and the right food adequately available. Further migration was restricted by natural barriers—high mountains and inland seas such as the Turgai Seaway, which divided Asia in two until about 60 MYA. Nevertheless, up to 50 MYA distribution of the more primitive holarctic mammals in North America, Europe, and Asia was remarkably similar [Essay 2].

First the Thulean Landbridge and then the DeGeer Landbridge (**14**) had been severed by the opening of the Nordic Sea. Only the far northern Bering Landbridge now remained, and mammal migration was therefore restricted to this natural filter. The crossing was 160 km (100 miles) closer to the North Pole 36 MYA than the Bering Strait is today, which may have limited the interchange of North American and Eurasian animals to those mammals that had adapted to cooler climates and arctic vegetation. The holarctic animal world became broadly segregated into regions of evolutionary influence, and these extended from north to south according to topography, climate, and food resources. This was equally true for North America, Asia, and Europe, where holarctic mammals were now beginning to assume their present appearance.

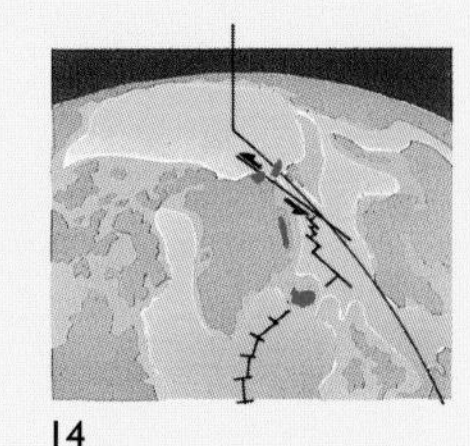

14

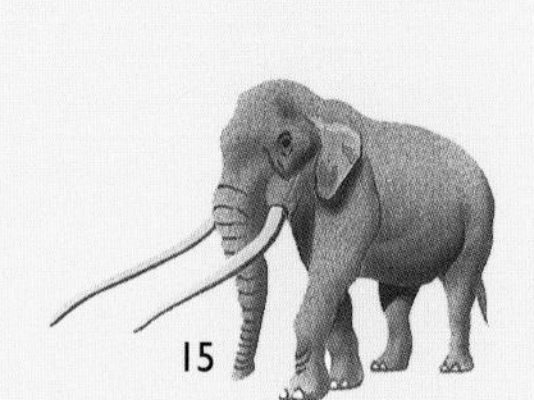

15

16

17

At first, all herbivores fed on leaves and herbaceous plants. There were few grasses before 36.6 MYA, and although they were well established by 23 MYA, they did not predominate in the holarctic and African world until after 15 MYA, when cooler, drier climates had become established. The most successful large herbivores evolved into the ungulates, which were either even-toed (like the camel and the antelope), or odd-toed (like horses). Long-nosed animals (like mastodons and elephants: **15**) evolved in Africa about 55 MYA and had migrated to Eurasia by 18 MYA. Mastodons were the most common migrants at a time when most herbivores were indigenous; they reached North America via the Bering Landbridge 11–5.3 MYA and South America via Panama by 1.6 MYA.

IN THE CONTEXT OF familiar modern creatures, the principal large animals on the North American scene during the period 23.7–15.1 MYA included members of the pronghorn antelope, camel, horse, and peccary families (peccaries are pig-like but not members of the pig family). The pronghorn antelope family (which are not true antelopes) are still exclusively North American. The modern genus of horse, *Equus*, migrated from North America during the last 2 MY. It included asses, which are broadly distributed, and zebras, which are now considered as African. Interestingly, the camel family (**16**) was also restricted to North America until about 4 MYA, after which it radiated widely to South America (where it became llamas and alpacas), Asia, and Africa.

The corresponding Eurasian mammals of the same period were members of the pig, deer, and bovine families (L. *bos*, cow). The pig family became Afro-Eurasian by 22 MYA, but true pigs did not reach North America at all. The deer family was indigenous to Eurasia until about 3 MYA; it now includes the Eurasian reindeer and the North American caribou, moose, and elk. The bovine family probably originated in Asia and had diversified across Eurasia by 17 MYA; it reached Africa about 15 MYA and North America by about 6 MYA. The bovines included sheep, goats, buffalo, and a remarkably varied range of antelopes (now mainly African).

All the principal holarctic carnivores evolved from a common ancestor, a ferret-like animal (*miacid*) dated 57 MYA: they diversified into weasel-like animals (*mustelids*), dog-like animals (*canines*: **17**), cat-like animals (*felines*), or omnivorous bear-like animals (*ursids*). Raccoons, pandas, hyenas, and civets are also included in this holarctic group: civets are from the same family as mongooses; they generally secrete musk—a substance used in perfumery.

All these holarctic carnivores had differentiated teeth that included pointed incisors for tearing flesh, and cheek teeth for slicing it. They also had a hardened bone in their ears that provided them with excellent hearing, and ultra-strong wrist bones, to keep their victims firmly held down with their five-toed and sometimes clawed feet. The ability to run swiftly was a huge advantage for the later carnivores over the more primitive carnivore groups that they replaced. The bear, cat, and dog families were habitual denizens of the Bering Landbridge, which acted as a filter for their prey. Carnivores spread out widely across their respective domains and adapted well to extremes of climate, because their food supply was available "on the hoof" irrespective of climate or season.

By 20 MYA some holarctic carnivores (*pinnipeds*, from the Latin *pinna pes*, winged foot) had adapted to a full-time marine existence in the North Pacific basin. Some experts think that weasel-like otters led to the evolution of the earless seal by 15 MYA, for example. In contrast, the eared marine carnivores—the walrus, sea lion and fur-seal families—are thought by some experts to be a branch of the bear family because of an anatomical similarity in skulls and teeth. Others believe that all pinnipeds came from the same lineage (**18**). However, one group of seals certainly originated in the North Atlantic and Western Tethys, later migrating to the Pacific and then to the Southern Hemisphere; conversely, the walrus migrated from the Pacific to the North Atlantic, and later reentered the North Pacific. The sea lion family, which evolved by about 11 MYA, has never left the Pacific.

The establishment of the Antarctic ice sheet had a desiccating effect on the global atmosphere—a similar effect to the defrosting device in a domestic refrigerator. This drier global climate affected the distribution of many plants and animals on the northern continents. Heavily forested regions became interspersed with grassland and herbaceous plants. After 15 MYA the landscape began to be dominated by savannahs, grassland regions with scattered trees blending into plains and woodlands [Essay 5]. Meanwhile, warmer-climate broad-leaved plants in Arctic regions had been replaced by temperate forests low in diversity. Between 15 and 5 MYA the Arctic climate remained comparatively mild, and although the variety of vegetation diminished, its extent remained roughly the same. However, northern climates continued to deteriorate; by 3.5 MYA permanent ice had formed in the Arctic Ocean (**19**).

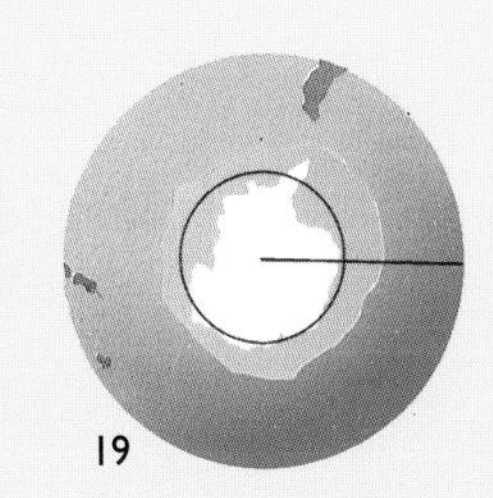

Modern-looking holarctic herbivores and carnivores expanded rapidly as open areas developed (**20**). Groups of grazing animals became more fleet of foot, putting greater pressure on the dependent carnivores—which had to respond in kind or become extinct.

Rainforest, Trinidad and Tobago, Lesser Antilles

Herbivores also had to adapt to new food sources, particularly grasses, as well as to the open spaces and increased danger from carnivores. Of all angiosperms the grasses—including wheat, maize, oats, barley, and rice—had the most profound effect upon the evolution of species.

Grass stems are constructed from cellulose reinforced by silica. As tooth enamel is softer than silica, herbivore teeth are worn down by grazing. Modern ungulates have evolved cheek teeth that are high-crowned, grow continuously, and have bands of enamel on their grinding surfaces. As a result, these teeth resist abrasion much better than the low-crowned teeth of early ungulates. Cellulose is a carbohydrate that is not easily digested and metabolized, so "ruminant" ungulates (caribou and wildebeest for example: **21**), which depend on grasses for food, have four-chambered stomachs to cope with this difficult-to-digest material.

About 40 percent of a ruminant's weight is taken up by the animal's complex digestive system. Bacteria break down the cellulose and partially absorb nutrients in the "rumen," the first stomach. The partially digested food is then returned to the mouth as cud for rumination. Often, the ruminant grazing strategy is to eat as much as possible as quickly as possible, and then retire to a safe place to "chew the cud." After the cud has been chewed, the food can then pass through a small orifice and into the acidic environment that governs the rest of the animal's digestive system. Particles of food that are not properly ground up are simply returned to the rumen for further bacterial processing. For all these reasons, ruminants have a limited capacity for processing food, so what they do eat must be nourishing. Ruminants digest slowly, and their blood sugar does not rise immediately after a meal. This accounts for their "bovine" natures, and generally slow response to provocation.

Non-ruminant ungulates (horses and zebras, for example) can thrive on poor quality food because their one stomach is acid based. —they are not wholly dependent upon bacterial processing to receive nourishment from food. Non-ruminants eat the stems of grasses (rather than the leaves as ruminants do) and they can increase their intake of food according to need, which ruminants cannot. Acids break down ingested food so that nutrients (particularly sugars) can be absorbed quickly into the bloodstream from the digestive tract. The general process of digestion is completed in the animal's bacteria-laden intestinal tract. Consequently, non-ruminant digestive systems account for only 15 percent of their body weight. Non-ruminants' blood sugar rises quickly after a meal, and the animals' reactions are immediate and lively, especially in rhinoceroses. Thus non-ruminant ungulates and ruminants often graze the savannah together. The non-ruminant ungulates eat the older, longer grasses and fibrous woody undergrowth. The ruminants eat the fresh, close-cropped grassy growth.

The effect of savannahs upon animal evolution was not limited to the ungulates and their attendant carnivores. As savannahs spread, and herbaceous plants began to color the grassy and occasionally wooded landscape, the length and complexity of the terrestrial food chain increased steadily. Insects of all kinds proliferated in number and variety. Plant-pollinating and parasitic insects encouraged the expansion and diversification of animals dependent upon *them* for food. Frogs and insectivores too diversified and increased as insect numbers multiplied. Savannah birds also ate insects, and berries whose seeds were distributed with droppings. Birds ate worms, grubs, small amphibians, and the snakes that now proliferated with the increased numbers of amphibians.

This was a time of major bird adaptation, diversification, and radiation. Birds that lived on the savannah and in the forests now staked out and defended their territories. They warned their kind of an approaching predator, and greeted the dawn with a chorus of song—songbirds now began to dominate the avian world. Owls and other birds of prey also began to appear. As over the millennia winters became more severe, migratory birds began to establish their seasonal flyways, while predatory birds stayed in place and fed on rodents. The latter, the most prolific of all mammals, had multiplied exponentially with the spread of savannah and grasses.

By 15 MYA animals grazed, carnivores hunted (**22**), bees built hives to store honey, and birdsong filled the air. The modern world was taking shape: a hectic, thriving, highly competitive, evolutionary scenario—the prelude to the appearance of "hominid primates."

The word "primate" (L. *primus*, foremost) was first used by the Swedish botanist-explorer Carl Linnaeus (1707–78), in the 1758

[*continued on page 232*]

Bamboo, Trinidad and Tobago, Lesser Antilles

5 Grasses and Savannahs

Grasses now dominate vast areas of the Earth's surface, and provide food and shelter for countless numbers of animals. They produce the staple grains people depend upon the world over, and their evolution played a paramount role in the domestication of grazing animals and the birth of civilization. Though their origin is not known, their rapid advance may have been due to the changes in global climate from about 35 MYA.

Once sea ice and then ice sheets began to form in the Antarctic, the global climate generally became cooler and drier. This cooling and drying process, together with the destruction of all the northern landbridges, had a considerable effect on the distribution of holarctic plants and animals.

Cuillin Hills from Loch Brittle, Isle of Skye, Scotland

The earliest known unequivocal fossil grasses date from about 30 MYA: bamboo (inset) is one of the oldest living members of this family of almost 10,000 species. Grasses are angiosperms that do not have attractive flowers; they depend instead upon the wind to disperse their pollen and seed. In most cases new growth stems from the base of the plant's elongated leaf—and this unique characteristic allows grass leaves to keep on growing even after severe grazing or cropping.

By 23.7 MYA many forested regions in northern continents had become interspersed with grassland, herbaceous plants, and lone stands of trees; by 15 MYA, such savannah landscapes dominated the ecosystem. Meanwhile, as grazing animals adapted to these new food sources as well as to open spaces, they became more fleet of foot. Their newfound speed put great evolutionary pressure on their dependent carnivores, and as a result, both predators and prey became "modern-looking." The complete holarctic ecosystem underwent a major evolutionary reorganization. The combination of cooler, drier climates with the rapid diversification of grasses and herbaceous plants resulted in an exponential evolutionary surge that began about 15 MYA and continues to this day.

[*continued from page 229*]

edition of his classic *Systema Naturae*. Linnaeus invented the modern system of animal and plant classification (taxonomy), particularly the custom of using two qualifying names to define a species. His expression "primate" classified mankind long before the extraordinary length and complexity of human evolution was appreciated.

The modern Order of Primates embraces several hundred species that include tarsiers, lemurs (**23**), monkeys, apes, and humans. In paleontology one method of classifying an animal as a "primate" is by its teeth. If the dentition of a particular fossil mammal complies with a particular formula, then it is a primate.

23

Teeth, the only parts of an animal really difficult to destroy, are the most frequently preserved fossils, and often provide the main clue to identification in primate research. For example, the statement "3–1–4–3 (upper jaw)" means that this or that animal had three incisors, one canine, four premolars, and three molars on each side of the jaw: the formula suggests a primitive placental mammal, although many modern mammals have retained this pattern. In accord with their behavior and feeding habits, primates developed dental formulas that modified the primitive layout, refining jaw shape and teeth arrangement. They suppressed the third incisor, and either the first, or first and second, premolars. The formula of more advanced primates, and anatomically modern humans, is 2–1–2–3.

There are other anatomical clues to taxonomic identity: all primates have five fingers and toes, mobile thumbs and big toes, and a tendency for claws to be replaced by flattened nails. The facial parts of more advanced primates became flattened as their sense of smell deteriorated; the shape of their brain cases altered accordingly and became enlarged as they gained intelligence. As primates continued to advance, their eyes tended to face forward; stereoscopic vision—the ability to judge distance and size began to evolve. More advanced primates had skulls with eye sockets facing forwards rather than to the side. But, because they are more often preserved, the type and disposition of primate fossil teeth generally decide the classification.

PRIMATES FIRST EVOLVED in North America, or so it is believed from the fossil record. The oldest known primate fossil genus, *Purgatorius* (**24**) (named after a locality in Montana) is known from a solitary tooth, a lower molar dated about 67–66 MYA. The tooth was found in the Western Maritime Province alongside fossils of third-regime dinosaurs. It is thought to have belonged to an animal that *looked* like a rodent: primates are of course "primate" and not at all related to rodents. Later North American primates were squirrel-like herbivorous animals (*Plesiadapis*: **25**). They died out in North America, but not before some of their kind had migrated to Europe before 50 MYA. Subsequent radiation of ancestral *Plesiadapis* in Europe led to the evolution of Eurasian, African, and South American primates.

24

25

The most important primate fossil site discovered in the Old World is dated 31–30 MYA and is located just southwest of Cairo near the Egyptian town of El Faiyâm, in the Faiyâm Depression (**26**). This remarkable site has produced the fossil remains of eight distinct primate lineages: three genera related to tarsiers and lemurs, and five of long-nosed primates called "catarrhinids" (Gr. *kata rhinos*, down nose, indicating that the nasal passages pointed down, not out). Of the five catarrhinid genera, three were advanced primates and two were more primitive and appear to have become extinct. Of the advanced groups, *Aegyptopithecus* (Gr. *pithekos*, ape: **27**) led to the evolution of hominoids (see below). However, it is not known whether *Aegyptopithecus* comes from a time before or after the divergence of hominoids from Old World monkeys, and New World monkeys. (As a rudimentary guide, modern hominoids do not have tails; all Old and New World monkeys do.)

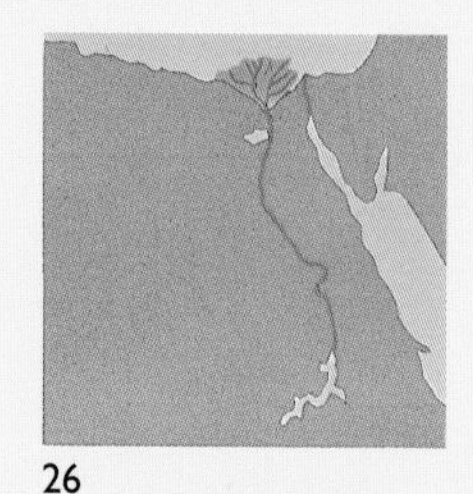

26

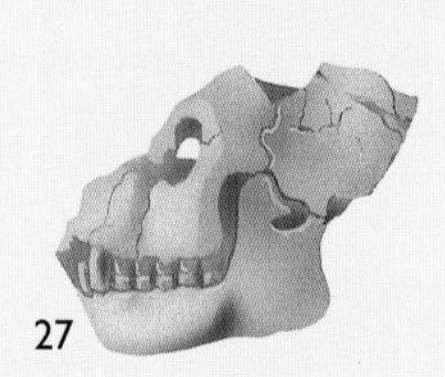

27

THE FIRST UNEQUIVOCAL HOMINOID to appear in the fossil record is *Proconsul*, and is dated 22–18 MYA. The term "hominoid" (meaning animals with a human-like appearance) is used to describe the superfamily of primates that evolved from *Proconsul*. This superfamily includes the *lesser* apes, the *greater* apes, and the *hominids*. The words "lesser" and "greater" have nothing to do with relative body size. They are used to imply that the "greater" apes (the chimpanzee, the gorilla, the orangutan, and other animals that are extinct) are more closely related to *hominids* (meaning animals with the basic characteristics of humans) than the "lesser" apes (the gibbon, and other now-extinct animals). The family Hominidae is the family of man, and that is how the human race is classified in the animal world.

There are *many* distinguishing anatomical characteristics among the three groups of this family. From an evolutionary point of view, the most important single difference between the hominids and the greater and lesser apes is that hominids had longer legs, stood upright with their arms free, and walked on the forest floors and across savannahs. Ancestral apes, meanwhile, were forest-dwellers that could not walk upright, and used forearms and knuckles to help movement. Bipedalism gave hominids the same advantage over other animals of their time as it had previously given the first-regime dinosaurs over the mammal-like reptiles 200 MYA. One could say that bipedalism appears to be the key to ascendancy in all animals.

And now we come to the main stumbling block that has inhibited research into the origin of humans [Essay 6]. There are major gaps in the African fossil record at a crucial moment in time. After *Aegyptopithecus*, no hominoid fossils have been found in Africa for the period 30–22 MYA, and the African record is also blank from 14–5 MYA.

The oldest known *Proconsul* fossils, found in Kenya and Uganda near Lake Victoria, are dated 22–18 MYA. These animals (**28**) were arboreal and are believed to have lived in regions dominated by forests mixed with open woodland and grassland: true savannahs—less than 20 percent canopy—did not develop until 10–5 MYA. By 16 MYA, *Proconsul*-like animals were widespread in Africa and Eurasia. One interpretation of the fossil record is that at some unknown time after 14 MYA they separated into African hominoids and Asian hominoids. The ancestral Asian hominoid is dated 12 MYA, and is called *Ramapithecus* (L. *ramus*, branch). This animal too was a forest-dweller, and the modern orangutan (a greater ape) is its surviving descendant. *Ramapithecus* was long thought to have been *the* ancestral hominid but, in accord with a consensus reached in 1984, it has now been reclassified as a greater ape. But in Africa no hominoid fossils have been found in deposits dated 14–5 MYA. There is simply no known fossil record of where and how and why hominids evolved out of the *Proconsul* line of hominoids [Essay 6].

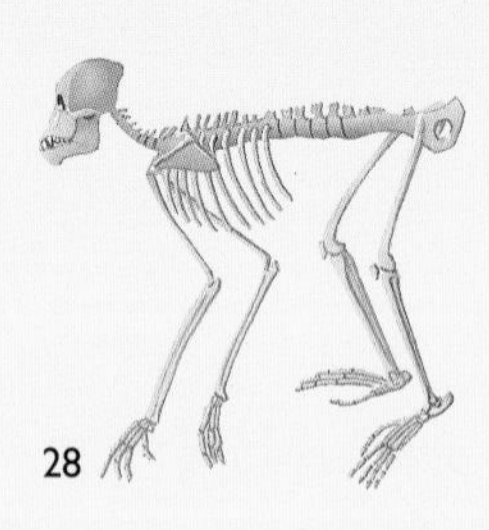

28

THE OLDEST KNOWN *hominid* fossils that have been found were discovered in Kenya—near Lake Turkana in 1970, and near Lake Baringo in 1984. Both of these rare finds are about 5 MY old. The first was a fragment of jawbone (5.0–5.5 MYA) and the second a larger fragment of a lower jawbone with two molars in place (5.0–5.25 MYA). Perhaps both the antiquity and the reality of these early ancestors of humankind seem too remote and fragmentary to bring home the fact that they belonged to the earliest hominids.

Hominids walk upright with arms swinging free; hominoids do not, and this is the prime factor that distinguishes them. Arising from this difference, perhaps the most evocative of all hominid fossil finds (by Paul I. Abell and Mary D. Leakey in 1978), was in the Laetoli area of Tanzania, 48 km (30 miles) south of Olduvai Gorge, at the southern end of the Ethiopian triple-rift system, the Afar Triangle. The fossils are not bones at all—they are hominid footprints.

From the size and depth of the footprint impressions at Laetoli, it appears that they were made by two males, and a female or a juvenile hominid, all walking upright. There are various interpretations of the prints. The most widely accepted is that the female walked slightly behind and to one side of the leading male hominid, while a second male followed behind, treading in the footsteps of his leader as he did so. The prints were made in a fresh but muddy cover of fine volcanic ash which fell 3.6 MYA. These are the first known unfaltering footsteps of the Family of Man.

6 Family of Man

The Laetoli footprints, reproduced below, are preserved in volcanic ash called the "Footprint Tuff"; they were found by P.I. Abell and M. Leakey in 1982. The character of the tracks is interpreted in various ways. The most commonly accepted interpretation is that a female hominid, or a juvenile, walked beside and slightly behind a larger hominid, a male. A second and smaller male hominid then trod in the prints formed by the leading male walking ahead of him. It is impossible to judge whether or not all three walked together or at different times. The exciting fact about this evocative scene is that the small creatures who left the trail—the tallest was a little short of 150 cm (five feet) tall—walked upright with their arms swinging free. They had "the rounded heel, uplifted arch and forward-pointing big toe typical of the human foot." And they had a natural bipedal gait.

Little is known about these creatures or about the divergence of the apes and the hominids from their common ancestors the homi*noids*. Experts agree that the division took place in Africa 14–5 MYA. There is, however, a major gap in the primate fossil record in Africa for this period of time.

Why the hiatus? And if part of that record still exists, where might it one day be discovered? To find a possible answer, we must turn to the tectonic scene in Africa during that period.

The Afar Triangle is the pivotal region of a triple-armed rift system that stems from an Ethiopian hot-spot dome. The three arms of the system form the Red Sea, the Gulf of Aden, and the East African Rift Valley—the site of the Laetoli footsteps and the Serengeti uplands. The mantle plume that caused the Ethiopian Dome to form is one of about forty hot spots resident below the slow-moving African continent, which are indicated by red dots on the globe above. African domes vary from modest uplifts to major features like the Ethiopian Dome, which encompasses Mount Kilimanjaro (5,802 m; 19,340 ft) and Mount Kenya (5,117 m; 17,058 ft). Mountainous country resulting from doming is drained by streams and rivers. Slight movements of the African continent over the plumes have determined the topography of much of the African landscape, and therefore the character and history of its river systems. Indeed, it is the now extinct and buried African river systems and their canyons that may provide some answers to the mystery of the missing hominoid-hominid record (see over).

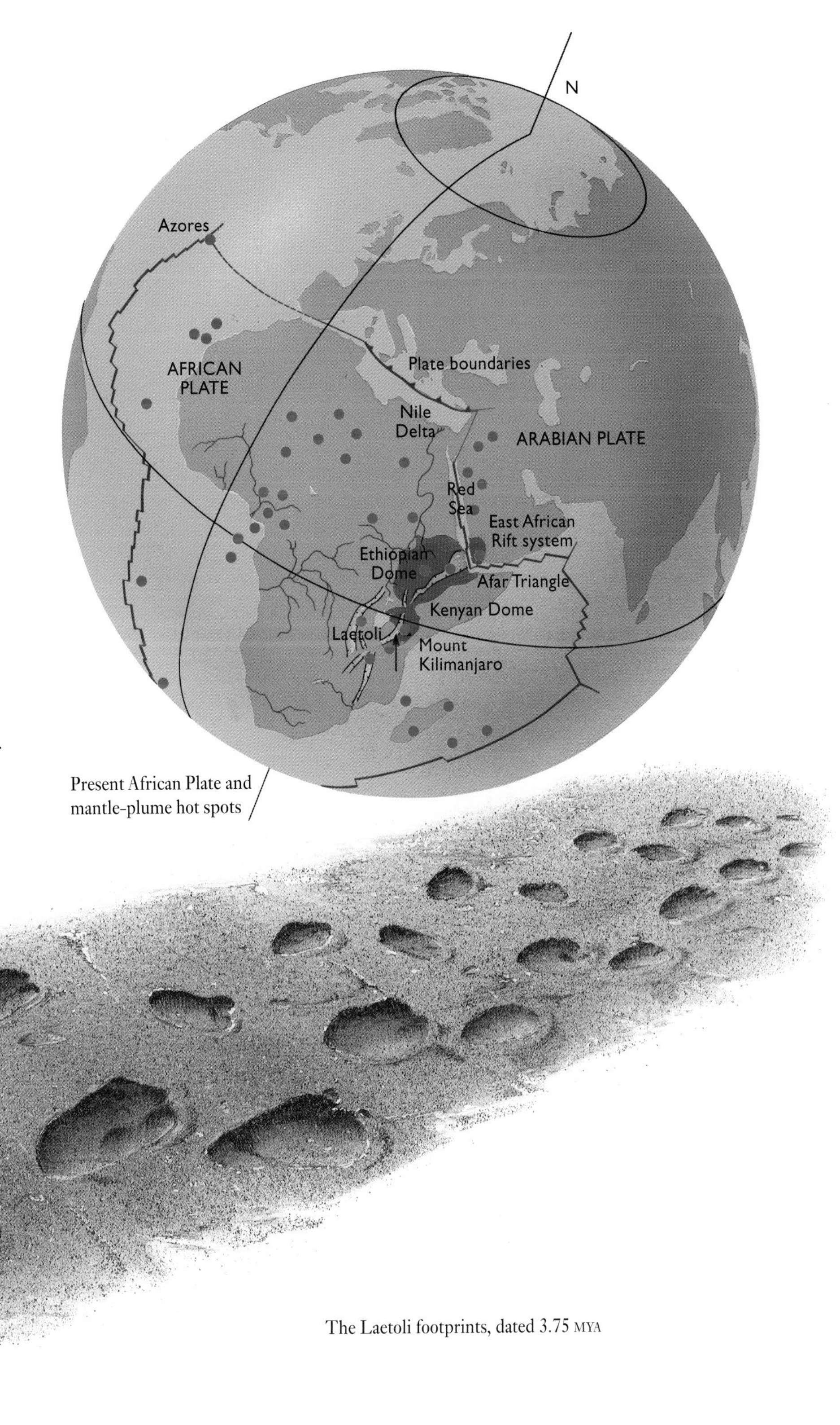

Present African Plate and mantle-plume hot spots

The Laetoli footprints, dated 3.75 MYA

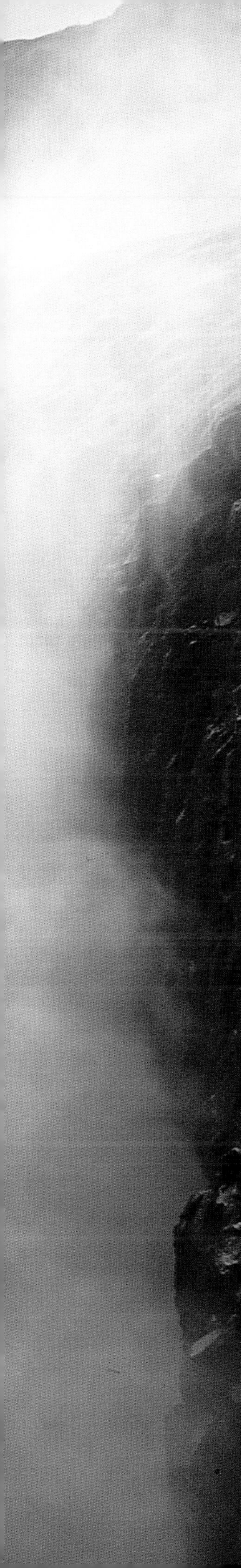

Gullfoss, Reykjanes Ridge, central Iceland

Mediterranean Crisis

Radar-imaging photographs of the Sahara Desert from NASA's space shuttles have revealed a sand-filled system of channels cut into the bedrock over vast regions of Africa. These are riverbeds, once part of several formidable river systems that drained much of the continent from about 18 MYA. At that time, Africa was lush with vegetation and free from the super-arid conditions of today. It was a period when active rifts were forming in domed uplifts above the many mantle plumes under the slow-moving continent [page 233]. Subsequently these rifts became inactive; they cooled and sank, forming river channels. In fact, most of the world's major rivers flow in now-inactive rifts. Sometimes rifts can capture other rivers—piracy of a kind demonstrated in this picture of Gullfoss pouring into an inactive rift in Iceland.

The now-buried river channels in Africa are the courses of tributaries leading from rift systems to an *ancestral* Nile River. Between 6 and 4.5 MYA this ancestral Nile cut a canyon into the African continent that may have rivaled Grand Canyon in depth and far outdistanced it in length. The formidable river that cut the canyon flowed into the Mediterranean basin at a point well below the present Nile delta and therefore below present sea level. The question is, how could this possibly have been the case?

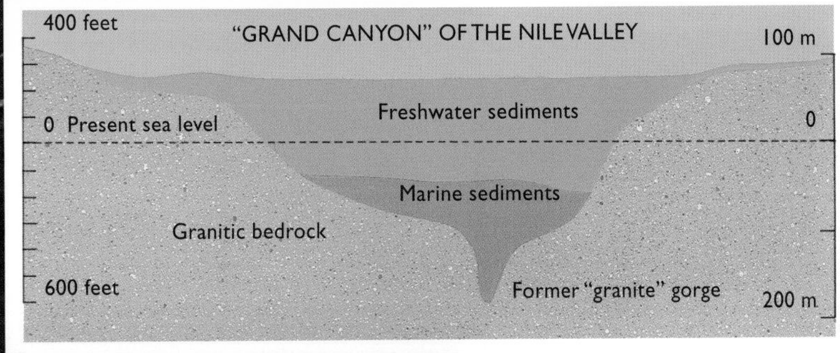

Between 6 and 4.5 MYA changes in global sea level, and the uplift of surrounding regions of the shallow Strait of Gibraltar [Essay 1], caused the Mediterranean basin to be periodically cut off from the Atlantic Ocean. Isolated, the enclosed sea lost most of its water through evaporation. This was in spite of the inflow of innumerable rivers and the drainage of the seas beyond the Dardanelles (the strait that joins the Black Sea to the Aegean Sea).

The amount of salt that collected on the Mediterranean seafloor during this time is estimated to have been in the order of several hundred thousand *cubic miles.* Extraordinarily this is more than thirty times the amount of salt contained in the present waters of the basin. From this (and much other evidence) it is implied that during times of low global sea level the Mediterranean basin dried up. During times of high sea level the basin was replenished by a gargantuan waterfall pouring over the Gibraltar sill from the Atlantic. Furthermore, this event happened *many* times during the course of 1.5 MY. It is also probable that other great conduits poured into the basin at such times—water from the ancestral Nile, the Rhone, and indirectly, the Danube.

Exploratory drilling in the Nile region during the construction of the Aswan Dam in Upper Egypt, and beneath the Nile delta 1,200 km (750 miles) downstream, revealed the existence of a deep granite gorge beneath the site of the modern river (inset diagram). This gorge is three times the length of Grand Canyon and possibly its equivalent in depth and width. Other canyons of the same age (6–4.5 MYA) are known to exist beneath the Rhone River in France and elsewhere around the Mediterranean. In addition, innumerable channels filled with river gravel and alluvium cut into the continental shelf off the coasts of France and North Africa. The common factor in all these features is that they lie well below present sea level. They were most likely formed when the Mediterranean basin was nearly empty, and its floor a salty desert.

"GRAND CANYON" OF THE NILE VALLEY

Desiccation of the Mediterranean Sea 6–4.5 MYA

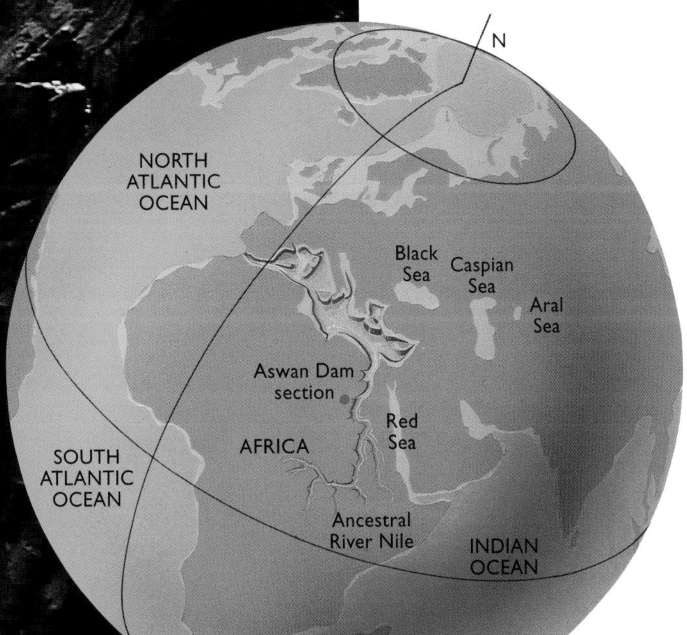

The Missing Record

The desiccation of the Mediterranean caused a radical climate change in surrounding regions. Cool, moist climates in Central Europe became temporarily warm and arid when the basin was dried up, and in North and Central Africa the climate turned extremely hot and arid. Before the time when the Laetoli footprints were made by early hominids, permanent flow between the Atlantic and Mediterranean was restored. The European climate reverted to its previous state but African climates tended to remain arid.

When the ancestral Nile River was cutting its "Grand Canyon"

Zaouïa-Sidi-Abdelâli,
Jebel Sahara, Morocco

between 6 and 4.5 MYA it was a time of drought. It is probable that African hominoids sought refuge along river banks and in canyons like the one pictured here at Zaouïa-Sidi-Abdelâli, on the edge of the Jebel Sahara of Morocco. Steep-walled, undercut canyons are characteristic land forms in arid country. The floors of such canyons are usually lush with riverside flora and fauna. Maybe the missing fossil record of the hominid transition from hominoids actually lies buried along the ancient river banks or in that "Grand Canyon" beneath the Nile?

XI FOUNTAINS OF YOUTH

This is the meeting place of two great ocean currents. The cold Labrador Current (left) flows southwards down the East Coast of North America to this point. The tropical waters of the Gulf Stream (right) flow northwards off the Florida and Carolina coasts and meet the Labrador Current here at Cape Hatteras. Although the picture was taken in a light breeze on a balmy day, the ten-foot-high cross-breaking waves testify to the meeting of leviathans. These currents are just two of many components in an extraordinarily complex three-dimensional ocean-river system. This network of surface and subsurface currents in the North and South Atlantic Oceans has largely governed the climate of the North Atlantic realm for the last three million years.

FOUNTAINS OF YOUTH
1.6 MYA–PRESENT

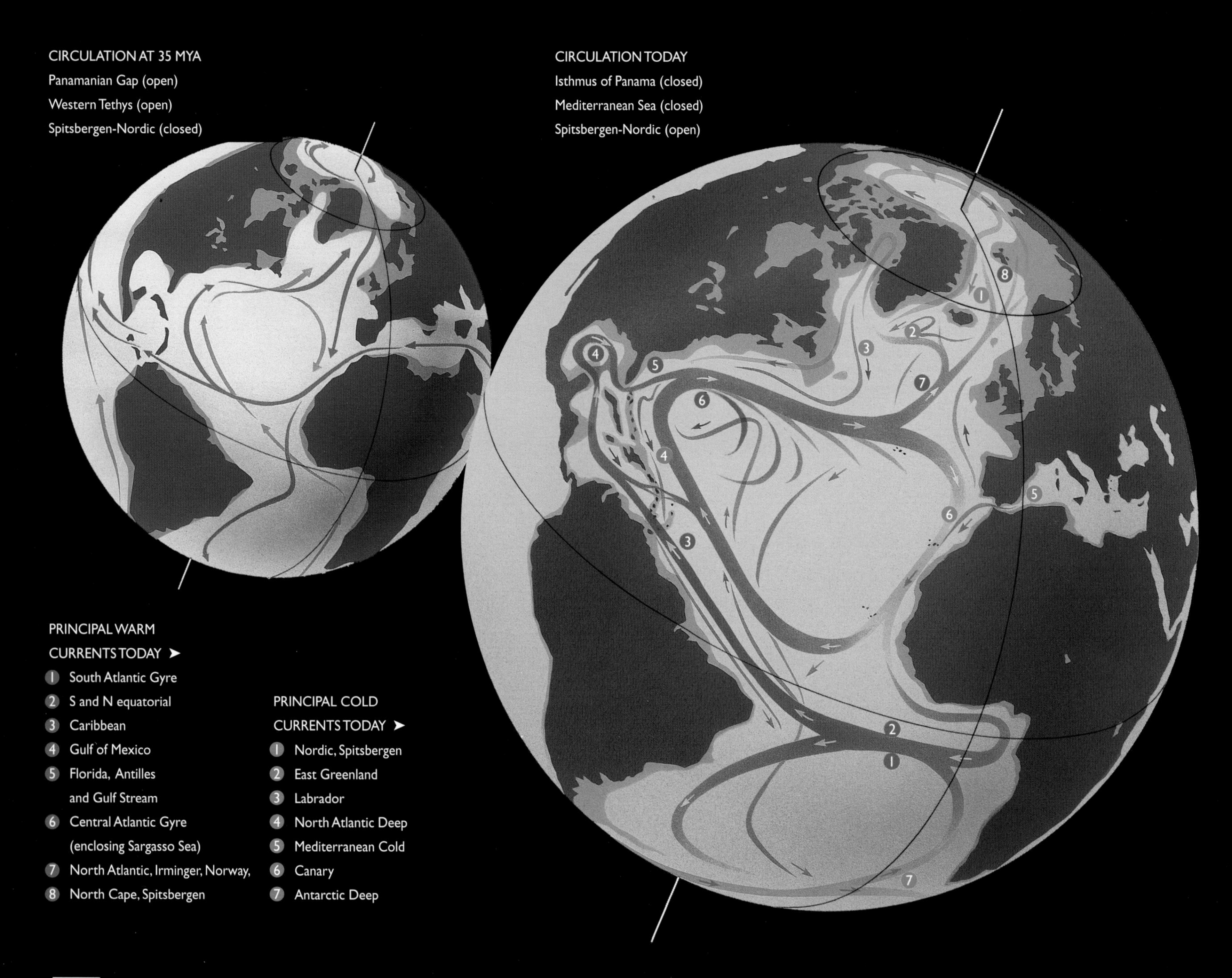

The smaller globe shows that 35 MYA the Central Atlantic circulation was still part of an equatorial system. At that time, South Atlantic currents were redirected into the Pacific, and North Atlantic currents could not freely interchange with Arctic waters. The larger globe illustrates today's interpolar circulation system.

The North and South Atlantic are no longer connected to the Pacific through Panama, or to the Tethys Ocean system, which has been reduced to the almost totally enclosed Mediterranean Sea. Instead, a proportion of wind-driven South Atlantic equatorial currents (red) join Caribbean currents that flow into the Gulf of Mexico. Here, heat lost by evaporation in colder regions is regained. A tropical ocean-river then spills out of the Gulf and flows along the East Coast to the far reaches of the North Atlantic. At this point Arctic water meets and mixes with the warmer northward-flowing current. Heat is released during this process and it is this heat that modifies northern winter climates.

The super-salty ice-cold sea water that results from this process sinks to the seafloor (blue) and flows back to the Southern Ocean. Similarly, ice-cold bottom-water circulating round Antarctica then flows back via the South Atlantic to the North Atlantic. This freezing bottom-current is called the "psychrosphere," and is the hidden factor in the three-dimensional, twin-ocean, circulating system that drives and controls global climate.

Christopher Columbus was the first European to see the Isthmus of Panama, during his fourth and final transatlantic voyage of 1503. However, it was to be another, later explorer who first fell foul of one of the many extraordinary consequences of the formation of that isthmus [Essay 1].

According to the 16th-century Spanish historian Antonio de Herrera, Ponce de León (1460–1521), who had been on Columbus's second expedition in 1493–94, completed a voyage from Puerto Rico in the Greater Antilles to an island in the southeastern Bahamas in March, 1513. The island, so Herrera records, was "the first State that the admiral Don Christoval Colón discovered, and where, in his first voyage, he went ashore and named it San Salvador." On Easter Sunday, March 27, Ponce de León sailed from San Salvador with a small fleet of three ships to discover and settle the fabled island of Bimini, which West Indians had told him lay to the north of Cuba.

According to Herrera, Ponce de León proposed this mission on hearing of "the wealth of this island and especially that of a particular spring, so the Indians said, that restores men from age'd men to youths." Ponce de León was acting in the name of King Ferdinand of Spain, but at the king's insistence he undertook the voyage entirely "at his own expense"—understandable in light of the explorer's prime objective, to quaff water from a West Indian equivalent of Arethusa's mythical fountain of perpetual youth, fed by the god Alpheus, who took the form of a river flowing beneath the sea.

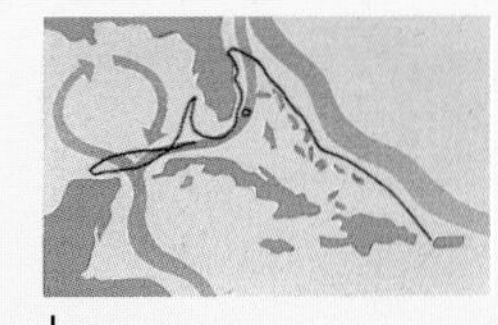

1

Ponce de León and his three ships sailed northwest (**1**), following the North Atlantic coastlines of Cat and Eleuthera Islands. However, they did not realize that they were being swept along by a strong current as well as by the wind. This current, coupled with bad weather, caused them to miss the Providence Channel that lay to the west between Eleuthera and Abaco Islands, a deep channel that would have taken them in the general direction of Bimini. So it was that they sailed beyond Abaco and into the northern extension of the Straits of Florida. On Saturday, April 2, they sighted land—what they thought to be a continuation of the Bahamian archipelago. They followed the shoreline for a considerable distance northwest, until deteriorating weather forced them to seek shelter. That night the ships anchored in an inlet and "believing this land was an island, they named it La Florida because they discovered it in the time of the Feast of Flowers"—that is, Easter time. Their anchorage is assumed to have been between Daytona Beach and Cape Canaveral.

They were greeted with vigorous hostility wherever they landed. This first time, having received a particularly hostile reception, the Spaniards put to sea again in their three ships and sailed farther north for a day or so before turning to sail southwestward. This was the point at which they discovered that "although they had a strong wind and put out all sails, they could not go forward, but rather backward. The current was so great that it was more powerful than the wind." Although they were quite unaware of it, they had indeed found an authentic "river" flowing beneath the sea—the Florida Current, part of a colossal multilevel ocean-river system that is far, *far* greater in volume than all the rivers in the world combined. This was a Fountain of Youth worthy of the name, for the ocean-river system is a key factor in the complex mechanism of global climate.

Some days later, after the scattered fleet had reassembled, Ponce de León was able to resume his southwestward course by sailing uncomfortably close to the treacherous coastal waters. By following the coast, and later the Florida Keys, the fleet entered the Gulf of Mexico and sailed some distance up the peninsula's west coast. For their return voyage in August, they decided first to head for Cuba, before embarking on a voyage to their home port at Puerto Rico. However, they set a generally southwestward course instead of due south, and this led them astray to the Yucatán peninsula, west of Cuba. After much argument about their whereabouts they decided to return to the northeast, and as they did so they were once again caught in the Florida Current. This time, it carried them past Cuba and Hispaniola, both now out of sight below the southern horizon, to Grand Bahama Island. After some delay and, one suspects, confusion, they clawed their way back through dangerous shallows towards Hispaniola. En route they "found Bimini, although not the spring." Ponce de León ordered one of his captains to land and search for the spring, but according to Herrera, Ponce de León did not himself set foot on any of the several islands that form the Bimini group.

The following year Ponce de León was appointed hereditary governor-conqueror of La Florida by Ferdinand of Spain. It was in this capacity that he later returned (in 1521) "in order to satisfy himself as to whether it was a mainland [but] the Indians sallied out to oppose him, and fighting stubbornly against him, killed some of his men, and he, wounded in the thigh, returned to Cuba, where he ended his days." The king's appointment was an empty gesture, for Ponce de León had suffered many indignities at the hands of the natives: he was hardly their "conqueror." The potential value of his discoveries was completely unappreciated.

Later in the 16th century, after the conquest of Mexico, Ponce de León's fellow countrymen were able to benefit from the Florida Current he had found. Sailing in galleons from the Isthmus of Panama with "all the diamonds, emeralds, amethysts, and gold" they could plunder, they used the Current to help propel their ships from the Caribbean Sea and the Gulf of Mexico up the East Coast as far as the latitude of Cape Hatteras [photo, pages 238–39, and Essay 2]. They then used the Gulf Stream, which flows from Hatteras to the Grand Banks of Newfoundland, to catapult them northeast, before sailing eastwards to find the Azores and Canary Currents, which in turn helped to convey them to their Iberian ports of origin.

2

The Canary Current, a cold southward-flowing current that joins the westward-flowing North Equatorial Current off the Cape Verde Islands, had aided "the admiral Don Christoval Colón" in his second and third voyages of discovery. The Antilles Current (in which Ponce de León was caught when he sailed past Abaco) completes the circuit to the Florida Current and Cape Hatteras. For centuries after Ponce de León's demise, this system, the North Atlantic gyre enclosing the Sargasso Sea, was the mainstream not only for the supply of wealth to Spain, but also for abject misery and disaster to the indigenous peoples of the Caribbean.

The currents that make up the central North Atlantic gyre (**2**), are parts of an extensive and generally clockwise-rotating system of currents. They are driven entirely by the action of wind on the surface of the sea, and primarily by the trade winds near the Equator. The trade winds are easterly (westward-blowing) winds that dominate the tropics, and were so called by merchant sailors from the 18th century onwards. The wind propels enormous volumes of generally northward-flowing warm water off the East Coast into the apex of the roughly triangular-shaped North Atlantic basin. The result is that surface currents are funneled towards Iceland, the Nordic Sea, and the Arctic Ocean. Although the Arctic Ocean does have an outlet into the Pacific, no significant volume of the Arctic's cold bottom-water can escape from the narrow and shallow Bering Strait. Since the volume of (lighter) warm water pumped into the North Atlantic and Arctic Oceans has to displace an equivalent volume of (heavier) cooled water, the latter is forced to return to the south along the Atlantic seafloor.

The influx of water into the North Atlantic gains heat from two sources. The first is the warm surface water driven by the easterly trade winds at the Equator. This is called the North Equatorial Current, and originates off the coast of West Africa, near the Cape Verde Islands. In part, the current links with the Antilles Current flowing north of the Leeward Islands and the Bahamas. As we have seen, beyond Abaco the combined flow merges with the Florida Current. Another strand of the North Equatorial Current flows closer to the Equator and drives into the Caribbean Sea between the islands of the

[*continued on page 246*]

Seafloor sediments, Scotland District, Barbados, Lesser Antilles

1 Panamanian Gap

The paleomaps at top left show that until the Isthmus of Panama was formed, the Atlantic and Pacific Oceans were connected through the Caribbean Sea. With the closure of the gap between South America and North America about 3.5 MYA, the pattern of Atlantic circulation switched from equatorial to interpolar. This change is thought to have been the key event that triggered the present Cenozoic Ice Age in the Northern Hemisphere. Other major contributory factors are discussed elsewhere in the chapter. Three stages in the crucial Panamanian event are illustrated in the paleomaps.

The Isthmus of Panama was formed from an arc of volcanic islands in the Pacific Ocean, and from accumulations of sedimentary rocks that were scraped from the surface of the Pacific Ocean floor as it subducted under the West Coast of America—called "accretionary prisms." The evolution of the modern Lesser Antilles island arc, of which Martinique and Barbados are part, can be followed on the right-hand side of each of the three maps. The Pacific island arcs formed in a similar way over 30 MYA during the early stages of the closing of the Panamanian gap.

The main picture is of Mount Pelée on the volcanic island of Martinique. Mount Pelée erupts and continues to build an island as an indirect consequence of the westward-spreading North Atlantic seafloor's subduction beneath the Caribbean seafloor. The inset-picture is of colorful sedimentary rocks that have indeed been scraped off the seafloor and are now exposed in the Scotland region of the island of Barbados. This island as a whole is an accretionary prism encrusted with corals. It was formed on the descending edge of the North Atlantic floor, east of the Antilles.

2 Central Gyre

The Isthmus of Panama moved into place around 3.5 MYA, and from this time on, equatorial currents from the Atlantic could no longer enter the Pacific Ocean. An increased volume of tropical water was deflected into the North Atlantic and distributed northward. As these tropical currents cooled by evaporation, they released moisture that ultimately precipitated as snow on northern continents in winter. From about 3.0 MYA snow began to accumulate year by year,. forming the nuclei of the ice sheets that were soon to envelop North America, Greenland, Iceland, and much of Europe from the British Isles to the Urals.

Because of the rotation of the Earth, ocean currents have a natural tendency to move to the west until they meet an obstruction. The currents then divert to the north in the Northern Hemisphere and to the south in the Southern Hemisphere. Accordingly, the mainstream of tropical currents in the North Atlantic—the Florida Current and the Gulf Stream—pass along the American East Coast between Cape Hatteras and the Bermuda Islands. At Hatteras the Gulf Stream is deflected away from the coast by the salty, and therefore dense, Labrador Current.

Other surface currents in the Central Atlantic, offshoots of the mainstream, contribute to the clockwise motion of the ocean gyre. The pressure of the winds that drive all surface currents also causes the sea to "pile up" in front of the wind. Water on such slightly sloping surfaces flows down-slope to the right at 45° to the direction of the wind, as shown in the diagram. This downslope water-flow drags the layer of water beneath it to the right at an angle a little less than 45°, and each layer of water acts on the one beneath it in a similar fashion. This results in a spiral-staircase effect, which contributes to both the formation and the clockwise motion of the gyre. The "dead" water inside the gyre is called the Sargasso Sea, an ocean desert that lies southeast of Bermuda.

The Bermuda Islands also mark the present northernmost limit of North Atlantic corals—determined by the temperature of sea water. The circular reefs in the inset picture are a rare form of micro-atoll. They were built by marine snails and algae on the tops of sand dunes as the dunes were gradually being submerged during a period of increasing sea level. The original dunes were formed during the last glacial maximum about 18,000 years ago when sea level was hundreds of feet below the present and Bermuda was fifty times larger as a consequence.

CENTRAL ATLANTIC GYRE

Red—warm surface currents
Blue—cold surface currents
Blue-grey—bottom currents

Potboiler reefs off Bermuda

Wind direction
Surface current
Net Ekman transport

EKMAN LAYERS

Nonsuch Island, Bermuda

[*continued from page 241*]

Lesser Antilles. Here it combines with the second source of warm-water flow, the South Equatorial Current.

South America's north coast is set at an oblique angle to the equator. Consequently, instead of continuing its counterclockwise gyre down the east coast of Brazil, a large portion of the South Equatorial Current (**3**) is diverted along the north coast of South America and into the Caribbean Sea. This flow crosses the Equator into the Northern Hemisphere near the Amazon Delta, where it combines with the Guiana Current and merges with the North Equatorial flow. As a result of the Coriolis effect, all western-boundary currents in the global ocean are forced to flow either north or south in accord with the geography of the continental margins.

Before the Panamanian gap between North and South America was closed by the Isthmus of Panama, this huge composite current flowed from the Caribbean into the Pacific Ocean. However, for the past 3 MY or so, since the formation of the Isthmus, its flow has been diverted into the Gulf of Mexico through the Yucatán Channel, vastly increasing the amount of tropical surface water conveyed into the North Atlantic region. The high humidity of this warm-water flow increases rainfall in the middle and high latitudes of North America and Europe—conditions which helped trigger the development of continental ice sheets, as we will see later.

THE GULF OF MEXICO, a huge oval basin roughly 970 km (600 miles) from north to south, and 1,600 km (1,000 miles) from east to west, is a natural heat-sink. Warm water flowing in from the Caribbean circulates with great volumes of warm water from the shallow shelves surrounding the basin. It then spills out through the channel between Cuba, the Bahama Banks, and Florida in a familiar guise—the Florida Current that so troubled Ponce de León. The Florida and Antilles Currents then combine to form the Gulf Stream off Cape Hatteras. Beyond the Grand Banks of Newfoundland (**4**), the Gulf Stream ends and the North Atlantic Current starts. The latter has a subsidiary, the Irminger Current, which splits off, turning counterclockwise south of Iceland, flowing round the southern tip of Greenland and into the cold Labrador Sea. North of Iceland, the Nordic Sea system of surface currents—including the Norway, North Cape, West and East Spitsbergen Currents—completes the surface distribution of warm waters that originate in the tropics.

This tropical ocean-river loses heat by evaporation as it crosses vast reaches of open ocean. It flows at the rate of more than 12 cubic miles a second and extends to a depth of two miles or more (3.2 km). As it evaporates and cools, its salt content becomes more concentrated. For example, the areas of greatest heat-loss in the Gulf Stream section are those east of Cape Hatteras and the Gulf of St. Lawrence. Here, relatively cold winds from the North American continent meet the moist, warmer air associated with the Gulf Stream. Much of the stream's dissipated heat is carried east by prevailing westerly winds, contributing to the humid but temperate climates of Western Europe.

The comparatively warm Irminger surface current enters the Labrador Sea, where it modifies the temperature of icy surface currents issuing from Baffin Bay, and from around the southern tip of Greenland (the East Greenland Current). During this process, water lost through evaporation of the Irminger contributes to the high humidity needed to maintain the Greenland ice sheet. The resulting modified, cold surface current—the Labrador Current—flows southward along the East Coast until it meets the Gulf Stream off Cape Hatteras [see pages 238 to 239]; it also contributes to a south-flowing subsurface current called "Slope Water." This moderately deep water issues primarily from the Sargasso Sea in small volumes, flowing as if in conduits beneath the Gulf Stream. It combines with water from the Labrador Current and modifies the temperature of waters in the Newfoundland Banks fishing grounds.

Most of the returning cold water, forced south by the influx of equatorial currents, issues from the Iceland and Nordic Sea regions

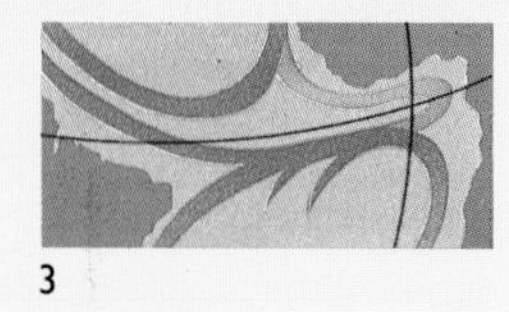
3

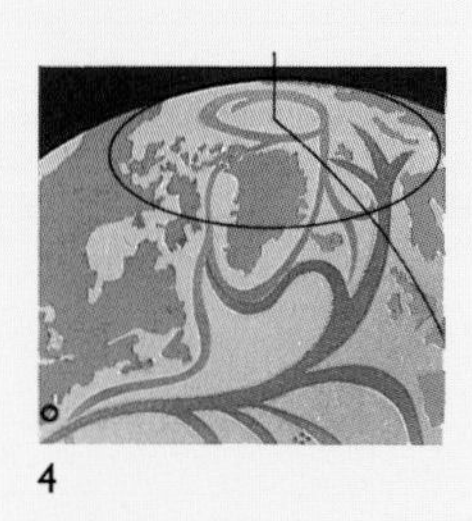
4

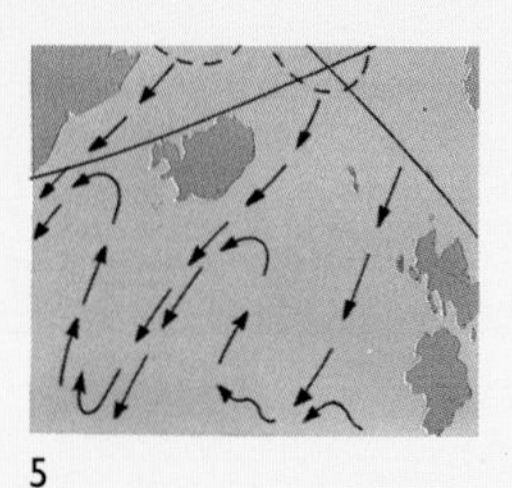
5

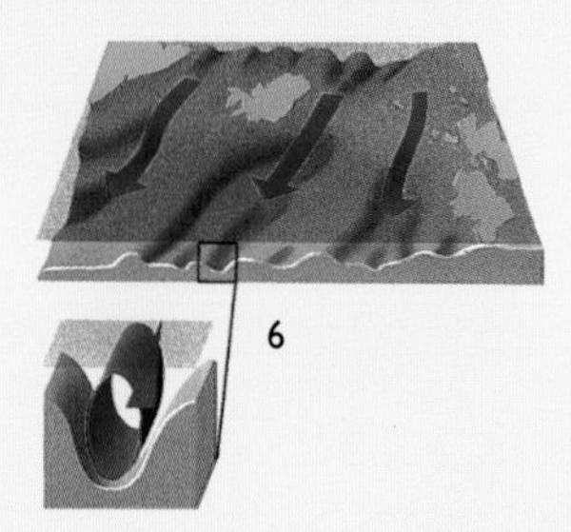
6

at great depth and in staggering volume. As we have seen, because of evaporation the north-flowing Gulf Stream becomes more salty and dense as it progresses. Its higher density should make it sink, but in the colder seas above Newfoundland, surface currents, although saltier, are still warmer and therefore less dense than the Arctic bottom-water flowing southward beneath them. They therefore remain near the surface at intermediate depth. In summertime, they allow the relatively warm North Atlantic Current (here no longer called the Gulf Stream) to reach far into the Arctic Ocean, up to the very edge of the permanent pack ice that surrounds the North Pole.

DAYLIGHT HOURS in the northern winter become steadily fewer and drop to nil above the Arctic Circle. Even so, the action of winds on the relatively "warm" surface of the Arctic sea causes rapid evaporation; the stronger the wind, the faster the rate of evaporation, and the greater the cooling effect on the sea. As the northern sea surface water evaporates and cools, it increases in salinity and therefore density, sinking to be replaced by marginally warmer water from beneath. But as surface water approaches freezing point, it loses its capacity to hold salt in solution; indeed, frozen seawater only retains about a third found in the sea. The excess salt released by freezing temperatures at the surface is absorbed by the upwelling warmer-water layer, increasing the salinity of that layer too. In turn, this water mixes with warm, intermediate currents introduced by the North Atlantic and Nordic Sea Currents.

The result of all this interaction is an extraordinarily powerful vertical-mixing process, releasing the remaining tropical heat in these far reaches of the ocean-river system to the surface. The heat is immediately soaked up by cold, dry, subpolar and polar winds, while the now ice-cold high-salinity water of the intermixed region sinks towards the ocean floor, where it displaces equally cold but *less* saline (and therefore less dense) bottom water. The high-salinity water accumulates in deep basins in both the Nordic Sea and south of Iceland (**5**). Ultimately it spills over its retaining basins and begins to move southward along the seafloor. The overflow, called "North Atlantic Deep Water," compensates for the influx of warm water into the North Atlantic Ocean at the Equator [see globe, page 240].

The total heat released during the vertical mixing process is estimated to be equivalent to one-third of the input of solar heat into the northern North Atlantic during the summer. Since this excess heat is released during autumn and early winter months, it moderates winter temperatures east of the areas where the mixing occurs—in Norway and the British Isles in particular. Contrary to popular belief it is the heat released during this heat-exchange process, and not the Gulf Stream, that accounts for the comparatively mild winters and ice-free ports of Northwestern Europe.

The magnitude of the vertical interaction of currents in the generating areas around Iceland was only appreciated with the discovery of enormous ridges on the seafloor. These dune-like ridges of disturbed and redistributed sediments are hundreds of miles long and tens of miles wide, and are created as turbulence produced by the mixing process scour the seabed (**6**). They are associated with the Denmark Channel between Greenland and Iceland, and the Iceland-Faeroe Ridge and Faeroe-Shetland Channel to the east of Iceland. The sediments on either side of these channels postdate the formation of the ridge between Greenland and Scotland, which developed during the separation of North America from Europe.

A ZONE OF RAPID TEMPERATURE CHANGE between ocean currents is called a "thermocline," and a zone of changing saline density a "pycnocline." Both act as barriers, separating currents of different temperature and salinity into gigantic but thin-walled conduits. This allows individual currents at different levels to move in different directions, altitudes, and speeds.

The North Atlantic Deep Water thus overflows the deep basins in which it accumulates, displaces surrounding water, and flows southward. It deflects to the west (**7**) in response to the Coriolis effect and then flows through the valleys of the mountainous seafloor. At the latitude of Nova Scotia (45°N), it meets the northern extension of the cold "Antarctic Bottom Water." Because this northward-bound current is even colder and more saline than its southward-bound counterpart, the Arctic water is pushed over the Antarctic stream as an intermediate-depth southbound current—a submarine river about 2,000 m (6,500 ft) thick.

This intermediate current flows south through the western North Atlantic, following (and to some extent modifying) the shape of the continental slope on North America's East Coast. Having reached the Equator, it flows down the western side of the South Atlantic towards the Southern Ocean. Part of its mass then flows into the region of the Weddell Sea off Antarctica, and joins the circumpolar Antarctic current. The rest of it flows around southern Africa, before percolating into the far reaches of the world ocean.

The North and South Atlantic interpolar circulation, caused by the formation of the Isthmus of Panama, is now thought to be one of several vital components of the global climate mechanism. It is, in effect, a "mainspring" of the heat-exchange mechanism between ocean and atmosphere. Interpolar circulation conveys more heat energy towards the North Pole from the tropics than the atmospheric circulation itself can carry. It returns an equivalent volume of cooled high-salinity bottom water towards the South Pole and ultimately into the Indian and Pacific Oceans.

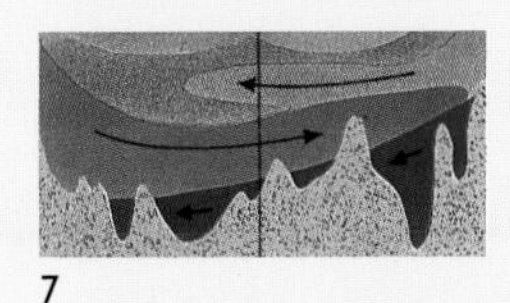
7

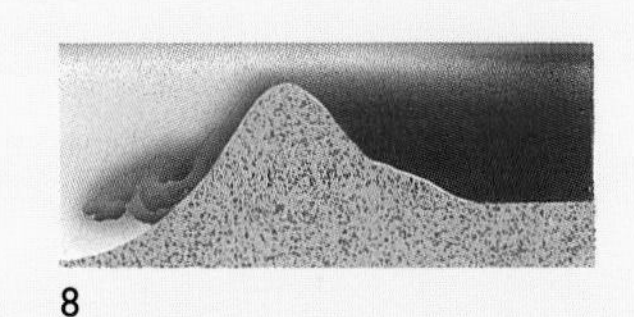
8

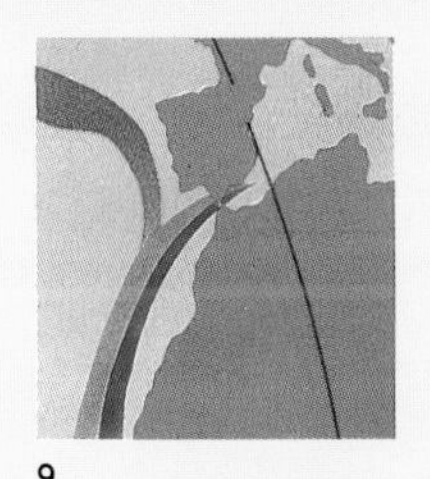
9

10

THERE IS, however, a notable aberration in the system. The high rate of evaporation in the almost enclosed Mediterranean basin causes comparatively warm but dense salty water to accumulate in great volume on the floor of that basin. As warm North Atlantic surface waters flow eastwards into the Mediterranean Sea, they displace an equal volume of accumulated Mediterranean saline bottom-water. This spills over the basin's containing sill between Morocco and Gibraltar (**8**), and then flows down the continental slope into the Atlantic where, because of its density, it sinks to the seafloor off Northwest Africa. The warm but dense Mediterranean bottom-water displaces cold but less dense bottom-water towards the surface. This upwelling contributes cold water to the Canary Current section of the North Atlantic gyre (**9**)—the gyre that helped to propel Columbus towards the North Equatorial Current and the New World during his second and third voyages.

The benefit of such favorable winds and currents was not felt so much on the first voyage of discovery (1492–93). This was made along an almost direct line from the Canary Islands (28°N) to the Bahama island of San Salvador (23°N). But on subsequent voyages, Columbus crossed the ocean on progressively more southward courses, evidently seeking ever more favorable sailing conditions ... as well as perhaps trying to ensure a more southwesterly landfall on what he still thought to be a coastal region of Asia.

The second voyage had a landfall in the Lesser Antilles at 15°N, on the same latitude as the Cape Verde Islands (Dominica, 1493–94). His third crossing (Trinidad and Venezuela, 1498–1500) was made south of 10°N. The admiral was a brilliant navigator—his principal talent, many historians would say. He was probably the first seaman to realize the connection between the Canary and North Atlantic Equatorial Currents and their associated easterly trade winds. As a matter of interest we should add that Columbus's fourth and last voyage of discovery to Mesoamerica—Honduras, Nicaragua, Costa Rica, and Panama—in 1502–04, is sometimes called his "High Voyage." Columbus deliberately returned to a higher latitude for the crossing—one approximating that of his second voyage.

WE GENERALLY TAKE CLIMATE for granted over the duration of our individual lifetimes. We may remember colder winters or warmer summers than we seem to have today, or vice versa. In fact climate varies tremendously even within the passing of a century. Importantly, in view of the present concern about climate change, such transformations have occurred many times in past centuries, long before the advent of fossil fuels. For example, Columbus had the good fortune to undertake his four great voyages of discovery during an interlude of favorable climate in an otherwise cold-climate period called the "Little Ice Age." This cold spell is one of the biggest known climatic anomalies since the end of the last glacial age. It began in the 13th century and lasted until the mid-19th century. Because of its historical relationship to this and the following chapter it is important to define the term in its original context, as the period in the last millennium when glaciers worldwide expanded and fluctuated around more advanced positions than before 1400 AD or after 1900 AD. (Some scientists use the term "Little Ice Age" for a specific period: from 1550 to 1850.)

During the Little Ice Age it was the custom in England to hold festivities on the Thames River whenever it froze sufficiently in wintertime, which was more often than not. This was in part because the Thames was not confined by embankments at that time: it was a wider and shallower river and therefore prone to freezing in cold winters. Tidal races that would have inhibited freezing in the Thames were also dissipated along the shallows near river banks. The first organized "Frost Fair" on the Thames was in the winter of 1607–08; the most protracted was in 1683–84, when the river-ice was over ten inches thick. The last Frost Fair was held in 1813–14, when an elephant was walked across the frozen river. From these and other testimonies, one might judge that temperatures had fallen much more than scientific evidence supports. This evidence suggests that a swing of only 2°C (3.6°F) in mean temperature was sufficient to cause major glacial fluctuations and the onset of the Little Ice Age in Europe, and then to provide a definite end to it. The return to "normality" was marked by a general retreat of glaciers worldwide that started in the 1920s and continues today.

PERTINENT CLIMATE INFORMATION is sometimes available from historical records: and is often used to support scientific data gleaned from the glaciological record. For instance, the annual dates of grape harvests in France, Switzerland, Germany, and elsewhere in Europe from the 14th century to the present have been researched, collated, and analyzed. These data have been prepared and averaged on the premise that an early harvest, measured from September 1 each year, indicates a warmer spring-summer period of growth, and that a late harvest implies a cold summer. For example, in 1488 the harvest was late (October 17), indicating a cool summer, but in 1495 the date was early (September 12), indicating a warm summer. The illustration here (**10**) is redrawn from the Peterborough Psalter, an early-15th-century celebration of a grape harvest, a regular event in England at that time. Similarly, in medieval times the annual price of wheat in Britain fluctuated according to available supply and demand. The low prices indicate warm, dry summers of plenty, and high prices indicate cold, wet summers of shortage and possible famine.

From analysis of harvest-date averages for vineyards in Europe it is evident that the Little Ice Age consisted of a series of frequent climate fluctuations throughout the six centuries or so of its duration. Each fluctuation deviated from the general trend of climate for individual years, for clusters of years, or sometimes even for decades—such as during the warm interval when Columbus crossed the Atlantic. The Little Ice Age is nevertheless classified as an ice age in miniature. This is because the incidence and frequency of severe winters and cool summers was such as to permit glaciers generally to advance all over the Earth, and also to remain far more advanced than any appear to be today. The Rhone glacier-snout in Switzerland has withdrawn a long way from the hotels built for 19th-century tourists. Navigation maps drawn after the Second World War show glacial fronts in northern fiords significantly more advanced than today—although this trend started to reverse in Norway and Spitsbergen at the end of the 20th century.

[*continued on page 254*]

3 Milankovitch Cycles

Julibreen is just one of Spitsbergen's innumerable glaciers. They stem from a central ice cap that is about 1,120 km (700 miles) from the North Pole. Like glaciers the world over, all Spitsbergen's glaciers advanced and retreated many times during a period called "The Little Ice Age," which lasted for most of the last millennium. During the last period of major glacial advance 20,000 years ago, the Spitsbergen ice cap was several miles thick and its glaciers proportionately much larger.

Such geologically frequent fluctuations in the extent and thickness of permanent ice are believed to be a result of eccentricities in the Earth's orbital geometry (see illustration). These affect the Earth's solar radiation "budget," and are known as Milankovitch cycles, after their discoverer, Milutin Milankovitch (1879–1958). Milankovitch's theory is based on the notion that the gravitational pull of the Sun, Moon, and planets on the Earth affects the shape of the Earth's orbit round the Sun. A near-circular orbit produces a steady stream of sunlight through the year. An elliptical orbit causes the intensity of sunlight to vary by as much as 30 percent during a year. The cycle from near-circular to most elliptical state takes 90,000–100,000 years to complete, and is just one of a number of eccentricities that Milankovitch took into account.

Over the course of a 43,000-year cycle, the Earth's axis varies in its angle of tilt to the Sun by almost 3°. The more upright the Earth is relative to the Sun, the smaller the circles of darkness encircling the Earth's poles on their respective midwinter days. Small-diameter polar circles lead to a warmer Earth, and larger circles infer a colder Earth. The Earth also gyrates and wobbles on its axis like an off-balance spinning-top. Gyration and wobble cause "precession of the equinoxes," as each hemisphere's moment of closest approach to the Sun regresses by a few minutes each year. The axial cycle takes 24,000 years to complete and the precessional cycle takes 19,000 years.

The graph is a reproduction of a classic trace produced from analysis of a section of seafloor core. It shows how the estimated global volume of ice has indeed varied in accord with the Milankovitch predictions over the last 500,000 years.

Julibreen, northwest Spitsbergen, Svalbard

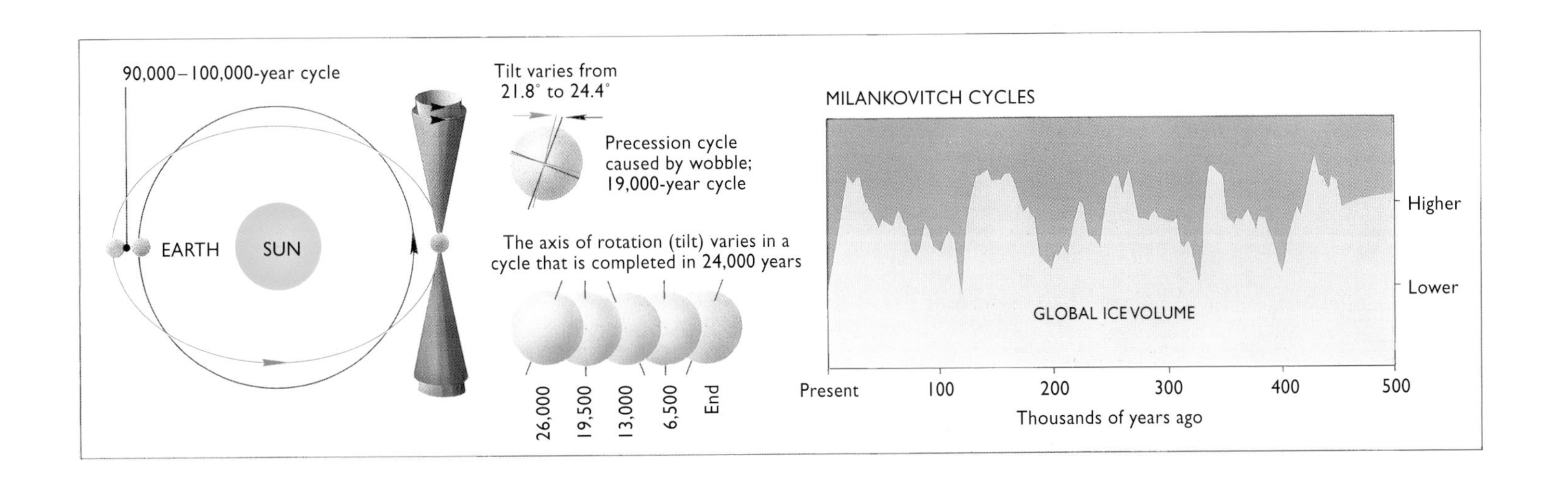

4 Glacial Maximum

By 3.0 MYA rocks carried by glaciers were being dropped on the seafloor by melting icebergs in the North Atlantic, which indicates that permanent ice caps had formed over the northern continents. Snow had turned to ice under increasing pressure of accumulation. Ice-sheets were forming and, where the masses of ice were carried downhill by gravity, glaciers were extending. However, much of Asia had remained free from ice; the region was too far removed from the moisture-bearing winds generated by North Atlantic warm currents. But prevailing westerly winds from Asia picked up moisture from the Pacific and conveyed it towards the Pacific Northwest. Here it precipitated as snow on the high coastal ranges of Alaska and British Columbia, and this region now began to form its own ice sheet and glaciers. By 1.6 MYA, extensive areas of North America and Europe had become glaciated. Apart from a number of interglacial interludes, such as the present one, North America and Europe have remained glaciated. According to theory, they will remain so until continents shift, ocean circulation changes, and the global climate responds.

The summit ice of Iceland's Vatnajökull pictured here is still 2 km (1.25 miles) thick, about the same thickness as the sheet of ice that enshrouded the whole of Iceland at the peak of the last glacial age. The once evenly distributed covering of volcanic ash on the ice cap's surface has been distorted by different rates of movement within the ice. The Sun's reflection can be seen glinting on the surface: the ratio between this reflected energy and the energy absorbed by the ice is termed "albedo." The albedo of ice sheets and marine ice fields is so high that, once commenced, glacial ages are inclined to be self-perpetuating. The increased amounts of ice reduce the amount of solar radiation being absorbed by the Earth's surface, cooling the planet still further, and leading to further expansion of the ice sheets in a feedback loop. Vegetation, rock, clouds, and water all have their own albedo ratings, which are much lower than those of ice and snow.

The two circumpolar projections show the Northern Hemisphere and its ice cover, permanent sea ice, and continental shelves, at present and at the last glacial maximum, 18,000 BP. During the maximum, the tropics were still tropical, but permanent North Atlantic sea ice extended considerably farther south. The North Atlantic gyre was therefore compressed between two extremes of climate and circulated more vigorously than now. The vast accumulation of terrestrial ice caused a major fall in sea level, at times 120 m (400 ft) below the present. Shallow shelves on continental margins were therefore exposed above the sea.

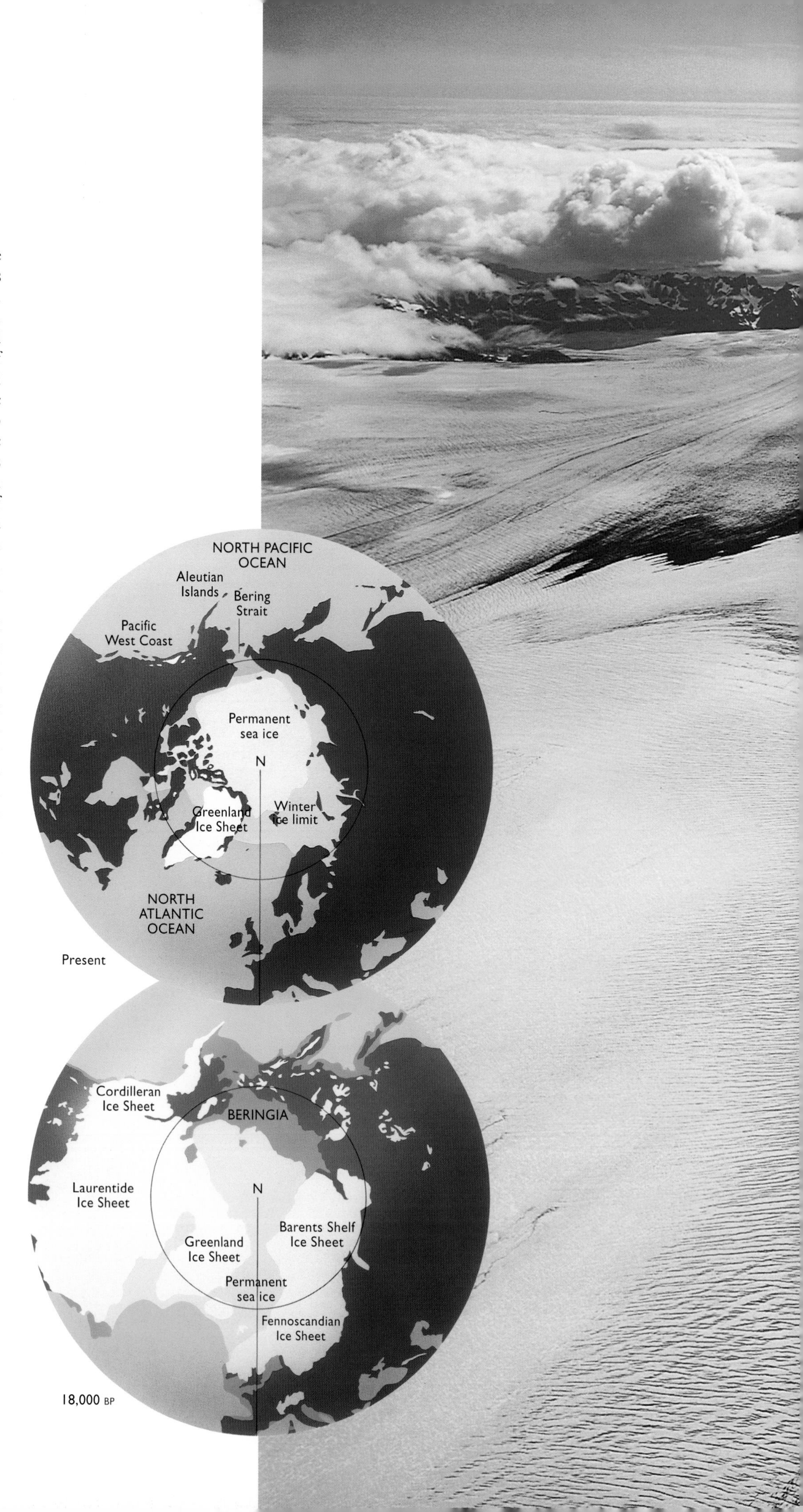

Summit of Vatnajökull
ice cap, Iceland

5 Glacial Retreat

Late Wisconsin terminal moraines (purple-brown shapes) and rivers about 10,000 BP

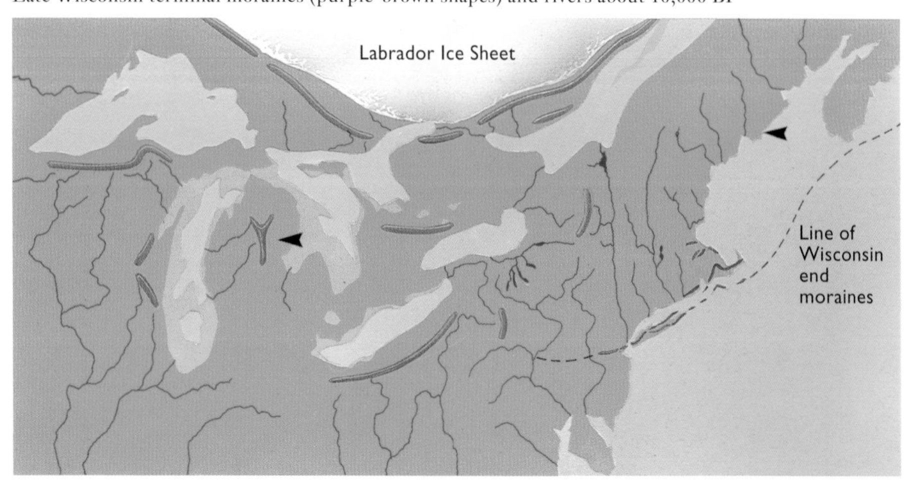

Each glacial age is named after the place where its principal ice sheet reached maximum extent—the point at which a terminal moraine was deposited. In North America the main glacial ages in the last 1.6 MY are called the Nebraskan, Kansan, Illinoian, and Wisconsin transgressions. In Europe there are numerous names for corresponding glacial ages; for instance, the four Alpine glacial ages are termed Günz, Mindel, Riss, and Würm after tributaries of the Danube River. Similarly, the names used for the glacial ages in Britain, Scandinavia, Italy, and elsewhere in Europe are all related to regional place names.

Glacial ages are followed by interglacials. In North America these are named after places in Iowa, and in Europe after each of several localities where fossils of broad-leafed trees and forests first reappear. Such places are also marked by signs of isostatic rebound by landscapes relieved of the weight of ice, or by the reappearance of marine sediments in regions that had previously been above sea

level. The final, unmitigated retreat is signaled by a growing web of established lakes and rivers such as those illustrated here.

Bass Harbor Lighthouse on the Acadian coast of Maine has some well-rounded visitors on its foreshore—the rocks in the foreground to the left of the picture, known as "glacial erratics." These rocks were gouged by glacier ice from a rocky outcrop hundreds of miles to the northeast. Then they were worn and rounded by water and ice before being deposited here as the Wisconsin ice sheet retreated about 10,000 BP. The inset picture is of plowed farmland in Wisconsin. The elongate hillocks are "drumlins," the compacted products of glacial grinding that accumulated under the margin of the ice sheet that once covered this region, and were shaped by its flow. The highway runs due north, so one can judge the direction of the retreat of the ice sheet.

Ice age landscape, Wisconsin

[*continued from page 247*]

Although they seem considerable, the glacial advances and fluctuations of 1.6 km (1 mile) or so during the Little Ice Age do not compare to the "great" glacial advances and retreats of the past several million years. These amounted to hundreds and thousands of miles. It is simply an illusion of our times that the Earth may be warming *irreversibly*, and that the Ice Age is at an end. The great glacial advances and retreats of the past will undoubtedly recur [Essay 3].

CLIMATE CHANGE OCCURS on different time scales. For example, lake sediments are deposited annually and accumulate in thin layers called "varves"—they can sometimes accumulate for centuries or even for thousands of years, depending on the lifetime of the lake. The varying thicknesses and characters of individual varves indicate short-term fluctuations in climate.

In contrast, bands of lithified seafloor sediments up to several feet thick can also indicate climatic change—but on a time scale of many thousands of years. Formations called "rhythmites" found in the Dolomite Alps, the Atlas Mountains of Morocco (**11**), and elsewhere, are all examples of rhythmic climate change. Each band is composed of carbonate rock that is discolored according to the degree of impurity in its carbonate content. The bands accumulate in sequences and thousands of feet thick. When uplifted and exposed by erosion, they display an uncanny rhythmic quality in their appearance, as the sequences are repeated time and again. In some instances it is possible to see a second rhythmic change of color intertwined with the first series. Each pair of carbonate bands may have taken 20,000 years or more to form. A second rhythmic pulse might occur at approximately 40,000-year intervals of rock formation. It is evident that there is something very unusual about the formation of a series of rhythmites in a cliff several thousand feet high. The regularity of deposition and change in character must relate to cyclic climate changes during formation.

So it seems that climate varies in some intricate but cyclic way that influences grape harvest times, frost fairs, and varve formation at the "microscopic" end of the scale, and the formation of post-Pangean rhythmites at the "macroscopic" extreme. Today it is known that there are indeed cyclic variations within variations of climate, and that these rhythms are as intricately interwoven as a Bach fugue is interwoven with variations on a single theme. In fact, many climatic variations respond to changes in the Earth's orbital geometry. They are called "Milankovitch cycles," after Milutin Milankovitch (1879–1958), a Yugoslav mathematician [Essay 3].

MILANKOVITCH WAS A CONTEMPORARY and associate of Alfred Wegener (in his profession of meteorologist) and Wegener's father-in-law, Wladimir Koppen (1846–1940). Koppen is considered by many to have been the most distinguished meteorologist and climatologist of his time—the father of modern meteorology. As a student and budding theoretician, Milankovitch had sought new realms to conquer in the world of mathematics. He recalled in his autobiography that after sharing several bottles of good wine with a poet friend in celebration of the publication of a book of poems, he had been "attracted by infinity." He had decided to "grasp the entire universe and spread light into its entire corners!" In his postgraduate days his intellectual ambition matched this early and perhaps slightly intoxicated vision. Milankovitch decided to calculate the temperature at the parallels of latitude on Earth and on other planets in their varying orbits around the Sun, and from these calculations then to attempt to describe the climates of the past—a major task that was to take him 30 years to complete.

In 1914 Milankovitch published a paper that addressed the problem of the *Astronomical Theory of the Ice Age*—a theory first advanced by a Scottish scientist, James Croll, in 1867. Croll (1821–90) had discovered that the eccentricity of the Earth's orbit changes cyclically, and had published the hypothesis that these changes were linked to the occurrence of ice ages. Earlier that century there had been decades of fierce controversy about even the possibility of ice ages; the question being ultimately resolved by Louis Agassiz in 1840. In 1920 Milankovitch published a formula for calculating the intensity of the Sun's radiation as a factor of latitude and season. He showed that changes in radiation due to changes in orbital geometry would be sufficient to cause ice ages. This paper caught the attention of Koppen, and in 1924 he published a joint paper with Wegener applying the Milankovitch formula to three different northern latitudes. They showed how summer solar radiation at these latitudes had varied over the last 650,000 years. Their findings were in general accord with predictions that Milankovitch had made from his calculations to that point. With Koppen's encouragement, Milankovitch then continued his work, and eighteen years later in 1938 he published the final version of his theory of the occurrence of ice ages.

11

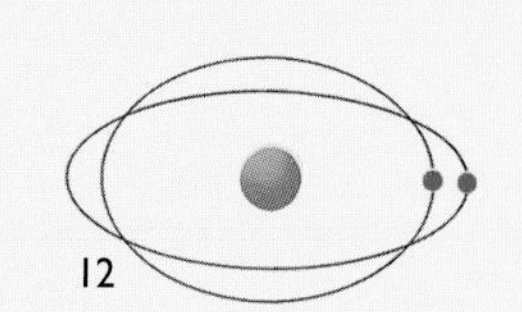
12

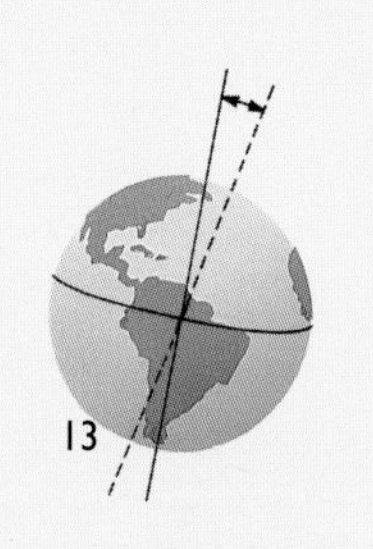
13

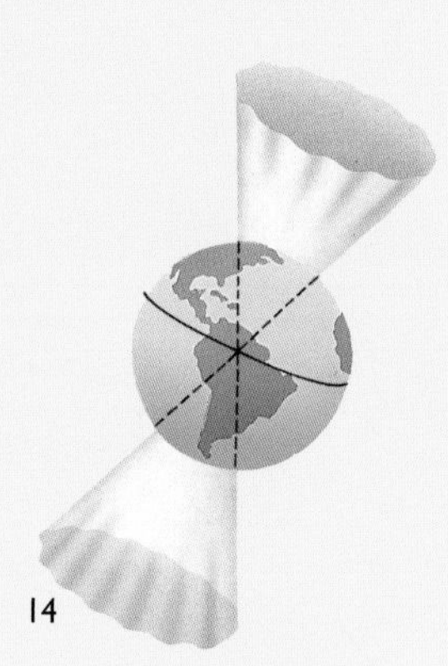
14

Milankovitch based his theory on the fact that the Earth's orbit varies from near-circular to a pronounced ellipse, and from elliptical to near-circular, in a cycle that takes 90,000–100,000 years to complete (**12**). During this time the total annual amount of radiant heat received by the Earth remains the same. However, the intensity of the sunlight received during a year varies with the Earth's distance from the Sun, unless there are fluctuations in the Sun's output of radiant heat from year to year—as indeed there are.

Milankovitch also knew that during the course of 43,000 years, the Earth's axis varies in its angle of tilt to the Sun. Since the Earth's income of radiant heat is constant each year, it follows that between subpolar and tropical latitudes an increased or decreased angle of tilt alters the magnitude of seasonal change from winter to summer (**13**). This is less severe when the angle is decreased and more severe when it is increased.

The Earth gyrates and wobbles as if off balance during this gradual change of axial angle. During these gyrations and wobblings (**14**), which have 24,000-year and 19,000-year cycles, the season of closest approach to the Sun gradually precesses from summer to winter and back to summer in each hemisphere. The summer and winter solstices simply occur a few minutes earlier each year. Thus, about 10,000 years ago, at the end of the last glacial age, the Northern and Southern Hemisphere seasons were the reverse of today. The Earth was closest to the Sun in northern midsummer, and farthest from it in northern midwinter [Essay 3]. All the planet's geometric eccentricities are superimposed one upon the other, and the near-miracle of Milankovitch is that he cataloged their combined effect upon the Earth without a computer.

In a sense, the cycles that Milankovitch demonstrated are passive. They did not and do not predict ice ages. Rhythmite rock formations indicate fluctuations in climate that match Milankovitch cycles. They register the seasonal and geographical distribution of the radiant heat that falls on the Earth's surface. But it so happens that the cycles also indicate optimum times on Earth for glacial advance or retreat during an ice age—such as the present Cenozoic Ice Age—when geographical conditions are *already* right for an ice age. Ice ages are rare in the Earth's history. When they do occur, Milankovitch cycles can predict the most likely times for the main glacial advances and retreats within that event. But Milankovitch cycles themselves have been continuous for billions of years.

ON A GEOLOGIC TIME SCALE it seems that there are at least two primary factors that trigger ice ages like the Cenozoic event. The first is the location of the drifting continents relative to the poles and the second is the pattern of global ocean circulation. On the Milankovitch time scale of glacial advance and retreat within a major event, other, more rapidly variable, factors may be at work. Such variables could include changes in the intensity of solar radiation and fluctuations in the level of carbon dioxide in the Earth's atmosphere.

Analysis of the CO_2 content of ice cores that date back tens of thousands of years before the present (cores drilled both in

Antarctica and in Greenland) has shown that CO_2 variations in the Earth's atmosphere in the immediate past happened in step with glacial advances and retreats. Atmospheric CO_2 levels were low during times of glacial advance and high during glacial retreat (the so-called greenhouse effect). They also show that the Northern and Southern Hemispheres have been in phase with each other during the last several glacial advances and retreats. This suggests a global link, such as the proportion of CO_2 in the atmosphere. There are several complex theories about how CO_2 levels could have been influenced before the burning of fossil fuels was a factor; but whatever the cause of CO_2 fluctuation, it has not yet been explained why the levels changed in rhythm with the glacial retreats or advances predicted by Milankovitch cycles. Are CO_2 fluctuations the cause of climate change, or are they a result of changes induced by other factors—such as variation in the intensity of the Sun's radiation? We may be in for a protracted academic argument before this issue can be finally resolved.

WHERE ARE WE TODAY in the present series of Milankovitch cycles? According to astronomical calculations, in the 100,000-year cycle the Earth's orbit is nearly circular and is becoming more so. Thus solar radiation is almost constant in intensity throughout the year. In the 43,000-year cycle the angle of the Earth's axis is decreasing to near minimum, and the gradient of seasonal change between subpolar and tropical latitudes is therefore decreasing. In the 24,000-year and 19,000-year cycles the Earth is closest to the Sun during Southern Hemisphere summers (which are therefore relatively warm), and farthest from the Sun during Northern Hemisphere summers (which are therefore relatively cool).

All this adds up to the conclusion that at present we are at, or near the end of, the Milankovitch cycles that led to the present interglacial warm period [Essay 5]. The Earth's climate should now be deteriorating over the centuries, and we should be heading into a glacial age. The possible irony of our time is that artificially induced greenhouse warming *may* be strong enough to postpone the inevitable decline. That consideration aside, once descent into a renewed glacial age begins, as it most assuredly will within time, a glacial advance with peaks and troughs of intensity should continue for the next 20,000 years or so. In terms of the present Cenozoic Ice Age as a whole, lasting perhaps another 20 MY, Milankovitch cycles will continue at a frequency of about ten complete cycles per million years. This will go on until the configuration of continents and the pattern of ocean circulation change the Earth's climate regime, and therefore its susceptibility to astronomical cycles.

FOR DECADES there were serious doubts about the validity of the Milankovitch theory in the scientific community, and particularly about the near-coincidence of northern and southern glaciations that the Milankovitch mechanism required. Doubts evaporated when it was shown that marine temperature variations on the Southern Ocean floor had corresponded to the 43,000-, 24,000-, and 19,000-year Milankovitch cycles.

Cores taken from the seabed revealed that the makeup of protoctista assemblages (foraminifera, radiolaria, coccoliths) changed in rhythm with the cycles predicted by Milankovitch. The characteristics of these minute organisms—their size, type, and direction of coiling—changed according to the increase or decrease in the salinity of the ocean, an indication of the amount of water locked up as ice on land. Also, variations in the levels of different oxygen isotopes (oxygen molecules with different masses) in a core can be attributed to glacial advances and retreats. Most convincingly, this pattern of change was repeated in sediments dating back 150,000 years [Essay 3]. Further work over many years has confirmed that both hemispheres do undergo near-synchronous glaciations, and have done so over millions of years. Even so, it seems that the glacial advances around Antarctica appear to lead the planet into and out of glacial

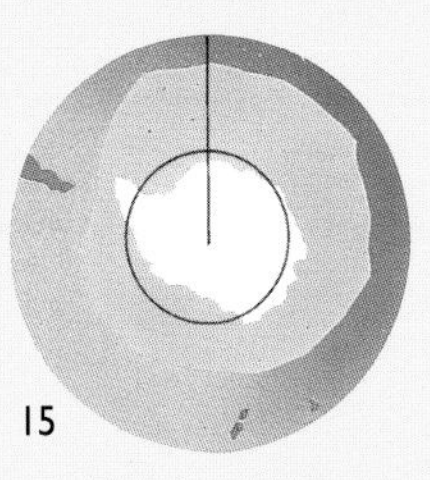
15

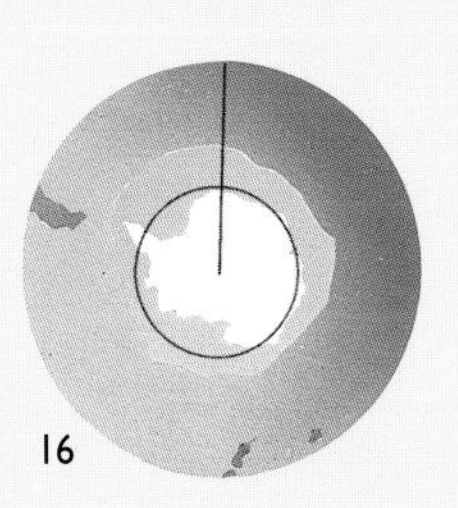
16

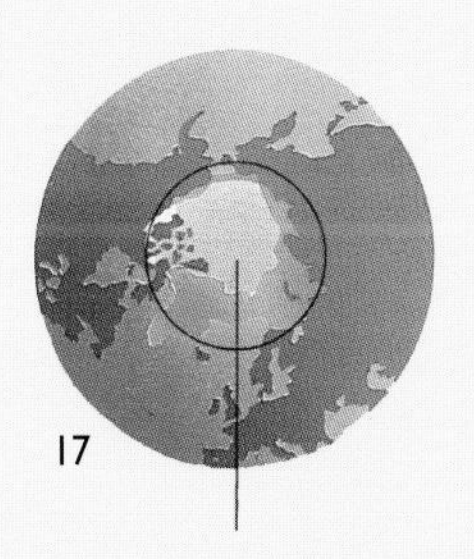
17

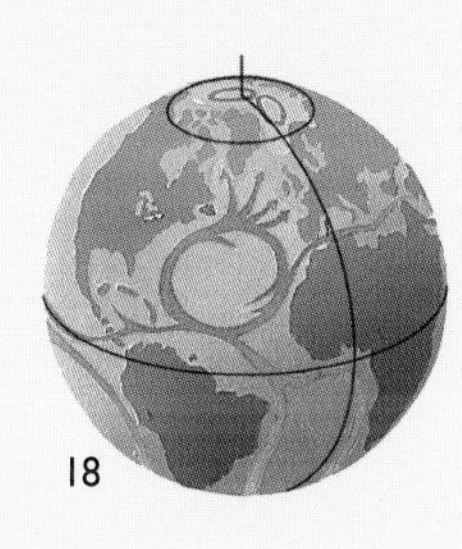
18

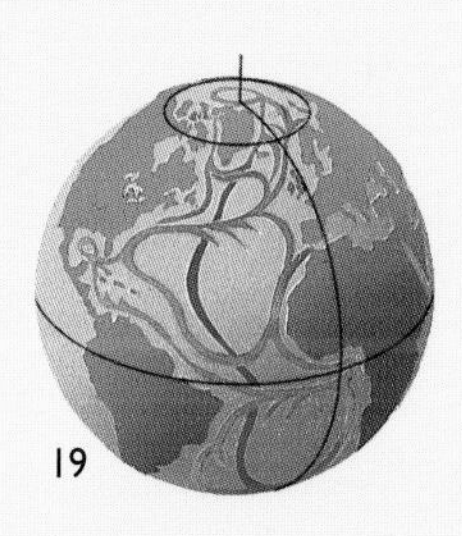
19

20

stages. This is so because of the present disposition of the continents [Essays 4 and 5: **15**].

The continent of Antarctica, stationed over the South Pole, is covered by a permanent ice sheet (**16**) that reflects sunlight in the polar summer. As it enshrouds a whole continent, the ice sheet has a fixed capacity to reflect rather than to absorb radiant heat (its "albedo"). Since the area of the continent is fixed, although the thickness of its land ice can vary, the net reflection of heat by the Antarctic ice sheet remains the same.

The variable albedo of the Antarctic therefore depends on the amount of sea ice that forms around the continent in winter. A severe winter enhances the formation of sea ice, which increases the area of ice to reflect sunlight the following summer. A series of severe winters and cool summers, induced by the progress of a Milankovitch cycle, increases south-polar albedo rapidly. Reflection of sunlight off clouds, volcanic dust, and other aerosols, also has a marked albedo effect, which further complicates the issues.

Conversely, northern continental ice sheets melt quickly—catastrophically, some think—for the same reason that they are slow to establish themselves (**17**): the continental surfaces on which they form retain their warmth [Essay 6]. Ice surrounding Antarctica dissipates rapidly at the end of a glacial cycle, but the vast south-polar continent retains its glacial cover long after northern continental ice sheets have disappeared. This is why Antarctica appears to lead both hemispheres into periods of glacial advance, and to lag behind the Northern Hemisphere during periods of glacial retreat.

By 23 MYA major deposits of rocks dropped by melting icebergs off Antarctica had accumulated on the surrounding seafloor in one locality. The presence of these "dropstones" shows that glaciers had reached the south polar sea by that time, though some think that an Antarctic ice sheet of modern proportions did not develop until 15 MYA or later. However, recent evidence of a grounded ice sheet in another locality at 35 MYA may change the overall view. As we saw in the last chapter, although the Earth was obviously undergoing a traumatic change in climate, the Northern Hemisphere remained comparatively warm, and the Arctic Ocean, with its surrounding continents, stayed free of perennial ice.

In the period 3.5–3.1 MYA the sediments deposited in the Yucatan Channel between the Caribbean Sea and the Gulf of Mexico show an extraordinary change. They indicate that the voluminous Pacific-bound Caribbean currents (**18**) had been redirected into the Gulf (**19**). The Isthmus of Panama had slotted into place, and the North Atlantic Ocean was no longer connected to the Pacific. The North Pacific Ocean surface circulation lost the benefit of warm surface currents from the South Atlantic Equatorial Current, and that Pacific heat loss was the North Atlantic's heat gain.

ALTHOUGH THE CLOSURE of the Isthmus of Panama is thought to have been the key event that triggered the Cenozoic Ice Age in the Northern Hemisphere, other tectonic events paved the way. For example, vast tracts of the American Southwest to the north of the Caribbean region had been elevated through many thousands of feet. The uplift caused the Colorado Plateau region to be raised to about 3,300 m (10,000 ft) above sea level. The Northern and Southern Rocky Mountains were also elevated by thousands of feet above their original height to their present 4,200 m (14,000 ft). The change in continental topography altered the pattern of distribution of moisture-laden air flowing from the Pacific to the east (**20**)—it was now deflected either south over the Gulf of Mexico or north towards Canada. The uplift of the Southwest and the docking of Panama share a common cause—the subduction of the Pacific Ocean seafloor beneath North and Central America. The combination of all three events may well have triggered the start of the present ice age in the Northern Hemisphere.

By 3.1–2.7 MYA glaciers had formed on West Coast mountain ranges from Alaska to the Sierra Nevada in California. In the far

[*continued on page 260*]

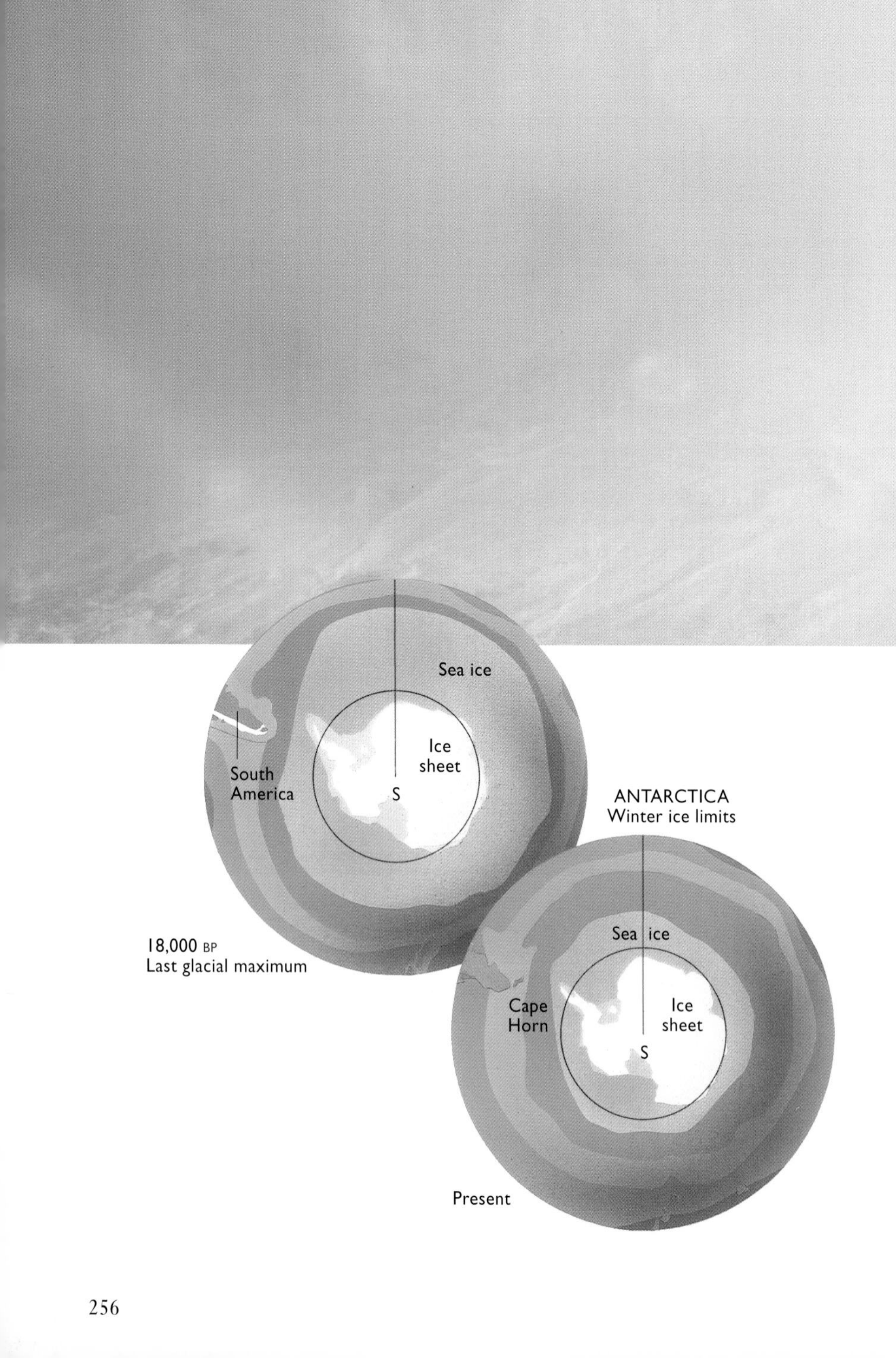

6 Northern Catastrophe

The Southern Hemisphere responds to changes in the global temperature with a rapid increase in the area of Antarctic sea ice at the onset of a glacial age, and an equally rapid reduction at the onset of an interglacial period. Meanwhile, the huge polar continent is kept refrigerated by the cold, windswept, and encircling Southern Ocean. The thickness of Antarctica's ice sheet may vary, but the area of the sheet, and therefore its albedo, remains relatively constant.

In the Northern Hemisphere, the position is almost exactly reversed. The Arctic Ocean is covered by relatively thin perennial sea ice that can spread or diminish in extent in direct response to climate change. It is surrounded by enormous continental surfaces that act as reservoirs for the Sun's radiant heat. Thus initial growth of ice sheets on northern continents is inhibited but, as ice-free continental surfaces begin to warm during an interglacial period, their retreat is rapid.

Laurantide Ice Sheet and ancestral rivers

Rapid melting of sea ice around Antarctica and in the Arctic Ocean can do no harm. But rapidly melting ice on northern continents had a catastrophic impact on their landscapes. For example, the retreat of the Wisconsin ice sheet left familiar-shaped depressions in the North American landscape, now the Great Lakes. The shaping of these, the largest freshwater lake system on Earth, might have involved *many* catastrophic events. The maps show just two stages of one such catastrophe.

After 14,000 BP the ice lobes that had formed the Great Lake basins, and that had filled and blocked the present Gulf of St. Lawrence, beat a sporadic retreat northwards. At some point during this retreat a vast flood of water burst from Lake Erie, pouring over a sill—now the site of the Niagara River and Falls pictured here—and into Lake Ontario. From there it flooded into the partially ice-blocked Champlain Sea, now the upper St. Lawrence River, and so into the North Atlantic.

7 Parallel Roads

The prime center of the last glacial maximum in Europe was the Fennoscandian ice sheet that formed in the Gulf of Bothnia, between Sweden and Finland. This sheet filled the Baltic Basin and spread over the adjoining region. One arm of the sheet reached across the North Sea Basin to Scotland, where it connected to a lobe that covered much of the British Isles. As in North America, there were many catastrophic ice bursts during the general retreat of the ice front, and one of them is illustrated here.

About 9,000 BP, glacier ice had advanced across the Great Glen of Scotland, from left to right of the main picture, and had dammed the entrance to Glen Roy (map and inset). During intermittent stages of glacial retreat and partial re-advance, a series of lakes accumulated in Glen Roy behind that dam, lakes about two-thirds the volume of Loch Ness. The ice dam burst several times during glacial retreats, leaving the shoreline of a former lake imprinted on the sides of the glen, and each time glacial debris was swept down the Great Glen. The debris accumulated into the mounds of rubble that now separate Loch Lochy, in the foreground of the panorama, from Loch Ness at Fort Augustus in the mid-distance; and Loch Ness from Beauly Firth at Inverness beyond. From this sequence it is difficult to understand how the Loch Ness Monster or its ancestors, having survived the glacial age, could have entered Loch Ness from the sea.

Glen Roy and its parallel "roads" have historic importance. The first correct explanation was given by Swiss naturalist Louis Agassiz (1807–73), the leading 19th-century protagonist of the concept of ice ages. He showed in an 1840 paper that they were "successive beaches of a receding glacial lake." This was at a time of fierce controversy over the very possibility of ice ages. The theory was thought too far-fetched by many Victorian scientists, including Darwin, Murchison, and Lyell. Agassiz returned the favor by bitterly opposing Darwin's concept of human evolution! It seems that fierce controversy is the essence of science ...

Glen Roy off the Great Glen, Scotland

Fennoscandia and Barents Shelf Ice Sheets 9,500 BP

[*continued from page 255*]

north and east of the Canadian Rockies, winter snow began to accumulate and to remain unmelted through the summer, eventually forming glaciers—slow-moving rivers of compacted snow. Cooling climate led to greater precipitation in normally arid regions in the south. A huge system of lakes began to occasionally flood landlocked regions of the Basin and Range Province to the east of the Sierra. One such lake was glacial Lake Bonneville (**21**), now vastly reduced in extent and called the Great Salt Lake of Utah.

During the course of the uplift of the Southwest, the Colorado Plateau had become an elevated desert region in the rain shadow of the Californian mountain ranges. Now raised in height, the Southern Rocky Mountains to the east had wrung moisture out of Pacific air currents redirected from north and south, and formed their own mighty glaciers. The Great Plains from the Rockies to the ancestral Mississippi, not yet the majestic river we know today, became semi-arid grassland. The prevailing westerlies that blew over the plains were loaded with moisture. This vigorous airstream billowed inland from the Gulf of Mexico and the Gulf Stream off Cape Hatteras. As the global climate continued to cool, increasing amounts of this moisture were precipitated as heavy snow to the north—snow that also did not entirely melt in summer months.

From 3.0 MYA to 1.6 MYA there were prolonged periods of glacial advance and retreat. After 1.6 MYA, the amplitude of fluctuations reached their maximum. It is thought that the increased albedo of the Northern Hemisphere contributed to a gradual increase in the extremes of global climate change. However, from the last 1.6 MY to the present, sea level has fluctuated by over 120 m (400 ft). The precise fluctuation is difficult to measure because the continents and their shorelines were, and sometimes still are, depressed into the asthenosphere by the weight of land ice. The surface of continental shelves now 90 m (300 ft) beneath sea level was often exposed for prolonged periods during times of glacial advance. During interglacial periods sea levels sometimes rose higher than today. For example, 125,000 years before the present (BP), sea level was on average about 6 m (20 ft) higher than now.

The degree of short-term sea-level fluctuation measured over hundreds and thousands of years was, and still is, affected by the amount of fresh water locked up in snow and ice on land, and this volume varies frequently. As a consequence, the sea level has also varied frequently, and will continue to do so. On the other hand, long-term sea-level fluctuations measured in millions of years are thought mainly to be the consequence of tectonic activity and the swelling and shrinking of ocean floors.

During interglacial periods ice melts and glaciers recede, the ice sheet's albedo decreases, rivers become swollen, continental sediments are transported into the ocean, sea levels rise, and continents rebound as they are relieved from the weight of ice. During glacial periods snow cover extends, albedo increases, ice sheets grow thicker, the annual flow of rivers is reduced, sediment transport is diminished, and continents sink under the weight of accumulating ice.

Between glacial extremes, specific climate zones can be displaced by as much as twenty to thirty degrees of latitude (roughly 2,400–3,200 km; 1,500–2,000 miles: **22**). So during a glacial age terrestrial life in the temperate zone, now squeezed between tropical and arctic extremes, has to suffer extreme swings between winter and summer conditions. During interglacial times (as now) the temperate zone is wider and summer-winter temperatures are not so extreme. For example, the present line of *permanent* sea ice in the Arctic Ocean lies just north of Spitsbergen. At the height of the last glacial advance 18,000 BP, permanent sea ice formed an arc in the North Atlantic, stretching from just south of Newfoundland in the west, to the south of Iceland in the north, and to the south of the British Isles in the east. But the tropics remained tropical. Between glacial peaks and interglacial troughs the ice front moves as if hinged from the southeastern tip of Newfoundland. It appears to rotate on its Newfoundland hinge through an angle of 45° to the north, to the permanent-ice front today. Because of their vulnerability to relatively fast changes in albedo, Northern Atlantic continents suffer the most rapid and most severe swings in climate on Earth as global climate switches from a glacial to an interglacial period.

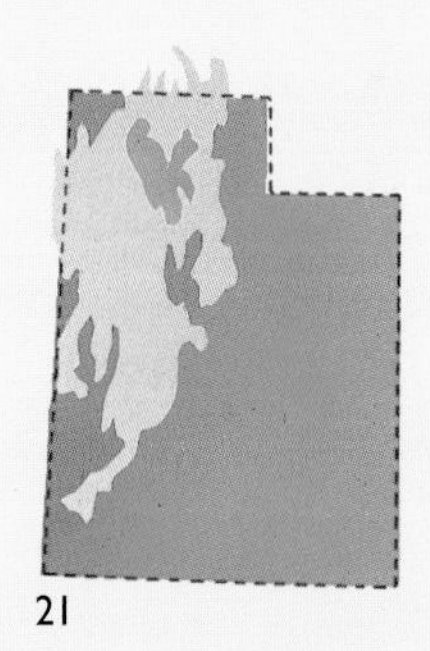

21

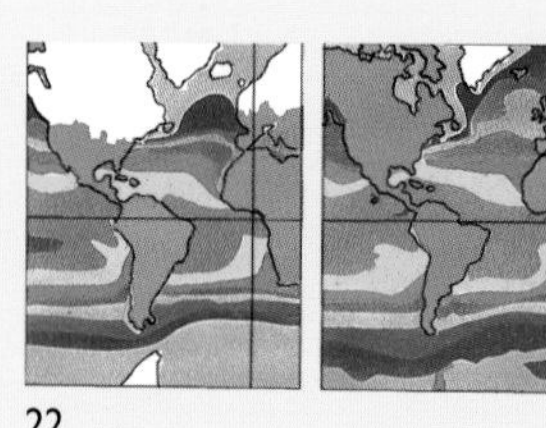

22

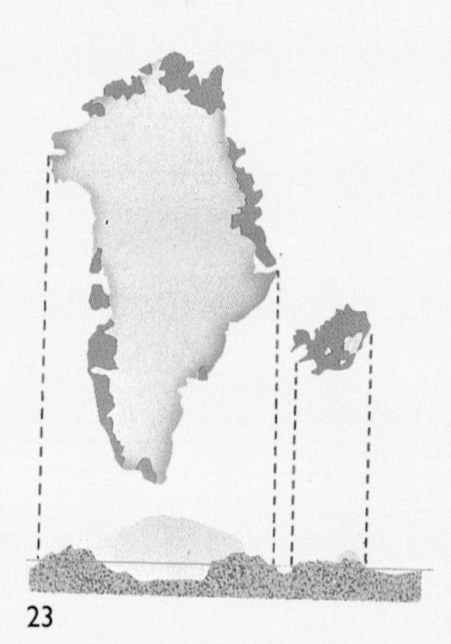

23

24

25

In the marine realm, changes in the latitude of the permanent-ice front cause a large shift in the pattern of North Atlantic currents. The northern and southern limits of permanent ice determine the extent to which warm saline currents can flow into Baffin Bay to modify the Labrador Current, and into the Nordic Sea to produce vertical mixing and release heat from the tropics. They also determine whether or not the warm saline currents are turned southeastward to be contained within the central gyre. It is only at peak interglacial periods such as the present that these currents can penetrate into the Labrador Sea and north of Iceland. There may have been only a few such occasions in the last 1.6 MY.

During Quaternary time, from 1.6 MYA to the present, there have been at least four glacial ages of great severity and varying duration in the Northern Hemisphere [Essay 5]. Each of the first three glacial ages lasted hundreds of thousands of years. (Future research may reveal that there were more than three; it is difficult to determine the duration of glacial ages in the distant past, as succeeding glacial episodes tend to eradicate traces of the previous major event.) However, the fourth and most recent glacial age is thought to have lasted around one hundred thousand years. The actual measured duration of a glacial advance of course depends on the latitude and geographical point from which the advance is gauged.

At the beginning of each glacial age, continental ice sheets developed from specific centers (**23**). We must explain that "ice sheets" are ice masses of considerable thickness and extent—over 50,000 sq km (20,000 square miles) in area. In contrast, "ice caps" envelop mountain ranges, and although large, are localized phenomena. So as ice sheets grew in thickness, they engulfed the local landscape, and then spread and radiated in every direction that did not offer obstruction. When piled in great mounds a mile or more high, ice tends to flow outward in all directions from its thickest points. A subtle shift in the region of heaviest snowfall caused individual sheets to change their outlines relative to neighboring or adjoining ice sheets. Because of this constant evolution, the shape and geography of continental ice varied substantially during the course of each glacial age.

In order to form ice sheets as large and thick as those found today in Antarctica and Greenland, prevailing winds had to supply sufficient moisture to permit the accumulation of massive amounts of snow, which consolidated into ice under their own weight. This only happened in regions of North America and Europe that were cold enough for long enough, and meant that the focal points of glaciation on each continent remained almost constant. Conversely, neighboring areas of continents surrounding the North Pole (**24**), starved of moisture by dry cold wind. became part-desert and part-tundra, and were far greater in extent than the ice sheets themselves.

In North America there were three prime ice-forming centers that overlapped. The very high coastal ranges of the Pacific Northwest (around 5,400 m; 18,000 ft) were the center of the Cordilleran ice sheet, which was 2.8 km (1.7 miles) thick at the last glacial maximum, *c.*18,000 BP, although glacial maxima varied from region to region. The Cordilleran sheet overlapped the Laurentide ice sheet, which was 3.8 km (2.35 miles) thick (**25**). The latter sheet, by far the most massive of the northern ice sheets, had several centers, but the main one was in the Hudson Bay region of Canada. The third North American center was the Greenland ice sheet (3.2 km, 2 miles, thick), which has almost the same elevation today as it had during the last glacial maximum. As the ice sheets melted during the present interglacial, the island's basement rocks isostatically rebounded as they were relieved of weight by melting ice.

Cape Hatteras lighthouse, North Carolina

In Europe, the Baltic Sea was the center of the Fennoscandian ice sheet (2.5 km, 1.5 miles, thick), which formed in the Gulf of Bothnia between Sweden and Finland. This sheet grew to fill the Baltic Basin and spread over adjoining continents. Associated ice caps were centered to the southwest in Scotland [Essay 7], and beyond the Ural Mountains to the northeast—land ice that spilled into the Arctic Ocean (**26**). Iceland too was once the center of a single ice sheet over the whole island. Because of the island's proximity to the North Atlantic's warm surface currents and their associated atmospheric moisture, this sheet may have been the first to develop at the start of the last glacial age. Today, the Vatnajökull ice cap, which now measures 2 km (1.25 miles) at its thickest, is the largest of four Icelandic ice caps.

26

THE MATTER OF GLACIAL ADVANCE and retreat is actually far more complicated and varied than even these major glaciations suggest. Scientists have identified thirty or more pulses of intermediate glacial advance and retreat during the course of the glacial ages reviewed above. Each intermediate pulse lasted in the order of 43,000 years, and within each pulse there are lesser pulses of about 24,000 and 19,000 years' duration. These intermediate pulses are called "stadials" and "interstadials" (a *stade* is a substage, a secondary advance within a glacial age). They are thought to coincide in overall length and intensity with Milankovitch cycles. When plotted on a graph, stadials and interstadials exhibit a saw-toothed geometry (**27**) that is superimposed on the generally upward or downward trend of a major glacial age. Within the saw-tooth pattern there are even finer gradations of climatic change of the kind exemplified by the Little Ice Age. Although centuries long, however, the Little Ice Age and perhaps many hundreds of similar interludes in the past amount to only minute fibrillations within the heartbeat of a Milankovitch cycle. The trace of these fibrillations could pass almost unnoticed on an oscilloscopic display of a procession of glacial ages, interglacial ages, stadials, and interstadials.

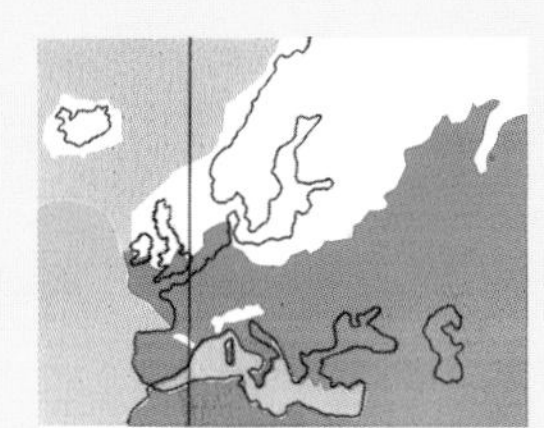
27

This is a sobering thought, for we are all inclined to make judgments about climatic events in relative terms—primarily the terms of our own lifetimes. The plain fact is that the technical difficulties that face paleoclimatologists in providing an *accurate* reconstruction of the progress of past glacial climates are formidable for any time span less than a few thousand years. This makes it doubly difficult to predict future climate trends.

WE BEGAN THIS CHAPTER with a description of the closure of the Isthmus of Panama (**28**), an event that had a profound and lasting effect upon climate in the Northern Hemisphere. We referred to Ponce de León's accidental discovery of one of several "Fountains of Youth" in the North Atlantic, and followed this with an overview of our present understanding of this ocean-river system. We have made passing reference to the great voyages of discovery that were made during a respite in the Little Ice Age at the turn of the 15th century—a late success in the progress of man. And we have delved into the intricacies of Milankovitch cycles and seen something of the effect of astronomical changes.

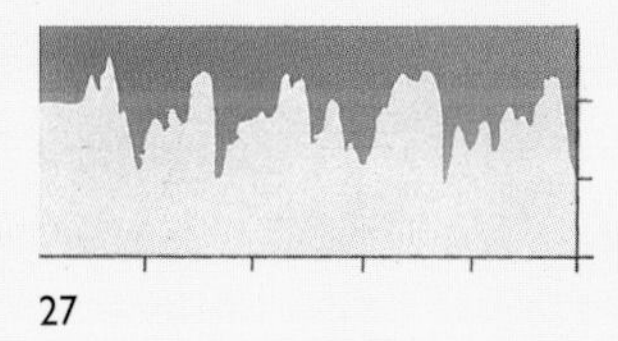
28

But through this story there runs yet another dominant thread—that of our own ecological and social evolution, which took place largely in the context of the Cenozoic Ice Age. In the chapter that follows we will retrace this same period of the last 3.5 million years. We will tell the story of the ascent of anatomically modern man in the setting of the last glacial age. Fernand Braudel (1902–85), one of the 20th century's greatest historians, put it this way:

"The world was 'discovered' a long time ago, well before the Great Discoveries . . . Europe's own achievement was to discover the Atlantic and to master its difficult stretches, currents and winds." Perhaps we should add " . . . and to develop the means to do so."

XII CHILDREN OF THE APPLE TREE

To reach the Orient from Europe one had first to reach the Canary Islands off the shores of Africa. Then one had to persist by sailing as close to due west as possible across the Ocean Sea—or so Christopher Columbus believed. This picture looking directly into the western Sun was taken on Gran Canaria in the Canary Islands. The small island of Gomera, Columbus's point of departure on September 8, 1492, lies just beyond the cloud-wrapped volcanic peak of Tenerife on the horizon. According to Columbus's account, the sails of his three becalmed ships caught the wind at 3 a.m. on that day. An enormous and irreversible change was thus set in motion, and the redistribution of human populations around the North Atlantic Ocean commenced.

CHILDREN OF THE APPLE TREE
1.6 MYA–500 BP

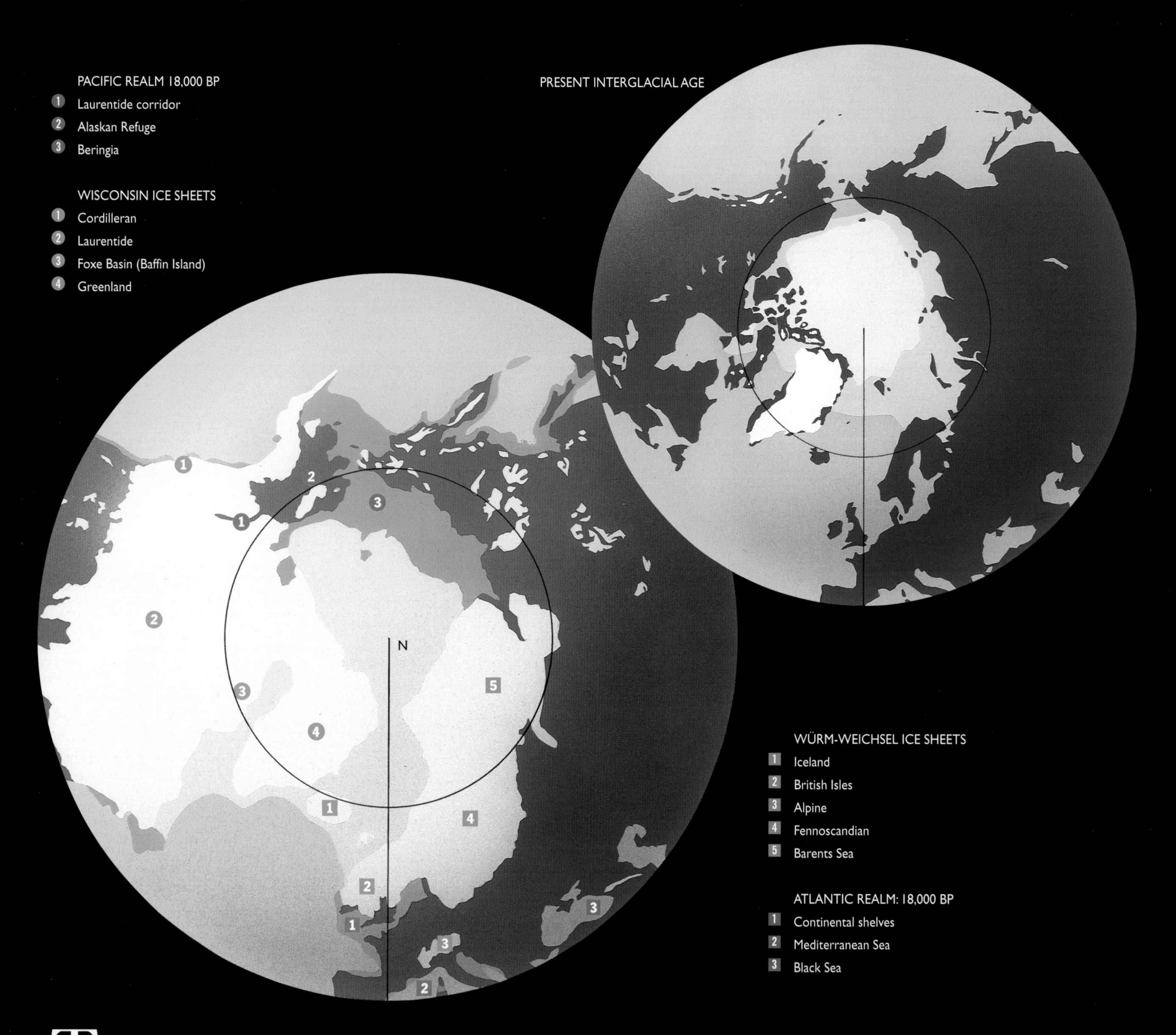

The two globes are circumpolar views of the Northern Hemisphere. The larger projection shows the extent of land ice, winter sea ice, and ice-free regions at the height of the last glacial age 21,000–17,000 BP. This glacial age had begun by 120,000 BP and concluded about 10,000 BP. The smaller modern globe is for comparison. At various times during the period of the last maximum glacial advance, sea level reached a low point around 120 m (400 ft) below the present level. At this time landbridges (and therefore terrestrial migration routes) between island continents were at their most extensive.

The Northern Hemisphere projections opposite have been redrawn from data produced by the Climap Project of 1981. The prime objective of this project was to complete a survey of the Earth's surface during the last maximum glacial advance—called the Wisconsin advance in North America, and the Würm or the Weichsel advance in Europe.

According to the rhythm of Milankovitch Cycles, in the third millenium we are entering, or are about to enter, the next 90,000–100,000-year-long glacial cycle of the Cenozoic Ice Age. In view of present publicity about global warming, we should remember that the great voyages of European discovery in the 11th, 15th, and early 16th centuries were made during warm periods of global climate. These periods punctuated centuries of glacial advance during the so-called "Little Ice Age" of the last millennium. The Little Ice Age began in the 13th century and lasted until the mid-19th century. During the last millenium sea level has fluctuated by ± 80 cm (3 ft) above and below present levels.

Perennial ice first began to form in the Northern Hemisphere in North America, Greenland, and northern Europe about 3 MYA. This climatic change signaled the start of the Cenozoic Ice Age in the north—it had already been in progress in the Southern Hemisphere for 15 MY or more. Since that time Afro-Eurasians have evolved from modest numbers of primitive people (*Homo erectus*) to the present teeming global population of anatomically modern man (*Homo sapiens sapiens*). In this final chapter we retrace the course of Cenozoic Ice Age events in terms of human evolution and experience, and particularly in the context of the origin of the North American peoples.

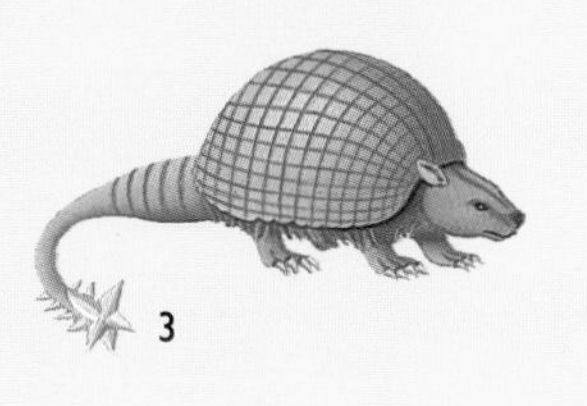

By 3.5 MYA the Isthmus of Panama, together with Honduras, Nicaragua, and Costa Rica, had jammed and fused, connecting Mexico (part of the North American tectonic plate) and South America. North American and South American terrestrial animals had not been able to migrate between the two continents, except in the latter stages by occasional island-hopping, as their opportunities to do so were largely dependent on fluctuating sea levels. Once the tectonic fusion was complete, faunal traffic became routine—between two continents that had been separated for at least 150 MY since the opening of the Americas Seaway during the Pangean breakup. The result was catastrophic for some mammalian species, primarily those of South America, yet opportune for others, mainly North American. The placentals simply overwhelmed the marsupials [Essay 1].

Mammals in South America had evolved in complete isolation from the Laurasian world, and from Antarctica and Australia for at least 35 MY. Although a large part of this fauna was marsupial (Gr. *marsippos*, pouch), the rest consisted of placental ungulates (L. *unguis*, nail) and edentates (L. *edentatus*, toothless). This population had been augmented from time to time by the arrival of small placental animals from other continents—waifs and strays that arrived by accidental rafting on masses of floating vegetation. This is deduced from the fact that the endemic faunas appear throughout the Cenozoic fossil record in South America, whereas the waifs do not: instead, they appear suddenly in the record and radiate rapidly once they have appeared.

Some South American marsupials evolved as opossum-like animals, while others were carnivorous. Many South American animals were remarkably similar in appearance to their northern counterparts, although completely unrelated—ideal examples of "parallelism" or convergent evolution. Among the carnivores were wolf-like animals and saber-toothed cat-like animals (**1**) that were as large as their placental counterparts in the northern continents. Some South American ungulates were camel-like and others were very large, with trunks (**2**)—they vaguely resembled the Holarctic mastodon and the African elephant. The largest and most varied group among the South American hoofed animals ranged in size from small rabbit-like creatures to "gigantic" forms—huge, heavy-bodied grazers. The early edentates in South America had evolved into the armadillos, sloths, and anteaters. Some of these creatures also evolved gigantic forms, such as the spike-tailed glyptodonts (**3**).

All these South American animals, marsupial, ungulate, and edentate alike, had evolved in isolation and for that reason were uniquely South American. Even the placental waifs developed differently. For instance, the first rodents and monkeys arrived in South America 30–24 MYA. From what must have been small numbers initially, waif-rodents bred on such a scale that they had become the most numerous mammals in South America by the time the Panamanian connection occurred; and by this time they too were quite different in appearance from rodents on other continents. Similarly, South American "New World" monkeys evolved in a distinctly different way to their Old World counterparts; they have prehensile tails (Old World monkeys do not), and they have flat-shaped noses and flared nostrils (Old World monkeys have elongate noses).

Once the Panamanian connection was made, the fossil record shows that there was a massive invasion of South America by mammals from the north. Transfers were hastened by the deteriorating northern climate after 3 MYA and they continued throughout the glacial ages of the northern Cenozoic Ice Age. Early arrivals in South America included insectivores, rabbits, squirrels, and mice. These were followed by members of the dog family, bears, weasels, large and small cats, and raccoons that supplemented the waif-raccoons that had arrived earlier. These animals were in turn followed by mastodons (a smaller relative of the mammoth), horses, tapirs, peccaries (**4**), llamas, and deer. The counter-migration was modest, but

[*continued on page 268*]

1 Missing Link

The Isthmus of Panama was formed about 3.5 MYA when elements of an island arc carried by the Pacific Ocean floor jammed between South America and North America. This blocked a seaway and caused the formation of a substantial landbridge between the two continents. The previously open Caribbean Sea was no longer an equatorial confluence of three great oceans, the Pacific, Central Atlantic, and Tethys Oceans. It became part of the North Atlantic realm, linked by innumerable channels that thread their way through a necklace of volcanic islands.

Before the Panamanian Landbridge formed, tropical marine organisms were free to move from the Caribbean and Atlantic realms into the Pacific realm, or vice versa. Meanwhile, land-bound North American and South American mammals had been separated by a broad seaway for many millions of years. After the landbridge formed, the evolutionary roles of both marine and terrestrial animals were reversed: previously similar marine organisms now became vicarian species, separated by physical barriers, while terrestrial animals became free to migrate to new regions. It is evident from the fossil record that after 3.5 MYA species of corals, tropical fish, and protoctists evolved along different paths east and west of the isthmus. At the same time, northern mammals largely replaced southern mammals.

The illustration shows a small number of the many mammals that made a successful crossing of the landbridge. They were "successful" in the sense that their genera overcame local competition, adapted to new environments, and evolved new species—many of which survive today. The North American immigrants included just one marsupial from South America, the opossum, but many edentates (strange-looking mammals like armadillos and ground sloths), rodents like porcupines and capybaras, and a few others. South American immigrants from North America included canines, cats, skunks, peccaries, camels, llamas, horses, deer, tapirs, and rabbits, to name a few of many.

But this exchange did not include representatives of the genus *Homo*. Indeed, there is no evidence to suggest that archaic *Homo sapiens* evolved anywhere in the Americas. It seems that the evolution of *Homo sapiens sapiens* (anatomically modern humans) was strictly an Afro-Eurasian affair, and that *all* Americans, both ancient and modern, are immigrants. It is just that some arrived earlier than others.

At a late stage in the closure of the Panamanian Gap, the scene may have looked rather like Drake Channel, pictured here.

Drake Channel, British Virgin Islands, Lesser Antilles

Left column: North American migrants
Right column: South American migrants

[*continued from page 265*]
it did include ground sloths, armadillos, porcupines, opossums, and a few other animals.

Although the long-established inhabitants of South America had grown used to crowded conditions long before the intercontinental linkage and invasion from the north, indigenous South American animals succumbed to the ecological pressures created by the placental immigrants from North America. Thus, the larger marsupial carnivores and ungulates gave way to their more adaptable placental counterparts—particularly cats, and browsers such as horses, tapirs, peccaries, and deer. The day of the dominant marsupial was effectively done in South America.

This evolutionary event, triggered by the tectonic closure of the Isthmus of Panama, was played out over a prolonged period, perhaps several million years. But only 500 years ago, the European discovery of the same locality at the turn of the 15th century was an equally momentous event with similar evolutionary consequences for indigenous peoples. At that time individual Caribbean and Mesoamerican populations, who had evolved in isolation and numbered many millions of people, were either extinguished altogether [Essay 15] or decimated in numbers, and superseded by an interbred population [Essay 14]. In the latter case, a new race and new culture resulted from displacement—the Mexican people and their "mestizo" culture. Almost five centuries after the event, 60 percent of the population of modern Mexico is still of Meso-Iberian descent. ("Mesoamerica" is a term used by anthropologists to describe those parts of Mexico and neighboring Central America that already had interactive societies, both civilized and uncivilized, at the time of the arrival of Europeans.)

Mesoamerican and Iberian cultures, although far distant from each other, had evolved and advanced towards their own forms of civilization in extraordinarily parallel ways [see Essays 14–15 in this chapter, and the Epilogue pages 304–323]. However, the Mesoamericans had evolved their way of life in isolation. Their advancement was the consequence of thousands of years without threat from outside influence. The Iberian culture, an ocean away, had evolved with comparative speed from primitive people to sophisticated nation. The speed of their advance had been driven by conflict and competition between cultural centers in Eurasia. When the Mesoamerican and Iberian cultures came face to face for the first time, naivety was quickly overwhelmed by sophistication. It seems that both the pace and character of the social evolution of humankind parallels that of the South American marsupials and North American placentals to a remarkable degree.

In regions of Iberia and France on either side of the Pyrenees, there is overwhelming evidence of the presence of human forebears dating back to about 500,000 BP. In the New World, no fossils or traces of hominids other than anatomically modern humans have been found anywhere. The evolution of anatomically modern humans seems to have been entirely an Afro-Eurasian process that took place in the present Cenozoic Ice Age.

As we saw in an earlier chapter, bipedalism is the prime physical signature of humankind—that is why the Laetoli footsteps in East Africa are so evocative [Essay, page 233]. Bipedalism, which is thought to have characterized hominids by 5 MYA, was accompanied by other rudimentary but novel physiological features. The size of the early hominid brain relative to body weight was one of these, and the brain itself had unusual properties—latent capacities for reason, compassion, aesthetic sense, and wisdom. The ability to oppose thumb and fingers was a second key hominid aptitude. This enabled hominids deliberately to shape a useful tool from stone with which to trim a stick to spear prey or to fend off a predator. With the passage of millions of years these properties and aptitudes, along with many other characteristics, were incorporated into the genetic code that distinguishes humankind, the genus *Homo* from all other mammalian species.

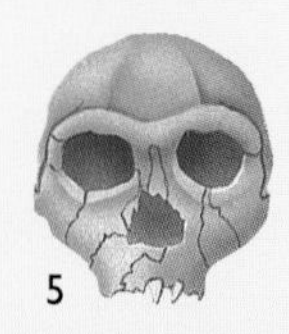
5

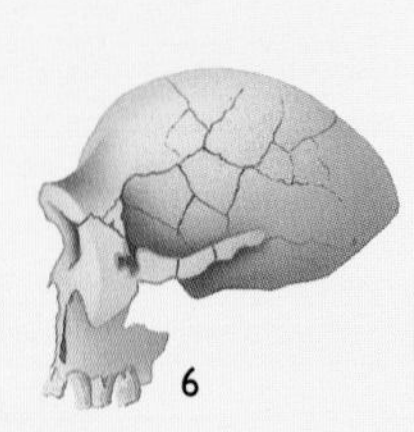
6

In Latin the word *homo* means "man," and *sapiens* means "wise" and the term *Homo sapiens* is often used in a general sense to describe humankind as a whole. In paleoanthropology, though, it is necessary to differentiate between archaic humans and anatomically modern humans. Archaic humans (*Homo sapiens*) first appear in the fossil record about 250,000 BP. Anatomically modern humans (*Homo sapiens sapiens*) first appeared in Europe 35,000 BP and in Asia 50,000 BP. In order to avoid confusion we will use the term "archaic humans" in place of *Homo sapiens*, and "modern humans" (meaning anatomically modern) in place of *Homo sapiens sapiens*. We will also use proper nouns for associated subspecies like "Neanderthal" and abbreviated versions of Latin names for predecessors of *Homo sapiens* —for example, the term *Homo erectus* will be abbreviated to "*erectus*"... and so on.

By at least 1.0 MYA the genus *Homo* (**5**) began to leave Africa and disperse thinly but widely throughout the habitable regions of Eurasia. This dispersal, and the ascent of humans that occurred thereafter, took place during a formidable phase of the current Cenozoic Ice Age, the "Pleistocene Epoch," (around 1.6 MYA–10,000 BP). By 1.0 MYA coordination of thought, dexterity, and inclination to experiment had led one species of the genus *Homo*, to create the first well-fashioned pear-shaped stone tools—this was *Homo erectus* (**6**). It seems that *erectus* may have already developed a sense of collective responsibility. This is judged by remains of what are thought by some to be *erectus* "base camps."— fixed short-term settlements to which these people may perhaps have returned nightly. By 500,000 BP *erectus* had also developed a means to cross from mainlands to islands, and in at least one case from continent to continent: they used raft-like log-floats to cross the Strait of Gibraltar from Morocco to Iberia [Essay 2].

By then, fire was in regular use in ice-age Europe. For example, charred bones in stone hearths dating from this time have been discovered at Escale near Nice in the South of France. Whether or not the means of deliberately *making* fire had been invented by this time is just not known. However, once started, perhaps with a smoldering remnant from a bush fire, fires could be banked in a hearth and kept alight for prolonged periods. The *erectus* tribal fire provided warmth, encouraged social activity, eased the winter night, and, quite importantly, offered protection from marauding animals.

Fire, no doubt, led to the accidental discovery of cooking. Roasted meat and vegetables baked in the embers of a fire, or perhaps skewered between two sets of antlers bridging a fire, more readily warmed the inner self. Cooked food was tastier, more tender, but not necessarily more nutritious than raw meat and vegetables. The switch from raw food to a regimen of cooked food required adaptation of the human digestive system. The vestigial human appendix may be a leftover from this period; we can no longer digest cellulose, which was the probable function of this now-rudimentary organ. Tastier and more easily eaten food no doubt won favor quite early in the evolution of ice-age peoples. One result was that molars became smaller, and other teeth changed in shape. This in turn led to changes in the shape of the jaw and the skull.

Fire also permitted improved methods of toolmaking—for it seems that *erectus* discovered, possibly from cooking food, that fire hardens organic materials like bone, antler, and wood. When charred these materials made improved tools and weapons. For instance about 400,000 BP a herd of over 60 mammoths was ringed by fire and stampeded in a marshy region near Torralba, Spain. Some experts interpret the remaining evidence to mean that the fire was set deliberately and that fire-hardened wooden spears were used to kill the ambushed animals. Others say the evidence indicates a series of smaller kills, and to others still, the scene suggests that *erectus* people were scavenging from animals that had met an accidental death.

However this may have been, from the production of the first deliberately hand-shaped tools it was but a short step in geological time, perhaps 100,000 years or so, to the appearance of archaic humans (*H.sapiens*). They made tools with special functions and displayed the first signs of artistry. Archaic *H.Sapiens* (**7**) evolved an artistically-inclined society and a non-utilitarian culture. The oldest known decorated artifact is an ox bone dated 300,000 BP (found at Peche-de-l'Aze, France). Swanscombe Man, dated 225,000 BP, is perhaps the best-known representative of archaic Europeans. Although there is a broad consensus that archaic humans evolved in Africa, surprisingly little is known about their origins. However, it is known that they were widely dispersed throughout Eurasia and Africa by a quarter of a million years ago. They led to the evolution of anatomically modern humans ... at least, in the traditional view.

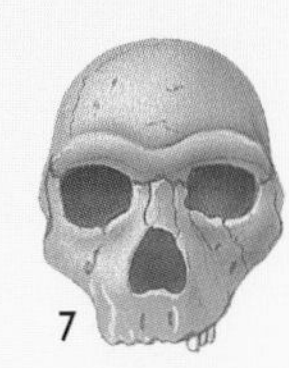
7

THERE ARE TWO BASIC PALEONTOLOGICAL THEORIES that attempt to account for the development of humankind from *erectus* through archaic to modern humans, and for the appearance of different regional characteristics during the course of evolution. These are called the "local continuity" and the "replacement" models. The first and earlier model is the traditional one, derived from the fossil distribution of *erectus* and archaic humans. The more recent and quite contradictory "replacement" theory is based on interpretation of the genetic code of modern humans.

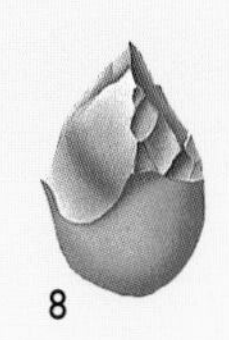
8

The main premise of the continuity model is the fact that *erectus* appears to have been widely distributed throughout Afro-Eurasia by 500,000 BP. As a species, Afro-Eurasian *erectus* is represented in western Eurasia by Arten Man, and in eastern Eurasia by Java Man and Peking Man. From this and much other fossil evidence, the model assumes that archaic humans evolved regionally from widely dispersed Afro-Eurasian *erectus,* who adapted to particular environments in particular localities. In western Eurasia this process led to the evolution of archaic humans from 250,000 BP to 35,000 BP, to Neanderthals from 120,000 BP to 35,000 BP, and to the first anatomically modern Europeans around 35,000 BP. In eastern Eurasia, so the theory suggests, the same species of Afro-Eurasian *erectus* led to Asian archaic humans by 250,000 BP, and to anatomically modern Asians by 50,000 BP.

9

The replacement model is based on genetic evidence alone. It suggests that archaic humans evolved only in Africa from African *erectus*, and resulted in the evolution of anatomically modern humans in Africa. According to the genetic model, modern humans first appeared 200,000–100,000 BP in Africa, migrating into Eurasia around 100,000 BP. Subsequently they dispersed throughout Eurasia with the consequence that all descendants of western Eurasian *erectus*, archaic humans, and Neanderthals were superseded by modern humans of African origin by 35,000 BP. Similarly, in eastern Eurasia, Asian subspecies of archaic humans were replaced by modern humans of African origin by 50,000 BP. The two theories converge at 35,000 BP. From this point in time both theories suggest that global distribution of modern humans depended upon the adaptation of regionalized peoples to specific environments.

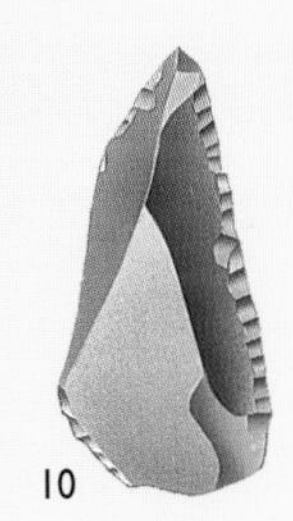
10

But in some important respects the models are diametrically opposed. The traditional continuity model appears to explain skeletal and ethnic differences between eastern and western Eurasians (and similar differences in Africans), but does not explain the universality of the human genome. This genome, the basis of the replacement model, indicates that anatomically modern humans and their ethnic differences are phenomena that developed in parallel in Africa from about 150,000 BP. Thus the replacement model satisfies the genetic criteria but does not satisfactorily explain why the earliest known modern humans—*H.sapiens sapiens*—are anatomically similar to their counterparts all over the world. There are vehement advocates for both models, but one model or the other must be incorrect—or both must be flawed.

THE ENVIRONMENTS THAT SHAPED AND CONDITIONED the character of modern humans was sometimes temperate, sometimes tropical and sometimes harsh. This was particularly the case during stadials and interstadials during the last glacial age (the Wisconsin advance in North America, and the Würm-Weichsel advance in Eurasia). This glacial age lasted from 120,000–10,000 BP and reached its maximum period of advance 21,000–17,000 BP. During the latter period, global sea level fell to its lowest levels since the appearance of modern humans, 90-120 m (300-400 ft) below the current level. Such low sea level exposed vast areas of continental shelving that became well vegetated with tundra, grasslands and/or forest, depending on latitude and local conditions. Exploratory oil-drilling cores containing timber shards have often been recovered from ice-bound sites in the Arctic, for instance. Continental shelves permitted widespread migration of both animals and people to sparsely populated or previously unpopulated localities. At the present sea level, we can now see islands or island-continents where once extensive landscapes existed. However, the weight of land-ice causes continents to sink, and to rise again as the ice melts. This process (called "isostatic" depression and rebound) complicates the issue of exposed continental shelves during a glacial advance, as we will see later in this chapter.

ONE FORM OF HARD EVIDENCE that allows paleoanthropologists to judge the relative advance of hominids is the comparative technological advance between one species or subspecies of the genus *Homo* and succeeding species. For instance, the first-known hominid tools (dated between 2.6 and 2.5 MYA) were found at Gona Kaba, in southern Ethiopia.They are little more than pebbles that were deliberately cracked and fractured to provide a hand-held cutting edge. It is believed that tools of this type were used as multipurpose choppers for cutting and pounding vegetables, butchering animals, and cracking bones to remove nutritious marrow. This ultrasimplistic form of tool was not refined until primitive hand axes began to appear from about 1.4 MYA.

The development of such axes for the dismemberment of animals, and later, axes for felling trees and other utilitarian purposes, remained a key technology. Hand-axe blades become progressively thinner, straighter-edged, and more effective, particularly after the introduction of soft-hammer percussion—the use of a piece of bone or antler to finish the shaping (**8**). They required less muscle and time to make, but greater skill and judgment in the selection of materials and their preparation. This progress suggests that the development of toolmaking skill was proceeding in parallel with the evolution of the brain. The neural tissues in the motor area of the human brain were increasing in size, allowing better coordination between thumb and forefinger, hand and eye.

Swanscombe Man, an archaic human, lived in England around 220,000 BP during an interstadial, when a comparatively mild climate permitted mammoth and hairy rhinoceros to range the Thames Valley. Swanscombe Man used sophisticated straight-edged hand axes (**9**) that were produced by shaping a suitably shaped rock with a smaller rock. The technique consisted of striking one rock with the other in order to detach flakes from either side of the larger rock, leaving a pear-shaped axe head with a rough edge. A piece of an antler or a bone from a butchered animal was then used to chip and dress the edge of the axe so as to produce a sharp-edged blade.

This technology was augmented by a new method of toolmaking called the "Levallois technique" (**10**), which was in general use by 200,000 BP. The technique was first recognized by archaeologists during the destruction of some old houses in the Anglo-Paris Basin in a region called Levallois-Perret—a suburb of present-day Paris. As the site was being cleared for new buildings, prehistoric artifacts were found that had required the toolmaker to use imagination in their manufacture. The Levallois toolmaker first had to select a flint nodule. Flint is a brittle but dense and impure quartz that is harder than most steels. The toolmaker had to visualize the potential axe blade, or other tool, resident within the nodule, then shaped it by

[*continued on page 274*]

2 Eurasians

Hominid evolution from ape-like ancestors to anatomically modern humans took place in Africa and Eurasia in the Cenozoic Ice Age. Continental ice sheets began to form in the Northern Hemisphere around 2.0 MYA, but did not reach the extremes of their several major advances until after 1.6 MYA. This latter stage is called the Pleistocene Epoch. It was during the Pleistocene that hominids moved out of Africa and into Eurasia, but it is difficult to reconstruct an unequivocal chronology because human bones were rarely fossilized. They were destroyed, or they decayed, or they were eaten by predators or ground up by glaciers. However, the four skulls reproduced here illustrate crucial stages in the evolution of the human brain case and facial features from about 600,000 BP.

H. erectus (upright man) was the widely distributed Afro-Eurasian genus that led to the evolution of archaic *H. sapiens*, which in turn gave rise to subspecies: *H. sapiens neanderthalensis* (Neanderthal) and *H. sapiens sapiens* (anatomically modern man). Both subspecies may have lived in the Middle East during the period 120,000–60,000 BP. Ultimately *H. sapiens sapiens* became established throughout the habitable Afro-Eurasian world, but Neanderthals remained restricted to the Middle East and Europe. The oldest known western Europeans are called Cro-Magnon (after a locality in France), and they appeared about 35,000 BP, perhaps several thousand years before Neanderthals disappeared from the scene.

The segment of the Northern Hemisphere, illustrated above right. shows the probable extent of both sea ice and exposed continental shelves in the North Atlantic, and terrestrial ice sheets in Eurasia at the last glacial maximum. Due to the growth of continental ice sheets sea level was at least 105 m (350 feet) lower than it is today. At that time *H. sapiens sapiens* probably lived on exposed areas of continental shelf off the coast of northern Spain and France in the Bay of Biscay, and in the Mediterranean basin. These

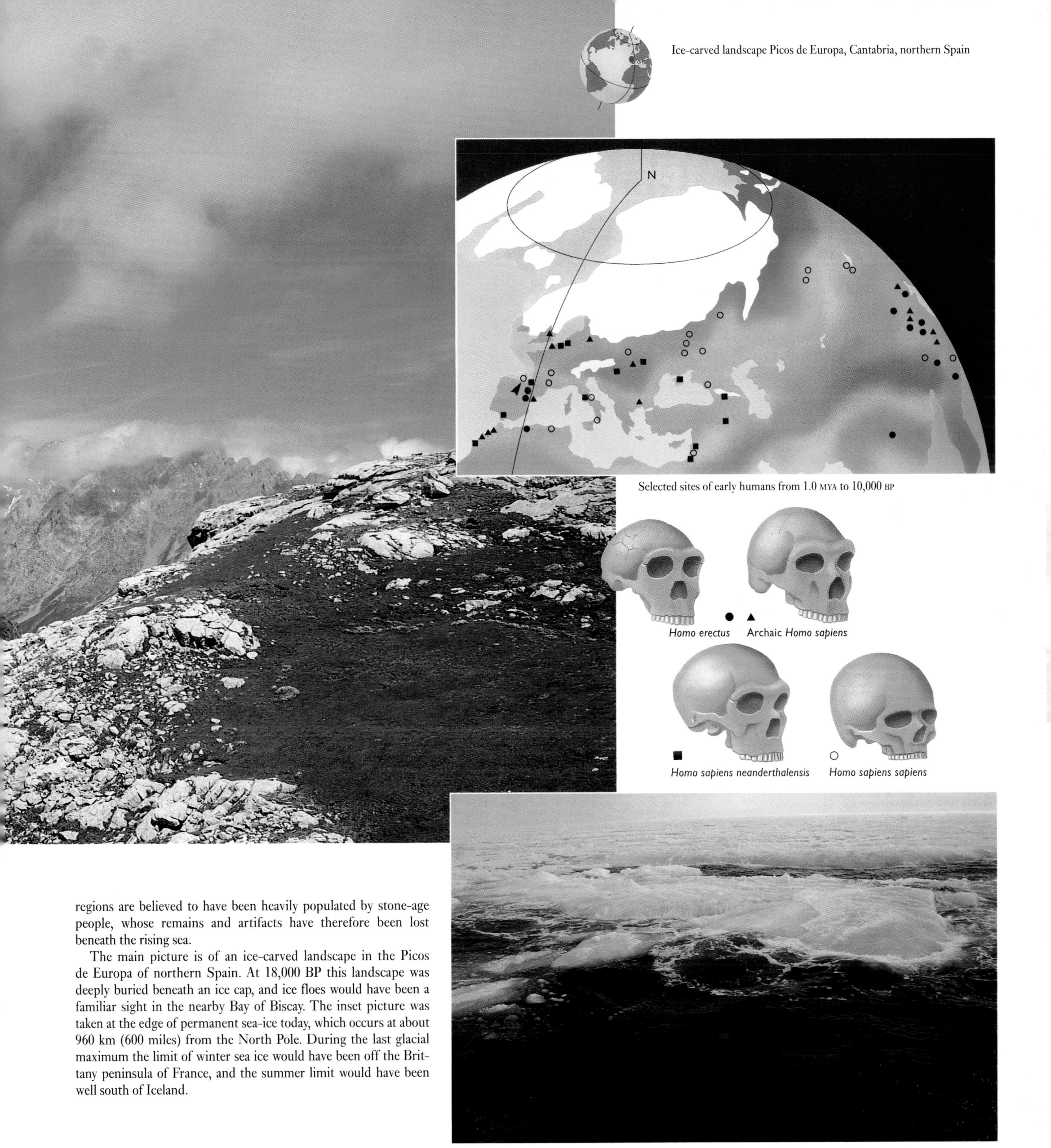

Ice-carved landscape Picos de Europa, Cantabria, northern Spain

Selected sites of early humans from 1.0 MYA to 10,000 BP

The summer limit of sea ice, southern Arctic Ocean

regions are believed to have been heavily populated by stone-age people, whose remains and artifacts have therefore been lost beneath the rising sea.

The main picture is of an ice-carved landscape in the Picos de Europa of northern Spain. At 18,000 BP this landscape was deeply buried beneath an ice cap, and ice floes would have been a familiar sight in the nearby Bay of Biscay. The inset picture was taken at the edge of permanent sea-ice today, which occurs at about 960 km (600 miles) from the North Pole. During the last glacial maximum the limit of winter sea ice would have been off the Brittany peninsula of France, and the summer limit would have been well south of Iceland.

Cirque de Gavarnie, Pyrenees, southern France

3 Europeans

Some paleoanthropologists argue that modern humans evolved from the Neanderthals. Others claim that our ancestors evolved directly from archaic *H. sapiens* in Africa. The former view is supported by strong circumstantial evidence and slim fossil evidence, and the latter by strong genetic evidence alone. Whatever the answer may prove to be, there is no doubt that the ecological and social evolution of hominids in Europe first gathered momentum during the time of the Neanderthal.

H. sapiens neanderthalensis and *H. sapiens sapiens* were both present in Europe around 35,000 BP. They were hunter-gatherers who employed identical toolmaking technologies; they hunted mammoth, bison, and deer, which they killed with flint spearpoints, and butchered with other specialized tools. Both peoples lived in small communities, cared for the old, and buried their dead—often with gifts of tools, food and drink, trinkets, and perhaps flowers. They also had middens (rubbish dumps). Indeed, both graves and middens are grist to the paleoanthropologist's mill.

Long after traces of Neanderthals vanish from the record, a vital technological advance was made by *H. sapiens sapiens.* This was the invention of the "burin," an elongated, multipurpose flint tool with a cutting edge at one end and sometimes a scraper at the other. Although other tools—special-purpose blades and double-edged knives—were in use at the time, the advent of the burin added a new dimension to the art of toolmaking itself. Burins made it possible to hone fire-hardened antler and bone and to fashion needles from splinters. These could be used to sew tight-seamed clothing with sinew, an insurance against frostbite. Splintered bone was also formed into harpoons and fishhooks (to extend the range of the food supply). Burins were also used to carve ivory into figurines such as the bison illustrated here.

By 15,000 BP the tools and artifacts of *H. sapiens sapiens* spanned a spectrum from utility to artistry, and their cave murals transcended mere craft—they could articulate an explanation or express a mystical viewpoint. Such paintings decorate innumerable cave walls and roofs in the Dordogne area of France, the French Pyrenees, and the Cantabrian mountains of Spain. They date variously between 17,000 and 12,000 BP.

A measure of the depth and extent of ice around 17,000 BP can be gauged from the main picture of the Cirque de Gavarnie in the French Pyrenees, just south of the Dordogne. The vertical wall beneath the largest waterfall and the U-shaped rock that frames the cirque were carved by ice during the last glacial maximum.

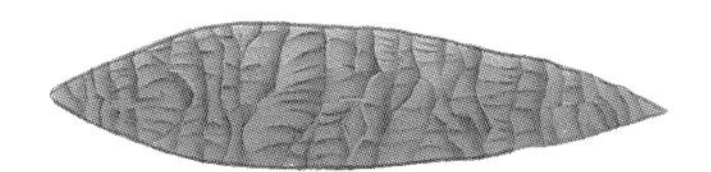

H. sapiens sapiens spearpoint

Three views of a burin

Ivory Figurine
La Madeleine,
Dordogne

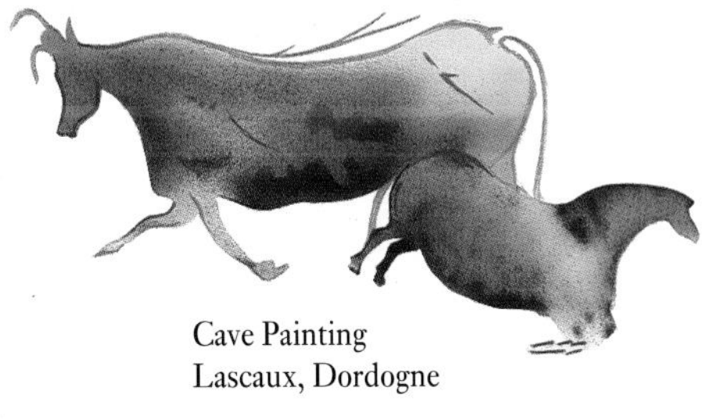

Cave Painting
Lascaux, Dordogne

Sites of cave paintings
and figurines

[*continued from page 269*]

flaking it with a hand-held pebble or rock fragment so as to produce a shaped core with a flat platform at the base [Essay 3]. One carefully aimed blow with maximum force (**11**) to the platform end of the core would now cause it to shatter into flakes whose shape had in effect been determined in advance. Only an expert toolmaker, a craftsman who could foresee how the flint would fracture, could achieve the hard, razor-sharp axes, cutting tools, scrapers, and spear-points that resulted. Since lives depended on this prowess, the toolmaker's standing in the hierarchy of a community was a high one.

By 35,000 BP several advances had been made in this struck-core technique, and these are named according to the locality in which they are believed to have developed. The basic technology remained the same, but the rate of production and the quality of the tools increased. The ultimate form of the Levallois technique in Europe emerged in the Dordogne area of southern France, where it has been most extensively studied. The Dordogne toolmakers discovered that the degree of efficient and exhaustive preparation of the core before the final blow was struck determined the eventual result—either a number of usable flakes (unifacial points) or one large pear-shaped flake (a scraper). Meanwhile, struck-core technology (**12**) had became widespread in Eurasia and Africa, where similar tools were made from locally available fine-grained and very hard rocks such as obsidian (volcanic glass). In Europe and the Middle East the technique was used by both anatomically modern humans and their contemporaries, the Neanderthals.

Neanderthals are named after the Neander valley near Düsseldorf in Germany, where the first fossil part-skeleton was found in 1856. Although they may have evolved in warmer climes long before 100,000 BP, Neanderthals (**13**) were not established in Europe much before that time. They long preceded anatomically modern Europeans, but disappear from the record after 30,000 BP (R. Prostch, 1989). It seems that there was a distinct overlap in time between these two hominid cultures in Europe, a theory that has been confirmed by the discovery of a classic Neanderthal site at St. Césaire in France dated around 34,000 BP.

The reason for the "sudden" disappearance of Neanderthals is unknown, one of the many unsolved mysteries of paleoanthropology. However, recent research in climatology has shown that there was a period of formidable glacial advance in the French Vosges between 50,000–30,000 BP. In this region, the advance was greater than that of 18,000 BP, the date generally accepted to be the time of the last *global* glacial maximum. This discovery could prove to be very significant: such extreme conditions could have contributed to the demise of some of the European Neanderthals by the combined effect of severity of climate and the reduction and changed distribution of flora-fauna on which Neanderthals depended for food. It is also thought that they were probably not as adaptable to severe conditions of cold as their anatomically modern human contemporaries.

Another theory accounting for the disappearance suggests that Neanderthals had more primitive tools than archaic humans. For this reason, it is suggested, they could not compete with more "advanced" modern humans in the harsh European climates of their time. The problem with this idea is that both cultures used similar tools throughout their coexistence, and that Neanderthals were not necessarily "inferior" beings. They are the victims of quite serious slander, often caricatured as savage cave dwellers brandishing clubs and dragging women by the hair. Indeed, Neanderthals were more robust and stronger than their contemporaries and in height and stature. They were within today's standards of size but were much stronger than we are. Their brutish reputation probably arises from the fact that they had very distinctive heavy-browed skulls, sloping foreheads, and protruding jaws (**14**)—*not*, incidentally, the weak chins often portrayed in drawings. In facial appearance, a few paleoanthropologists suggest, they were not so different from some people of European origin today. However, facial features may be

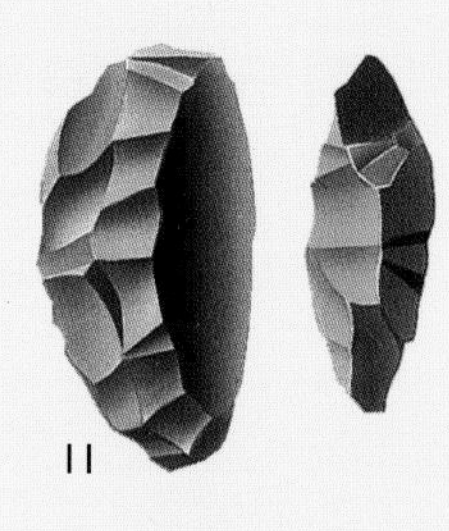
11

12

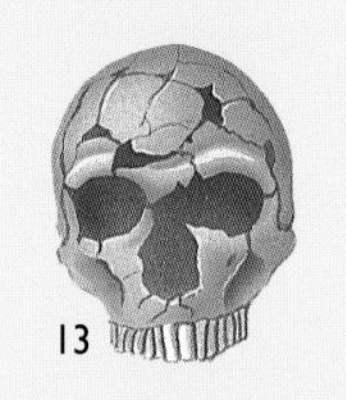
13

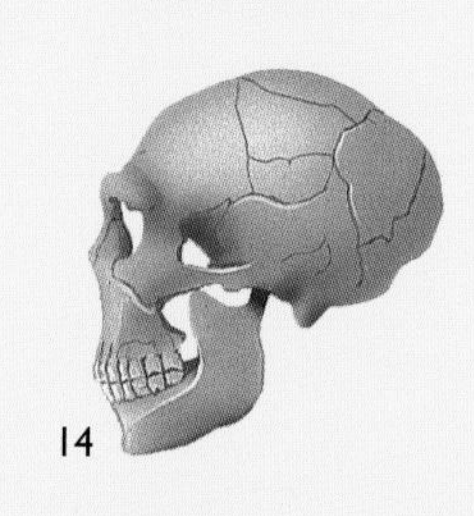
14

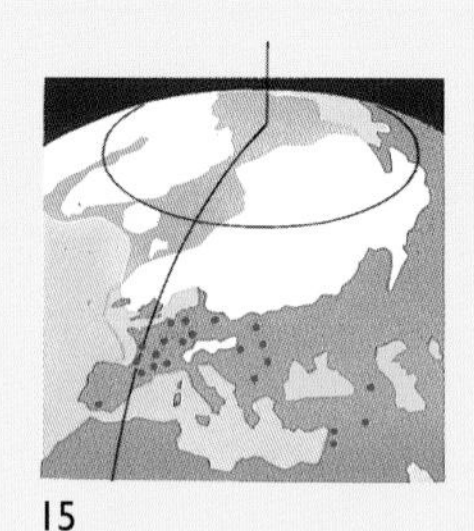
15

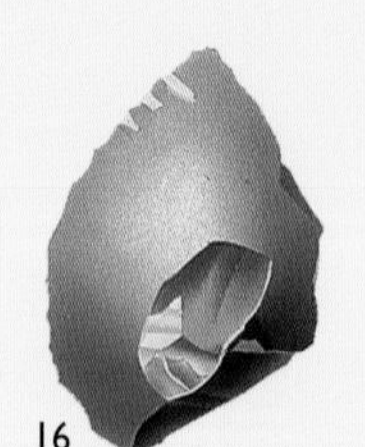
16

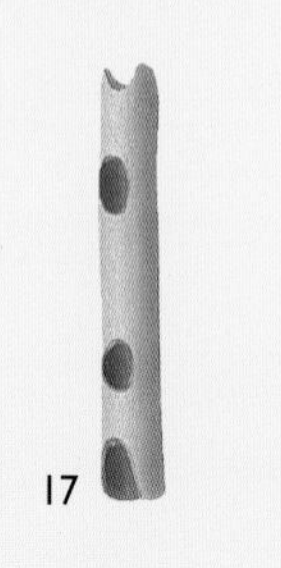
17

related to functional cause as much as to inheritance. For example, chewing hides to soften leather from an early age would build up massive jaws and jaw muscles and thus give an unreliable impression of genetic cause. In any event, Neanderthals lived in contrasting environments (**15**) that varied in extremes from the arid Middle East to the cold, treeless tundra of Central and Eastern Europe. Their remains are most frequently found in the more equable but still inclement ice-age conditions of Western Europe, in Iberia and France. It follows that there were several cultures of Neanderthal and local skeletal variations suggest that there were tribal differences.

They wore skins, decorated their bodies with ocher, and lived both in caves and in open-site shelters constructed from mammoth bones and covered with skins. In Central and Eastern Europe Neanderthals employed cooperative hunting strategies. Groups joined to hunt for mammoths, which provided meat and hides for all. The bones of the butchered animals were used as fuel for fires in areas beyond the tree line. In Western Europe cooperative hunting was not always necessary because people could take advantage of chance encounters with deer, horse, bison, and other ungulates. They may also have speared or trapped salmon in the well-stocked ice-age rivers of France and Spain.

Neanderthal living quarters were a hive of activity. Animal hides were dressed and some of the tools that were used for this process resembled modern implements that are used to make suede today (**16**). Judging from the wear on their cutting edges, specialized tools were often used for chopping large tree branches. Others were used for shaping smaller pieces of wood, and still others for the preparation of meat and vegetables before cooking on a hearth. Tools were often made from material that was obtained locally. Sometimes flint or flint cores were imported from places 32 km (20 miles) away. This shows that Neanderthal toolmakers prized nodules for their toolmaking qualities. The prepared core technique was an important technological advance of Neanderthals over archaic humans. Not least of the Neanderthals' technical accomplishments, although contentious, is that they may have made the first known musical instrument. This is a flute (**17**) made from a young cave-bear femur. It is dated between 43,000 and 82,000 BP—an unusually large margin for error. All that remains of the original bone is a segment with four holes and broken ends. The remnant flute, if indeed it is a flute, was found at a Neanderthal campsite in 1997 by Dr. Ivan Turk of the Slovenian Academy of Sciences in Ljubljana. The find triggered the intense interest of musicologists everywhere. The line-up and distance-apart of the finger-holes indicates that the instrument when complete was capable of playing a full diatonic scale, a near replica of the scale that can be played on any modern instrument. This find, if authenticated, has profound implications for both the Neanderthal and for the technical evolution of music.

Neanderthals also buried their dead, and are the first people known to have practiced funeral rites. Judging from skeletons found in graves, the old, and the seriously injured or deformed members of a clan, received care and attention. Some skeletons have arthritic deformities, or missing or injured but healed limbs. The old and young alike were often buried with beautifully crafted tools beside them—perhaps specially made or selected—and some graves may have been scattered with flowers. Compassion and grief are hardly present in the commonplace cartoon portrayal of Neanderthals.

Most experts now also accept that Neanderthals had some form of language. A throat-bone, called the "hyoid" has been found in a Neanderthal skeleton dated 60,000 BP (Kebara Caves in Israel). The site of the hyoid relative to the rest of the throat either permits or prevents the articulation of a systematic language, but unfortunately the original anatomical position of the bone is unclear. However, Neanderthal brain cases have a small depression in the region next to that part of the brain that would have controlled speech—a depression that also exists in modern brain cases. Furthermore, Neanderthal brain cases were as large or larger than our own today. Whatever the argument for and against articulate speech, with the degree of culture and technology that Neanderthals had evolved in their time, they must surely have shared some form of elaborate communication.

A ROCK SHELTER containing the remains of four skeletons, some skeletal fragments, stone tools and a hearth, was discovered accidentally by laborers in the Vézère Valley in the Dordogne in 1868. The remains were very old and appeared to have been deliberately buried. The site is at Cro-Magnon, near Les Eyzies, France, hence "Cro-Magnon" has become the name used to describe these people, once the oldest-known representatives of anatomically modern humans in Europe (**18**). (This distinction is now credited to human remains found in Moravia, Czech Republic, which are dated between 30,000 and 35,000 BP.)

The oldest Cro-Magnon-type tools, found elsewhere in the Dordogne, are reliably dated to about 35,000 BP. This is the date normally ascribed to the first appearance of anatomically modern humans. The best preserved skull at the original site belonged to a tall, muscular, adult male. He is affectionately known to paleoanthropologists as the "Old Man from Cro-Magnon," although his age was probably no more than fifty years when he died.

From Cro-Magnon time, technological advance in the stone and bone industries in Europe and elsewhere escalated steadily. Much of Cro-Magnon and subsequent technological progress must have been the result of experiment, but discovery was often incidental to the intended line of research and development—a methodology that has not changed all that much in tens of thousands of years. One could speculate that the real change was in attitude of mind: anatomically modern humans were able to sense a need for technical improvement and to have ideas about how this might be achieved—and then to progress by experiment. Neanderthals may also have possessed this aptitude, but with the appearance of Cro-Magnon, the development of technology came into full play. This is exemplified by the development of the "burin" [Essay 3].

A BURIN IS AN ELONGATE, dual-purpose stone tool (**19**), with an incisor at one end, sometimes a scraper at the other, and a spine, for added strength, from one extremity to the other. It was made from a blade already removed from a carefully prepared cylindrical core. As many as 250 chisel-like strokes were made around the edges of the core platform (**20**), each stroke producing a potential burin. The prime use of the burin was for the manufacture of other tools—it was a toolmaker's tool. By 15,000 BP, burins had allowed the development of the "groove and splinter" bone industry. Bone and antler (and wood), although often deliberately splintered, had previously resisted being worked. But now, however, splintered bones could be fashioned into hammers to dress flint cores; into needles to sew tight-seamed clothing with sinew and grass fibres; into harpoons and fishhooks; and into triggers for snare traps (**21**), thus extending the range of the food supply.

One particular product shaped with a burin is an ivory figurine from France, called the Venus of Brassempouy (**22**), dated to 25,000 BP. It will serve to demonstrate the degrees of technological advance and product development that the burin made possible in the hands of a skilled craftsman. It will also show something of the mental attitude and mode of dress of our early predecessors. The figurine, a carving of a woman's head, was fashioned from mammoth ivory, a material that could not be worked unless it had previously been heated. The head was carved out of solid material with a burin and burnished with a polishing instrument, possibly made from sandstone. The figurine is so well carved that it could be mistaken for a Greek classical-period miniature: long hair in a netted "snood" flowing back over the shoulders, almond-shaped face, wide eyes, firm

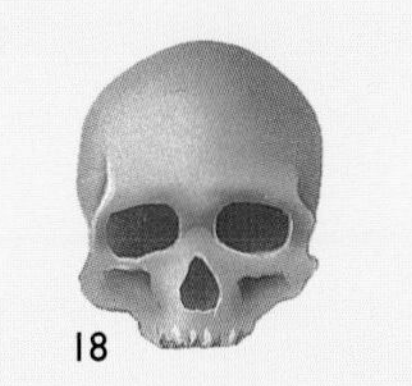
18

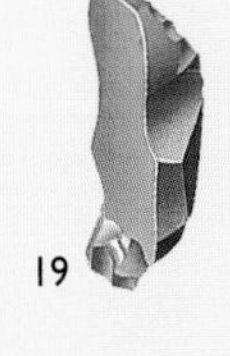
19

20

21

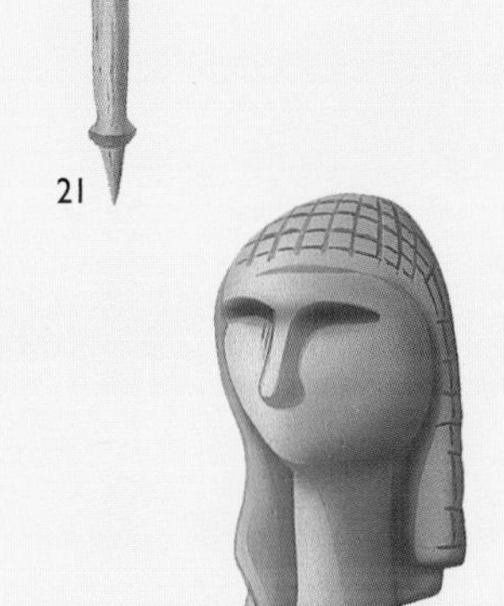
22

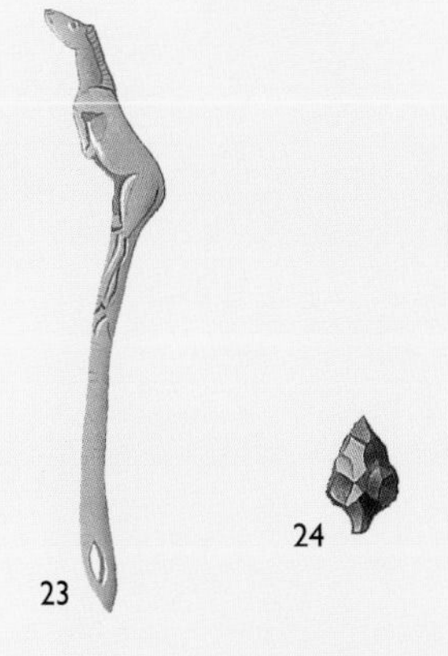
23

24

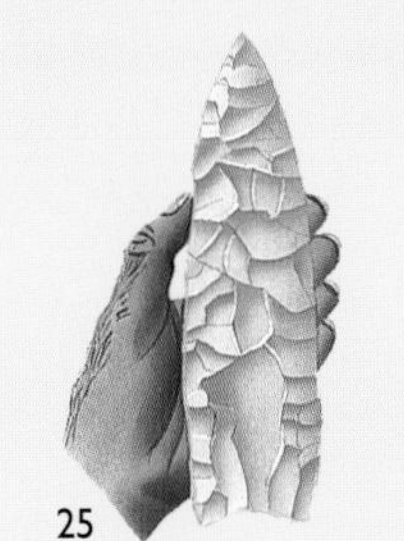
25

chin, and long neck. It is quite unlike the (to us) grotesque, faceless fertility figurines of the time, which emphasized heavy breasts, large belly, and fat buttocks. The Venus of Brassempouy was fashioned with feeling, perhaps with passion. The artist could visualize the shape of the head and facial features he wanted to reproduce.

Another technological advance made possible by the burin required the toolmaker to appreciate some principles of physical science: levers, tension, and centrifugal force (**23**). Hand-held spear-throwers made from antler, bone, or wood appeared on the scene about 17,000 BP and ultimately were used throughout Eurasia, North America, and Mesoamerica. These instruments extended the exact length of a hunter's arm, and were counterbalanced to give the user the right "feel" when hurling the spear. They were designed to maximize the conversion of his wound-up muscle tension, and the shift of body weight from one foot to the other, into distance; accuracy of course, depended on the thrower's skill. Spear-throwing devices were just about as carefully crafted as titanium-shafted golf clubs are crafted today. They no doubt involved the same degree of heated discussion, experiment with materials and design, and in the end the personal preference of the user. Modern testing has shown that if correctly designed, as these spear-throwers undoubtedly were, the flight of a spear can be extended from 39 m by hand to 69 m with the thrower (130 ft–230 ft). Its penetrating ability at close quarters also increased dramatically. (230 feet is about as far as an average golfer can hit a distance-designed golf ball with titanium shafted metal-headed clubs!)

ULTIMATELY THE SPEAR THROWER was largely replaced by the bow and arrow, a considerable technological advance and surely the most enduring mechanical invention of all time. Artifacts that could be interpreted as arrow points (which would have required arrow shafts and bows) have been found in North Africa and Eurasia, and date from 24,000–11,000 BP. In effect, the bow stored the energy expended in drawing back its sinuous bowstring. When this energy was released into the small cross section at the base of a flighted arrow, it was almost instantaneously converted into momentum. This momentum projected the arrow further and with more velocity and accuracy than a projectile could be thrown or directed with a spear-thrower. The triangular-shaped arrowheads that made the arrow most effective for killing prey (**24**) were at first the specialty of the Aterian culture at Bir-el-Ater in Tunisia, but the technique of making such arrowheads spread along the Mediterranean coast of North Africa, from eastern Libya to the North Atlantic.

The craft of making bifacial spear points appears to have been established in Europe no later than 12,000 BP. By about 11,000 BP an advanced version of a bifacial spear point was being produced by toolmakers half a world away. These were beautifully crafted tools that are unique to North America: there are no known Eurasian intermediates between the Aterian style and the North American style of manufacture. The North American products are called "Clovis bifacial points" (**25**): they are truly wonderful examples of neolithic craftsmanship. The people responsible for this industry were hunters of mammoth and long-horned bison. They fashioned their remarkably efficient projectile points at Blackwater Draw, near Clovis in New Mexico—the spear-point's "type" locality. But who, if anyone, had preceded the Clovis people into North America? This remains a controversial issue in the field of paleoanthropology.

FOR THIRTY YEARS there have been two conflicting and often acrimonious schools of thought on this matter. On the one hand there is the traditional pro-Clovis school of distinguished paleo-archaeologists, and on the other, the innovative pre-Clovis school of equally distinguished paleo-archaeologists. Until 1997, based on the dating of Clovis points, the pro-Clovis school could not accept an age earlier than 11,200 to 10,900 BP for arrival of humankind in North America. After the discovery of a site where human artifacts have

[continued on page 284]

Arctic potentilla

4 Siberians

The first Americans originated in West Beringia, a permafrost and mountain tundra region of Siberia neighboring the present Bering Strait. At the height of the last glacial advance it resembled the scene in the main picture—mountain tundra in the Canadian High Arctic. The inset picture is of a subspecies of potentilla, a common arctic-tundra plant that might have relieved the bleakness of the scene. During the last glacial age, little snow fell in Siberia in winter, but the short, dark days and long nights were severely cold. In contrast, the long summer days were warm, dry, and sunny. These conditions were the result of a resident high-pressure weather system, formed as a consequence of reduced sea levels that exposed continental margins around the Arctic basin, and a perennially frozen ocean with high albedo.

Tundra and ice cap on Ellesmere Island, Canadian High Arctic

Early *Homo sapiens sapiens* sites in Asia to West Beringia relative to the last glacial maximum at 18,000 BP

The consensus is that nomadic groups of *H. sapiens sapiens* were living in West Beringia before 36,000 BP, although the majority of the widespread sites so far discovered date only to 16,000–12,000 BP. But why should people have been drawn to such an inhospitable place?

Much of Siberia was covered by "steppe-tundra," an ecosystem that has no modern equivalent. Steppe and tundra are usually separated by taiga, swampy areas of coniferous forest. But in Siberia "Steppe" is a dry, grassy, and treeless plain, more arid than North American prairie. "Tundra" is also a treeless domain, but it is underlain by permafrost and its surface is often marshy in summer, when it supports the growth of lichens, mosses, and low-growing shrubs such as arctic willow. The lush summer growth of all forms of steppe-tundra vegetation attracted large numbers of ice-age mammals—woolly mammoth, bison, horse, musk ox, camel, arctic fox, and enormous numbers of caribou. These animals attracted hunter-gatherers.

In summer, these people subsisted on fresh meat and the partly-digested stomach contents of slaughtered ruminants, which together provided all the vitamins essential for survival. In winter, hunter-gatherers ate meat from storage pits dug into the permafrost during summer. They used mammoth bones as frameworks for skin-covered shelters, burned bone for fire, and used animal skins for clothing. This way of life seems to have been so attractive that the steppe-tundra was never abandoned for long—even at the time of the last glacial maximum 21,000–17,000 BP.

5 Beringia

During the period 75,000 to 14,000 BP in the Wisconsin-Würm glacial age (120,000–10,000 BP), Beringia existed at least from 40,000 to 36,000 BP and from 32,000 to 17,000 BP. Outside these times, the Beringian sub-continent either was either near its threshold level (45 m or 150 ft below present sea level), or actually submerged. At this point, Asia and North America were linked by a tenuous isthmus. But, as reconstructed in the sea level curve and globe here, there were prolonged intervals when the sea dropped well below the threshold point. Central Beringia then extended to about 1,600 km (1,000 miles) or more north-south and east-west of the present strait. At such times small numbers of Siberian hunter-gatherers, following the animals on which they depended, would have wandered from Asia to North America.

Although the permafrost landscape pictured here is a relict of extreme northeastern Beringia, it is nevertheless analogous to a Central Beringian landscape. Here, the depth of permafrost extends 300 m (1,000 ft) or more beneath the surface. In low-lying regions "ice lenses" developed in winter as a consequence of sub-surface melting in summer. This cycle caused frost heaves to form—mounds of uplifted subsoil with ice lenses at their center-some of which still exist beneath the sea. Each year ice lenses were fed from beneath by percolating meltwater from moats that formed around the mounds, thus causing both lens and mound to grow in size. Such "pregnant hills" are called "pingos" by the modern Inuit people and were possibly used by Central Beringian hunter-gatherers to sight mammoth and other animals.

Central Beringian winters were dark, frigid and very windy—but relatively snow-free; the brief summers were warm, sunny, and dry. The tundra-type landscape was virtually treeless, although wood fragments are sometimes brought to surface during oil drilling in the Beaufort Sea (in what was East Beringia). The wood debris in the foreground of the picture is a mixture of birch and

Pingos and birch-forest debris, Tuktoyaktuk near Mackenzie River delta, Northwest Territories, Canada

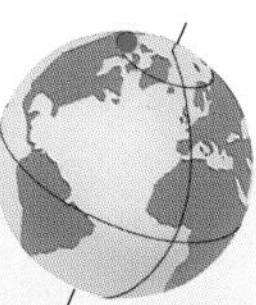

GLOBE: All known sites of early *Homo sapiens sapiens* in Central Beringia relative to the last glacial maximum at 18,000 BP

spruce—trees that grew in this region of East Beringia during a period of environmental change that was to mark the end of the Wisconsin-Würm glacial advance. This interval is accordingly called the "birch zone."

Calculating the possible timing of opportune crossings is fraught with difficulty, particularly for the potential "first people-crossing" periods of 40,000 to 36,000 BP and from 32,000 to 17,000 BP. Global sea level fluctuations did not affect Central Beringia in the same way as other exposed or flooded continental shelves. The weight of ice on the North American continent depressed the continent rather like an overloaded ship in the sea, changing the *relative* sea level, and therefore the times in which Beringia was above its threshold. [This subject and alternative access to North America by maritime-culture people is discussed in the text on page 284.]

All known sites of early *Homo sapiens sapiens* in the Alaskan Refuge relative to the last glacial maximum at 18,000 BP

6 Alaskan Refuge

During the last glacial advance the Old Crow River basin pictured here was part of the "Alaskan Refuge"—the region of East Beringia in which the first Americans found refuge. Old Crow lay to the northeast of the region within an ice-sheet-free zone that straddled the Arctic Circle from Siberia to the Mackenzie River delta on the edge of the Beaufort Sea. The refuge was bordered by a perennially frozen Arctic Ocean to the north and by the northernmost limit of the coastal range ice sheet to the south. A wall of continental ice lay to the east of the Mackenzie.

The V-shaped notch visible at center-right on the Old Crow River bank is the site of a key archaeological dig that has yielded bone artifacts made by early humans in North America. But a glance at the shape of this meandering river basin will explain why it is difficult to date such artifacts. Sediments laid down by the river, and any artifacts they may contain, have been reworked by river ice and water time and time again. Archaeologists here have therefore relied upon the radiocarbon dating method [see page 285]. However, bone is porous, and it is now known that in conditions such as those around Old Crow, the likelihood of contamination, and therefore inaccurate results, is high. Indeed, Old Crow bone artifacts, supposedly dating from 29,000–26,000 BP, were recalibrated to only 1,300 BP some time after their original discovery and preliminary dating. Subsequently, mammoth-bone flakes and cores dated 40,000 BP were discovered at Old Crow, but have been largely ignored or explained away because of the previous dating errors (C.V. Haynes, Santa Fe, N.M. *Clovis and Beyond* conference: 1999 address). Old Crow is still a vitally important site.

The Old Crow region was only one of many sparsely populated regions in Eastern Beringia, where generations of hunter-gatherers lived in permafrost environments (inset, near right) before wandering southward. Their ultimate migration was both opportunistic and spasmodic. Some nomads may first have followed migrating herds through the unglaciated Richardson Mountains (inset, far right) east of Old Crow to the Mackenzie River. Humans and migrating animals would then have been forced to turn upstream towards the ice-free Mackenzie Corridor, which led to the northern limit of the open prairies 2,250 km (1,400 miles) away. Part of the timing problem is that when Central Beringia was fully exposed above the sea during the periods of severe glaciation, migration directly to the southeast was blocked by swollen ice sheets.

Permafrost polygons, Old Crow basin

Richardson Mountains between Old Crow basin and the Mackenzie Delta

St. Elias Mountains, Alaska-Canada

7 Corridor South

The St. Elias Mountains in southern Alaska (left) are permanently snow-covered, and are cut by the largest glaciers outside of Antarctica and Greenland. The high mountains in this range (the highest, Mt. Logan, is 5,955 m; 19,850 ft) wring most of the moisture out of prevailing Pacific westerlies. As a result, valleys east of the range are in a snow shadow (below); they remain warm and dry in summer and are almost free of snow and ice in winter.

The same conditions applied during the Wisconsin-Würm glacial age from 120,000 to 10,000 BP. Relatively warm moist air associated with the North Pacific gyre created heavy snowfall on an unbroken 2,250-km (1,400-mile) chain of coastal ranges. This chain stretched from the Aleutian Islands to Alaska, and from Alaska down the West Coast to the region of Vancouver in British Columbia. These coastal ranges defined the western limit of the

Cordilleran ice sheet. In the north, the ice sheet bordered the southern edge of East Beringia (the region north of the Aleutians, now submerged beneath the Bering Sea), and from the Anchorage region to Vancouver it reached inland for an average distance of about 960 km (600 miles). During this time much of the Alaska-Canada coastal region was obstructed by 1600 km (1000 miles) of repeated and extensive glacier-snouts that would have been difficult to by-pass, if not entirely insurmountable, for humans.

To the east of the Cordilleran ice sheet there was another sheet of even greater extent—the Keewatin, one of three sectors of the Laurentide ice sheet that covered much of continental North America. The western Keewatin ice front stretched from near the Mackenzie Delta in the Beaufort Sea southeast to the region of present-day Edmonton.

At times of severe glacial advance the Keewatin and Cordilleran ice sheets coalesced in northern Alberta. But in times of temporary retreat an ice-free corridor may have developed between the two sheets. If and when the corridor was open before and after a glacial advance, it was a potential migration route for animals and hunter-gatherers alike. Like many other facets of the timing for human migration into the Americas, the existence of an ice-free corridor at times between 36,000 and 22,000 BP and after 17,000 BP is challenged by some experts and thought probable by others. But whenever the corridor existed, the first Americans must have used it as the gateway to a lush new world of plants and animals hitherto unseen by human eye—a *Shangri-la*.

Sheep Mountain near Mount Wrangell east of St. Elias Mountains, Alaska

[*continued from page 275*]

been found and dated to 12,500 BP, the Clovis school now accept this as the earliest known date for human habitation in the Americas. But the site is just the youngest of *two* sites at Monte Verde, at the southern end of Chile's Pacific Coast (**26**).

The pre-Clovis school, using apparently overwhelming evidence, accepts dates of arrival in North America going back to 17,000 BP at Meadowcroft Shelter, Pennsylvania. Tentatively, they also accept a date as far back as 33,000 BP for human artifacts at the second site at Monte Verde, 24,000 km (15,000 miles) from Alaska and the Bering Strait—more than two-thirds of the way around the Earth. The only universally accepted method for humankind to have gained access to the Americas, is for Asian nomads to have crossed the Bering Landbridge from Siberia to Alaska on foot. A minority view is that in-shore but not maritime skin-covered kayaks and umiaks might have been invented by this time.

Migrants could have been members of one of two fundamental cultures. They were either hunters and gatherers, adapted to a continental life-style, or maritime people adapted to coastal regions. However, there is no trace anywhere in North America of a culture old enough to have made the journey from Alaska to the Monte Verde site by 33,000 BP [Essays 4 to 7]. Even so, the Bering Landbridge was "open" between 40,000 and 36,000 BP and nomads of one culture or the other could have crossed from Asia to North America during this period of time. Indeed, the evidence for there being humans at Monte Verde 33,000 or more years ago is so compelling to paleo-archaeologists of both schools that a major and costly dig is planned for early in this century. The older Monte Verde site is in a water-saturated peat bog rich in tannin, a natural preservative. If the peat-bog artifact-dates are validated, it is likely that the pro-Clovis school will accept the findings and join in a new era of paleo-archeological exploration in America. This dig will be exhaustive, expensive, definitive, and closely watched.

The issue of dates before 12,500 BP for the arrival of humans in the Americas raises questions about access from Asia, via the ephemeral Bering Landbridge, to the American Southwest, Mesoamerica, and South America. It also raises questions about the timing and progress of hunter-gathers from a refuge in Alaska along the intermittent Mackenzie Corridor to the heart of the continent [Essays 6 and 7]. Conversely, for supporters of maritime colonisation, it raises issues about the extent of the continental shelf exposed during the intermediate periods when the Bering Landbridge would have been partly submerged (**27**). These questions in turn relate to the timing of stadials and interstadials during the Wisconsin-Würm glacial age (120,000–10,000 BP), fluctuating sea levels, and not least to the whole issue of dating techniques and their dependability. The origin of the endemic languages and modern genetic interpretations of the relationship between native peoples just add to the complications.

The main change between the world of 33,000 BP and the present is the vast reduction in the extent and volume of continental ice in the Northern Hemisphere. Another difference is in the disappearance of an immense volume of sea ice that formed in the Southern Hemisphere during these distant times. Although the extent of sea ice made little difference to sea level, it would have impeded migration. The accumulation of continental freshwater ice resulted in a considerable fall in the level of the global ocean and the exposure of continental margins above the sea. But at the same time, the buildup of land ice depressed the continental crust considerably, and simultaneously the level of the ocean floor level rose as it was relieved of a vast weight of water. So we cannot assume that, just because continental ice volume was sufficient to lower sea level by 120 m (400 ft), continental shelves of that depth today were necessarily exposed. One factor counteracts the other.

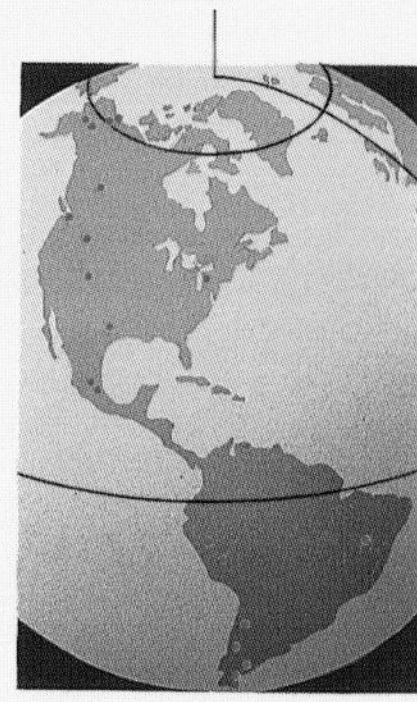

26

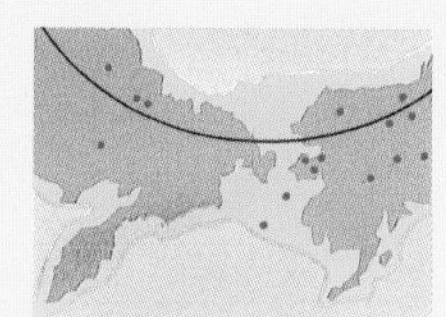

27

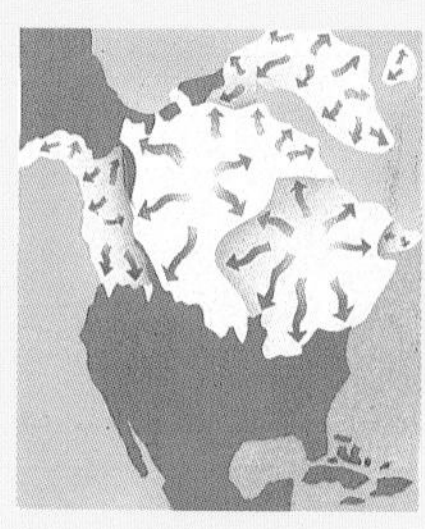

28

The extent of exposure of continental shelves during a glacial advance, that might have permitted human advance from Alaska to the south, also depended on the nature of the continental shelf itself. Intertidal flats would become high, dry, and extensive with only a modest drop in sealevel. But steeper shelves would only expose a small additional area, even with a sharper drop in sea level. Additionally, extensive stretches of Alaskan-Canadian coastline north of Vancouver and Seattle would have been impassable because of glacier snouts from mountain ranges protruding into the Pacific Ocean.

The Wisconsin-Würm landbridge connection across the Bering Strait required a *relative* fall of at least 45 m (150 feet) in sea level below the present global level to reach a "threshold" stage—the level at which land would have appeared above the sea in the earliest stage of forming a landbridge. Although potentially rich in marine resources, ther thrshold landbridge would have been a very wet and miserable place—but acceptable to a maritime culture. There were prolonged periods during the last glacial age when global sea level dropped to a point that where the landbridge grew to become subcontinental, about 1,600 km (1,000 miles) wide. When exposed to this extent the region is called "Beringia," and it was Beringia that was home to the hunter-gatherers during their traverse from one continent to another. Both Beringia and much of Asia were free of ice sheets even at the height of the last glacial advance.

All the available information, much of it contradictory, some of it still uncomfirmed, adds up to the conclusion that Beringia existed above sea level at various times from 40,000 to 36,000 BP and from 32,000 to 17,000 BP (**27**). Outside these times, the Beringian subcontinent either was either near or below its threshold level from 70,000 BP onwards (though it may still have been traversable at times). It follows that the opportunity for human migration from Siberia to the region known as the Alaskan Refuge, was infrequent but prolonged. Conversely the coastal route, where and if practical, was available as an option much of the time. But what of the other side of the equation? Once in the Alaskan Refuge, what were the windows of opportunity to reach the heart of the continent via the Mackenzie Corridor to the American Southwest and beyond?

During much of the Wisconsin-Würm glacial age, North America was enveloped by two primary ice sheets (**28**). The first was the West Coast ice sheet, termed the Cordilleran ice sheet: this was sustained by snowfall from the North Pacific gyre. The second was the Laurentide continental ice sheet.

The Cordilleran ice sheet reached its maximum advance about 18,000 BP, and by 10,000 BP it was not much more extensive than it is today. The Laurentide sheet to the east was a much more complex and long-lived structure. It was one of the Earth's largest-known combinations of ice sheets, covering two-thirds of North America. The thickness and extent of these sheets depended on the amount of precipitation produced by the moist, warm air associated with both the North Pacific and North Atlantic Ocean gyres. The shape, speed, and direction of flow of the Laurentide ice-sheet sectors altered as regional snowfall patterns changed from year to year.

The advance of the Cordilleran ice sheet eastward, and the Laurentide ice sheet westward, proceeded simultaneously from about 24,000 BP onwards [Essay 7]. The two sheets coalesced along the line of the Mackenzie River, and then southeast to the region of Calgary-Edmonton in Alberta. The actual timing of the final meeting is difficult to assess because the advance and retreat of glaciers during interstadials and stadials simply destroys the evidence needed to make such calculations. However, according to S. Dyke and V. Prest of the Canadian Geological Survey (1987), it is certain that the merger was absolute by 18,000 BP, and unlikely that a corridor formed between the two sheets until some time between 14,000 and 13,000 BP, . However, some experts disagree and point to pollen evidence that suggests an earlier date. But there is no disagreement that, whenever it opened, the corridor was rock-strewn, and often flooded by extensive glacial lakes and dangerously swollen rivers—it would have been a thoroughly inhospitable and uninviting place.

We have discussed some of the physical evidence that suggests that people could have crossed from Siberia to the Alaskan Refuge on occasions before 14,000 BP. There is evidence of their presence in Siberia before and after this date, but there is no evidence of them in the Alaskan Refuge before 13,000 BP, except some that is currently considered unacceptable. The earliest accepted artifacts are dated 13,000-11,500 BP (Bluefish caves; **29**). So there is an important missing link in the chronology still to be proved, which is why the redating of a caribou tibia (**30**), from Old Crow River, from 27,000 BP to 1,300 BP added fuel to the fire of a heated debate [Essay 6].

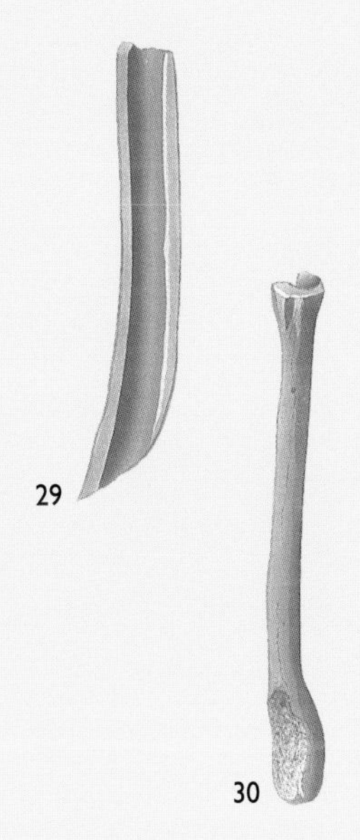

RADIOCARBON DATING OF ORGANIC MATERIAL is the most frequently used method for dating archaeological sites back to 40,000–50,000 BP. This method measures the proportion of a radioactive form or "isotope" of carbon (denoted ^{14}C) as a dating tool. During a plant or animals lifetime, it is constantly exchanging carbon with its environment, and so the ratio of ^{14}C to normal carbon remains constant. After death, however, carbon exchange ceases, and the ^{14}C in the organic matter starts to radioactively decay into other atoms. Therefore the proportion of ^{14}C left in a sample provides an indication of the length of time since the sample material died.

For years, the radiocarbon method was accepted without question because it agreed (at least at the lower end of the scale) with dates obtained from the record of ancient tree growth-rings, which provide an unbroken dating sequence back to almost 10,000 BP (a method called dendrochronology). But it was later found that ^{14}C dates did not always conform to the tree-ring record; the isotope was somehow being "contaminated."

The main flaw was a simple one—radiocarbon dating was based on the assumption that carbon dioxide (CO_2) levels in the environment (and therefore in living organisms) had not changed—when in fact they had. Variations in the composition of the atmosphere during the last 50,000 years could affect the accuracy of dating of many organic materials—bone and ivory as well as wood.

Some way of calibrating radiocarbon results against the changing levels of atmospheric CO_2 was now urgently required, and this was accomplished in 1990 by Richard Fairbanks of the Lamont-Doherty Geological Observatory of Columbia University. The new method was based on dating of cores from ancient corals drilled in Barbados, matched against analysis of ice cores drilled by other scientists in Antarctica and in Greenland. Using a sophisticated dating technique based on another radioactive isotope, the Lamont team were able to obtain unequivocal dates for corals built up since 10,000 BP, and compare them with their carbon dates.

Meanwhile, the annual layers from the ice cores (analogous to tree rings) were analysed. Each layer contains measurable amounts of gases, including CO_2, that were present in the Earth's atmosphere at the time that the snow which formed the ice was precipitated, and together they showed that CO_2 in the atmosphere had varied considerably over time. The CO_2 content was appreciably higher during warmer interstadial interludes and lower during the colder stadial periods. These variations also matched the patterns of tree-ring and coral growth during the last 10,000 years. From all this evidence, it became clear that ^{14}C dates for materials prior to 10,000 BP are most probably older than stated, perhaps by several thousand years. [Even so, we must continue to quote dates used by authors in their papers, as listed in the bibliography.]

Changing CO_2 content is also a means of judging the past condition of the global climate and highlighting potential periods for glacial advance (low CO_2 content) and retreat (high CO_2 content). It is a means of judging periods of time when Siberians could have crossed the Bering Landbridge, during glacial advances, as well as later opportunities for southward migration during interstadials. In this context, Antarctic ice cores show that there was a global CO_2 high on either side of 30,000 BP, and a low about 28,000–15,000 BP. After 15,000 BP, CO_2 content of the atmosphere increased rapidly for several thousand years before assuming a slower and more fluctuating increase to the present level.

ALTHOUGH PALEOANTHROPOLOGISTS AGREE that Native Americans derive primarily from Siberian stock, the origin of Native American languages is enigmatic. Over a thousand such languages have been described, and many more are believed to have become extinct. The vocabularies of the known languages divide into about 150 families of mutually unintelligible tongues. Only a very few of these have even a remote similarity to Eurasian languages.

The Arctic people have had free rein around the northern shores of Asia and America; they have a common-language denominator that is not to be wondered at. More surprisingly, the Navajo and Apache family of languages, like that of the coastal peoples of the northwest, is thought to have a tenuous affinity with certain Siberian tongues—both tribes are believed to have migrated from Athabasca, Canada to the south only seven centuries ago. But other Native American language families, although they appear to be generically related to each other, are mutually unintelligible and have no likeness to other known languages—they appear to be collectively unique. This diversity supports the notion that the paleo-Native Americans were completely isoolated throughout their cultural evolution—and support for such great antiquity arises from recent genetic research.

ALL MEMBERS OF AN INTERMARRYING POPULATION possess certain genetic markers that undergo minute changes with succeeding generations. Should an original population split into several groups, each producing a distinctly separate population, then the pattern of the common genetic marker in each population will begin to diverge from the time of separation. Thus the genome markers of several populations derived from a common stock will have markers that are genetically distant from one another today. Knowing the approximate rate at which such genome markers change in character has enabled geneticists to work out the "genetic distance" between Native American populations and therefore the approximate time of their divergence.

The main conclusion of a twenty-year-long program of research to discover such genetic origins, completed in 1985, was that Native Americans derive from an Asian common stock that produced three distinct populations. Furthermore, the team of researchers, drawn from several North American universities and institutions led by R.C. Williams of Arizona State University, concluded that the initial migration of the core populations took place in waves commencing around 40,000 BP at the earliest.

More recently, Douglas.C.Wallace and associates of Emory University, Atlanta, completed a similar but independent study. They relied specifically on a small fragment of genetic code containing only 37 genes. This fragment of DNA is called the "mitochondrion" and its genes are called "mitochondria." It is passed on to the next generation by the mother of a child alone: all other genes in human DNA molecules are inherited equally from mother and father. Through this feminine fragment, ethnic groups that intermarry can be identified and linked to their continent or region of origin by the mutation patterns in their mtDNA. Each continent or region has its own mtDNA identity. From years of research, Wallace and his associates concluded that the original peopling of the Americas may have taken place in a single wave between 20,000 and 40,000 BP—not three. Furthermore Dr.Wallace found that some of the immigrants into the Americas possessed mtDNA which had more in common with the aboriginal populations of China, and very little to do with their Siberian counterparts.

ALTHOUGH IT RUNS CONTRARY to Dr. Wallace's conclusions, the general view of many paleoanthropologists (other than the pro-

[*continued on page 286*]

[*continued from page 285*]

Clovis school), has been formed from the Williams research. The consensus is that the first immigrants appear to have arrived in North America from Asia in several stages between 40,000 BP and 16,000 BP. Genetically these first-wave people were ancestral to nearly all subsequent Native American populations. A second wave of migrants entered North America between 14,000 BP and 12,000 BP and resulted in the evolution of at least two Canadian (Athabascan) cultural groups. A third migration occurred about 9,000 BP—the Arctic cultures whose descendants are the present Aleuts and Inuits of North America and Greenland. The consensus also agrees with Williams view that all three paleo-American populations originated in northeastern Siberia, which has "the genetic variation, the geographic proximity, and geological history required of such a homeland."

The best known of a number of exhaustively studied pre-Clovis sites is at Meadowcroft at Avella in Pennsylvania [Essay 9: **31**]. The ^{14}C dates at this site were established from charcoal remaining in hearths that appear to have been deliberately built. According to the archaeologists who have excavated the site, the radiocarbon dates of these charcoal remains conform to the date suggested by the older methods of stratigraphy. The dates for Meadowcroft and its artifacts (**32**), challenged unsuccessfully by the pro-Clovis school, are the most secure for any site in the Americas, and place the site at 17,000 BP and more. As we have seen, Monte Verde dates (vigorously challenged) are older—and range from 12,500 back to 33,000 BP.

Yet another site, at Pedra Furada, Brazil, has produced artifacts stratigraphically dated from 5,000 to 50,000 BP (*Nature* Vol. 362: 1993) but dates remain contentious, and it has yet to receive the vigorous attention it will ultimately get if Monte Verde is proved beyond question to be aged 33,000 BP. With these major sites in mind, one can take the previously summarized linguistic and genetic research into account. To it one should add the physical and archaeological research that we have outlined in this chapter. From this conjunction it should be possible to assess the merits of the pre-Clovis and pro-Clovis points of view for oneself. This author is irresistibly reminded of what Sherlock Holmes is reported to have said to Dr. Watson in *The Sign of Four*, "How often have I told you that when you have eliminated the impossible, whatever remains, *however improbable*, must be the truth?"

During 9,000 years of generally retreating ice sheets, ice caps and glaciers, human populations in Europe had also increased from thousands to millions. Europeans had advanced from primitive hunter-gatherers, to farmers and traders, townspeople and city folk, and from tribal domains to empires. Europeans evolved political, religious, and military organizations, on a much greater scale than the parallel advances that had been made in a few isolated areas of the Americas. And Europeans had mastered the principles of ocean navigation while the Americans had not.

When, as a consequence of this mastery, between 1,000 and 500 years ago Europeans discovered the existence of two new continents in addition to their own, they were unaware of the geographic extent of their discoveries. They were ignorant of the fact that these new-found lands were already populated by thirty to forty million people, roughly 10 percent of the world population at that time. What followed was catastrophic for the indigenous peoples, opportune for the Europeans, and predictive for the future of humankind in an overcrowded world.

8 First Americans?

In this picture we are looking down at the Colorado River of the Grand Canyon from an Anasazi cliff dwelling and granary at Nankoweep. The dwelling was in use about eight centuries ago. The inset picture is of rare colored pictographs painted on the walls of a sandstone amphitheater, up-river in the Maze region of Canyonlands. The frieze depicts priestly fertility rites. The ghostly figures are about ten feet high and were painted on a prepared surface by people of the Fremont Culture about 1,100 BP.

The oldest known perishable artifacts in this Colorado River region are found on the Kaibab Plateau. they date from 8,000–10,000 BP and consist of basketry, cordage, sandals and related materials (J.M. Adavasio; personal communication). But by about 12,000 BP, a sophisticated technique for making bifacial spearpoints for killing large animals (termed "megafauna") had been developed by Native Americans elsewhere in the Southwest. These spear-points are unique: the technology of their manufacture had no counterpart in Eurasia.

There were two basic spearpoint designs, named after their respective discovery sites in Folsom and Clovis, New Mexico. The

Nankoweep, Marble Gorge, Grand Canyon, Arizona

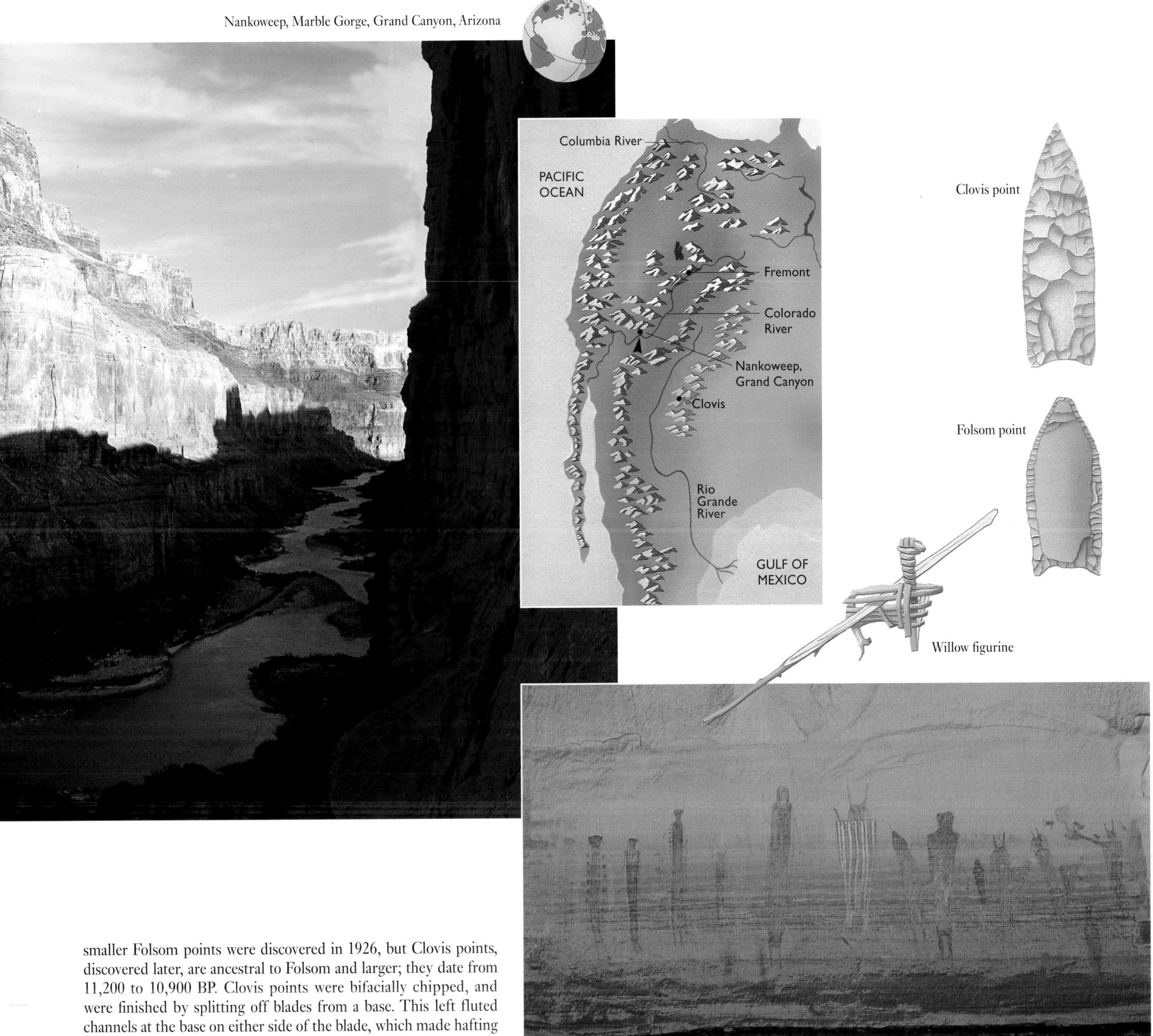

Clovis point

Folsom point

Willow figurine

Fremont Culture frieze, the Maze, Canyonlands, Utah

smaller Folsom points were discovered in 1926, but Clovis points, discovered later, are ancestral to Folsom and larger; they date from 11,200 to 10,900 BP. Clovis points were bifacially chipped, and were finished by splitting off blades from a base. This left fluted channels at the base on either side of the blade, which made hafting them onto a wooden shaft easier and more effective. Bindings were protected by grinding down the sharp edges of the flutes. The makers of these points hunted herds of woolly and imperial mammoths as well as other long-extinct Wisconsin-Würm glacial-age animals, including long-horned bison, North American camel, tapir, horse, and other animals that roamed the western plains.

Clovis points are pivotal to a continuing controversy over the timing of the arrival of the first Americans. For decades a pro-Clovis school of paleoarcheologists has maintained that people arrived in the Americas only a thousand years or so before the Clovis people. Pre-Clovis archeologists argue that the first Americans arrived at least ten thousand and possibly twenty thousand years before Clovis points appear in the record. The pro-Clovis school now accepts a slightly older figure for artifacts found at a site at Monte Verde, Chile, but a second site at the same location is claimed to contain artifacts 33,000 years old, and this date is not generally accepted. The latter site will be the subject of a major research project early this century. This investigation will hopefully resolve past differences and promote a new era of research in the field of North and South American antiquity.

9 River People

During periods of glacial advance, the difference in temperature between the warmest and coldest months of the year was at its greatest in areas between the leading edge of an ice sheet and the subtropical regions to the south. Thus areas of North America free from ice experienced the greatest changes between extremes of climate, while subtropical and tropical regions to the south tended to remain steadily warm.

The first North Americans are thought to have been hunters of megafauna, although there is a growing view that there was also a strong maritime culture on continental shelves now flooded by the sea. Even so, for day-to-day sustenance, northern people were also dependent upon the gathering of wild fruits and nuts, and the killing of small animals. According to biologists, an area of about 65 square kilometers (25 sq miles) of tundra-taiga-steppe would have been required to support each individual. It follows that population density in North America remained low and campsites few and far between until the ice sheets had retreated, and more bountiful varieties of flora and fauna had become established.

In strong contrast to this set of conditions, Mesoamerican hunter-gatherers, from the time of their first arrival in that subtropical region, could rely more upon plants and plant-derived foods than upon animals for sustenance. The ready availability of edible plants led to early plant domestication, and therefore to concentrated population centers and the earliest steps in the evolution of the many Mesoamerican civilizations.

Meadowcroft

This is the exterior of the famed Meadowcroft Shelter, "famed" because the interpretation of the massive data excavated here in 1973–78 has survived all criticism. Although there are several other sites for which greater antiquity is claimed, particularly in South America, Meadowcroft has proved to be the strongest candidate for the pre-Clovis presence of humankind in the Americas. It has yielded evidence of periodic use from 17,000–1,600 BP, and possible evidence dating back to 19,000 BP.

Meadowcroft was a "motel," an intermediate stop for travelers. The shelter is perched on the side of a deep ravine that was at least partially filled by ice at times during the Wisconsin-Würm advance. At its maximum advance the Laurentide ice sheet itself was only a few miles north of here, and the nearby Ohio River and its upstream tributaries, the Allegheny and Monongahela Rivers, were fierce glacial streams that drained from the edge of the ice sheet.

The figure at the now-covered entrance to the dig is Dr. James Adovasio (Director of Mercyhurst Archaeological Institute), who led the Pittsburgh University interdisciplinary team of scientists who excavated the shelter.

18,000 BP

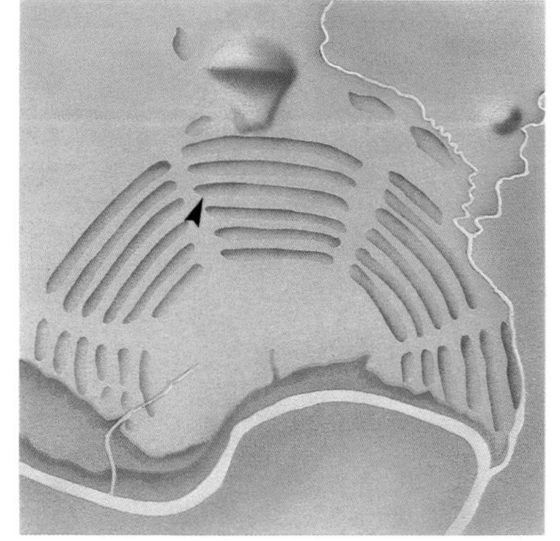

Poverty Point site plan

Poverty Point

The picture above is of Poverty Point, Louisiana—a great earthwork by the Mississippi River that was inhabited from about 3,500–2,700 BP. We are looking across a shallow pool-filled depression between the now-eroded ridges of nested octagons (inset, near left), and towards the large mound shown in the plan. This complex is 1,300 m (0.8 miles) in diameter, and was the earliest of its kind in North America.

10 Woodlanders

The general retreat of the Laurentide ice sheet was not a quiet affair: it was violently active. Burst ice dams and catastrophic water surges were frequent. For example, about 8,000 BP, as the Laurentide ice sheet began to disintegrate, an enormous lake about 315 m (1,000 ft) deep formed in an ice basin in the general region of Hudson Bay. This was drained "instantly" when the sea penetrated the edge of the basin. Such ice-dammed lakes were fed by gigantic, ever-shifting river systems on the surface of the remaining ice. Rivers also flowed in tunnels within and beneath the ice. The interaction of surface and subsurface rivers cut through melting ice sheets to produce sudden and tumultuous surges.

There was water everywhere on North America as the Wisconsin glacial period came to an end: swollen lakes and rivers, raging streams, rising sea levels, and flooding continental shelves. On the eastern coast, terminal moraines that marked the maximum advance of the ice sheet's Labrador sector into the open sea were at first left abandoned on exposed continental shelves. As sea level rose, these untidy piles of glacial debris became islands and peninsulas, of which Long Island and Cape Cod are examples. The Cape region had once extended three or four miles to the east, but now marine erosion began to shape the fishhook appearance of the Cape's present headlands, beaches, spits, and sand bars. As shorelines shrank, the energy of ocean waves constantly drove half-drowned dunes and offshore sand bars landward. The swirl and wash of the tide formed them into barrier islands parallel to the coast. Fire Island, pictured here, on the ocean side of Long Island is one example, but there are hundreds of others, stretching from the Gulf of Mexico to Maine.

The first hint of human presence on Cape Cod is dated around 9,000 BP. From about 3,000 BP to the time of the first European contact, the Archaic and Woodland Period peoples (as they are now called) made use of a wide variety of exotic materials that had been deposited by ice sheets. These included flint, quartz, and other minerals used for tools, and, later, glacial clays which were used for making ceramics.

Nauset Bay, Cape Cod, Massachusetts

Nauset Bay

In 1605 AD Samuel de Champlain and Pierre du Guast, the Sieur de Monts, made several voyages of exploration down the coast of New England from bases in the Bay of Fundy. The French were searching for a place to establish their first settlement in North America, one that would avoid the kind of extremely harsh winter they had experienced the previous year. They chose Port Royal (now Annapolis Royal, Nova Scotia), and Champlain later detailed his first impressions of the Woodland peoples on the coast of Maine and here in Nauset Bay on Cape Cod. He wrote of the lush beauty of the summer landscape, the plentiful food and game, the cultivated fields, the general way of life, and the usually warm reception that his small party received. The best-known European encounter on the Cape occurred much later, in November 1620 AD; it was an unfortunate brush between the Woodlanders and the Pilgrims from the *Mayflower*.

River systems *c.*10,000 BP

Flooded Continental shelf today; Fire Island, New York

Soapstone quarry, Fleur de Lys, Newfoundland, Canada

11 Inuits

To this point we have followed the progress of people who made their way from the Alaskan Refuge through an ice-free corridor into regions of the continent south of the Laurentide ice sheet. We turn now to the Alaskan-Canadian coastal cultures—the people who stayed either beyond or near the Arctic Circle. They lived in bleak maritime zones like Ellesmere Island (main picture), just south of the permanent ice. They also lived near the southern limits of drifting ice floes, such as the Aleutian Islands in the North Pacific, and in northern Newfoundland at the entrance to the Labrador Sea.

By about 6,000 BP the coastal hunter-gatherers who had remained in the Alaskan Refuge had adapted to a maritime life style. By 4,000 BP the Aleutian tradition to the south had developed into an isolated culture. The first purely Inuit (Eskimo) culture, the so-called Small Tool Tradition, had by then appeared in the far north. The Small Tool people produced sophisticated microblades, the stone equivalent of modern razor blades, which were designed to make larger tools from bone, ivory, and wood.

As the main sectors of the Laurentide ice sheet contracted into individual sheets, these early Arctic peoples dispersed rapidly throughout the maritime zone, from Alaska and Canada to

Distribution of early Arctic cultures

1 Bering Strait
2 Skraeling Island
3 Thule
4 Fleur-de-Lys
5 South Greenland

Greenland. They made ever-increasing use of marine resources, and as they became more proficient, their "small tools" gave way to a new technology, the Dorset tradition of toolmaking, named after Cape Dorset in Baffin Island. The distinction is that by 2,500 BP the Dorset people had developed harpoons and other advanced weapons, and were now largely dependent upon the hunting of seals and walrus for survival.

Around 2,700 BP a third basic culture began to appear in the Bering Strait region, and by about 1,100 BP it had evolved into the modern Thule culture, so-called after Thule in northwestern Greenland. The Thule way of life spread rapidly; its technology included large umiaks and small kayaks, bows and arrows, dog sleds, harpoons of advanced design, whale-oil lamps, and snow igloos—ideas that were quickly adopted by the neighboring Inuit communities. Ultimately, the Dorset people of Ellesmere Island and Nares Straight were either replaced or absorbed by the Thule. By the time these early North Americans made their first contact with Europeans, the Norse, about 750 BP they were indeed Thulian. They spoke dialects of a mutually understandable language that is still used in areas ranging from Greenland to northern Alaska and coastal Siberia.

The main picture shows one of many finger-like peninsulas that protrude from Ellesmere Island into the Kane Basin of Baffin Bay. This is a region where tools of all three Arctic traditions have been found. The inset picture is of a soapstone quarry worked by Dorset people 2,000–1,500 BP at Fleur de Lys in northern Newfoundland. It is thought that the soft and easily worked rock face was first carved to leave a stubby and still-attached cylindrical shape. The cylinder was hollowed out, and the bowl-like shape was then cut from the cliff face and used for cooking. The Dorset used such soapstone utensils at sites in the Ellesmere region, and the Thule seem to have reverted to their use after experimenting with pottery.

12 Brief Encounter

At 79°N, this little-known region is at once the most northerly and one of the most noteworthy of all prehistoric sites in North America. The first crossing of Arctic peoples from Ellesmere Island in the Canadian High Arctic to Greenland took place from the region of the Bache Peninsula on the horizon. Evidence of the earliest certain contact between native North Americans and Europeans has been found here. It was the existence of "polynyas" in this region of the Kane Basin that first attracted ancestral Inuit people to the area. "North Water" polynya can be seen trailing out of the panorama to the right, off the end of the Bache Peninsula.

A polynya is an open channel in an otherwise frozen sea that either opens early and closes late in the summer or stays open permanently. Polynyas form when powerful currents pass over a shallow seafloor, in this case from Nares Strait over the Kane Basin floor. They attract every kind of terrestrial and marine fauna within reach—prey irresistible to arctic hunters. Skraeling Island in the left mid-distance is well placed for access to North Water, the largest of all known polynyas. The island was a winter settlement for Thule hunters in the 12th and 13th centuries. There are twenty-seven ruined stone structures on the island. All have deep foundations, tunneled entrances, and roofs that were supported by whalebone and covered with turf sod or animal skin.

Excavation of the Thule sites on the island has yielded a variety of Norse artifacts. These include a small piece of chain-mail of European origin, iron boat rivets, and a piece of woven cloth made with wool dated 1200–1250 AD. The cloth is characteristic of clothing worn by Norse settlers in southwest Greenland seven centuries ago. Norse artifacts dated from 1190–1390 AD have also been found at nearby locations. It is thought that the latter were either acquired by Thule traders in Greenland and brought here, or they were traded by visiting Norse.

Navigational cairns at L'Anse aux Meadows

13 Vinland Norse

This is the only authenticated Norse settlement in North America. It is at L'Anse aux Meadows in northern Newfoundland and is dated around 995 AD. Unlike Skraeling Island, described in the previous essay, there is no evidence of contact with Native Americans here. But there can be no doubt about the presence of Norsemen. In addition to many small artifacts, the foundations of buildings have been found and excavated here. Iron slag and furnace remains show that the Norse also built a smithy to smelt bog iron—technology that was unknown to Native Americans.

The Norse also constructed navigational aids such as the cairns pictured in the inset. When returning to this bay, they probably

lined up such pairs of cairns from out in the bay, in order to guide their boats safely ashore. The outline of the original Norse beach can be seen quite clearly; its elevated level is due in part to the continuing rebound of the region after the retreat of the last ice sheet, but mainly to cooler climates and lower sea level today.

Global climate averaged around 2°C (5°F) higher for centuries before and after the Norse made their first exploratory voyage from Iceland to Greenland about 980 AD. As a consequence, sea levels were several feet higher and permanent ice in the Arctic lay far to the north. Grain could be grown in both Iceland and Greenland, and there are even records of imported orange trees. In Europe vineyards grew 300 miles north of their present limit. But these warm centuries were followed by centuries of glacial advance, termed "The Little Ice Age."

Sea ice and stormier seas began to interrupt Norse communication with Greenland by the early 13th century. Fish migrations took different tracks. The last reported Norse expedition from Iceland to Greenland was made in 1347 AD—after that time the Little Ice Age tightened its grip. Life in Greenland now became impossible for Europeans. Grain would not grow; food became scarce; farm animals died. By the mid-to-late 15th century the Norse saga came to an end—almost 500 years after it had begun.

14 Mesoamericans

Tulum is a small Mayan outpost set on the edge of the Yucatan peninsula, beside the Caribbean Sea. Its setting is a startling contrast to the Kane Basin and L'Anse aux Meadows. But the story behind this tranquil scene is very long and about as unexpected as the discovery of a human population on another planet. Indeed, the only common factor between the Mayan civilization

represented here, the Thule people of the Arctic, and the Greenland Norse of European origin, is that all these three cultures existed simultaneously.

Each civilisation was essentially the product of its environment. Thus Thule people lived in small, scattered communities that had adapted to the harsh Arctic conditions. The Norse, however, failed to adapt their European culture successfully to the privations of the Little Ice Age in Greenland. In contrast, the degree of civilization achieved by the Mayan nation—and the others that existed in parallel or long preceded it in Mesoamerica—rivaled in sophistication and complexity that of contemporary Old World nations to be found in Eurasia and Africa.

Lowland Swamps to City Streets

Wild Maize

During the Wisconsin-Würm glacial age, Mesoamerica's rugged topography and tropical maritime climate produced environments ranging from lowland rainforest and grassland to arid high plateau and snow-capped mountains (Map 5). This diversity resulted in a correspondingly wide range of edible plants growing in different areas. As a consequence, early Mesoamericans were more dependent upon plants for food than their North American cousins were.

The natural human inclination was to collect the more abundant wild-grass seeds for processing into food, and to select the larger, plumper soft fruits and edible roots to eat. Highland people had a preference for teosinte, a wild grass that is thought to have mutated into wild maize; lowland people preferred cassava, a root plant that yields tapioca. Both highlanders and lowlanders gathered and ate the fruit of other wild plants wherever they could be found—beans, pumpkins, squash, and peppers in particular. Such preferences caused all these staple-diet plants to change in character as they were regularly and *naturally* selected for domestication—a process that has been described as "evolution directed by the hand of man."

Maize and cassava are the most important examples of such domestication in Mesoamerica—the equivalent of the domestication of cereals in Eurasia. The small wild wild-maize "corncob" illustrated here may have derived from teosinte. During natural domestication its kernels gradually became larger but completely enclosed by husk. The husk then needed to be stripped off before the kernels could seed, so they had to be planted by hand to ensure germination. Similarly, cassava evolved so that it produced no seeds at all at lower altitudes, and it then had to be propagated from cuttings. Continued selection of the more productive varieties of early cultivated maize and cassava led to a steady increase in food supply, which in turn both encouraged centralized populations and permitted them to grow in numbers.

Hunting and gathering was thus gradually replaced by a more sedentary life style. It is thought that pottery may have appeared in Mesoamerica by about 2,300 BC, a sure sign of community life. Although present much earlier, by 2,000 BC the crafts of cordage and basketry had evolved highly elaborated designs, and a wide variety of groundstone tools were available for food processing. Villages grew in size. The larger communities swelled into towns and finally into cities that offered a civilized urban life style with a political and religious structure. The best regional example of this process began as a settlement on the northeastern side of the Valley of Mexico near Lake Texcoco.

By the 1st century AD this river plain and lake region had become the site of the future city of Teotihuacan. By 400 AD Teotihuacan was a metropolis larger by far than most towns and cities in the Old World (Map 1). It rivaled Rome and Athens in its town planning, architecture, and population. Its rulers wielded enormous economic and political influence, mightier than that of the neighboring Mayan city-states, or the Toltec and Aztec empires that were to follow. By 500 AD, Teotihuacan was perhaps the fifth largest city in the world. Even so, the ruins of Teotihuacan today are just one shining example of a civilisation that has yet to be fully revealed by archaeologists.

The principal Mesoamerican cultures and their regional centers as they are presently understood are shown on the maps below. The Olmec are thought to have evolved the first civilization in Mesoamerica. This culture evolved spontaneously on the Gulf Coast of modern Mexico (Map 3) from about 1200 BC, long before the Maya appeared in Guatemala and Yucatan around 400 BC (Map 4). Although the Olmec people were ancestral to the Maya, they were probably unrelated, except culturally, to other and later civilizations: the Teotihuacan, 1st–7th centuries AD, the Toltec, 10th–13th centuries AD, and the Aztecs, 14th–15th centuries AD (Map 2).

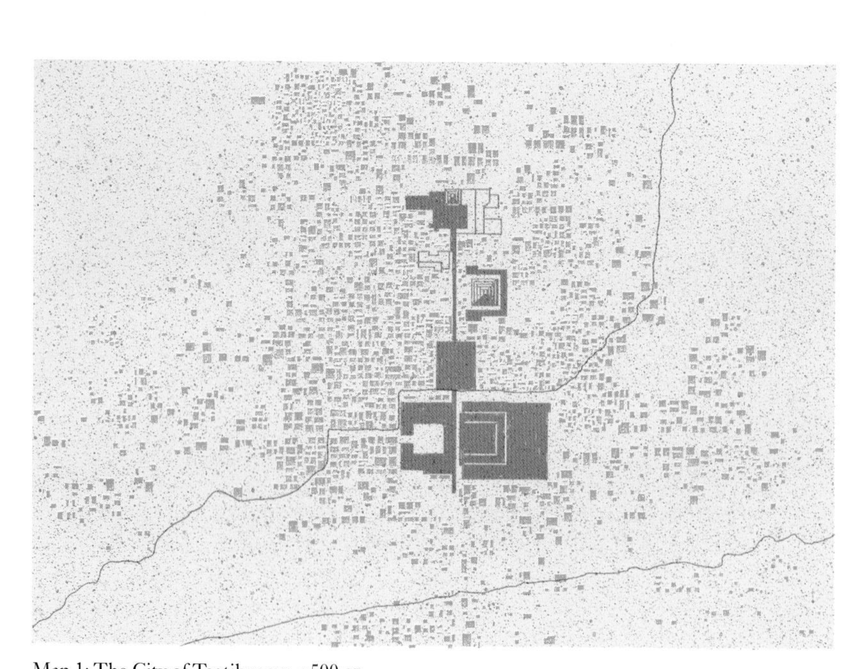

Map 1: The City of Teotihuacan *c.*500 AD

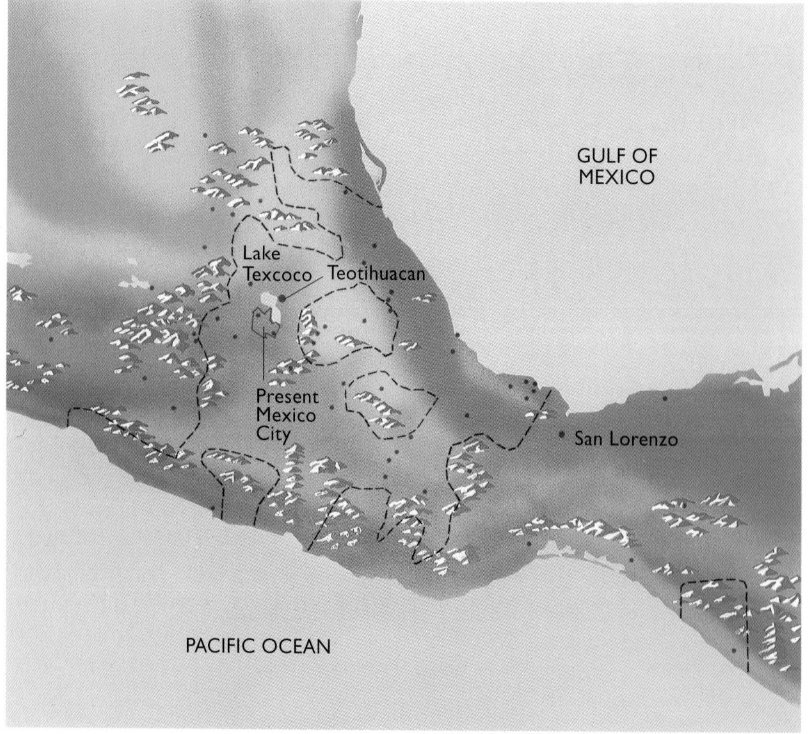

Map 2: Site of Teotihuacan *c.*500 AD and map of the Aztec Empire [14th to 15th centuries]

Olmec and Rameses II

The Olmec culture began to evolve in a swampy lowland region of southern Veracruz and Tabasco about 1200 BC (Map 3). This was about the time of the burial of the Egyptian Pharaoh Rameses II at Abu Simbel. There is a parallel between the Olmec and Egyptian cultures that is perhaps more than just a coincidence of timing. These civilizations may well be an example of the vicarian evolution of cultures—the notion that *Homo sapiens sapiens* living in widely separated places with similar environments and ecologies will tend to adapt to such conditions in similar ways. As they evolve they then develop parallel but unrelated cultures.

The parallel between the evolution of Egyptian and Olmec culture starts with the Nile-like seasonal flooding that took place in the Veracruz and Tabasco swamplands. Flooding produced rich and fertile deposits of mud along river levees. Fertile soil attracted a large and increasing population of early Egyptians in the Nile Valley and Olmec farmers in Veracruz and Tabasco. The Egyptians quarried sandstone, limestone, and granite to carve into figures with hard stone tools. Centuries after their first establishment, the Olmec sculptured likenesses of their rulers from basalt boulders with hard stone tools—their usual obsidian tools were sharp and hard, but too brittle for this task.

Like the Egyptians, the Olmec built temples, and later temple pyramids of increasingly complex design to match their intricate ceremonial rites. Both introduced a system of hieroglyphics inscribed in stone. The Maya descendants of the Olmec developed this craft into an art comparable in required skill to the finest Egyptian examples. During the development of their cultures, both the Olmec and the Egyptians established distinctive styles that characterized their regional architecture and artifacts for thousands of years. Yet there could not have been a diffusion of cultures: ancient Egyptians and ancient Olmecs were separated by a sea, an ocean, and a distance of 12,000 km (7,500 miles).

The Olmec head and the Egyptian head reproduced here are comparable in relative size, in quality of execution, and are reproduced to the same scale. The smaller portrait is of an unknown Olmec ruler, one of eight such heads discovered at San Lorenzo, Mexico. The Egyptian sculpture is of Rameses II, and was redrawn from one of the four figures guarding the entrance to Abu Simbel at Aswan.

Olmec ruler
Approx: 2.5 m (8 ft)
*c.*1,000 BC

Rameses II
(head only)
Approx. 2.7 m (9 ft)
*c.*1,200 BC

Map 3: Olmec civilization 1200–300 BC

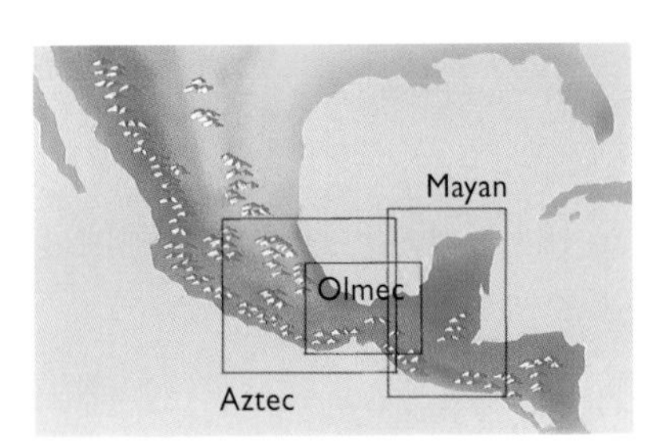

Map 4: Location key to maps 2, 3, and 5

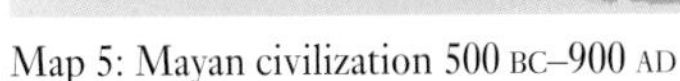

Map 5: Mayan civilization 500 BC–900 AD

15 Arawaks

The people who first began to populate the Caribbean islands lived on Cubagua Island, off Venezuela, from about 7,000–3,000 BP. These Ciboney are thought to have migrated to the Lesser and Greater Antilles from about 4,500 BP. To cross the open sea, they may have used balsa-wood sailing rafts (inset), perhaps considerably larger but nevertheless similar to rafts used today on the Island of Dominica. A large palm frond placed in a hole drilled in the central balsa log of the modern raft acts as an effective and easily managed sail.

Between 3,000–2,500 BP a second wave of people, the Arawak-speaking people of the Orinoco, migrated from their riverside retreats in South America to the Orinoco delta. From here they moved north along the mainland coast to the adjacent island of Trinidad. Over a prolonged period they crossed the open sea in dugout canoes or rafts to populate the islands of the Lesser Antilles island arc, and ultimately the Greater Antilles and the Bahama Islands. These Island Arawak largely replaced the Ciboney, who appear to have withdrawn to the far western extremities of the islands of Cuba (in the neighborhood of modern Havana) and Hispaniola as the Arawak advanced.

The Island Arawak evolved a number of subcultures, of which the most advanced, and most numerous, were the Taino-Arawak of

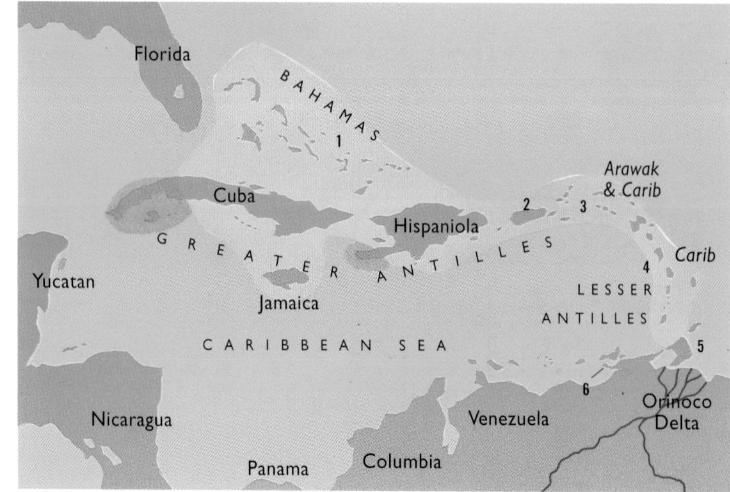

The Ciboney (blue) and Arawak (yellow-green) cultures' migration from South America from 2,500 BP

1 San Salvador
2 Puerto Rico
3 Virgin Islands
4 Dominica
5 Trinidad and Tobago
6 Cubagua Island

Cuba, Hispaniola, Jamaica, and Puerto Rico. The Taino were farmers who planted the staple foods—maize, cassava, beans, and squash—but also cultivated sweet potatoes, peanuts, pineapples, and planted fruit trees. They also grew cotton for making cloth and produced a great variety of fine-quality decorated ceramics.

The less culturally advanced Island Arawak adapted mainly to a maritime life. For instance, the Arawak who populated the small and comparatively infertile Bahama Islands depended upon canoes and rafts for fishing off coral reefs. They also used them to exploit the offshore resources of the myriad uninhabited islands of the kind portrayed in the panorama. It was these Island Arawak, living on a Bahama island they called Guanahani, who greeted Christopher Columbus and his companions when the discoverers arrived literally out of the blue on October 12, 1492. Columbus, who believed he had landed on an island off the shores of Japan (then called Cipangu) renamed Guanahani "San Salvador."

Although estimates vary considerably, the population of the Caribbean in 1492 may have been as high as six million people. A century after the arrival of Columbus, the Ciboney, the Taino-Arawak, and most of the Island Arawak were extinct.

They died from a variety of causes, but principally through virulent European diseases against which they had no immunity, through harsh labor practices inflicted upon them, and as victims of an illegal slave trade. However, a few Arawak descendants still live in a remote region of the Island of Dominica, in the Lesser Antilles' Windward Islands.

Until recently such Windward Islanders were called "Caribs," and considered to be descendants of a culturally war-like South American people of possibly cannibalistic habit. Research has now shown that they are indeed descended from Island Arawak. Columbus was initially responsible for an incorrect notion about such islanders through his transcription of the Taino-Arawak word for the more aggressive islanders (*kanibna*) into *caniba* (Sp. *canibalis*, cannibal). This was later further corrupted into the general term *Caribes,* and used loosely in the 16th century as a general term of description to describe all aboriginals who resisted Spanish conquest. Aboriginal people were widely proclaimed to be of "human flesh-eating habit" and "enemies of Christians"—a specifically political, not a cultural or factual, viewpoint. In turn, the word "Caribbean" was derived from *caribe* and in contemporary usage this word has also become a synonym for "tropical paradise." In an ironic twist "paradise" is exactly what the Caribbean became for the hundreds of thousands of Arawaks in Columbian times, as they died at the hands of their genocidal conquerors.

EPILOGUE

BETWEEN TWO WAVES OF THE SEA

THIS IS SAGRES POINT ON CAPE ST. VINCENT IN PORTUGAL, AND WE ARE LOOKING INLAND TOWARDS THE SMALL GARRISON TOWN OF VILA DO INFANTE. WHEN THE FORTRESS SCHOOL OF NAVIGATION ON THE CLIFF-TOP TO THE RIGHT WAS CONSTRUCTED BY HENRY THE NAVIGATOR IN 1418, THIS COASTLINE WAS BELIEVED TO BE THE WESTERN LIMIT OF THE WORLD. TO HENRY AND HIS NAVIGATORS THE NORTH ATLANTIC ROLLERS WE SEE BENEATH US MARKED THE EDGE OF THE UNKNOWN AND FEARFUL OCEAN SEA THAT ENCOMPASSED THEIR WORLD.

Sagres Point and
Vila do Infante,
Cabo de São Vicente,
southwestern Portugal

BETWEEN TWO WAVES OF THE SEA
c. AD 1500

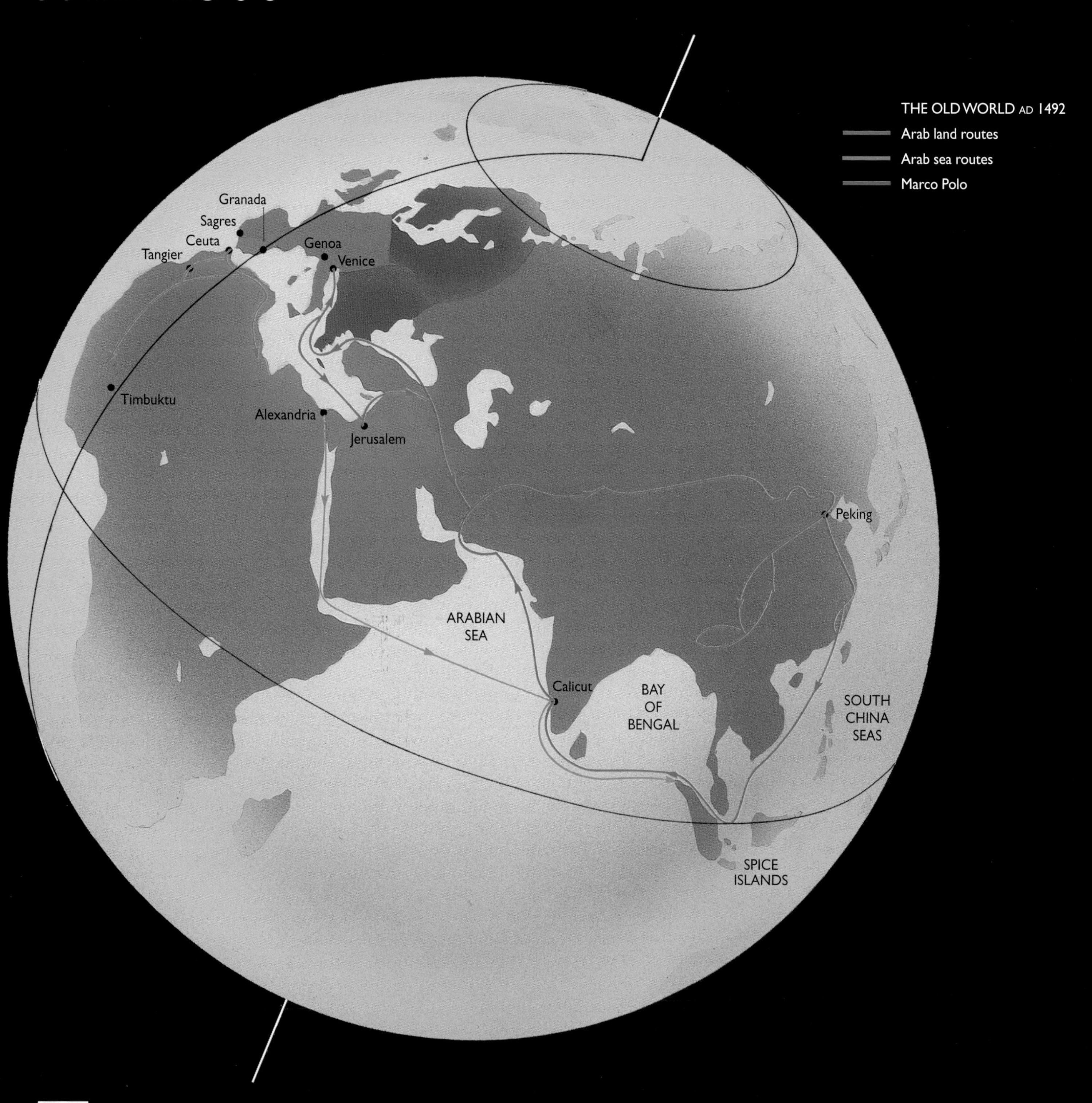

This globe shows the Old World in modern guise. Those parts of the Earth's surface that were generally known to Europeans by the 15th century are detailed, and those generally unknown are simply shaded out. However, the cities, localities, and routes identified are limited to those discussed in this introduction and in succeeding essays.

Five Centuries Ago

To this point we have followed the 700 million years of tectonic and biological changes that led to the geological and geographical present. The time scale involved is virtually unimaginable. If we were to assign one inch to every million years, the scale would require almost sixty feet. In the last chapter we focused down to human technological adaptation to the conditions of an ice age, a span that barely stretches a tenth of an inch on the one-inch per million-year scale. The last five hundred years from the Great Discoveries to the Third Millennium would occupy just one two-thousandths of an inch on the scale. To view this in more philosophical terms, perhaps we should paraphrase part of T.S. Eliot's *Little Gidding* a second time. Maybe the whole human experience of shaping the modern world to our liking took place "between two waves of the sea" in an ocean of time?

By the 13th century AD Afro-Eurasians in the Mediterranean sphere of influence visualized the world as the single continent described by Claudius Ptolemaeus of Alexandria (*c.* AD 90–168). Ptolemy was a Roman geographer and astronomer-extraordinary whose work was sacrosanct in medieval times. It remained sacrosanct until the Ptolemaic concept of a spherical universe with a spherical Earth at its center was ultimately displaced by the Copernican system, which recognized that the Earth moved round the Sun. Such acceptance, almost 150 years after Copernicus's death, followed the publication of Sir Isaac Newton's *Principia* (1687). People had found it easy enough to visualize the Ptolemaic universe; they could see it rotating over their heads from the supposedly fixed surface of the Earth on which they stood. But they found it difficult to visualize the geography of the Earth's surface—even as we find it difficult to imagine the Universe as a whole. Although Ptolemy had suggested how the difficulty of the Earth's geography could be overcome by disciplined mapping, his concept of latitude and longitude was not revived until the 16th century.

Nevertheless, by the beginning of the 14th century the East-West extent of the Afro-Eurasian Old World had been defined by the limit of Marco Polo's travels from Venice to the Orient 1271–95. As a result, Ptolemy's single continent was now visualized more like a table-top Pangea. The northern extent of the landmass was sketchy and the southern limit of Africa was completely unknown to Europeans—literally *terra incognita*. But it was generally assumed that the shape of the supercontinent, although amorphous, was roughly disk-like. It enclosed the Mediterranean Sea to the west and the Arabian, Bengal, and China Seas to the east. The contiguous whole was surrounded by a much feared ocean of unknown size—the *only* ocean on Earth, in essence a Panthalassa. Many names were used to describe this ocean: "Oceanus" (from Greek mythology), the "Green Sea of Darkness" (at Sagres), or the "Ocean Sea" (the term used by Columbus).

Despite the enormous distances, East-West overland trade across Africa and Eurasia was brisk: from China via Mongolia and Turkestan to Persia, and so to the Mediterranean and Southern Europe; and from Turkestan via Russia to Northern Europe. Brisk, that is, until after 1368, when Mongol rulers began to lose control of their Far Eastern Empire, and the trade route pioneered by Marco Polo was gradually abandoned.

By the early 15th century Europeans, particularly the Portuguese, confined to the Iberian North Atlantic seaboard, were increasingly preoccupied with the mounting Muslim threat to their future prosperity. Muslims dominated southern Iberia from their Kingdom of Granada. They held the fortress Rock of Gibraltar, and on the southern side of the narrow strait the Moors (Berber-Arab Muslims) had fortified Ceuta (adjacent to Mount Hacho in northwest Africa). Thus both Pillars of Hercules at the entrance to the Mediterranean Sea, and the Atlantic coast of Morocco to Tangiers and beyond, were under Muslim authority. Furthermore, the Moors, who ruled Morocco and populated North Africa, also governed the profitable trade routes across the Sahara Desert to the Ivory Coast in equatorial West Africa. Islam thus dominated the Mediterranean and controlled the heart of the known world. In addition, Arabian traders had a stranglehold on Far Eastern sea trade. They monopolized the sea routes to the East from the Red Sea (via Alexandria) and the Persian Gulf (via Damascus) across the Arabian Sea and the Bay of Bengal to the South China Sea.

There were just two ways for the Christian world to circumvent the geographical strictures of the Islamic world and to trade directly with the Orient. One was to follow the coast of Africa all the way around to the Arabian Sea. The other was to sail due west across the Ocean Sea from Iberia to the Orient. But there could be no significant ocean exploration of any kind until suitable vessels were designed to ensure a safe return of venturing ships and their crews. Nor could long voyages of exploration be undertaken until there were reliable techniques for charting unknown coasts and for navigating an unknown ocean.

As a result of the advances that were made in the 15th century, the true nature and geography of the Earth's surface began to dawn upon Europeans. This awareness led to the discovery of a totally unexpected extension of the previously known limit of the Western World. It was to give Europeans an enormous advantage over other peoples in the Old World community.

1 Gauntlet Thrown

This 14th-century Dominican abbey at Batalha in Portugal contains the tomb of Henry the Navigator (1394–1460) and that of his parents, King John I of Portugal and Philippa of Lancaster, his English Queen. The inset is of a Manueline-style window at the Convent of Christ at Tomar, the headquarters of the chivalrous Portuguese military Order of Christ of which Henry was Grand Master from 1418 until his death. These places are commemorative of one of the most significant periods in the history of the Western World.

From 1095 to 1291 Christians of Western Europe had fought for, but had failed to capture, the Holy City of Jerusalem from the Muslim powers. The hated "infidels," already controlling trade routes to the Orient, were now free to dominate the Mediterranean world and to threaten Europe. The Order of Christ was founded in Tomar (1314–20) to replace the ancient Order of the Knights Templar of England, who had sought refuge there after their resounding defeat in the last Crusade: the new holy order inherited the Templar wealth and tradition. Subsequently, in 1418, and in keeping with that tradition, the order was able to finance its new grand master's innovative scheme to break the strictures of Muslim dominance: the founding of a school of navigation at Sagres.

Since wresting Portugal from the clutches of Castile in 1385, King John of Portugal had become ever more concerned about the threat from neighboring Granada, a Muslim kingdom. Significantly, Hispanic Muslims controlled Gibraltar and Arab-Berber Muslims held Ceuta in Morocco—they thus commanded both Pillars of Hercules at the entrance to the Mediterranean. In 1415

Prince Henry of Portugal led an attack on Ceuta and won his knightly spurs for its conquest. He was appointed governor and was astonished to discover the prosperity of the place. This was the consequence of a lucrative overland trade in gold, ivory, and slaves across the Sahara Desert between Ceuta and West Africa.

In 1418 the Muslim rulers of Fez in Morocco and the Alhambra in Granada attempted to recapture the fortress city. Although they were repelled, it was obvious that the Christian foothold on the African continent was tenuous. Spurred on by his success at Ceuta, Henry evolved an ambitious scheme: to explore the coast of Africa so as to outflank his enemies by attacking from the rear; to divert the trans-Saharan trade at its source for the benefit of Portugal; and, later, to find a route to the Orient by circumnavigating Africa.

Early 15th-century Manueline window, Convent of Christ, Tomar, Portugal

Batalha Abbey, Batalha, Portugal

Early bitácula of the type used at Sagres

Mariner's astrolabe (early 17th century)

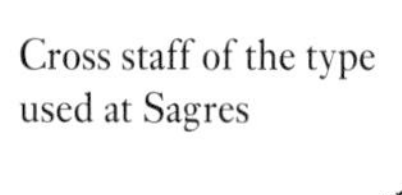

Cross staff of the type used at Sagres

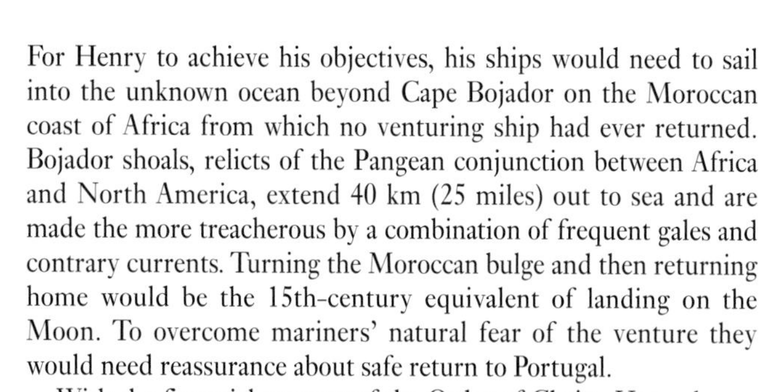

Turning Point

For Henry to achieve his objectives, his ships would need to sail into the unknown ocean beyond Cape Bojador on the Moroccan coast of Africa from which no venturing ship had ever returned. Bojador shoals, relicts of the Pangean conjunction between Africa and North America, extend 40 km (25 miles) out to sea and are made the more treacherous by a combination of frequent gales and contrary currents. Turning the Moroccan bulge and then returning home would be the 15th-century equivalent of landing on the Moon. To overcome mariners' natural fear of the venture they would need reassurance about safe return to Portugal.

With the financial support of the Order of Christ, Henry began to construct a wall across Porta de Sagres, the narrow, windswept, southwestern tip of Portugal: the structure was fortified in case of Muslim attack. Behind the wall Henry built a chapel, school buildings, and barracks to house military and students (top left, panorama). He added teaching aids such as a giant wind rose (foreground), which indicated prevailing winds at sea and marked it with the points of the compass. To simulate seagoing conditions in this windswept location, Henry built a platform on a high building overlooking the wind rose (top right).

According to archivists at Vila do Infante, the platform was fitted with the stem of a ship rigged with lines and pulleys, and could be made to pitch and roll. From this structure it is thought that seamen were taught how to take wind direction into account when using a bitácula, a primitive tub-like water compass set upon the rolling deck. It is also thought that the marine astrolabe was developed here, an adaptation of the more complex astrologer's astrolabe. The simplified instrument was used for measuring the height of stars in the night sky—particularly the "fixed" star over the then theoretical North Pole, the Pole Star. The instrument was suspended by a ring in one hand so that the fixed crosspiece stayed level, while the movable crosspiece was adjusted to line up with the Pole Star. The nearer one's position was to the Equator, the smaller the angle of the adjustable crosspiece relative to the fixed crosspiece, and the nearer one's position was to the North Pole, the greater the angle. Thus, from the angle measured on the astrolabe one could judge a ship's relative distance from the Equator or North Pole in degrees of latitude.

The mariner's astrolabe was later replaced with the simpler cross staff, a graduated staff fitted with a moving crosspiece. The crosspiece could be slid along the staff and locked in place. The staff was held to the eye, and the crosspiece moved until its extremities coincided with the horizon and the Sun. In this way the midday altitude of the Sun could be measured on a particular date. Longitude could not be measured except by dead reckoning. In this method, a bitácula was used to record direction, and distance was estimated by plotting the direction of travel for 24 hours on a chart. Time was measured by turning narrow-necked hour glasses. By estimating the average speed of the vessel one could calculate the distance sailed by the ship from one day to the next. Such linear progress was recorded by projecting successive lines on a chart, lines that took the ship's latitude into account.

Sagres students were trained in the use of charts for recording wind direction and currents. They learned to estimate distance and

Wind rose and compass points, Henry the Navigator's school of navigation, Sagres

The Azores Arc and Sahara trading routes (15th century)

1 Cape Verde Islands
2 Canary Islands
3 Azores
4 Sagres
5 Marrakesh
6 Timbuktu

record direction, and to use mathematical tables for measuring latitude from the angles of declination of the Sun or the stars. Such charts were taken on every expeditionary voyage that departed from the nearby port of Lagos or from Lisbon, then returned with notes added for master cartographers to interpret and to register data. Slowly, a picture of the African coast, offshore islands, North Atlantic currents, and prevailing winds began to emerge. Courses were set using the Canary Current and easterly winds for outward-bound ships, and the Azores Current and westerly winds for ships returning to Portugal from the mid-Atlantic. This route became known as the "Azores Arc."

The long reach to mid-ocean, during which the Azores were discovered in 1432, and the return voyage to Lisbon or Lagos, was made necessary by the cumbersome but seaworthy merchant ships of the day. The heavy, square-rigged carrack could not sail close enough to the wind to explore the coast of Africa and to turn Cape Bojador. As exploration proceeded, ship design was continuously studied, with the result that a ship able to tack across winds and currents was developed from a Mediterranean fishing-boat design. That ship had lateen sails and was called a "caravel."

The caravel was an explorer's ship, the key to turning the Cape. The Sagres school modified the original foremast and mainmast lateen-rigged Mediterranean vessel. They fitted a deck, a low sterncastle, and either a combination of foremast, mainmast, and mizzenmast (as shown on some reconstructions), or just a mainmast and mizzenmast alone. However, although the caravel could sail close to the wind and could gain more seaway than any other ship of its time, it was cramped. It simply did not have the cargo space for prolonged oceangoing voyages.

Henry ultimately failed in his attempt to outflank his enemies: his attack on Tangiers in 1437 was a disaster. But Cape Bojador was turned in 1434. By reaching beyond Cape Verde on the West Coast of Africa in 1456, Henry's ships were able to divert the Guinea trade in dyes, wood products, sugar, wine, spices, gold, ivory, and slaves. They switched this ancient trade from its traditional trans-Saharan route to Morocco and conveyed it directly to Portugal by sea. The implication of this initiative went far beyond securing the West African trade. Portuguese mariners rounded the horn of Africa in 1487 and reached India in 1498. This was followed by Malaysia, China, and Japan and even, due to an error in navigation, a first sight of the coast of Brazil in 1500.

Since the school at Sagres was run under the strictures of a holy order it offered monastic conditions of discomfort and discipline. Even so, it remained the center of maritime science until Henry's death there in 1460 (it closed in 1470). Together with the associated town of Vila do Infante, Sagres became a repository for sea and coastal charts, and it provided a means of training navigators, astronomers, cartographers, instrument makers, and shipbuilders. Its graduates and their pupils changed the course of human history and the destiny of the Western World.

A certain irony threads through all this achievement. Many of the technical advances gained at Sagres actually had their roots in Islamic science and technology—both were far more advanced than their Western counterparts at that time.

Reconstruction of a 15th-century caravel—one of many interpretations

2 Muslim Province

By the early 9th century AD the Muslim Empire extended from Middle Asia to the Atlantic coasts of Iberia and Morocco. Ultimately, the Kingdom of Granada at the foot of the Betic Alps became the focal point of Muslim power in the West. A fortress, palaces, and gardens to the left and to the rear of the Alhambra, pictured here, were built by Ibn al-Ahmar and his successors between 1238 and 1358. Architecture and interior design, in particular decorative tiles and elaborate carvings (inset), were the most prized expressions of Islamic art, and the Alhambra is the finest Western example. The Moorish style was combined with the Romanesque to form the modern expression of Spanish Mediterranean architecture in Europe—and in the New World.

Far left: Three-masted ship *c*.1425, reconstructed from Hispano-Moorish bowl at the Victoria and Albert Museum, London.

Left: Early use of the triangular sail, tomb of Amenophis II, 14th century BC

Above: Arabic astrolabe, by Ibrahim al Sahli, Toledo, Spain, AD 1068

Arabian Influence

The Islamic world contributed far more to Western culture and technical advance than a style of architecture. By the 5th century seamen of the Red Sea and Persian Gulf were traversing the Arabian and Bengal Seas to the China Seas and exploring the coast of East Africa [see page 306]. Muslim scientists systematized their knowledge of astronavigation, sea currents, tides, prevailing winds, and seasonal weather, as well as their daily observations of star altitudes and solar declination, recording information in astronomical tables and almanacs for navigators. Mathematical tables required a practical form of numbering, and this led to the development of arithmetic, algebra, trigonometry and logarithms. Although Arabic numbers and methods of calculation were available in the Mediterranean from about the 9th century, Western mathematicians were still using the less practical Roman method of calculating numbers.

Arising from Arabian advances in mathematics and astronavigation, an astrolabe for reading off the position of the constellations was developed from the original Greek astrolabe in AD 771 by Al Fazaf. The instrument was graduated according to the Arabic calendar. The advanced and complex Hispano-Moorish astrolabe illustrated above was developed in Toledo in 1068. This was matched by yet another very early Arabian navigational aid, Al Khwaflzomi's Staff—a form of quadrant for use at sea where the astrolabe was difficult to use. The quadrant was abandoned for a logarithmic method used widely in the East but not adopted by the West. Nevertheless, Arabian numerals and navigational aids were later adopted or independently developed by Henry the Navigator at Sagres. Possibly these included the bitácula: the compass was in use in the China Seas from about AD 300 and the Muslims traded with China.

The earliest-known Arabian sailing vessels were Egyptian: the illustration is from a hieroglyphic on the tomb of Amenophis II and is dated 14th century BC. The raft was fitted with what appears to be a triangular sail, perhaps a forerunner of the lateen-type rig that became characteristic of craft on the Nile and in the Mediterranean Sea. This was the style that Henry the Navigator adopted for his caravels almost 3,000 years later. However, the square-rigged *three*-masted ship proved to be the greatest maritime invention of all time, with a square-sailed mainmast for power, a foremast fitted with a spritsail, and a mizzenmast fitted with a lateen rig for controlling and steering the vessel. Remarkably, one of the earliest-known illustrations of such a ship appears as a decoration on the interior surface of a Hispano-Moorish bowl. This does not mean that the Moors developed the three-masted ship, but it does infer that they were at the forefront of sailing technology. The curvature of the bowl's interior distorts the original Moorish illustration and has been removed in this reproduction.

Western Gateway

The Alhambra was captured by the Spanish in January 1492. Five years before, in 1487, the Portuguese had rounded the Cape of Storms (renamed the Cape of Good Hope) and thus had beaten Castile in the race to discover a gateway to the Far East. The Portuguese had promptly lost interest in Christopher Columbus, who had offered them his ideas for a western route to the Indies, and his services to put them into effect. Having been refused by the Portuguese and other European courts, Columbus tried to sell the Spanish court the prospect of his exploring his optimistically short route to the Far East at modest financial risk to his sponsor—but Queen Isabella and her advisers were also skeptical.

One of many authoritative views of the cause of the Queen's sudden change of mind in 1492 is that the capitulation of the Alhambra and the end of Muslim dominance of Andalusia had provided a great cause to be celebrated by some such grand gesture. By his own account (reproduced opposite) Columbus was in Granada at the time of the capitulation. He saw the royal banners of King Ferdinand and Queen Isabella placed by force of arms on the towers of the Alhambra. Later, Columbus described the background to this view in a prologue to his journal of the first successful crossing of the Atlantic. The journal was completed while homeward bound off the Azores in mid-February 1493.

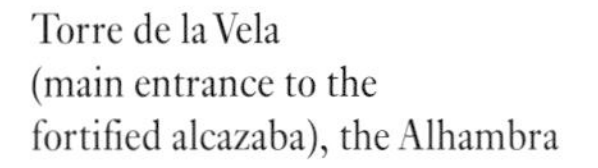

Torre de la Vela
(main entrance to the
fortified alcazaba), the Alhambra

"Most Christian and most exalted and most excellent and most mighty princes, King and Queen of the Spains and of the islands of the sea, our Sovereigns: Forasmuch as, in this present year of 1492, after that Your Highnesses had made an end of the war with the Moors who reigned in Europe, and had brought that war to a conclusion in the very great city of Granada, where, in this same year, on the second day of the month of January, I saw the royal banners of Your Highnesses placed by force of arms on the towers of the Alhambra, which is the citadel of the city, and I saw the Moorish king come out of the gates of the city and kiss the royal hands of Your Highnesses and of the Prince, My Lord, and afterwards in that same month, on the ground of information which I had given to Your Highnesses concerning the lands of India . . . our Highnesses, as Catholic Christians and as princes devoted to the holy Christian faith and propagators thereof, and enemies of the sect of Mahomet and of all idolatries and heresies, took thought to send me, Christopher Columbus, to the said parts of India ... [You] ordained that I should not go by land to the eastward, by which way it was the custom to go, but by way of the west, by which down to this day we do not know certainly that any one has passed; therefore, after having driven out all the Jews from your realms and lordships, in the same month of January, Your Highnesses commanded... [and] I departed from the city of Granada on the twelfth day of the month of May in the same year of 1492, on a Saturday, and came to the town of Palos, which is a port of the sea, where I made ready three ships, very suited for such an undertaking, and I set out from that port, well furnished with very many supplies and with many seamen, on the third day of the month of August of the same year, on a Friday, half an hour before the rising of the sun, and I steered my course for the Canary Islands of Your Highnesses, which are in the Ocean Sea, thence to set out on my way and to sail until I should arrive in the Indies ..."

Christopher Columbus
Ship's Journal, 1493

Aboard the brigantine *Romance*, Lesser Antilles

3 Ship Shape

In the prologue to his journal, addressed to Ferdinand and Isabella in 1492, Columbus had written of "three ships, very suited for such an undertaking [his forthcoming voyage]" and of steering a course for the Canary Islands, "which are in the Ocean Sea, thence to set out on my way and to sail until I should arrive in the Indies." The ships were of course the *Santa Maria*, the *Pinta*, and the *Niña* illustrated here. The *Santa Maria* was built as a square-rigged ship. Columbus converted the *Niña* from a lateen-rigged caravel to a square rig in the Canary Islands; the *Pinta*, a second caravel, already had been converted to a square rig when she was purchased.

By committing himself to square-rigged ships, Columbus was being prudent. Before launching into unexplored reaches of the Ocean Sea beyond the known Azores Arc he was ensuring the best chance for a safe return and, like other oceangoing mariners of his day, he obviously had greater faith in the stable center-powered square rig than in the less stable off-center lateen rig—rather like the modern preference for four-wheeled over three-wheeled vehicles.

Columbus was later very critical of the *Santa Maria* as having too deep a draught and for being unweatherly. But this might just have been a jaundiced view. The *Santa Maria* struck a reef off Hispaniola late at night on Christmas Eve 1492 after the crew had been celebrating ashore. The ship was lost and 39 crew members put back ashore to establish an improvised settlement; there were no survivors when Columbus returned a year later. After the wreck of his flagship, Columbus transferred to the caravel *Niña*, on which later he lavished praise. Nevertheless, it was the three-masted and the square-rigged ship that revolutionized the maritime world and not the easily maneuvered yet unstable lateen-rigged caravel.

It was the Mediterranean climate with its generally light and variable winds that permitted the broad use of the lateen sail. The great advantage of this rig was that it permitted a ship to be sailed much closer to the wind than was possible with a square sail. It also

Santa Maria

gave mariners a wider range of options in choosing a course—as Henry the Navigator had discovered. The sail was set from a lateen yard suspended at an angle on the leeward side of a mast and could receive the wind on either side. But to change to another tack, the yard had to be swung almost vertically: in rough weather this was a hazardous, even dangerous operation. The lateen provided "off-center" power to drive the vessel, so that when running before the wind, there was always a danger that a sudden heavy gust would capsize the vessel.

In direct contrast, the square sail rig was developed in northern seas where gale-force winds are commonplace and often prolonged. Under such conditions it is essential to ensure that as much of the wind's force as possible is applied over the ship's center of gravity. A square sail is therefore set from a yard suspended from the forward side of a mast and receives the wind evenly. Although the angle of the sail may be changed relative to the angle of the ship's progress so as to continue to receive the wind squarely, wind is only received on one side of the sail. The center of power in the boat is always directly above the ship's keel.

It follows that the main difference between northern and Mediterranean ships in the early 15th century lay in the convenience and power of the square sail in heavy weather, and the easy general management of ships fitted with lateen rigs. But the shape of the northern carrack hull allowed larger and bulkier cargoes. Their size varied but the principle of their building remained practically unchanged through the centuries; their bottom planking was laid edge to edge over a basic structure, much like the framework of a Newfoundland fishing boat pictured here. Their steep, high sides were then built with overlapping planks from the turn of the bilge. Carracks, therefore, cut through the waves and the ship flexed with the battering of the sea. They sailed faster than caravels and were generally more seaworthy, but these advantages were reduced by their inability to sail close to the wind.

Ship's structure, St. Anthony, Newfoundland, Canada

Pinta

Niña

Illustrations of Columbus's ships based on the work of José Maria Martinez-Hilago, Museo Maritime, Barcelona

Sailing Fair and Square

At some unknown point in the 15th century, the square- and lateen-rig designs were merged—possibly an accomplishment of Genoese shipbuilders. The latter catered to shipowners who traded in bulk cargo—grain, alum, or wool—and needed capacity. Competitive Venetian shipbuilders catered to merchants who traded in luxury goods—spices, silks, and gemstones—shipowners who were not interested in capacity. Whatever the origin, the result is evident: merchant ships were now designed with a square-rigged mainmast and with a single lateen sail on a small mizzenmast. The earliest known illustration of such a ship appears as a decoration on the interior surface of a Hispano-Moorish bowl dated *c.*1425 [page 314]. The driving power of this ship and its successors, including the brigantine in the panorama, which is about the size of the *Santa Maria*, was derived from the square sail on the mainmast—which was stepped on the keel. The foresail and mizzensail produced leverage fore and aft of the ship, and in early ships are thought to have been aids to steering.

The cargo capacity of three-masted square-rigged ships must have proved attractive: shipowners now demanded even more cargo space. As mainsails grew larger the mainmast had to be lengthened. Since there is a limit to the height of trees suitable for making masts, extra length was achieved by binding together a number of shaped mast timbers of sufficient thickness and length. To allow for the arc of the tiller, the mizzenmast had to be placed forward of the stern post, and this too was stepped on the keel (see picture on previous page). The mizzen could therefore be a substantial mast that could be made to carry a larger lateen sail—larger than that required for steering alone. But the foremast could only be a light spar bearing a small square sail, because it had to be stepped as far forward as possible. The curve of the ship's stem necessitated that this be between decks on a beam below the forecastle.

All masts were supported by stays and shrouds, but the fashion of building forecastles and sterncastles caused major problems. Their wind resistance produced a drift to leeward that even the

lateen sail had difficulty in countering. The solution was to make the mizzen even larger relative to the size of the foresail. This is thought to have upset the power balance of the ship, and it was corrected by the invention of the spritsail, a square sail set from a yard slung on the underside of the bowsprit. The spritsail was found to exert a leverage disproportionate to its size, which was a bonus; but it also had the effect of pulling the bows lower into the water, consequently reducing the speed and the maneuverability of the ship. This in turn was countered by the introduction of the main topsail,which lifted the bows slightly and eased the passage of the ship's hull through the water. No doubt to the original designer's surprise, the topsail was found to produce more continuous power than the mainsail. This was because the wind was steady 12 to 15 m (40 to 50 ft) above sea level, but variable nearer to sea level in an ocean swell or in choppy seas.

With the invention of the topsail in the late 15th century, the primary evolution of the three-masted square-rigged ship was complete in all essentials. Maritime historians consider this one of the most significant technical advances there has ever been. It enabled the global ocean to be thoroughly explored; it permitted the development of world trade; and it shifted the balance of population around the Atlantic. The *principle* of the three-masted square-rigged ship remained almost unchanged for 350 years after the time of Columbus and his discoveries. It was indeed this design that had made his ocean crossing and safe return possible. After the careless wrecking of the *Santa Maria*, if Columbus had become dependent upon the customary *lateen*-rigged caravel instead of the *square*-rigged *Niña*, on his return voyage he may not have survived the North Atlantic gales of January-February 1493. In the event, he and his crew may well have owed their lives to foresight—the rerigging of the *Niña* before embarkation from the Canary island of Gomera on September 6, 1492.

4 False Premise

It is thought likely by most maritime historians that Columbus chose his heading from the Canary Islands to the west after reading *The Travels of Marco Polo*. Columbus concluded that the Canary Islands are on the same parallel of latitude as Cipangu, the name for Japan in the 15th century. In fact, Tokyo is near 35°N latitude and the Canaries are near 28°N latitude. From his other studies it seems that Columbus also formed an extremely optimistic view of the narrowness of the Ocean Sea that he believed separated the two island archipelagos. He calculated this to be about a quarter of its actual distance—a little less than the width of the North Atlantic Ocean at 28°N latitude.

According to his journal, Columbus sailed due west from the Canary Islands until he was blown off course by northwesterly winds, and a few days later followed a false sighting of land. After compensating for loss of heading, he resumed a more southwesterly course than previously, as if trying to ensure a landfall in the Indies, south of Cipangu. Having reached Asia by crossing the narrow Ocean Sea, his intention was then to return to Iberia using the favorable winds and currents of the Azores Arc. And if we take Columbus's viewpoint we can see that everything worked out more or less as he had imagined. The North Atlantic was wider than his expectation of the width of the Ocean Sea, and he worried about this. But Cipangu, in the guise of the Bahama Islands, was roughly where he expected Cipangu to be. Indeed, a Spanish map, dated 1500, shows two separate islands in Cipangu, Samana to the south and Guanahani to the north: either could have been his point of landfall. But which? Or perhaps some other Bahama island was the sight of that memorable first footing? The subject is extremely controversial because none of the candidate islands exactly fits contemporary description. However, the majority scholarly view favors the island the Arawak called Guanahani, pictured here, as the island on which Columbus landed and renamed San Salvador.

Possible first sighting of the New World on Friday, October 12, 1492, the island of Guanahani, renamed San Salvador by Christopher Columbus

Columbus memorial, San Salvador, Bahama Islands

This view is supported by the account of Antonio de Herrera, a 16th-century Spanish court historian, which describes the renaming of Guanahani in 1492. Columbus himself wrote of a calm harbor on the lee side of the island capable of holding "all the ships in Christendom"; although in this author's opinion exaggerated, this is a true description of the lee side of San Salvador. In fact, glass beads, a copper-alloy coin of the times, and a bronze buckle have been found inland from the memorial cross (inset) that now marks the shore where Columbus is thought to have landed. Columbus also described the "many waters" of San Salvador's interior, such as the salina that can be seen in this aerial panorama. The aerial panorama was taken approaching the point thought to be that first sighted from the *Pinta* at 2 am on Friday, October 12, 1492—a few hours before the course of history in the Western World was to change irrevocably.

5 Ymago Mundi

It is difficult to erase the picture of the globe we have in our minds and see it through 15th-century eyes. We live on a much-traveled Earth; we take its geography for granted—indeed, we even know how the Western World may have been formed during the last 700 MY! But for a moment we have to set our sense of the modern world aside so that we can imagine the New World discoveries in the perspective of their time.

To assist with this quite considerable undertaking, the modern coastlines portrayed on the small globe on the right have been drawn so that the coasts of Asia and Europe are opposed across a supposed Ocean Sea, with Cipangu on the same line of latitude as the Canary Islands. This projection corresponds to a Western view of the famous Martin Behaim globe of 1492, now at the Germanisches National Museum, Nuremberg, and shows how Columbus viewed the geography of the Ocean Sea. It also helps us to see why no one in Columbian times could have visualized two major continents stretching down the center of the Ocean Sea from pole to pole. And, as we shall see, even if their existence had been postulated, there was simply no room for them in people's minds.

Columbus approached various European monarchs, in addition to those of Portugal and Castile, in order to obtain backing for a transoceanic expedition to the "Indies," meaning, vaguely, the general region of the Indian Ocean and the South China Seas. He based his arguments on statements he had read and marked in a number of books he owned, particularly *Historia rerum ubique gestarum* (Aeneas Silvius Piccolomini: 1471) and *Ymago mundi*, a world geography written about 1410 by the French theologian-astronomer Pierre d'Ailly. Columbus made his most extensive annotations in his copy of the latter. He underlined passages with different pens and inks and added comments by the text.

Pierre d'Ailly based his ideas about the extent of Eurasia on the work of Marinus of Tyre (*c.* AD 100). Marinus accepted the idea of the Earth as a sphere and had estimated the extent of Eurasia to be 225° out of the encompassing 360° of a line of latitude on a globe. Ptolemy (*c.* AD 150) had estimated only 177° for the extent of Eurasia. Thus Marinus's model suggested a *narrow* ocean between Western Europe and East Asia, whereas Ptolemy's conclusion implied a *wide* ocean: therefore Ptolemy was a strong voice against Columbus's proposal for a western voyage, for it would be too hazardous if the Ocean Sea was so wide. It is significant that Columbus did not write *any* notes at all in his copy of Ptolemy's *Geography* (a newly translated edition dated 1479), although this book was held in high regard by the very court advisers he had to convince. On the other hand, Marinus of Tyre and d'Ailly's *Ymago mundi* not only supported Columbus's intentions, they demolished Ptolemy's science.

In one margin of a heavily annotated page of *Ymago mundi* Columbus wrote, "The eastern extremity of the habitable earth and the western extremity of the habitable earth are near to each other. The part that separates them is a small sea." An abbreviated section of that heavily marked page in Columbus's copy of *Ymago mundi* reads:

In his book of the disposition of the sphere, Ptolemy holds that about a sixth part of the earth is habitable with respect to water… Aristotle… towards the end of his book of the sky and the earth, [says] that the habitable area is greater than a quarter ... [and] that the sea is small between the western extremity of Spain and the eastern part of India ... Moreover, Seneca in the fifth book of the things of nature says that this sea is navigable in a few days if the wind be favorable. And Pliny teaches in the second book of natural history that one can navigate from the Arabian Gulf to the Columns of Hercules in no very great time ... For these and many other reasons ... some conclude that, apparently, the sea is not so great that it can cover three quarters of the earth. We can call on the authority of Esdras in his fourth book to support this, who says that six parts of the earth are inhabited, and the seventh is covered with the waters; the authority of which book the saints held in reverence and confirmed the sacred truths by it.

The last sentence expresses a key medieval ecclesiastic doctrine and belief: it refers to Esdras 6:4 in the Hebrew Bible, which was incorporated in the Christian Bible's Old Testament, but which is not included in modern Bibles. The verse reads (translated into modern English):

On the third day you commanded the waters to be gathered together in the seventh part of the earth, but six parts you dried up and kept them so that some of them might be planted and served before you.

But there was another book, Marco Polo's *Il milione* (first printed in 1477 and known as *The Travels of Marco Polo*); Columbus owned a copy of a later edition of this book which he marked extensively. Marco Polo's book was the source generally used for estimating the eastward extent of China. Even before Columbus, one Paolo dal Pozzo Toscanelli (1397–1482) had already confidently promoted the concept of a sea route to the East by sailing west from Iberia, basing his argument mainly on Marco Polo's report of the extent of Asia and the location of Cipangu (Japan). Marco Polo had estimated Cipangu to be some 1,500 miles off the China coast. The actual distance averages much less than a third of this, but Marco Polo's erroneous estimate had the effect of further reducing the theoretical width of the Ocean Sea from Cipangu to Iberia, and therefore encouraged Columbus in his point of view. Furthermore, it shows us why having found the Greater Antilles and Mesoamerica, throughout his lifetime Columbus believed that he had discovered a western route to Japan off the mainland of China. And thus why Columbus searched in vain for a strait between these supposed Japanese Islands that would take him across the Sea of Japan and so to the coast of China.

In his classic *The Southern Voyages*, the eminent marine historian Samuel Eliot Morison wrote of a friend who had said of Amerigo Vespucci (1454–1517) "you only have to mention the name to another historian to see his face grow red, his eyeballs protrude, his blood pressure to rise, and his voice become emotional." The reason for this view is that although Vespucci was a Florentine of noble birth who was said to be the man "who by the discovery of America rendered his own and his country's name illustrious; the Amplifier of the World," his reports of his voyages almost to the tip of South America and to the mainland of North America are considered by many to be a web of fact interwoven with fiction. But there are three very good reasons for bringing Vespucci into the conclusion of this story.

One is that Vespucci was indeed the first to realize that there are two continents stretching continuously down the center of the Ocean Sea, and the first to conclude that Ptolemy and everyone else were wrong about the juxtaposition of East and West. He thereby threatened the greatest inconvenience to publishers of maps and books, and to philosophers and governments, most of whom were quite happy to stay with the world as they knew it. The second is that he lent his name to the Americas.

The name "America" was coined by a clergyman by the name of Martin Waldseemüller (1470–1518), who was fond of inventing nomenclature. Waldseemüller, who had studied at the University of Freiburg, ran a small printing publishing house in the village of Saint-Die in the Vosges Mountains and was an admirer of Vespucci. In 1507 he had woodcut blocks produced for the first map, and woodcut gores for the first globe ever to depict two north-south continents on the Earth's surface and to use the term "America." Martin Waldseemüller also published a slim but important volume

The larger globe is reproduced at 90 percent of its actual size—the first known reconstruction of the Martin Waldseemüller globe of 1507 (for details see pages 341–42). The globe shows two continents stretching down the center of the Ocean Sea and uses the name "America" for the first time. Interestingly, both this projection and the corresponding large-size wall map of the world, show a gap in the region of Panama—the elusive gateway to the Orient sought by Columbus.

The small globe is a modern geography drawn so that the coasts of Asia and Europe are opposed across the Ocean Sea, with Cipangu on the same line of latitude as the Canary Islands. This rearrangement shows how Columbus is thought to have visualized the geography of the Ocean Sea between Iberia and Asia.

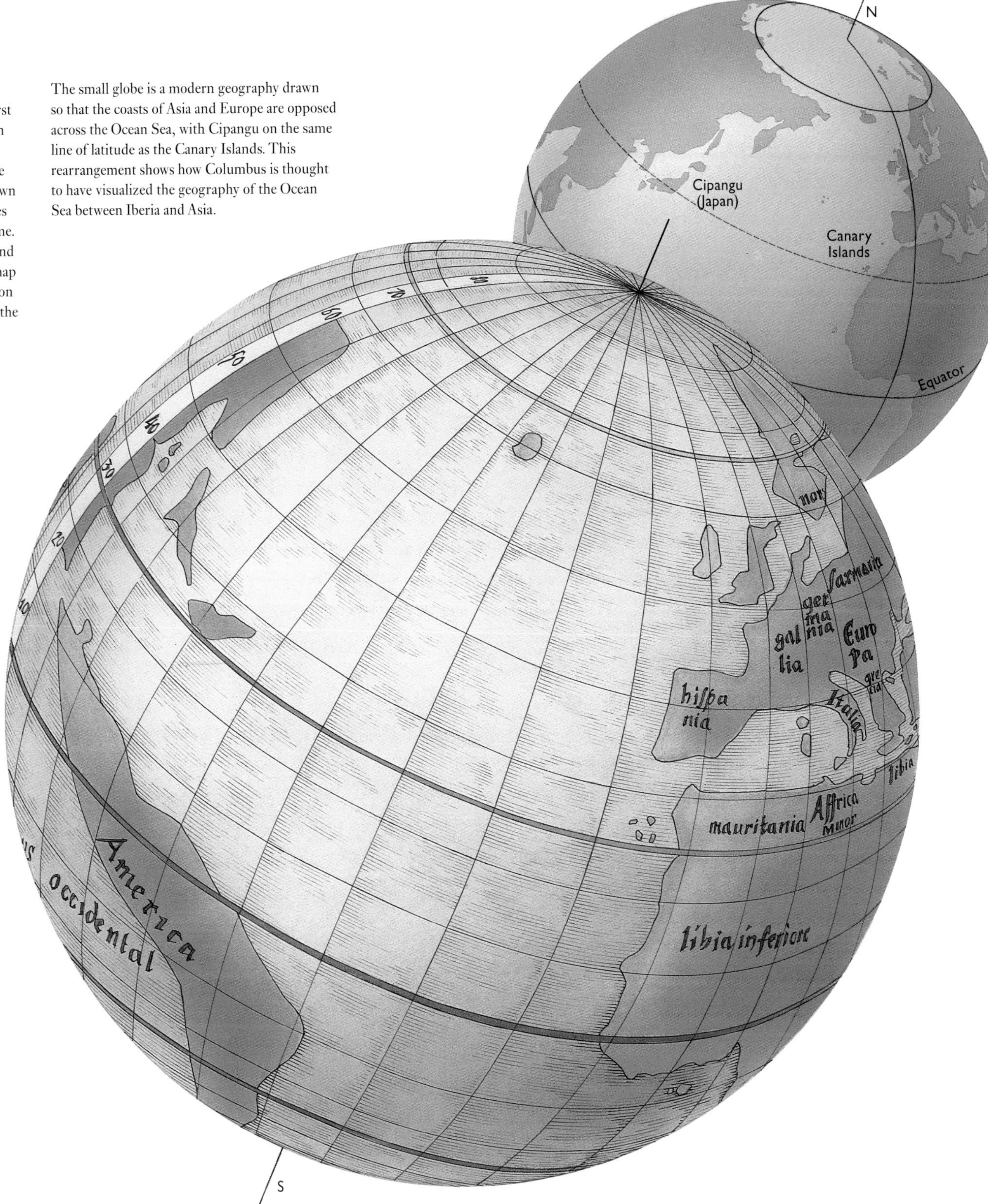

rejecting Ptolemy and promoting Vespucci's ideas. He wrote, referring to South America alone:

Inasmuch as both Europe and Asia received their names from women, I see no reason why anyone should justly object to calling this part Amerige, i.e., the land of Amerigo, or America, after Amerigo, its discoverer, a man of great ability.

Waldseemüller later changed his mind about Vespucci's "ability," but not before Gerardus Mercator had published a map of the world in 1538 in which he showed both a "South America" and a "North America." The die was cast.

The third reason is that Amerigo Vespucci in the 16th century and Alfred Lothar Wegener in the 20th century were viewed very much in the same light by their contemporaries. Although not an untruthful man, which Vespucci is considered to have been by many historians, Wegener's theories of "continental displacement," caused a similar disruption of the establishment view in a discipline not his own. They were to meet with similar rejection for some not dissimilar reasons. However, through his persistence, like Vespucci four centuries before him, Wegener succeeded in causing a permanent change in the way all of us view and understand the Earth.

Glossary

The use of technical language in a book about science poses a problem. My policy has been to use the technical term where it is sensible to do so, and to avoid it when it is not strictly necessary. But I must admit I have little patience with media people who cause confusion rather than provide help when they avoid using perfectly understandable and specific technical words like "fault" or "fracture" or "rift" by substituting vague words like "crack" or "break" or "split." But when a less obvious technical term *is* necessary, I have explained what it means and how it was derived, and soon after I use it a second time in a self-explanatory context.

The absolute time scale used in this book is based upon a revised version (1999) of that prepared for the Decade of North American Geology in 1983 and published by the American Geological Society. Absolute rather than relative time has been used in order to present a continuing story, and to avoid the many and sometimes confusing names used to describe the divisions of relative time. An abbreviated chart relating absolute time to relative time is included here. But there is a snag. There are controversies about some of the absolute dates applied to certain relative divisions of time. The absolute dates quoted throughout are indeed *mean* dates between extreme points of view.

Two key diagrams are included here for the interest of biologists. These are for *Life Begins* (pages 48–49) and *Crisis and Diversity* (pages 50–51), which are there described in nontechnical terms. Here there is no compromise: the language is indeed technical and intended for those who have particular interest in the subject of cellular biology. The graphic art is original, but the individual figures are based upon the work of Drs. Lynn Margulis and Karlene V. Schwartz in their book *Five Kingdoms 1998* (revised edition).

The purpose of this glossary is to refresh the reader's memory of terms used in the book, and, where possible, to add more information for the reader's interest. The terms were extracted from the text, and rewritten for the glossary by my research assistant Dr. Linda G. Martin. The text was read by Dr. R.D. Hamilton, adjunct professor of the Colorado School of Mines. While I am most grateful to both for completing this exacting task, the responsibility for the content remains mine alone. Readers should refer to the index if they wish to follow a particular subject through the book.

absolute dating The actual number of years that have elapsed from any given event expressed within defined limits. This is often calculated by measuring the half-life decay of radioactive isotopes such as carbon-14 (^{14}C).

abyssal plain The generally flat regions of the ocean floors more than 914 m (3,000 ft) deep.

Acadian orogeny A period of mountain building that occurred along the margin of what is now Maine and adjacent areas about 350 MYA. Caused by the subduction of the Rheic Ocean beneath Laurentia during its closure, and characterized by the emplacement of granite plutons.

accretionary prism Ocean-floor crust and sediments that are overthrust onto the margins of island arcs or onto continental margins. These prisms frequently include rock sequences called "ophiolites." See also *obduction; ophiolites*.

active margins Boundaries of colliding plates characterized by volcanoes and violent earthquakes, like those in the Mediterranean and Aegean Seas. See also *passive margins*.

Aegyptopithecus A fossil primate dated 31–30 MYA thought to be ancestral to hominoids. Found at the Faiyûm depression in Egypt.

aerobes Organisms that require a supply of oxygen to live. See also *anaerobic*.

aerosols Microscopic dust or droplets. Aerosols produced from a bolide impact would include pulverized rock from the bolide and the Earth's surface, evaporated water, compounds such as nitrous oxide, and possibly smoke and soot from forest fires.

ages The smallest divisions of geological time, which together make up an epoch.

albedo The capacity of an object to reflect rather than to absorb radiant heat. The ratio between the amount of the Sun's energy that is reflected by ice, snow, vegetation, rock, clouds, or water, and the amount absorbed by those surfaces.

Aleuts An Arctic culture that inhabits the Aleutian Islands.

algae The name applied to any of several protoctist phyla with plant-like properties. These phyla include green algae (Chlorophyta), red algae (Rhodophyta), brown algae (Phaeophyta), and golden algae (Chrysophyta). The name was previously applied to the blue-green algae (Cyanophyta), which have been reclassified as prokaryotic bacteria.

allantois The membrane within an amniote egg that collects the waste materials produced by the embryo. See also *amniote egg*.

Alleghenian orogeny The orogenic event that formed the Allegheny Mountains 330–300 MYA. This event was the result of collision between North American Laurentia and North African Gondwana.

alluvial fan A fan-shaped deposit of sediments that forms at the mouth of a stream valley as it opens onto a plain.

alluvium Water-transported, unconsolidated sediments including boulders, cobbles, pebbles, gravel, sand, silt, and mud.

Alpine orogeny The mountain-building event that occurred as Africa collided with Eurasia. It resulted in the formation of the Alps and other mountain ranges on either side of the Mediterranean Sea, from Morocco and Spain in the west to the Balkans in the east.

alpine-type mountains Mountains that result from the destruction of margins during continent-to-continent collision. See also *andean-type mountains*.

ambient Enclosing or surrounding.

ammonites A group of shelled cephalopods that are important index fossils throughout the late Paleozoic Era and all of the Mesozoic Era. They became extinct about 65 MYA. They had coiled shells and resembled the modern *nautilus*, but ammonite shells had convoluted septa and their siphuncle lay along the ventral (inside) margin of the shell.

amniote Any animal that produces an amniote egg which is either laid by or retained within the mother. All reptiles, birds, and mammals are considered amniotes.

amniote egg Any egg, whether contained in a shell (which is oxygen-permeable) or within the female of the species (which supplies oxygenated blood to the embryo), that has three enveloping membranes. These membranes are

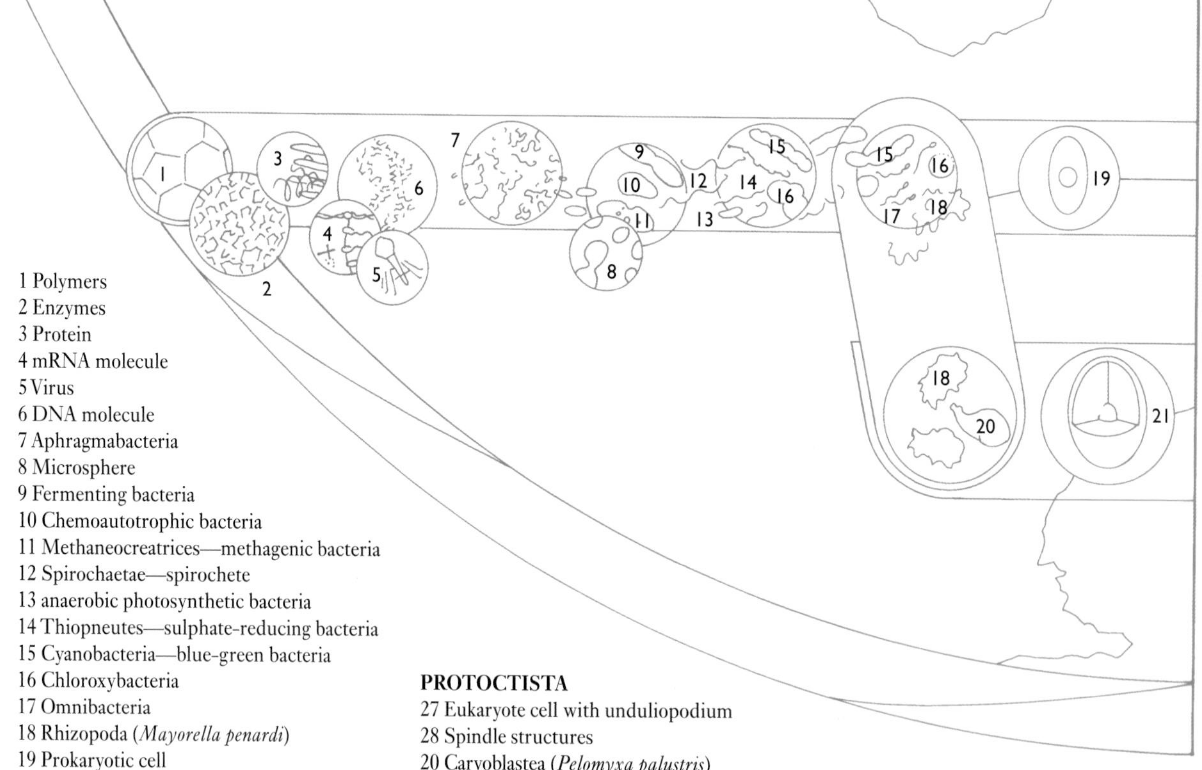

1 Polymers
2 Enzymes
3 Protein
4 mRNA molecule
5 Virus
6 DNA molecule
7 Aphragmabacteria
8 Microsphere
9 Fermenting bacteria
10 Chemoautotrophic bacteria
11 Methaneocreatrices—methagenic bacteria
12 Spirochaetae—spirochete
13 anaerobic photosynthetic bacteria
14 Thiopneutes—sulphate-reducing bacteria
15 Cyanobacteria—blue-green bacteria
16 Chloroxybacteria
17 Omnibacteria
18 Rhizopoda (*Mayorella penardi*)
19 Prokaryotic cell
20 Caryoblastea
21 Eukaryotic cell

MONERA
22 Prokaryotic cell with flagellum
23 Omnibacteria
24 Myxobacteria (*Stigmatella aurantiaca*)
15 Cyanobacteria (*Anabaena*)
25 Actinobacteria (*Streptomyces*)
26 Micrococci
12 Spirochaetae

PROTOCTISTA
27 Eukaryote cell with unduliopodium
28 Spindle structures
20 Caryoblastea (*Pelomyxa palustris*)
18 Rhizopoda (*Mayorella penardi*)
29 Chlorophyta (*Acetabularia mediterranea*)—green algae
30 Phaeophyta (*Fucus*)—brown algae
31 Actinopoda—actinopod
32 Dinoflagellata (*Gonyaulax tamarensis*)
33 Zoomastigina (*Joenia annectens*)
34 Cnidosporidia (*Glugea stephani*)—an animal parasite
35 Bacillariophyta (*Thalassiosira nordenskjøldii*)—diatom
36 Haptophyta (*Emiliania huxleyi*)—coccolithophore
37 Cryptophyta (*Cyathomonas truncata*)
38 Rhodophyta (*Polysiphonia*)—red algae
39 Myxomycota (*Echinostelium*)—plasmodial slime mold
40 Acrasiomycota (*Dictyostelium*)—cellular slime mold

outgrowths of the embryo itself: the amnion (which protects the embryo), the allantois (which collects its waste), and the chorion (which encloses the yolk sac, embryo, and allantois). This triple-membrane system encloses the embryo, which develops without an aquatic larval stage in its own self-sufficient but watery world.

amphibians Phylum of vertebrate animals that can live on land but must return to water in order to reproduce. They evolved about 365 MYA. This phylum was diverse and abundant from about 340 to 206 MYA. Today they are represented by frogs, toads, salamanders, and newts.

anabolism The process of absorbing digested food and transferring its latent energy into the bloodstream through the walls of the upper intestine. One part of the process of metabolism. See also *metabolism*.

anaerobic (1) A condition in which free oxygen is absent. (2) A term applied to an organism that can exist without the presence of free oxygen. See also *aerobe*.

anapsid A group of reptiles that have no apertures in their skulls other than eye sockets and nostrils. The group includes turtles. See also *diapsid; synapsid*.

andean-type mountains Mountains that are formed during the subduction of ocean crust beneath a continental margin. Mountains that generally include volcanoes and batholiths. They form on active margins. See also *alpine-type mountains*.

andesite A type of extrusive volcanic rock of intermediate composition, meaning that it contains more iron and magnesium than rhyolite but less than basalt. It is the extrusive equivalent of diorite. The characteristic rock type formed in andean-type mountains.

angiosperm Any seed-bearing plant that produces flowers. A plant that produces seeds that are enclosed by pulp, meat, or fruit. Some angiosperms are wind pollinated and others are pollinated by insects, birds, or bats. See also *dicotyledon; gymnosperm; monocotyledon*.

Animal Kingdom Multicellular eukaryotes that are all characterized by the fact that following fertilization their embryos form a blastula. They obtain their food from other living organisms or by ingesting organic materials. Their cells are enclosed by a cell membrane, not a cell wall.

ankylosaurs Armored dinosaurs that grew up to 10 m (33 ft) in length and to over 3,600 kg (4 tons) in weight. They lived only in western North America and western Asia 90–65 MYA.

Annelida Phylum of segmented worms.

anoxic Environment that is completely devoid of free oxygen.

anthracite Highest grade of coal. Virtually pure carbon, it is produced by intense compression and heat during the burial and folding of rocks.

anticlines Folded structures in which the youngest rocks comprise the outside layers of the structure and the oldest rocks form the inside layers or core of the structure. See also *synclines*.

Apache A North American cultural group that migrated to the Southwest around 600 to 700 years ago from Canada. Believed to be closely related to the Athabascan and Navajo cultures.

apatite Calcium phosphate mineral that hardens bone and forms enamel on teeth.

aragonite A mineral that is one crystalline form of calcium carbonate; sometimes called "mother of pearl."

Arawak A group of people that migrated from the Orinoco region of South America to the Caribbean islands 3,000–2,500 BP.

Archaeopteryx Oldest-known fossil that is generally accepted to be a bird, dated about 150 MYA.

Archean Eon One of two divisions of the Precambrian, dated 3.75–2.5 BYA. See also *Proterozoic Eon*.

archosaurs Carnivorous diapsid reptiles that evolved about 235 MYA. They diversified into several lines and led to the pterosaurs, dinosaurs, and phytosaurs.

Arctic Small Tool culture An Arctic culture that predates the Thule culture.

arthropods Phylum of animals that have exoskeletons, segmented bodies, and jointed legs. Arthropods include lobsters, crabs, insects, spiders, trilobites, etc.

asthenosphere The lower semiplastic layer of the two layers that compose the tectosphere. See also *lithosphere; mantle; tectosphere; upper mantle*.

asteroid A metallic or rocky object that orbits the Sun.

astrobleme An impact structure formed by a bolide striking the Earth.

astrolabe, marine An archaic instrument used to measure the height of stars, particularly the "fixed" Pole Star in the night sky in order to determine the relative distance from the Equator or the North Pole (degrees of latitude). It was suspended by a ring in one hand so that a fixed crosspiece stayed level, while a movable crosspiece was adjusted to line up on the star.

Athabascan culture A Canadian cultural group that arrived in North America 14,000–12,000 BP, considered the progenitors of the Navajo and Apache cultures.

Atlantic realm Term applied to the region of Western Europe in which the trilobite *Paradoxides* is normally found.

atoll A structure formed by coral reefs growing around the base of a half-submerged volcano in tropical seas. As the seafloor sinks, the corals grow higher, building upon those that have died in deeper and cooler waters out of reach of sunlight. The process continues, until the remnant volcano completely sinks out of sight, leaving a circular reef that continues to grow and which contains a shallow, sandy-bottomed lagoon without an island at its center.

aulacogen A failed rift. The East African Rift Valley is the failed rift of a triple-rift system. See also *triple rift*.

autotrophs Type of bacteria capable of metabolizing inorganic materials such as methane, sulphur, or nitrogen.

Avalonian Ediacara Fauna of soft-bodied animals that lived about 620 MYA. Their fossils are found on the Avalon Peninsula, southern Newfoundland.

background extinction The occasional permanent disappearance of one or more species during relatively stable periods.

back reef The region behind a fringing reef. It includes patch reefs located in the relatively calmer lagoonal areas behind the wave-swept fringing reefs.

Baltica Name for the plate that contained much of proto-Northern Europe throughout geologic time, including Scandinavia, European Russia, and northern Germany.

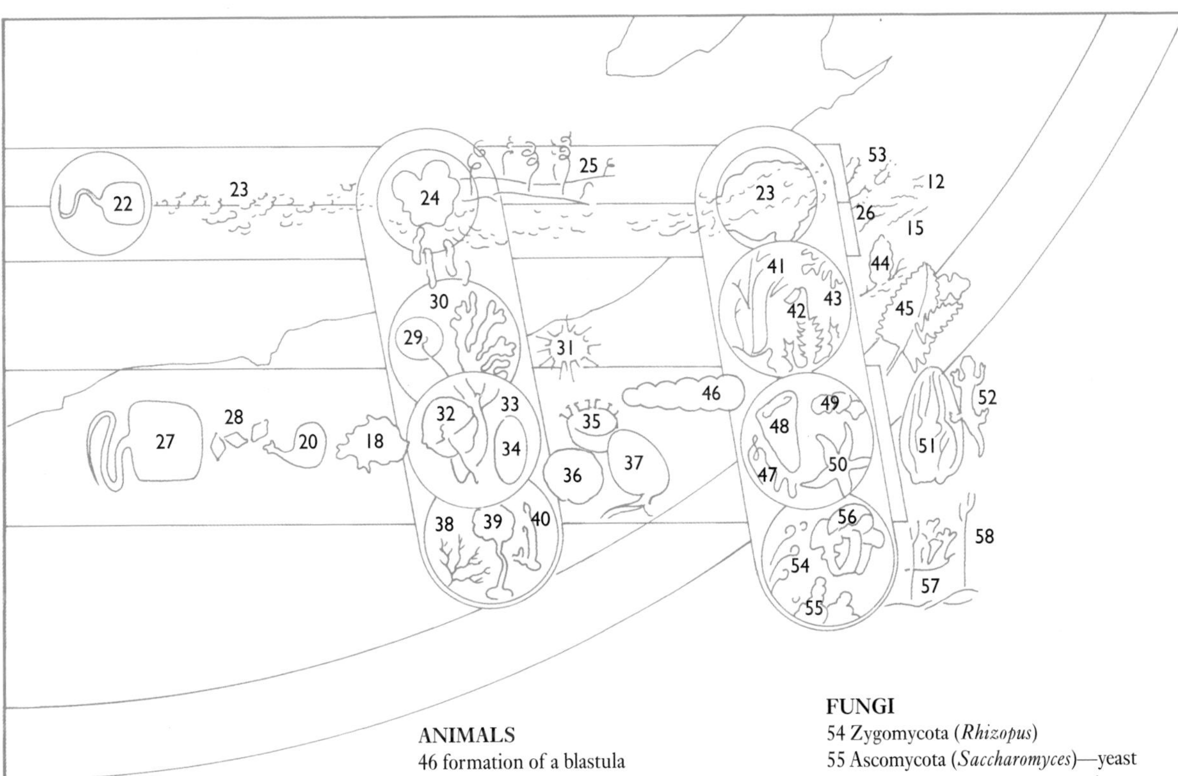

PLANTS
41 *Rhynia*
42 Bryophyta (*Polytrichum juniperinum*)—bryophyte
43 Lycopodophyta (*Lycopodium*)—club moss
44 Coniferophyta (*Pinus*)—conifer
45 Filicinophyta—fern gametophyte and adult sporophyte

ANIMALS
46 formation of a blastula
47 Nematomorpha (*Gordius villoti*)—horsehair worm
48 Porifera (*Gelliodes digitalis*)—sponge
49 Brachiopoda (*Terebratulina retusa*)—brachiopod
50 Echinodermata (*Asterias forbesi*)—star fish
51 Ctenophora (*Bolinopsis infundibulum*)—comb jelly
52 Vertebrata (*Ambystoma tigrinum*)—salamander
53 Arthropoda (*Pterotermes occidentis*)—termite

FUNGI
54 Zygomycota (*Rhizopus*)
55 Ascomycota (*Saccharomyces*)—yeast
56 Basidiomycota (*Boletus*)—mushroom
57 Mycophycaphyta (*Cladonia cristatella*)—lichen
58 Deuteromycota (*Penicillium*)

All organisms in these illustrations are identified either by general description or by phylum followed by genus and species name, followed by common name wherever appropriate. Classification is according to Margulis and Schwartz in *Five Kingdoms*.

barrier islands Long, narrow islands that are parallel to the mainland but separated by a lagoon.

basalt A type of extrusive volcanic rock that is dense, dark-colored, and iron-rich. It has the same composition as gabbro but instead of being intruded into other rocks, it is extruded onto the Earth's surface, often at spreading-centers such as the Mid-Atlantic Ridge. It is the major rock type that comprises oceanic crust. See also *gabbro*.

basement Undifferentiated crystalline rocks beneath sedimentary rocks and sediments. It extends to the base of the continental crust.

basin A low-lying, generally depressed, or down-folded region that may contain lakes or seas. The weight of the water, and sediments, may further depress the center of the basin, which through time will become filled with sediments, including evaporites.

batholith Collective name for a series of large plutons that were intruded in the same region but at various times.

bathypelagic zone The deep sea where no sunlight penetrates. It starts at 1,200 m (4,000 ft) beneath the surface and continues into the abyss. See also *epipelagic zone; mesopelagic zone*.

bedrock Any solid rock lying beneath poorly consolidated surficial sediments.

benthic fauna Bottom-dwelling fauna. Organisms that live on the seafloor regardless of ocean depth. Collectively called the "benthos."

benthic Pertaining to the seafloor.

Beringia The continental shelf area between western Alaska and northeast Siberia that emerges as a landmass when sea level is sufficiently reduced below present-day level. This is caused by the accumulation of ice on the continents during glacial advances and the melting of ice sheets during glacial retreats. The appearance and disappearance of Beringia determined the timing of human and animal migrations across the Bering Strait in prehistoric times.

biogeography Study of the pattern and significance of the distribution of species.

biomass (1) The total weight of organic life that the Earth's resources can support at any one time. (2) The total weight of organic life contained in a defined area or region at any one time.

biosphere The realm of living organisms on Earth. The envelope of life.

biota Organisms living in a geographic region or period of time.

bipedal Having the capacity to walk on two legs. See also *quadrupedal*.

Birch Zone A time interval from the last glaciation that is identified in part from the mixture of birch, spruce, alder, balsam, and poplar that grew during that time in arctic and subarctic regions.

bitácula A primitive tub-like water compass set upon a ship's deck. It was used to record direction. By plotting the direction of travel together with estimated speed for 24 hours, it was possible to chart the estimated distance traveled.

blastula A hollow ball of cells formed from a fertilized animal cell that will develop into an embryo.

block faulting Occurs in an extensional tectonic province in which the region is being stretched. Roughly parallel faults develop in response to the pull-apart tension, and some blocks of rock are down-dropped along those faults forming a series of grabens and horsts (valleys and ridges).

blue-green algae Morphologically simple organisms without nuclei; also called "cyanobacteria." They once were considered plants since they undergo photosynthesis, but now are classified as prokaryotic bacteria in the kingdom Monera. See also *cyanobacteria; Monera*.

bolide Any extraterrestrial object that hits Earth whether on land or in the ocean.

bottom currents Water masses that flow along the bottom of ocean basins.

bottom-water Any water mass that occupies the bottom of an ocean basin. The water is generally very cold, may be nutrient-rich, and may or may not be oxygen-deficient.

BP Number of years before the present.

brachiopods Phylum of marine animals with two shells (valves). They have bilateral symmetry across an imaginary plane that passes through the center of each shell.

brain corals Corals that grow into rounded, globular masses that resemble brains. The dominant corals that form patch reefs on the modern back-reef floor; these are more resistant to turbid tidal water than corals in fringing reefs.

bristlecone pine One of the oldest living varieties of tree on Earth. Those in Colorado have been dated 2,500 years BP and in the Sierra Nevada, California, 6,000 BP. The growth rings of these and other long-life trees are used in the science of dendochronology—for instance, in the dating of archeological sites and artifacts.

brown algae Protoctist kelp and common tide-line seaweeds formerly classified as plants. Members of the phylum Phaeophyta.

bryophytes Plants, such as mosses and liverworts, that do not have capillary tubes in their stems. This lack of a fluid transporting system ties them to a watery or very moist environment in which to live and reproduce. See also *nonvascular plants*.

burin An elongate, spined, dual-purpose chert tool with an incisor at one end and a scraper at the other. It was produced from a prepared core, and was used primarily for the manufacture of other tools, including bone and antler tools.

BYA Used in this book for billion years ago.

^{14}C dating Dating technique that calculates absolute age using the radioactive decay of carbon-14 atoms.

Cadomian orogeny Period of mountain building along the southern margin of Gondwana in a region, Armorica, dated 650–550 MYA. Remnants of this episode can be found in Normandy, Brittany, and the Channel Islands.

calcareous Of or pertaining to calcium carbonate materials, sediments, or processes.

calcareous green algae Protoctists that produce calcium carbonate skeletal frameworks. They are the major carbonate producers in modern carbonate shelf or platform environments. They belong to the phylum Chlorophyta.

caldera Large, roughly circular sunken volcanic basin that may contain several volcanic vents.

Caledonian orogeny Mountain-building episode that occurred when Laurentia and Baltica collided 360–250 MYA. This was the first mountain chain created during the assembly of Pangea. It was originally recognized in Scotland. Mountains from this episode still exist in Greenland, Spitsbergen, and Norway.

Cambrian Period Division of geologic time dated 543–490 MYA. The period originally thought to represent the beginning of life, since the oldest fossil shells were discovered in rocks of this age.

cap rock Impermeable rock, generally shale or an evaporite such as rock salt, that prevents hydrocarbon migration from the rocks underlying it. See also *seal; trap*.

caravel A moderate-sized, sleek ship with lateen sails that tacks well across winds and currents. Maneuverability made it an ideal explorer's ship, but it had limited cargo space.

carbonate A mineral that contains CO_2, such as calcium carbonate ($CaCO_3$), for example, calcite, aragonite, or dolomite. Minerals that are used by most marine organisms to make their shells or skeletons. Predominant mineral in limestone and marble.

carbonate bank (1) Ridge or mound made up of carbonate skeletal materials. (2) Flat-topped elevated submarine surface largely composed of and covered by carbonate sediments.

carbonate compensation depth Depth of the ocean at which seawater completely dissolves all carbonate material. The depth is variable, depending on ocean temperature and the amount of carbonate being produced in the epipelagic zone by protoctists. Often called the "CCD"; in this book referred to as the "carbonate snow line."

carbonate platform A large, flat or gently sloping elevated submarine surface composed of and covered by carbonate sediments. See also *carbonate bank*.

carbonate rocks Rocks composed primarily of calcium carbonate minerals or skeletal materials, i.e., limestone, chalk, and marble.

Carboniferous Period The European division of geological time, dated 354–290 MYA. The period is characterized by innumerable coal deposits. Equivalent to the Mississippian and Pennsylvanian Periods used in North America.

carpel That portion of a flower that envelops the ovules and thus the seed. It develops into a "fruit"—a pod, nut, burr, or a soft-fleshed fruit.

carrack Heavy, square-rigged, seaworthy merchant ships with limited maneuverability.

cartilaginous fish A group of fish whose skeletons are composed of cartilage. They first appeared about 400 MYA, and today are represented by sharks, dog-fish, and rays.

catabolism The biochemical process of breaking down food that occurs in the stomach and upper small intestine. Part of the process of metabolism. See also *metabolism*.

cell The rudimentary organizational unit of life, composed of protoplasm surrounded by a cell membrane. It may or may not contain a nucleus and other organelles.

Cenozoic Era Division of geological time dated 65 MYA to the present. The name means recent life. It is divided into the Tertiary and Quaternary Periods.

Cenozoic Ice Age The most recent ice age. It is dated 3.5 MYA to the present in the Northern Hemisphere, but is thought to have started in the Southern Hemisphere 30–15 MYA. We are currently in an interglacial period of the Cenozoic Ice Age.

Central Pangea Term used in this book to describe that area of Pangea that was roughly equatorial and encompassed most of the modern "Western World."

cephalopods Phylum of animals whose heads are surrounded by tentacles. Includes ammonites, *nautilus*, octopus, and squids.

ceratopsians Suborder of beaked dinosaurs generally with horns and a frill around the edge of their skull. They lived from about 80 to 65 MYA, and were confined to western North America and western Asia.

chalk A type of carbonate rock composed primarily of the microscopic shells of coccoliths.

chemosynthesis The process employed by bacteria in vent and seep communities to produce the primary organic compounds that form the base of the food chain in these environments.

chert An extremely hard, microcrystalline rock composed of silica (SiO_2). It was often fractured into tools such as arrows and axes during the stone ages. Sometimes called "flint."

chevron folds Rock units that have been compressed into V-shaped folds.

china clay Particular quality of clay (kaolin) formed from the chemical alteration of granite. It is used to make the thin-walled translucent ceramics called porcelain or china; also used in the manufacture of high-quality paper.

Chicxulub An area of the Yucatán Peninsula, Mexico, in which a subsurface impact crater measuring 170 to 300 km (102 to 180 miles) has been discovered. The bolide that formed the crater is estimated to have been from 6 to 16 km in diameter (3.6–9.6 miles). This may have been of sufficient size to have caused the mass extinction that ended the Cretaceous Period and Mesozoic time—most famously, the extinction of the dinosaurs.

Chlorophyta Phylum of protoctists known as green algae, which includes both unicellular and multicellular forms. The few genera that are calcareous are the prime sediment contributors on carbonate platforms. Previously classified as plants, since they undergo photosynthesis.

chorion The layer of an amniote egg that encloses the yolk-sac, embryo, and allantois. See also *amniote egg*.

Ciboney The first people to populate the Caribbean islands. They lived on Cubagua Island off Venezuela from about 7,000 to 3,000 BP and are thought to have migrated to the Lesser and Greater Antilles about 4,500 BP.

circumpolar ocean Ocean with unrestricted circulation that surrounds a polar continent, such as Antarctica.

class Division in the hierarchy of classification of organisms. One or more orders comprise a class. One or more classes comprise a phylum.

clay (1) A silicate mineral made up of multiple layers of alumino-silicate weakly bound together with water. (2) Any particle that is of a specific microscopic size (less than 1/256 mm [0.00016 inch] in diameter).

Clovis point Bifacial projectile points made by North American cultural groups that hunted mammoth and the now-extinct long-horned bison. Named after their type locality at Blackwater Draw, near Clovis in New Mexico, they date from 11,800 to 11,200 BP.

Cnidaria Phylum of relatively simple, tentacled animals that includes jellyfish, coral, and other similar animals. Also called "coelenterates."

coccoliths Short for coccolithophores. A group of microscopic, pelagic protoctists whose calcareous skeletons are made up of disk-like overlapping plates called coccoliths. They are major producers of carbonate sediment in warm and temperate oceans. Their skeletal fragments are the primary components of chalk.

coelacanth Primitive lobefin fish whose fins have hand-like bone structures. They were long thought to be extinct but are occasionally caught off Madagascar. See also *lobefin fish*.

coevolution Occurs when two or more organisms evolve in a mutually beneficial manner with one another, such as trumpet-shaped flowers and insects with long sucking mouth parts—the former is benefited by specific cross-pollination, and the later by a reliable food source.

cold seeps Deep-water openings on or near the seafloor from which cold water emerges, such as those at the foot of the Florida escarpment. Warm seawater from the surface of the continental shelf seeps downward through thousands of feet of porous rock, dissolving sulphide compounds along the way, and finally percolates from openings near the base of the escarpment.

comets Extraterrestrial objects covered by ice and frozen gases that orbit the Sun in long elliptical paths. They are thought to originate in the Oort Cloud, which lies about one light year from Earth.

conodonts Phosphatic microfossils that resemble teeth but whose function is unknown. They were produced by extinct, small, free-swimming animals that vaguely resemble arrowfish. They are important index fossils for most of the Paleozoic and the lower Mesozoic.

continent–continent collision The collision of two continental plates. Folded mountains with granite intrusions are formed along the margins of both continents during the collision,producing alpine-type mountains like the Himalayas.

continent–ocean boundary The boundary between a region of continental crust and a region of oceanic crust.

continental crust The portion of the Earth's crust beneath continental masses. It is far thicker but less dense than oceanic crust because it contains a higher proportion of aluminum and a smaller proportion of magnesium and iron.

continental drift Theory proposed by Wegener in 1912 that through time continents move, or drift, across the surface of the Earth. Wegener called the process "continental displacement." Now part of the science of "plate tectonics."

continental separation Term used in this book to describe the stage of continental rifting which follows crustal extension. Continental separation occurs when grabens become sunken, thinned, and then ruptured by volcanic activity and eventually form a continuous spreading axis along which oceanic crust is formed.

continental slope Submarine area surrounding the continents beyond the continental shelf and extending to the deeper parts of the sea. The area has an average slope of 3–6°.

convection Movement throughout a mass caused by the sinking of cooler, denser material and the rising of warmer, less dense material. Gases, liquids, and solids contract as they cool, expand when they are heated, and thus change in relative density according to their temperature; this causes the coolest regions of liquid or plastic material to sink, and the warmer regions of the same mass to rise.

convection cells Elongated, counter-rotating paired bodies in the outer core that transfer heat to the underside of the surrounding mantle.

convergent evolution Occurs when unrelated organisms develop similar physical characteristics in response to similar life styles or environmental pressures. Sometimes called "parallel evolution."

coral polyps The myriad, tiny individual animals that build and inhabit a coral head or branch.

coral reef A mass of individual coral heads or branching corals that form a mound or ridge. Generally includes other marine organisms with similar growth habits such as sponges and red algae.

core The central region of the Earth which is divided into an inner core and an outer core. The inner core is a solid sphere of iron-nickel alloy, and its radius is almost 1,600 km (1,000 miles). The outer core is a white-hot semi-liquid mix of mainly iron, nickel, and sulphur about 2,250 km (1,400 miles) thick, and contains circulating convection cells.

Coriolis effect A physical law that explains that on a rotating surface, there is an inertial component which acts upon the body at right angles to its direction of motion, resulting in a curved path for an object that would otherwise travel in a straight line.

cotyledon In angiosperms, the seed's leaf-like organ which, upon sprouting, becomes the plant's primary leaf or leaves.

cotylosaurs The primitive reptilian stock or stem-reptiles from which all other reptiles are descended. Cotylosaurs produced the first amniote eggs, i.e., eggs enclosed in leathery shells.

country rock Older rocks that are melted or intruded by younger plutonic magma as it rises.

Cretaceous Period Division of geologic time dated 144–65 MYA. Named for the numerous chalk deposits in this time period. Cretaceous means "chalky" in Latin. The third and final period of the Mesozoic Era.

Cro-Magnon Man Name applied to anatomically modern man in Europe, dating to about 35,000 BP. Named for the site at which the first skeletons were found at Cro-Magnon, near Les Eyzies, in the Dordogne, France.

cross staff A device used to measure latitude at sea that consisted of a graduated staff fitted with a moving crosspiece, which slid along the staff and could be locked in place. The staff was held to the eye, and the crosspiece moved until its extremities coincided with the horizon and either the Sun or Pole Star. Latitude was obtained from this information with the aid of an almanac.

crust The outermost layer of the Earth comprising the rigid lithosphere. Two types of crust are recognized: continental crust composed of granitic material, and oceanic crust composed of basaltic material. The crust and uppermost portion of the mantle together comprise the lithosphere. See also *continental crust; lithosphere; ocean crust*.

crustal extension The process of the stretching and pulling apart of a continent that causes block faulting and initial triple-armed rifting to occur. It is caused by crustal thinning over a domed region of the lithosphere.

cryosphere That portion of the Earth which is frozen, i.e., all ice and permafrost regions.

cyanobacteria Scientific name for blue-green algae, which are now recognized as prokaryotes. Formerly classified as Cyanophyta.

cycad Type of gymnosperm that is palm-like in appearance but which produces cone-like reproductive structures.

cyclothem A rock sequence in which stream sediments are overlaid by coal-forest deposits (peat), followed by a marine invasion, which is followed by a return to terrestrial stream deposits. This results in a sequence of rocks consisting of sandstone, shale, coal, sandstone, shale, marine limestone, shale, sandstone. These sequences are generally cyclic and result in the accumulation of coal and other sedimentary rocks, layer upon layer.

cynodonts Therapsids with canine-like teeth. They were carnivorous mammal-like reptiles.

delta fan A fan-shaped deposit of sediments that forms at the mouth of a stream valley as it opens into the sea.

dendochronology The study of the growth-rings of trees, particularly ancient and long-lived trees, for the purpose of dating the recent past (as distinct from the geological past).

depressions Down-warped areas or basins.

detrital sediments Particles formed from the erosion of pre-existing rocks.

devolatilization The process of the degassing of coal during burial and deformation, which causes the expulsion of volatile compounds such as water, carbon dioxide, and methane (natural gas).

diapirism The tendency of less dense materials to rise; e.g., molten plutons are liquid and are hotter and lighter than surrounding rocks, and thus rise towards the Earth's surface. See also *salt dome*.

diapsid Reptiles with four cranial openings, two on either side of the head, in addition to the eye sockets and nostrils. They first appear in the fossil record about 285 MYA, and are the largest group of living and fossil reptiles, including dinosaurs, pterosaurs, lizards, crocodiles, and snakes.

diatoms Single-celled, marine and fresh-water protoctists with two siliceous shells. They are primary food producers in cool-water masses. They were formerly classified as plants, as part of the phytoplankton.

dicotyledon One of two classes of angiosperm, generally called "dicots." Their embryos have two leaf-like organs that store food for the embryo. These leaves become the plant's two primary leaves upon sprouting. Dicots have flower parts (petals, stamens, etc.) that are usually divisible by four or five, and net-like veins in their leaves. In their stems, which may be woody, the vascular bundles are in a ring-like form. Their pollen grains have one to three furrows. See also *monocotyledon*.

dicynodonts The dominant group of therapsid herbivores. They had two long canines in their upper jaw but no other teeth.

dinosaurs Name coined by Richard Owen in 1842 for an order of extinct reptiles. Today the dinosaurs are classified into two separate orders because it is thought that the two groups evolved separately from each other. The term dinosaur, however, continues to be applied to all members of both orders. It is also often misapplied to other extinct reptile groups such as flying reptiles, the pterosaurs, and aquatic reptiles like plesiosaurs, and even to mammal-like reptiles, like the pelycosaurs. In this book dinosaurs have been classified into three overlapping regimes.

diorite A type of intrusive igneous rock that is intermediate in composition between granite and gabbro. It is the same composition as andesite, but instead of being extruded onto the Earth's surface, it cools in a pluton beneath the surface.

dipole A magnet having two ends with opposite polarity, such as a bar magnet.

dispersalism A short-term biogeographic process that involves day-to-day or seasonal migration and radiation of species to the limits of their natural environment on a contiguous landmass; e.g., the development of the Isthmus of Panama allowed migration to occur between North and South American mammals. See also *vicarianism*.

displaced terranes Fault-bounded regions of modern continents that are thought or known to have originated elsewhere. See also *terrane*.

displacement theory Wegener's term for the theory of continental drift—now called "plate tectonics." It explained anomalous fossil faunas and rock formations that are climatically misplaced by suggesting that the continents "drifted" into their present relative positions. See also *continental drift*.

diversity The variety of different types of interrelated organisms that exist at any given time.

divergent evolution This process occurs when a lineage develops into two or more different species in response to adaptation to fill new niches or habitats. See also *vicarianism*.

DNA Deoxyribonucleic acid. The complex molecules that make up genes.

DNAG (Decade of North American Geology) DNAG dating is a geological time scale established by the Geological Society of America for a multivolume synthesis of North American geology in the decade 1980–90. The revised version (December 1999) is used in this book.

doming The swelling beneath the weakest parts of the overlying crust caused by the upwelling and accumulation of molten mantle material above a hot spot in the mantle. This can result in crustal extension of the surface and the formation of triple-armed rift systems.

Dorset tradition An Arctic culture dated around 3,600–1,500 BP, which followed the Arctic Small Tool culture and predated the Thule culture. It is characterized by harpoons and other advanced weapons for hunting seals and walrus, and is named after Cape Dorset on Baffin Island.

drift Archaic term coined to describe the movement of continents through time into their present configuration. This process is now known as "plate tectonics."

dropstones Rocks or boulders generally transported by ice and deposited in a lake or sea by a melting iceberg, thus disrupting the soft, laminated sediments on the lake bottom or seafloor. They may also be transported in the roots of floating trees.

drumlins Elongate hillocks that are the compacted products of glacial grinding that accumulated as a result of subglacial flooding beneath an ice sheet.

East Central Pangea Term applied in this book to that portion of Central Pangea which includes present-day Western and Central Europe and formerly adjacent parts of eastern North America.

echidna The Australian spiny anteater. A type of monotreme that has an egg pouch in which to lay and incubate its eggs, but suckles its young like all other mammals.

ectothermic A condition generally correlated with cold-blooded animals, in which the body temperature depends on heat obtained from outside the body. See also *endothermic; heterothermic; homeothermic*.

edentates An assemblage of ungulates belonging to two orders of very different animals that are all characterized by the reduction, or loss, of all teeth. It includes armadillos, sloths, anteaters, and pangolins, plus several extinct forms.

Ediacaran fauna Diverse and complex soft-bodied animal fossils that are Precambrian in age (650 MYA or before), particularly those discovered in the Ediacara Hills of northern Australia.

Ekman transport A phenomenon in which the wind blowing across the ocean piles up surface water, which then flows downslope at an angle of 45° to the direction of the wind (to the right in the Northern Hemisphere and to the left in the Southern Hemisphere). This phenomenon deflects the layer of water beneath it to the right (or left) but at an angle less than 45°. Thus succeeding layers of water in the water column act upon underlying layers with ever-reducing effect. The result is a spiral staircase of diminishing force called an "Ekman spiral," which can cause a deep current to flow in the opposite direction to the wind-generated surface current.

embayment A downwarped area or bay filled with sediments or sedimentary rocks that may have accumulated at the mouth of a river.

endopterygotes Insects whose eggs hatch into grubs which go through a pupal stage. During the pupal stage there is a complete transformation of body structure, including the development of wings (from chrysalis to butterfly, in effect). This category of insects includes bees, wasps, ants, butterflies, and moths. See also *exopterygotes*.

endosymbiosis A symbiotic relationship at the cellular level with anaerobic bacteria.

endothermic A condition generally correlated with warm-blooded animals, in which a controlled, high body temperature depends on internally generated heat which is maintained by a high metabolic rate. See also *ectothermic; heterothermic; homeothermic*.

Eocene Epoch Division of geologic time dated 54.8–33.7 MYA. It is the second epoch of the Tertiary Period.

eolian Wind-formed or deposited sediments or structures.

eon The highest division of geological time. There are only three eons: the Archean Eon, the Proterozoic Eon, and the Phanerozoic Eon. The latter dates from the beginning of Cambrian time and is comprised of the Paleozoic, Mesozoic, and Cenozoic Eras. The Archean and Proterozoic Eons belong to the Precambrian interval.

epicontinental seas Seas that inundate continental shelves or interiors.

epipelagic Those organisms that live in the upper 200 m (675 ft) of the sea.

epipelagic zone The upper 200 m (675 ft) of the sea. This is the well-lit zone of the sea. See also *bathypelagic zone; mesopelagic zone.*

epochs Divisions of geologic time that make up a period, which in turn is divided into ages.

era Division of geologic time that is shorter than an eon but much longer than a period, e.g., the Paleozoic Era.

erratics Rocks that were gouged from rocky outcrops by glacier ice, worn and rounded by water and ice while being transported many miles, and finally deposited as the glacier retreated.

Eskimos Name often applied to the Inuits of North America and Greenland.

eukaryote Any organism whose cell or cells are nucleated. This applies to all plants, animals, fungi, and protoctists.

Euler's theorem Postulated by Euler in 1775, it proved that a section of rigid surface of a sphere can move only by rotating about an independent axis. When translated into terms of the Earth's spherical surface, this means that as continents drift and separate, they can only swing away from each other at an ever-increasing angle; i.e., on the surface of a sphere, segments cannot separate and still maintain a parallel relationship.

euphotic zone The extreme depth of the sea at which photosynthesis can take place. In ideal conditions of clarity this zone reaches down to 150 m (500 ft).

eustatic In geology, pertaining to a global change in sea level often caused by glacial fluctuations.

evaporites Minerals that precipitate from evaporating water, often from sea water. These minerals include rock salt and gypsum.

exopterygotes Insects that do not go through a pupal stage. The young are born resembling the adults, but without wings. The wings develop with adulthood. This group of insects includes dragonflies, mayflies, grasshoppers, locusts, and earwigs. See also *endopterygotes*.

extrusive Igneous rocks that have been erupted onto the Earth's surface either as lava flows or as ash. See also *intrusive*.

failed ocean An ocean-filled region in which a spreading-center developed for a limited time, then ceased to propagate, e.g., the Bay of Biscay.

Famennian Age Geological time dated 364–354 MYA. The uppermost age of the Devonian Period. The base of the period is the upper boundary of the Frasnian-Famennian mass extinction.

family Division in the classification of organisms. One or more genera comprise a family. One or more families comprise an order.

Farallon Plate That portion of the Pacific Ocean floor which descended beneath the West Coast of North America and South America and extended into the Caribbean.

fault A fracture or group of fractures along which there has been movement or displacement of adjacent regions.

faunal realm A geographical region characterized by a particular fauna.

fecal pellets Excrement of organisms, particularly of marine invertebrates, that may occur fossilized in sedimentary rocks.

first-regime dinosaurs Term applied in this book to the early dinosaurs that lived 230–190 MYA. They are all thought to have been bipedal. These dinosaurs were diverse, and were generally smaller than many later dinosaurs.

fissure swarm A series of roughly parallel volcanic fissures (long fractures from which lava may flow).

five kingdoms Classification system dividing all organisms into five kingdoms: Monera, Protoctista, Fungi, Plantae, and Animalia.

flood basalt Volcanic eruption on an enormous scale, generally along a fissure swarm. Lava in such an eruption is normally very fluid, always basaltic in composition, and flows across considerable distances before solidifying.

fold Geological structure that forms in deeply buried, layered rocks as they are compressed from opposite sides, causing the rocks to bend.

fold-belt Region that has been compressed forming parallel folds and folded mountains.

Folsom point Artifact from a North American tool-making technique which produced bifacial spear points, dating from around 11,000 BP.

foraminifers Type of mostly single-celled protoctists that generally make calcareous shells. They can be useful either as index fossils or as paleoecological indicators. The organisms resemble amoebas but construct elaborate shells. They can be either planktonic or benthonic in habit; a few are encrusting forms.

foredeep basins Elongated troughs or downwarped areas that form from the accumulating weight of mountain ranges depressing a subducting plate boundary.

foreland The leading edge of a continental plate margin, or the area in front of an adjacent area.

foremast The forward mast of a sailing vessel. The mast nearest the bow of a ship.

fossil fuels Hydrocarbons (organic compounds of carbon and hydrogen) that can be used for fuel, such as natural gas, petroleum, and coal.

fossil record The history of life as recorded by fossils.

fractionation (1) The separation into different parts or divisions, which in solids and semisolids, occurs mainly as a result of density differences, i.e., denser materials sinking and lighter materials rising. (2) A process that occurred in the early stages of the Earth's history that resulted in the separation of the Earth into the various layers of the core, mantle, and crust.

fracture zone A fault zone along which transform faulting occurs at right angles to a spreading-center. Also a site of volcanic activity.

Frasnian Age Geological time dated 370–364 MYA. The next-to-last age of the Devonian Period noted for a mass extinction that occurred 364 MYA called the Frasnian-Famennian extinction.

fringing reef Elongated reef track that forms on the seaward edge of an island, shelf, or platform. The organisms that form these reefs can tolerate high-energy waves. The reef breaks the waves, thus protecting the lagoon or bay behind it.

Fungi Kingdom Comprises all eukaryotic organisms that produce spores and do not develop embryos at any stage in their reproduction. Fungi absorb their food, and their cell walls are composed of chitin.

gabbro A dark, coarse-grained intrusive igneous rock that has the same chemical composition as basalt, i.e., it is iron-rich. Instead of being extruded and cooled on the Earth's surface, it cools below the surface and thus has a coarsely crystalline appearance. See also *basalt*.

gamete Mature reproductive cell, such as egg, sperm, ova, or pollen.

gastroliths Swallowed stones used in an animal's gizzard to help grind up food and thus aid digestion. Gastroliths are associated with some birds, dinosaurs, plesiosaurs, and crocodilians.

genome, human The genetic code that distinguishes humans from all other species.

genus Division in the hierarchy of the classification of organisms. A genus may be composed of one, several, or many species. One or more genera comprise a family. In scientific classification, the genus is the first name used for any species; for example, in *Homo sapiens*, *Homo* is the genus name, *sapiens* the species name.

geoid The general term used to describe the imperfect surface shape of the planet.

geologic time scale A representation of the divisions of geologic time, in terms of both relative and absolute time. Since the divisions and boundaries of geologic time are subject to varying interpretation, numerous time scales have been proposed through the years. The GSA 1999 time scale is used in this book.

geomagnetic (1) Pertaining to the Earth's magnetic field. (2) Magnetic characteristics that can be measured and which are exhibited by rocks containing certain iron-bearing minerals. These include paleomagnetic directions and relative magnetic strength which is related to the amount of iron in the rock.

glacial age A period of relative global cooling that is measured in thousands of years. It is characterized by glacial advances, an increase in the albedo, isostatic depression of the continents, and by lowered global sea level. May contain stadial and interstadial fluctuations. See also *interglacial age*.

Glossopteris A genus of primitive seed-bearing plants, a gymnosperm, that lived in temperate regions of Gondwana during the late Paleozoic.

gnetales A small group of rare, modern gymnosperms that exhibit some features in common with the angiosperms.

Gondwana A linked assembly of Southern Hemisphere continents: Africa, South America, India, Australia, and Antarctica. Named after a locality in India and meaning the "land of the Gonds."

graben A valley formed by a down-dropped fault block. See also *horst*.

granite A type of intrusive igneous rock that is light-colored, coarse-grained and contains a relatively high proportion of aluminum. It is the same composition as rhyolite, but instead of being extruded onto the Earth's surface, it is intruded and cools beneath the surface. The major rock comprising continental basement rocks is granitic in character. The main constituent of continental crust. See also *rhyolite*.

grapestones Sand-sized carbonate grains often formed from fecal pellets that become cemented together.

graptolites Fossils having a saw-tooth appearance that were made by tiny, extinct, colonial, marine animals classified as hemichordates. Important Paleozoic index fossils especially for the Ordovician Period (490–443 MYA).

Great Unconformity The nearly worldwide boundary between Precambrian and Cambrian time, in which there is a major gap in the rock record. It is very well expressed in the Grand Canyon of the Colorado.

green algae Members of the protoctist phylum Chlorophyta which includes both unicellular and multicellular forms. The few genera that are calcareous are prime contributors on carbonate platforms. Previously classified as plants since they undergo photosynthesis.

greenhouse age Term used in this book for a time of general global warming, with decreased polar ice and increased carbonate production. The opposite of an ice age. Often corresponds to times of high global sea levels. Greenhouse gases are released by volcanic activity, rice-paddy fields, the digestive tracts of animals, and by the burning of rain forests and fossil fuels.

Grenville orogeny The mountain-building event that occurred 1,000–800 MYA when Baltica (Scandinavia and associated regions of Russia) collided with Laurentia (North America). Associated with a locality in Quebec, Canada.

guyots Flat-topped seafloor mountains formed along mid-ocean spreading centers and transform faults. See also *seamounts*.

gymnosperms Plants that produce seeds that are not covered by a carpel as angiosperm seeds are. They reproduce via wind-blown pollen; many produce cone-like structures for seed protection. The group includes seed ferns, conifers, yews, ginkgos, and cycads.

gyre An extensive circulation system of either water or air, comprised of numerous currents. An ocean gyre is driven by the action of wind on the surface of the sea and is augmented by the Ekman spiral effect. The center of an ocean gyre contains little if any current action.

hadrosaurs A family of diverse, bipedal, duckbilled dinosaurs that existed 85–65 MYA in western Asia and western North America. Many genera had elaborate crests on their heads that varied in shape and function according to species.

hard rock Term applied to any rock formed from intergrown crystals, i.e., all igneous and metamorphic rocks. See also *soft rock*.

herbage A growth of plants. Used in this book to denote a phytoplankton bloom.

herbivores Animals that feed upon plants to obtain the materials they need for metabolism.

Hercynian orogeny Period of mountain building in East Central Pangea (Western Europe) dated 345–280 MYA. Generally considered to be the same event as the Variscan orogeny. See also *Variscan orogeny*.

heterothermic A condition in animals that have imperfectly controlled blood temperature and that possess some degree of warm-bloodedness. Occurs in certain "cold-blooded" animals, including some insects, some sharks, all tuna, and in some mammals like the modern armadillo. See also *ectothermic, endothermic, homeothermic*.

holarctic Pertaining to the region surrounding the Arctic, its landmasses, Greenland, North America, and/or Eurasia, and any animal or plant that was confined to that region.

Holocene Epoch Division of geologic time covering the last 10,000 years. The upper division of the Quaternary Period dated from the last glacial retreat. The Holocene is the division of time in which we are living.

homeothermic Condition in which an animal's body temperature is controlled by its great size alone, i.e., the mass of the animal's body acts as a giant heat sink. This may have been the condition of many of the really large herbivorous dinosaurs. See also *ectothermic; endothermic; heterothermic*.

hominid Having the basic characteristics of humans. Any fossil that is considered a modern human or a primitive human, or that is considered to have walked upright with its arms free and with the big toe in alignment with the other toes. The oldest-known fossils of this character are dated about 4.5 MY.

hominoid Animals with a human-like appearance. A superfamily that includes hominids (humans), the greater apes (gorillas, orangutans, chimpanzees, and extinct apes), and the lesser apes (gibbons and other similar animals). The oldest fossils date to 22 MYA.

Homo erectus Hominids that lived 1.6–0.5 MYA and that migrated from Africa throughout Eurasia. They made well-shaped, stone hand tools, used fire, and may have constructed raft-like log floats.

Homo sapiens Term used in this book to refer exclusively to *archaic* humans. Fossil humans dating from about 500,000–200,000 BP who made specialized tools that were more advanced than those of *H. erectus*. They made the first decorated portable artifacts.

Homo sapiens neanderthalensis Scientific name for Neanderthals, who lived in Europe and the Middle East 100,000–*c.*30,000 BP. The name means "wise man from Neander" (Germany). See also *Neanderthal*.

Homo sapiens sapiens Scientific name for anatomically modern humans, who are thought to have evolved around 180,000 BP, but first appear in Europe about 35,000 BP. They developed advanced tool-making techniques around 32,000 BP.

horn corals Common name for horn-shaped rugose corals. A group of extinct corals that lived 500–248 MYA. They are index fossils for most of the Paleozoic.

horsts Name for upthrown blocks or ridges in block-faulted regions. See also *graben*.

hot spot Stationary volcanic center formed at the top of a rising mantle plume. As plates move across this feature, chains of volcanoes form on the seafloor. Doming above a hot spot may form triple-armed rifts.

hyaloclastites Term used to describe the texture of glassy angular grains of lava formed during explosive, shallow-water, volcanic eruptions.

hydrobleme A crater formed by a bolide striking the Earth in any of its oceans or seas.

hydrocarbon Any organic compound made of hydrogen and carbon only; includes coal, natural gas, and petroleum.

hydrolyze Decomposition of an organic substance that occurs as the result of exposure to and reaction with water. Specifically applied in this book to the process of breaking down protective compounds in the seed coat so that moisture is able to penetrate the seed, activate the embryo, and thus enable the seed to sprout.

hydrosphere All of the water on or near the Earth's surface, including groundwater, rivers, ice, oceans, etc.

hydrothermal vents Chimneys on the seafloor through which extremely hot, mineralized seawater is released. The water, which begins as cold seawater, percolates through hot volcanic rocks (pillow lavas), dissolving a variety of

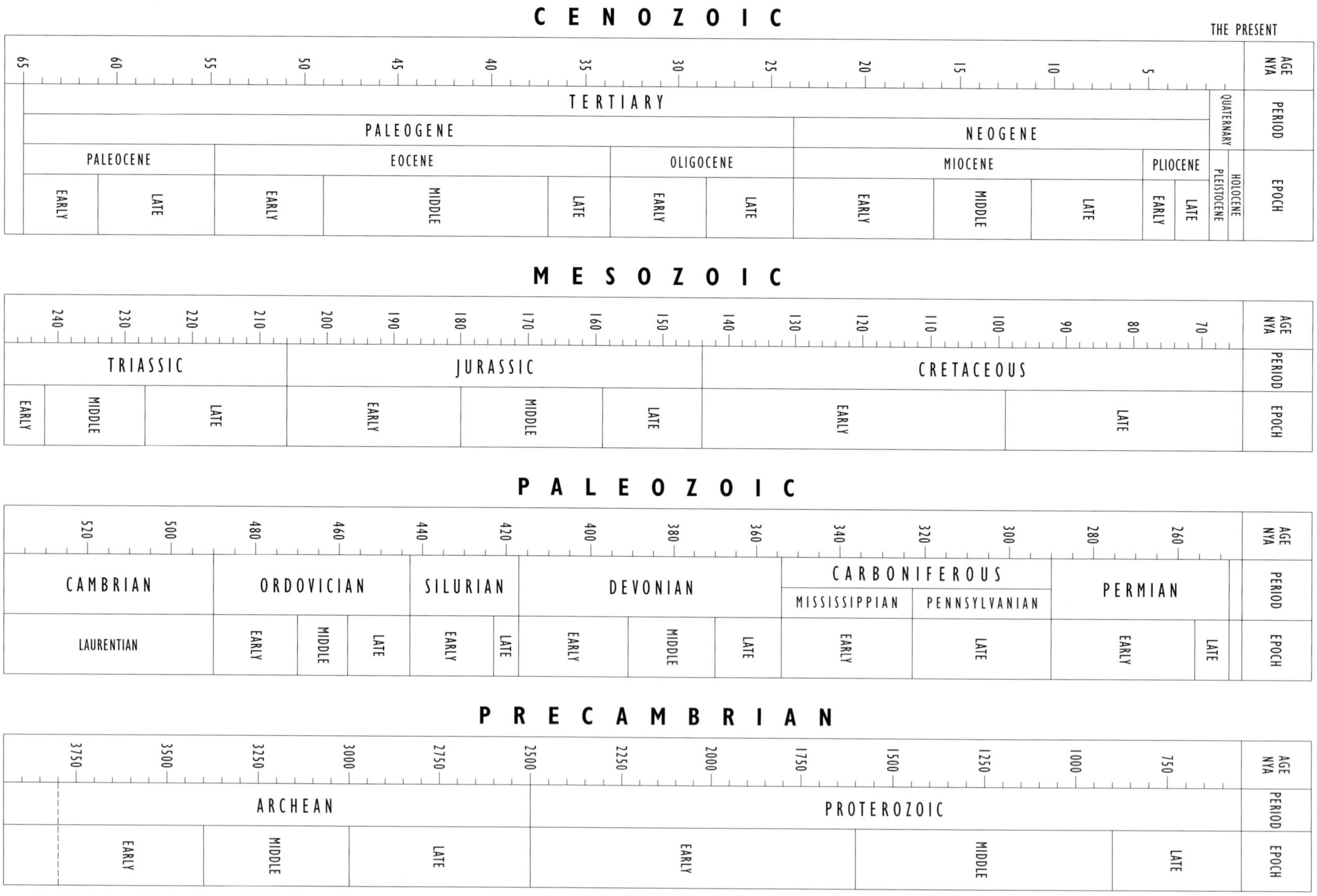

CENOZOIC

THE PRESENT

AGE NYA: 65 · 60 · 55 · 50 · 45 · 40 · 35 · 30 · 25 · 20 · 15 · 10 · 5

PERIOD		EPOCH	
TERTIARY	PALEOGENE	PALEOCENE	EARLY
			LATE
		EOCENE	EARLY
			MIDDLE
			LATE
		OLIGOCENE	EARLY
			LATE
	NEOGENE	MIOCENE	EARLY
			MIDDLE
			LATE
		PLIOCENE	EARLY
			LATE
QUATERNARY		PLEISTOCENE	
		HOLOCENE	

MESOZOIC

AGE NYA: 240 · 230 · 220 · 210 · 200 · 190 · 180 · 170 · 160 · 150 · 140 · 130 · 120 · 110 · 100 · 90 · 80 · 70

PERIOD	EPOCH
TRIASSIC	EARLY
	MIDDLE
	LATE
JURASSIC	EARLY
	MIDDLE
	LATE
CRETACEOUS	EARLY
	LATE

PALEOZOIC

AGE NYA: 520 · 500 · 480 · 460 · 440 · 420 · 400 · 380 · 360 · 340 · 320 · 300 · 280 · 260

PERIOD		EPOCH
CAMBRIAN		LAURENTIAN
ORDOVICIAN		EARLY
		MIDDLE
		LATE
SILURIAN		EARLY
		LATE
DEVONIAN		EARLY
		MIDDLE
		LATE
CARBONIFEROUS	MISSISSIPPIAN	EARLY
	PENNSYLVANIAN	LATE
PERMIAN		EARLY
		LATE

PRECAMBRIAN

AGE NYA: 3750 · 3500 · 3250 · 3000 · 2750 · 2500 · 2250 · 2000 · 1750 · 1500 · 1250 · 1000 · 750

PERIOD	EPOCH
ARCHEAN	EARLY
	MIDDLE
	LATE
PROTEROZOIC	EARLY
	MIDDLE
	LATE

This **time scale** is based upon a revised version (1999) of data published by the Geological Society of America for their Decade of North American Geology (DNAG) project 1980–90. The chart represents time in both relative and absolute terms. Relative dates are established by the chronological appearances and disappearances of fossils. Absolute time is calculated by measurements of the half-life decay of radioisotopes, in numbers of years elapsed from the present. It should be noted that in order to accommodate a time span of almost 4,000 MY, the divisions of geologic time are here progressively compressed. Thus the Cenozoic Era in which we live here covers a period of 65 million years, while the Precambrian time and its eons and their subdivisions cover a period of 3,250 MY within a similar space. It should also be noted that some of the absolute dates are very controversial—different isotopes can give very different results. In those cases this scale represents a midpoint between extreme views. [Note varying scale.]

chemicals contained in the seafloor, including metallic sulphides, compounds of carbon, sulphur, and oxygen. These vents may serve as oases for specialized forms of life on the seafloor.

hyoid bone A bone in the throat that is essential for articulating the vowel sounds *i*, *u*, and *a* and the sound *s*.

Iapetus Ocean Ocean that began to form about 650 MYA between ancestral North America (Laurentia) and the Baltic countries of Northern Europe (Baltica). An early ocean, thought to have been the ancestral, or proto, North Atlantic Ocean. The Iapetus was the first ocean to form during the breakup of a Precambrian continental clustering. It closed about 417 MYA, when Avalonia collided with Laurentia.

ice age Period of general global cooling lasting for millions of years that is characterized by episodes of increased glaciation and increased polar-ice accumulation, and normally lowered global sea levels. During an ice age, numerous glacial to interglacial cycles may occur.

ice cap An accumulation of ice that envelops mountain ranges, and although extensive it is a relatively localized phenomenon, e.g., Vatnajökull, in Iceland.

ice sheet An accumulation of ice that envelops part or all of a continent. May cover over 50,000 sq km (20,000 square miles), e.g., the Antarctic ice sheet.

ichthyosaur One type of marine reptile that lived about 248–100 MYA. They vaguely resembled porpoises with paddle-like flippers and a fish-like tail, and probably gave birth to live young. Their entire life cycle was in the sea.

Ichthyostega One of the oldest-known amphibians, an ancestral amphibian. A carnivore that lived about 365–354 MYA.

ignimbrite Compacted volcanic ash formed from hot, rapidly moving, dense ash flows that are released during violent volcanic eruptions.

impact theory A theory for the formation of the Moon which proposes that a Mars-sized object struck the Earth, causing the melting of both objects and the ejection of a huge filament of gaseous material, similar in shape to the plume of a solar flare; part of this eventually coalesced to form the Moon. It is also suggested that such an impact caused the Earth to tilt on its axis and to fractionate into core, mantle, and crust.

index fossils Fossils used by paleontologists to deduce the relative age of sedimentary rocks. Index fossils are normally distinctive and very widespread; their species are limited to a relatively short period of geological time.

infusoria Term used during Darwin's time to define microscopic "animals and plants," now known to be various types of protoctists.

inner core The innermost part of the Earth. It is a solid sphere of iron-nickel alloy with a mushy irregular surface and a radius of almost 1,600 km (1,000 miles).

interglacial age An interval of relative global warming measured in thousands of years and characterized by glacial retreat, a decrease in the albedo, rebounding of continents, and generally by a rise in global sea level. Occurs between glacial periods, and may contain stadial and interstadial fluctuations. See also *glacial age*.

intermediate-depth water Any mass of water which, because of relative density (which is a factor of salinity and/or temperature), occupies an area above the bottom waters but below the surface waters in the ocean.

interpolar circulation Ocean circulation flowing from pole to pole, i.e. from north to south, and south to north.

interstadials Relatively short pulses of warming that occur within a glacial or interglacial age. They are thought to coincide with Milankovitch cycles. See also *stadials*.

intrusive (1) Term applied to igneous rocks that intrude pre-existing rock beneath the Earth's surface and solidify. (2) Term applied to the process of the emplacement of molten rock into pre-existing rock. See also *extrusive*.

Inuit The preferred name of the native Arctic peoples commonly called "Eskimos."

invertebrate Any animal that does not have a backbone. See also *vertebrate*.

iridium Element of the platinum group found in relatively high concentrations in some metallic bolides. Though it is rare on the Earth's surface, it may be extruded from the mantle during volcanic eruptions. Iridium is also associated with certain types of bacterial activities.

iridium anomaly A zone with an anomalously high concentration of iridium. The best-known anomaly occurs at the Cretaceous-Tertiary (K/T) boundary.

Island Arawak An Arawak subculture in the Caribbean that adapted mainly to a maritime life and largely occupied the Bahama Islands.

island arc A substantial chain of volcanic islands that forms on a plate margin above a subducting plate, e.g., the Lesser Antilles and Japan. They are bounded on one side by a trench formed by the subducting plate.

isostasy An adjustment process that causes a continental mass to sink or rise in order to compensate for the weight of sediments, or seas or ice, as they increase or decrease in depth.

isostatic depression Relative sinking of a continental mass beneath the accumulating weight of ice sheets or epicontinental seas. See also *isostatic rebound*.

isostatic rebound Relative rising of a continental mass that occurs when the weight of an ice sheet or epicontinental sea is removed. See also *isostatic depression*.

isotope Various forms of an element which have the same number of protons in their nucleus but differing numbers of neutrons.

jawless fish A class of fish that evolved about 500 MYA. They have cartilaginous skeletons, are toothless, and lack jaws, and thus feed by sucking or scavenging. The sole survivors of this class are hagfish and lampreys.

joint (l) A fracture in a rock. (2) A hinged structure such as an insect's leg.

Jurassic Period That period of geological time dating 206–144 MYA, named after the Jura Mountains in Switzerland. The middle period of the Mesozoic Era.

K/T event A proposed bolide impact or impacts corresponding with the end of the Cretaceous (K) Period and the beginning of the Tertiary (T) Period. This event is suggested as the cause of the Cretaceous mass extinction. Evidence for the impact includes a clay layer called the "iridium layer," which contains a very high proportion of iridium, along with shocked quartz grains and tektites (glass-like spherules).

kerogen A black-colored, greasy solid formed by heat and pressure from preserved organic matter. Additional heat and pressure over millions of years may convert kerogen into oil and gas.

kingdoms The highest division of the classification of living things. Today five kingdoms are recognized: Monera, Protoctista, Fungi, Plants, and Animals.

landbridge A low-lying broad expanse of land that usually appears at times of low sea level, tens to thousands of miles wide, connecting one continent to another, or to another landmass such as a large island.

Laramide orogeny Mountain-building period 80–40 MYA, which formed the second generation of Rocky Mountains.

lateen rig Triangular-shaped sails that can be maneuvered to receive the wind on either side of the sail. The sails are attached to tilted masts, which allow for considerable maneuverability.

Laurasia A northern megacontinent consisting of most of modern-day North America, Greenland, and most of Eurasia.

Laurentia A continental plate consisting of most of modern North America, Greenland, and the Barents shelf.

Laurussia A megacontinent that existed before 700 MYA consisting of most of North America, Greenland, Siberia, the Baltic countries, and the Barents shelf (Svalbard).

leeward Of or pertaining to the side that is sheltered from the wind.

lenses Relatively thin bodies of rock that are confined by other layers of rock and that have an overall shape resembling glass lenses.

Levallois technique An advanced, stone tool-making technology in which a chert nodule was first shaped into a core with a flat platform at the base. The platform of the core was then struck to produce axes, knives, etc.

limestone A type of sedimentary rock made of calcium carbonate that often forms in warm shallow water. It may be formed as the product of shell or algae debris, but can also be formed by inorganic processes.

lithified sediments Sediments that have been cemented and/or subjected to enough pressure (burial) to turn them into sedimentary rocks.

lithosphere The thin, rigid outer shell or crust of the Earth that is fractured into seven large plates and numerous smaller plates. It is composed of the uppermost portion of the upper mantle and the continental and ocean crusts. This layer overlies the asthenosphere, and together these two layers make up the tectosphere. See also *crust; tectosphere*.

Little Ice Age A period of relative cooling and glacial advance that lasted about 500 years during historical times. This period began in the 13th century and continued to the mid- to late 19th century. Sometimes specifically used to describe the period 1550–1850.

lobefin fish An order of fish whose fins have a hand-like bone construction and who had internal air sacs for enhanced buoyancy. They include lungfish, coelacanths, and rhipidistians. This group of fish is ancestral to all tetrapodal vertebrate animals—amphibians, reptiles, birds, and mammals. This order evolved about 390 MYA.

Lystrosaurus An amphibious mammal-like reptile that lived about 240 MYA. Its fossils are found in Africa, India, and Antarctica and are considered part of the proof that those continents were connected at that time.

magma chamber A large reservoir of molten rock not far beneath the Earth's surface.

magnetic anomaly A magnetic reading in rocks in which the magnetic minerals in the rock do not align with the existing magnetic field.

magnetic pole Either end of the Earth's magnetic dipole, especially applied to the positive pole or to the Earth's present north magnetic pole, as distinct from its north geographic pole. See also *north magnetic pole*.

magnetic quiet An extended period of time, 120–84 MYA, during which the Earth's magnetic pole did not reverse except for brief intervals.

magnetic reversal The Earth's magnetic field reverses polarity episodically, so that north and south magnetic poles switch position. A period when the north (positive) magnetic pole is approximate to the north geographic pole is considered a normal interval, whereas a period when the positive magnetic pole approximates the south geographic pole is called a reverse interval.

mainmast The larger mast in a ship on which the mainsail is mounted.

mammal-like reptiles A group of fossil reptiles called "synapsids," which are considered ancestral to the mammals. These animals lived about 286–221 MYA and included the pelycosaurs and therapsids.

mammals A class of vertebrates that are generally endothermic, with hair or fur, that nurse their young on milk produced by mammary glands. Modern mammals are classified as either monotremes, marsupials, or placentals.

mantle The very hot but solid layer of the Earth between the core and the upper lithosphere, which is made up of iron- and magnesium-rich oxides and silicates. It is divided into a thicker layer, the lower mantle, and a thinner layer, the upper mantle. Heat from the Earth's core is transferred to the Earth's surface through the mantle by way of slow-moving convection cells, a movement termed "mantle creep." See also *asthenosphere; lithosphere; upper mantle.*

mantle plume A region in the mantle where hot, circulating currents meet and rise to the surface. The surface manifestation of a mantle plume is called a "hot spot." See also *hot spot*.

Manueline style A style of architecture named after King Manuel I of Portugal (1495–1521).

margins The edges of the continents. They may be passive, in which oceanic crust is joined to the continental crust (like those on either side of the Atlantic Ocean); or they may be active, in which the oceanic crust is being subducted beneath the continental crust (like those surrounding the Pacific Ocean).

marine reptiles Any of several groups of air-breathing reptiles that adapted for life in the sea. Many of these animals including ichthyosaurs and plesiosaurs are now extinct.

marine snow A term used in this book for the whitish-colored, minute particles of carbonate material that is either suspended in or slowly sinking through sea water. It consists of microfossil shells, fecal pellets, and precipitated calcium carbonate particles.

maritime province An area whose generally equable climate is largely regulated by its proximity to a sea or an ocean.

marsupial Any mammal that has a combined urinary and anal canal but a separate reproductive tract. A mammal that gives birth to very immature young, which are transferred to a pouch to be suckled and continue to develop.

mass extinction Period of higher than normal extinction or background extinction, generally much higher. The period may be nearly instantaneous by geologic standards, lasting for less than a few thousand years, or prolonged, lasting up to 10 MY. Five major mass extinctions, along with numerous smaller events, are recognized in the fossil record.

mass spectrometry Method of measuring isotope ratios, used for computing absolute-age dates based on radioactive isotope content. Considered to be the most accurate method today.

matrix Any generally finer-grained component rock that contains larger particles, fossils, or even blocks of a different kind of rock.

maturation The conversion of organic material, often kerogen, into natural gas or petroleum. There is a maturation envelope based on heat and pressure within which hydrocarbons develop. Anything outside that envelope is either still kerogen or is overcooked (overmature) so that all volatile material has been destroyed.

Meadowcroft At present the oldest -dated, best-authenticated, and longest continually used human habitation in North America: levels are confidently dated at not less than 14,000 BP and may date back to 19,000 BP.

megacontinent An assemblage of two or more continents, such as Gondwana or Eurasia.

megafauna (1) In paleontology, an assemblage of animals that are larger than microscopic size. (2) In paleoanthropology, a term used to denote larger mammals such as mastodons and bison.

megashear A major strike-slip fault that extends for hundreds or thousands of miles.

meiosis A special type of cell division in sexual reproduction, in which only one chromosome of each chromosomal pair is contained in the newly formed cells. These cells, gametes (sperm, egg, ovum, pollen, etc.), recombine during fertilization to form a new individual.

Mesoamerica Term used by anthropologists to describe those parts of Mexico and neighboring Central America which were civilized at the time of the arrival of Europeans.

mesopelagic zone The middle zone of the sea where sunlight can be detected only by using delicate instruments. This zone is below the epipelagic zone. Organisms that float or swim in this zone are called mesopelagic organisms.

Mesozoic Era Division of geologic time dated 248–65 MYA. The middle era of geologic time. It is divided into three periods: Triassic, Jurassic, and Cretaceous.

messenger RNA (mRNA) Messenger ribonucleic acid. Complex molecules that ensure that the proper genetic material matches up with a DNA strand when a cell is duplicating itself, or following fertilization.

mestizo A person whose parents belong to different racial groups, such as a mixture of Caucasian and Native American ancestry. Applied specifically to a mix of Spanish and Meso-american ancestry.

metabolism The complete process of the breakdown and absorption of food and the transfer of its latent energy.

metalliferous Containing or bearing metals. For example, a metalliferous rock is a rock that contains metals, such as gold, copper, lead, or iron, that may be extracted by heating, leaching with chemicals, or some other technique.

metamorphism The process of altering the mineralogical composition of a rock through exposure to heat and/or pressure, either by being adjacent to heated fluids or through deep burial and/or folding of the rock.

metasedimentary rocks Rocks of sedimentary origin that have been partially altered (metamorphosed) by heat and pressure.

meteor A stone or metallic space object that makes a streak of light in the sky as it enters a planet's atmosphere. Commonly called a shooting star.

meteorite A remnant of a meteor that strikes a planet's surface.

microcontinent A relatively small continental mass generally on its own tectonic plate. It may assemble with others to form continents; e.g, South China, North China, Tibet, Siberia, and others to form Asia.

microfossil Any fossil that is microscopic in size, such as a foraminifer or radiolarian.

micropaleontology The study of microfossils.

microplates Very small tectonic plates, each bearing its own oceanic or continental mass and behaving independently from other plates/microplates, e.g., remnant Caribbean plate.

microspheres Term specifically applied in this book to microscopic orbs that spontaneously form when dry amino acids are wetted out in the laboratory.

Mid-Atlantic Ridge (MAR) The spreading-center that bisects the Atlantic Ocean from the north of Iceland southward to near the Falkland Islands. It is characterized by a range of flat-topped submarine mountains, and is offset by numerous transform faults.

Mid Central Pangea In this book, that portion of Central Pangea between the Straits of Gibraltar and the Florida-Bahama block, and which corresponds today to the Eastern seaboard of the United States and northwestern Africa.

migration Movement or seepage of hydrocarbons through the pores in rocks and along fractures into neighboring porous rocks.

Milankovitch cycles Cyclic variations of climate that respond to eccentricities of the Earth's orbital geometry, including the angle of tilt of the Earth's axis, the shape of the Earth's orbit and the amount of wobble in the Earth's spin. The cycles occur in increments of 100,000, 43,000, 24,000, and 19,000 years, and are named after the Yugoslavian mathematician Milutin Milankovitch, 1879–1958, who first proposed them.

Miocene Epoch The division of geologic time dated 23.7–5.3 MYA. The fourth epoch of the Tertiary Period. Grasslands and savannahs became widespread during this time, along with an explosive radiation of insects, rodents, frogs, snakes, and birds.

Mississippian Period Division of geologic time used in North America, dated 354–323 MYA; corresponds to the Lower Carboniferous Period in Europe. The fifth period of the Paleozoic Era.

mitosis A process of cell division in which the chromosomal pair separate, and then duplicate themselves. Following chromosomal duplication, the cell divides into two cells, each of which are identical to the original parent cell.

mizzenmast The third or sternward mast on a three-masted square-rigged ship. It is stepped on the keel forward of the stern post and carries the mizzen sails, which may be lateen-rigged and which are partially used for steering.

mollusks A very large and diverse phylum of invertebrate animals both with and without shells, that includes, clams, snails, octopi, and nautilus. The name means "soft-bodied."

Monera Kingdom The kingdom that contains all prokaryotes—organisms without nuclei, i.e., the bacteria and cyanobacteria.

monocline A large, asymmetrically folded geologic structure in which the rocks appear to have been folded on only one arm or side of the structure, i.e., one side is steeply dipping while the other is gently dipping.

monocotyledon One of two classes of angiosperm, generally called "monocots," whose embryos have only one leaf-like organ that becomes the plant's primary leaf upon sprouting. Monocots have flower parts (petals, stamens, etc.) that are usually divisible by three, parallel veins in their leaves, and non-woody stems. Their vascular bundles are scattered at random throughout the stem and their pollen grains have only one furrow. See also *dicotyledon*.

monotreme An order of mammals that have a single opening which serves as a combined reproductive tract, urinary tract, and anal canal. Modern monotremes (duckbilled platypuses and echidnas) lay eggs but suckle their young.

Moon capture theory A theory that proposes that a smaller planet was captured by the Earth's gravitation and began to orbit the Earth. See also *impact theory*.

motility The ability of some micro-organisms to move by themselves.

mountain belts A chain of mountains or mountain ranges thought to have been formed from the same tectonic event or series of events.

mud A sediment made mostly of clay-sized particles—particles of a specific size, in geology.

MY General term for million years.

MYA General term for millions of years ago.

nappe A geologic structure in which blocks of layered rock units are overthrust and often folded back on themselves.

natural selection A theory of evolution that proposes that individuals who are better adapted or more vigorous will survive and thus reproduce, whereas less-adapted or less-vigorous individuals will not be able to compete as well and will be eliminated from the population.

nautilus A cephalopod with a coiled shell that has smooth, gently curved septa and a central siphon conduit, a siphuncle. They are marine predators. Ancestors of nautilus evolved about 500 MYA. Fossil animals with similar characteristics are called "nautiloids."

Navajo A Native American culture thought to be derived from the Athabascan culture, which may have arrived in North America 14,000–12,000 BP. The Navajo, along with the Apache, migrated to the Southwest about 700 years ago.

Neanderthal (*Homo sapiens neanderthalensis*) A subspecies of humans that lived in Europe and the Near East 100,000–35,000 BP. Named after the site of discovery in the Neander valley near Düsseldorf in Germany. They used fire, sewed clothing, made well-formed tools, and probably had the advantage of speech.

nectary Plant glands which secrete either sugar-based solutions and perfumes to attract certain insects or other pollinating animals, or strong poisons such as alkaloids to repel unwanted predatory insects.

nekton Animals that swim in the sea, such as fish or squid, and that are not hostage to the winds and currents. See also *plankton*.

Neogene Period of geological time, dated 23.7–1.8 MYA. The upper division of the Tertiary as used in some geological time scales. It is time equivalent to the Miocene and Pliocene Epochs.

nonvascular plants Land plants that have only a rudimentary system for the transport of fluids and nutrients. They are inevitably small and are tied to a moist environment in which to live and reproduce. They are rootless and leafless, but have creeping stems with short branches. First appeared around 423 MYA, and include mosses and liverworts. See also *bryophytes*.

Nordic Sea The northernmost extension of the Atlantic Ocean that lies between Greenland and Europe and includes the Greenland, Lofoten, and Norway Basins.

North Atlantic deep water The cold, dense, highly saline water mass that overflows from the Nordic Sea basin and flows southwards along the bottom of the ocean basin until it meets with the even denser Antarctic Bottom Water flowing northwards. The North Atlantic deep water then overrides the Antarctic Bottom Water, continues southward, and eventually rises to the surface near Antarctica.

North Atlantic Ocean That portion of the Atlantic Ocean which includes the Nordic Sea, the Iceland-Faeroe Ridge, and the Labrador Sea.

north magnetic pole The positive pole of the Earth's geomagnetic dipole where the magnetic force is strongest. It is distinct from the Earth's north geographic pole. First located in 1831, it has since slowly changed location through time.

obduction An occasional tectonic process in which part of the ocean floor, along with its sedimentary cover, overrides the edge of a continent. See also *accretionary prism; ophiolites*.

oblate The shape of an orb that is flatter at its poles and distended at its equator.

obliquity A variation in the tilt of the Earth's axis from a 21.8° minimum to 24.4° maximum that occurs in a 43,000-year cycle; it affects the geographical distance of the polar circles from the poles, i.e., the area of winter darkness, and thus affects climate. An increase in the radius of the polar circle causes global cooling; a decrease causes global warming.

ocean crust That portion of the Earth's crust beneath the oceans that was initially formed at mid-ocean spreading centers. It is thin, averaging 5–10 km (3–6 miles) in thickness, and is composed of three layers: pillow lavas on top, vertical dikes between, and gabbro (intruded basaltic rock) beneath. It is denser than continental crust, since it contains a higher proportion of magnesium and iron.

ocean cycle Term applied to the formation and destruction of oceans. See also *Wilson cycles*.

Olenellus Genus of trilobites that lived around 540–520 MYA. Important index fossils whose presence on both sides of the Atlantic provided partial evidence of a proto-Atlantic Ocean.

Oligocene Epoch Division of geologic time dated 33.7–23.8 MYA; the third or middle epoch of the Tertiary Period.

omnivore An animal that eats both vegetation and flesh.

oolites A sediment of egg-shaped grains normally formed from successive concentric layers of calcium carbonate. The grains can be formed by algal coatings or from inorganic precipitation on rounded grains that are rolled about by wave action on shallow shelves.

ooze Soft, watery sediments that form as pelagic skeletal material and clay settle to the seafloor.

ophiolites Rock sequences that are portions of ocean-floor debris and sedimentary coverings that have been thrust up and onto continental margins or other oceanic plate margins. These rock sequences frequently include metalliferous strata. See also *accretionary prism; obduction*.

order Division in the hierarchy of the classification of organisms. One or more families comprise an order, one or more orders a phylum.

Ordovician Period A division of geologic time dated 490–443 MYA. The second period of the Paleozoic Era.

orogeny Episodes of mountain building, often resulting from continental collision, which cause chains or belts of mountains to form often thousands of miles long.

osmosis The process of transmission of fluids through a semipermeable membrane caused by variable concentrations of solutions on opposite sides of the membrane.

outcrops Surface exposures of rock units.

outer core The outer layer of the Earth's core, which is a white-hot semiliquid mix of mainly iron and other elements, perhaps nickel and sulphur, about 2,250 km (1,400 miles) thick. It has turbulent, paired convection cells that transfer heat from the inner core to the mantle. The rotation of the cells is thought to generate the Earth's magnetic field. See also *core; inner core*.

overprint A process in which a later tectonic event superimposes new structures on top of older structures which formed in an earlier tectonic event.

overthrust A structural process in which one set of rocks is "pushed" up and over another set of rocks along a fault, as the result of compression on the plate.

oviparous Term applied to animals that lay their eggs, which then mature and hatch.

ovoviviparous Term applied to animals that give birth to live young from eggs that mature and hatch within the female's body.

ovule A female part of a flower that contains the ovum which, following fertilization, becomes the developing seed.

ozone A specific molecular form of oxygen (O_3). The ozone layer is an ozone-rich region that forms in the upper atmosphere and which partially absorbs the Sun's ultraviolet radiation.

pachycephalosaurs A suborder of bipedal "helmet-headed" dinosaurs which measured 2–4.5 m (6–15 ft) in overall length and weighed from a hundred (45 kg) to several hundred pounds. They were part of the third dinosaur regime and lived around 110–70 MYA.

Pacific realm A term applied to the region in which the trilobite *Olenellus* is normally found, i.e., much of North America.

pack ice Any area of sea ice, generally more or less a cohesive mass. May include areas that contain icebergs that are being moved by currents.

Paleocene Epoch A division of geologic time dated 65–54.8 MYA. The first epoch of the Tertiary Period.

paleoclimate A term applied to prehistoric climatic conditions.

Paleogene Period of geologic time, dated 65–23.8 MYA. The lower division of the Tertiary as used in some geological time scales. It is time equivalent to the Paleocene, Eocene and Oligocene Epochs.

paleogeography The study of ancient geography.

paleontology The study of ancient life through fossils and trace fossils.

Paleozoic Era A division of geologic time dated 543–248 MYA. It is divided into six periods (seven in North America): Cambrian, Ordovician, Silurian, Devonian, Carboniferous (Mississipian and Pennsylvanian in North America), and Permian. The name means "ancient life."

Pangea The name applied to the ever-shifting loosely amalgamated supercontinent that existed about 250–200 MYA. It was a clustering of all the larger continental plates and most of the smaller plates. A few of the smaller plates formed a shifting chain around the Tethys Ocean.

Panthalassa The name applied by Wegener to the ocean that surrounded Pangea.

Paradoxides A genus of trilobites that lived around 520–512 MYA. Important index fossils whose presence on both sides of the Atlantic provided partial evidence of a proto-Atlantic Ocean.

parasitic protoctist A protoctist that lives on or in another living organism and often causes damage to its host.

Paratethys An inland seaway that was a remnant of the eastern Tethys Ocean, which was closed off during the collision of India and other smaller plates, such as Turkey and Iran, with Asia. Today the Paratethys is reduced to the Black, Caspian, and Aral Seas.

passive margins Boundaries on separating plates that are free from volcanic activity and earthquakes, such as those on either side of the Atlantic Ocean. See also *active margins*.

patch reefs Patches of coral, algal, and sponge growth often on the floor of a lagoon or platform behind a fringing reef. Modern patch reefs are dominated by brain corals and sponges, which are resistant to more turbid tidal water.

peat Organic debris that accumulates from leaves and wood impregnated with tannin, in swampy areas where little or no decay occurs.

pelagic (1) Pertaining to the complete aquatic environment or region of the open ocean that is subdivided into life zones based on the amount of light penetrating the water. (2) Refers to any swimming or floating organism that lives in the sea (plankton or nekton). See also *epipelagic; mesopelagic; bathypelagic.*

pelletoidal Of or pertaining to excreted fecal pellets or other small, rounded grains.

pelycosaurs The earliest mammal-like reptiles. They first appeared about 300 MYA and became extinct about 250 MYA. Some of them had sails on their backs. They are the ancestors of the therapsids and are *not* related to the dinosaurs.

peneplain see *peneplane*

peneplane An extensive, nearly flat, undulating landscape produced by erosional forces. [Originally spelled 'peneplain' (Davis 1889): this form was replaced (Johnson 1916) to avoid giving the expression unintended connotation: Bates and Jackson—AGI *Glossary of Geology.*]

peneplanation Process of forming a nearly flat landscape through erosion.

Pennsylvanian Period Division of geologic time used in North America, dated 323–290 MYA; equivalent to the Upper Carboniferous of Europe. The sixth of seven periods of the Paleozoic Era. Named after the state of Pennsylvania, where thick and extensive coal beds that characterize this period were first recognized.

periods Divisions of geologic time, which together equal an era. Periods in turn are subdivided into epochs.

permafrost A phenomenon that occurs in arctic, subarctic, and alpine environments, in which a portion of the soil or subsoil remains permanently frozen year-round. In summertime, it may have an "active layer" at its surface, which melts or softens before refreezing in winter.

Permian Period Division of geologic time dated 290–248 MYA. The last period of the Paleozoic Era, named after the Perm province of Russia. This period ended with a prolonged mass extinction, the largest recorded in the fossil record (255–250 MYA).

Phanerozoic Eon Division of geologic time dated 543 MYA to present. Consists of the Paleozoic, Mesozoic, and Cenozoic Eras.

phosphates Chemical compounds of phosphorus and oxygen (phosphoric acid, PO_4^{-3}) that provide basic nutrients to plants and plant-like protoctists. In the epipelagic zone of the ocean these compounds are quickly removed from organic matter, mixed with seawater, and provide immediate nourishment for the primary producers (algae).

photophyta Archaic term used to describe plant-like protoctists.

photosynthesis Process that occurs in some prokaryotes and protoctists, and all plants, in which chlorophyll reacts with sunlight to convert carbon dioxide into carbohydrates. In this process oxygen is produced as a waste-product.

phylum Division in the hierarchy of the classification of organisms. One or more orders comprise a phylum; one or more phyla comprise a kingdom.

phytosaurs A group of early, predatory aquatic reptiles dated 230–206 MYA, that resembled crocodiles in overall appearance.

pillow lavas Lava takes this form when extruded directly into water below a critical depth and, therefore, under a critical pressure. The lava occurs as rounded to flattened masses, each enclosed by a rind that forms upon contact with the water. Pillow lavas comprise the top of three layers of oceanic crust. Though generally formed deep beneath the sea, they can also develop during subglacial eruptions.

pingos Frost-heave mounds of frozen ice and soil that assume a conical shape as the ice lenses within them grow in size and heave up the surface of the summer-softened active layer in permafrost conditions. The ice lenses are fed by percolating meltwater that forms moats around the pingos in summertime. Some pingos contain ice lenses that are many thousands of years old.

pinnipeds The general name applied to carnivorous marine mammals whose legs are modified into flippers. They are divided into the earless seals, like the harbor seal, and the eared pinnipeds, including walruses, sea lions, and fur seals. Pinnipeds first appear about 24 MYA. There is disagreement as to whether the two groups evolved from a common ancestor, or whether they are the result of parallel evolution: separate lines assuming similar form in response to similar ways of life.

placentals Mammals having separate body openings for the anal canal, urinary tract, and reproductive system, and in which a placenta envelops the fetus in a watery environment throughout gestation. The placenta permits oxygen and nutrients to be transferred to the fetus from the mother and allows for the removal of waste products from the fetus during gestation.

planetesimals Small, rotating spheres amassed during the formation of the solar system from amalgamated nodules of matter. Planetesimals are thought to have aggregated to form both planets and/or moons.

plankton Term used to describe the collection of both minute and relatively large organisms that populate the epipelagic zone. Plankton have either seriously limited powers of self-propulsion or have none at all. They are swept along by the wind or ocean currents.

Plant Kingdom Multicellular, sexually reproducing eukaryotes, all of which contain chlorophyll and undergo photosynthesis; all organisms that produce an embryo within maternal tissue at some stage in their reproduction and do not produce blastulas. They have cell walls made of cellulose.

plate Any of the many mobile parts into which the lithosphere is fractured. There are seven huge plates and numerous smaller ones. Each is autonomous but reacts to the movement of all other plates. Plates can separate, collide, shear by transform faulting (slide past each other), subduct (slide beneath each other), or obduct (thrust up over one another).

plate boundary The margin or boundary between two plates, generally located on ocean floors. Plates interact with one another along plate boundaries.

plate tectonics The study of the movement of plates through time. Formerly called the "theory of continental drift."

Pleistocene Epoch Division of geologic time dated 1.8–0.01 MYA; the first epoch of the Quaternary Period. The division of time that corresponds with the later advances and retreats of the Cenozoic Ice Age in the Northern Hemisphere. It is followed by the Holocene, the division of time in which we are living.

plesiosaurs Carnivorous marine reptiles that lived about 206–65 MYA. These animals had flippers on all four limbs, large rounded bodies, and short tails; some had long necks. It is thought that they laid their eggs on shore but spent the rest of their lives at sea.

Pliocene Epoch Division of geologic time dated 5.3–1.8 MYA; last epoch of the Tertiary Period. The Cenozoic Ice Age is thought to have begun in the Northern Hemisphere during this epoch.

pluton An intrusive igneous rock mass, generally very large. Several plutons that repeatedly intrude a given region at various times comprise a batholith.

polar wandering Term used to describe changes in the position of the magnetic poles relative to the rotational (geographic) poles.

polymer A large, complex molecule formed by the linkage of many identical smaller molecules, which spontaneously pair and bond to form a variety of complex compounds. These include carbon-based compounds such as nucleic acids, proteins, and other building blocks of life.

polynya An open area in an otherwise frozen sea within the Arctic Circle that either thaws early and freezes late in the summer or stays open permanently. Polynyas form when strong currents pass over a shallow seafloor. They attract both terrestrial and marine fauna and therefore drew the attention of early Arctic people.

polyps In biology a term applied to (1) the individual soft-bodied animals that together build coral heads by secreting carbonate in the form of aragonite; (2) any simple, hollow, tubular-shaped animal whose body terminates with an anterior mouth surrounded by tentacles.

Precambrian Division of geologic time dated 3,800–543 MYA that originally was considered to represent the time before life began. However, several soft-bodied fossil faunas dating back to 680 MYA have now been discovered in the upper Precambrian; stromatolite reefs date back to 2,500 MYA; and the earliest life forms may date to nearly 3,800 MYA. See also *Archeozoic Eon*; *Proterozoic Eon*.

precession of equinoxes Phenomenon in which the season of closest approach to the Sun gradually precesses from summer to winter and back to summer in each hemisphere—the spring and fall equinoxes occur earlier each year. About 10,000 years ago, at the end of the last glacial age, the seasons in the Northern and Southern Hemispheres were reversed. January would have been midsummer in the north and midwinter in the south. This process is caused by the gyration or wobble in the Earth's axis of rotation and occurs in 24,000- and 19,000-year cycles.

primate Any mammal belonging to the order that includes tarsiers, lemurs, monkeys, apes, and humans. It is thought that primates evolved in North America about 65 MYA.

Proconsul The first unequivocal hominoid fossils dated 22–16 MYA; found in East Africa.

prokaryotes The oldest-known life forms. All nonnucleated organisms, including bacteria and cyanobacteria, most of which are single-celled, but some form multicellular colonies. They reproduce by binary fission. All belong to the Monera Kingdom.

Proterozoic Eon Division of geologic time dated 2,500–543 MYA; one of two divisions of the Precambrian interval. See also *Archeozoic Eon*.

proto Prefix meaning "first" or "earliest form of."

proto-Atlantic Ocean Term used by J. Tuzo Wilson for an ocean he hypothesized had preceded the Atlantic Ocean, an ocean now called the Iapetus Ocean.

proto-Central Atlantic Name applied in this book to the narrow, elongate seaway that formed between North American Laurasia and West African Gondwana as Pangea began to break apart. It extended from Nova Scotia and Morocco, to the Florida-Bahamas and Senegal.

Protoavis Fossil found in West Texas, dated 225 MYA, that may replace *Archaeopteryx* as the oldest-known fossil bird.

The animal appears to have had a reptilian brain, scaly head, feathered body, and an avian wishbone.

Protoctista Kingdom Any eukaryote that does not reproduce by either producing a blastula or an embryo within maternal tissue, or by producing spores. This kingdom includes twenty-seven phyla. Members of this kingdom, called "protoctists," include foraminifers, radiolarians, coccoliths, green algae, and many other types of both single-celled and multicellular organisms.

protozoa Archaic term used to describe protoctists that were animal-like.

psychrosphere (1) The mass of cold bottom water that forms off Antarctica and slowly circulates (over centuries) along the bottom of the Atlantic, Pacific, and Indian Oceans. (2) The masses of cold bottom water that occupy abyssal ocean basins.

pterodactyls Suborder of pterosaurs that lived about 150–67 MYA. They had long faces and necks and short tails. Some had large, elongated crests on their heads.

pterosaurs The order of flying reptiles, with membranous wings, that evolved before 215 MYA and became extinct about 67 MYA. They were diverse and widespread animals, varying in size from small sparrow-sized animals to those as large as hang gliders. Early pterosaurs had short necks and long tails. Late pterosaurs are called "pterodactyls."

pycnocline Zone of changing water density in an ocean that acts as a barrier for currents of different densities and thus contributes to the stratification of the ocean.

quadrant A navigational instrument used to determine position of latitude by measuring altitudes of the Sun or of a star.

quadrupedal Characterized by walking on four legs. See also *bipedal*.

Quaternary Period Division of geologic time dated 1.8 MYA to the present. The most recent period of the Cenozoic Era. It is divided into two epochs, Pleistocene and Holocene, and encompasses most of the last ice age in the Northern Hemisphere and the time that has elapsed since the last glacial retreat. Name was proposed by Jules Desnoyers in 1829 for very young strata in the Paris Basin.

radiation (1) In biology, the dispersion or expansion of organisms into new regions or new habitats. (2) In physics, the process of transmitting energy via waves or particles.

radioisotopes Isotopes of elements that undergo spontaneous radioactive decay, which results in the emission of radiation and the formation of a new isotope or element. Measurement of the half-life decay of some radioisotopes, such as ^{14}C, can be used to determine the age of rocks and organic materials.

radiolarians Planktonic protoctists with siliceous shells that are used as index fossils. They have helmet-shaped, bell-shaped, or spherical shells formed from an intricate lattice-like network of bars and spines. As a group they are more abundant in cooler marine environments.

rain shadow Desert region that develops on the lee side of a high mountain range. Precipitation occurs on the windward side of the range, as clouds rise up and cross over the mountains. As a result, the leeward side receives little, if any, precipitation. See also *snow shadow*.

Ramapithecus Ancestral Asian hominoid dated 12 MYA. Formerly considered a possible direct ancestor of humans, but now considered to be the predecessor of the orangutan. Closely related to or possibly identical to *Sivapithecus*.

rayfin fish General term applied to what are called bony fish, including the teleosts, which evolved over 400 MYA and today are the most diverse group of marine and freshwater fish. All have a swim bladder used for buoyancy control, and paired and median fins supported by parallel fin rays. They comprise all fish not jawless, cartilaginous, or lobefin fish.

red algae Members of the protoctista phylum Rhodophyta. They are common "seaweed" inhabitants of the shore and near the shore. Some genera are calcareous and are prime contributors to reefs and other carbonate environments. Previously classified as plants, since they undergo photosynthesis.

reef A ridge-like or mound-shaped wave-resistant buildup of carbonate material secreted by various organisms throughout geologic time that provides an oasis for fish, clams, urchins, and other animals. Today reefs are primarily composed of living and dead corals, red algae, and sponges. Reefs generally develop at the edge of a platform, or shelf, or on the slope of a volcano (fringing reefs); or they develop in a lagoon behind the fringing reef (patch reefs). The animals that construct the reef require clear, warm (not less than 18°C; 65°F), well-oxygenated water.

relative dating A method of dating rocks or geologic features in which time is related either to the progressive and chronological appearances and disappearances of fossils, or to the relative relationships of geologic boundaries, such as cross-cutting relationships or the fact that younger rocks generally overlie older rocks. Relative dating has resulted in the establishment of the eons, eras, periods, epochs, and ages that comprise the geologic time scale.

reservoir rocks Porous and permeable rock units into which hydrocarbons have migrated and become trapped. They can be anywhere relative to the source rock—immediately above, beneath, or adjacent to it.

reversal A periodic change in the Earth's magnetism that results in a switch of the Earth's positive magnetic pole, from its normal position approximating the geographic north pole to a position approximating the geographic south pole. See also *magnetic reversal*.

Rheic Ocean Ocean that formed between Avalonia and Armorica-Iberia-Gondwana. It began to form about 500 MYA and closed about 250 MYA, with the final assembly of Pangea.

rhipidistians A group of lobefin fish considered to be ancestral to all tetrapods—amphibians, reptiles, birds, and mammals. They had fins with hand-like bone structures and are dated 390–270 MYA. See also *lobefin fish*.

rhyolite A light-colored, fine-grained extrusive volcanic rock with the same composition as granite. It is erupted onto the Earth's surface, rather than being cooled beneath the surface. See also *granite*.

rhythmites Regularly layered series of lithified marine sediments up to several feet in thickness that indicate the rhythms of cyclic climatic change. These bands can accumulate into formations that are thousands of feet thick over periods of millions of years. See also *varves*.

ridge (1) An elongate, narrow, raised or uplifted feature or region. See also *graben; horst*. (2) General term applied to a mid-ocean spreading-center, along with its long range of flat-topped mountains, such as the Mid-Atlantic Ridge.

rifting General term applied to a stage of tectonic extension and pull-apart that results either in crustal extension (the formation of grabens and horsts), or in continental separation (development of a seaway), and eventually the development of seafloor spreading-centers.

rift system A series of interconnected and related rifting valleys that develops during crustal extension.

rift valley A graben or valley formed during crustal extension.

RNA Ribonucleic acid. Any of several complex molecules that transfer information from a cell's DNA to other parts of the cell. RNA molecules are components of ribosomes, and are involved in protein synthesis and DNA duplication. Also called ribose nucleic acid.

rock cycle A complex process of formation, alteration, and subsequent destruction of all three categories of rocks—sedimentary, metamorphic, and igneous. Multiple pathways are possible within the cycle, so that any given rock may undergo a complete cycle of igneous formation, erosion, sedimentary deposition, and subsequent lithification, followed by deeper burial and metamorphism, followed by melting into an igneous rock. Or a rock may undergo only part of the cycle. About half of the mass of sedimentary rock on Earth today was formed from rock eroded within the last 600 MY.

Rodinia A name coined from the Russian word for motherland—*Rodina*. Other than Pangea there is much controversy in geoscience about the occurrence of supercontinents during the Earth's long history. However, most paleogeographers agree that a supercontinent existed about 1,100 MYA and this is called "Rodinia," but some scientists insist that it was more of a continental clustering than a unified whole. It is believed that North America formed the core of this assembly in similar fashion to Africa as the core of Pangea.

rudists Specialized bivalve mollusks that were major reef builders in warm, near-equatorial waters. They evolved about 150 MYA and became extinct at the end of the Cretaceous (65 MYA).

rugose corals Group of extinct corals that lived 485–248 MYA. Many of these corals were horn-shaped and are commonly called "horn corals"; others formed columnar, interconnected masses.

ruminants Even-toed ungulates (hoofed animals) having a rumen (first stomach) in which bacteria break down cellulose and partly absorb the nutrients. Ruminants are therefore pot-bellied, cud-chewing animals and include cattle, antelope, deer, and camels. See also *ungulates*.

salina A body of supersaline water, such as a salt lake, salt pond, salt spring, or salt marsh.

salinity Total amount of dissolved salts in seawater, measured by weight in parts per thousand.

salt dome A large plug, column, or mushroom-shaped structure formed as less dense evaporitic sedimentary strata such as salt, pierce, disrupt, deform, and rise through denser overlying sedimentary strata. Salt domes may form traps for hydrocarbons. See also *diapir*.

sandstone A type of sedimentary rock composed primarily of sand-sized grains of quartz, feldspar, and other minerals that are products of land-surface erosion (detrital sediments). Sandstones may be the depositional products of wind or streams, or they may be shoreline deposits.

sauropods Very large herbivores with long necks and tails and relatively small heads, e.g., *Brontosaurus* (now called *Apatosaurus*). Major representatives of the second-regime dinosaurs as categorized in this book.

savannah A grassland region with scattered trees that grades into plain or woodland.

schist A type of metamorphic rock characterized by thin layers of parallel-aligned minerals, often with a high mica content. The rock generally breaks or cleaves along these layers.

seafloor spreading Rifting of the seafloor, which results in the formation of new oceanic crust along the rift. The process occurs along mid-ocean ridges, i.e., seafloor spreading-centers.

sea ice Ice formed by the freezing of seawater, and therefore denser than fresh-water ice.

seal A barrier preventing further migration of hydrocarbons. May result from contact with an impermeable rock layer such as shale or salt, or from contact with a structure such as a fault. See also *cap rock; trap.*

sea level Relative average height of the world's oceans after discounting tidal effect.

seamounts Seafloor mountains formed along mid-ocean ridges and transform faults. May be either flat-topped or peaked. See also *guyot.*

seaway In tectonics, the term applied to the early stage in an ocean's development.

second-regime dinosaurs Term applied in this book to the dinosaur groups that dominated the fauna about 180–90 MYA. Comprised of sauropods and stegosaurs, along with their attendant carnivores and other smaller herbivores.

sedimentary Anything characteristic of or pertaining to sedimentary rocks or the processes of sedimentation.

sedimentary rocks Classification of rocks lithified from sediments including sandstone, shale, conglomerates, gypsum, chalk, and limestone.

sediments Particles formed from erosion, or from chemical precipitation from water, or by organisms. They include detrital sediments such as sand, silt, and mud; evaporites such as rock salt; and carbonate deposits such as chalk. Sediments are deposited by wind, water, ice, and organic processes. Most accumulate into unconsolidated layers that may eventually be lithified to form sedimentary rocks.

seed fern An order of seed-bearing plants with fern-like foliage that lived from about 370 MYA to about 210 MYA, and were common Carboniferous coal-swamp plants. They were gymnosperms not ferns, technically called "Pteridospermales," and include the important genus *Glossopteris.*

seismic reflections Reflections of seismic waves as they pass through the interfaces of various types of rocks. Different minerals, as well as various fluids in the rocks, produce dissimilar reflections than adjacent units or fluids. These differences are recorded, charted, and used to produce subsurface maps for exploration of hydrocarbons and minerals.

seismic waves Elastic waves of energy produced either naturally by earthquakes or artificially by explosives, airguns, or other means, and which are capable of penetrating rock layers. Parts of the waves are reflected back to the Earth's surface where they can be measured and recorded.

sepals The outer layer of special leaves that covers the bud until the flower opens. Sepals may retain their chlorophyll, or they may be colored like the petals of the flower. They may also be tougher than normal leaves. Collectively, the sepals form the calyx.

septa (1) General term applied to partitions in organisms. (2) The calcareous partitions inside a cephalopod shell. Each septum is built at the back of the living chamber as the cephalopod grows. In nautiloids the septa are smooth and gently curved. In ammonites the walls are more, often very, convoluted. It is thought that this difference made the ammonite shell stronger and the animal possibly less vulnerable to attack.

Sevier orogeny The mountain-building period in western North America dated about 150–60 MYA that resulted in deformation on the east side of the Great Basin (eastern Nevada, western Utah). It predated but overlapped the Laramide orogeny and the formation of the Idaho Batholith, and was caused by subduction of part of the Pacific (Farallon) Plate beneath western North America.

sexual reproduction Type of reproduction that involves segregation of DNA strands through meiosis to produce gametes (i.e., sperm and egg), fusion of the gametes (fertilization), and production of a new individual that has genetic material from two parents. This type of reproduction occurs in many eukaryotes—all plants, all animals, all fungi, and many protoctists.

shale A type of sedimentary rock consisting of finely laminated, lithified silt and clay-sized particles. See also *clay.*

shelf The water-covered area surrounding continents that lies between the shoreline and the continental slope, generally up to a depth of 200 m (600 ft). The shelf area has a very gentle slope on an average of 0.1°. Also called the "continental shelf."

shocked quartz Grains of crystalline silica with parallel shock lamellae (veins) formed by extreme pressure. Such grains are very common in impact craters but are very rare elsewhere. Grains of this type are found in the K/T iridium layer.

silica Compound of silicon and oxygen (SiO_2), which together with other elements and compounds form the group of minerals known as silicates, the most common materials that make up the Earth's crust and its overlying rock and sediment cover.

sill An igneous intrusion that forms a tabular elongated layer between the flat-lying layers of pre-existing rock.

silt Detrital particles that are smaller than very fine-grained sand but larger than clay-sized particles. Silt particles are 1/256 mm (0.00016 inch) in diameter.

Silurian Period Division of geologic time dated 443–417 MYA; third period of the Paleozoic Era.

siphuncle In cephalopods, both the shelly tubular openings in the septa and the membranous tube that runs through it, used by the animal to adjust for buoyancy.

slope Submarine area surrounding the continents beyond the continental shelf and extending to the deeper parts of the sea, having an average slope of 3–6°. See also *continental slope.*

slope water A south-flowing subsurface current that flows along the continental slope of northeastern North America. A mixture of cold water from the Labrador Current and small volumes of warm water from the Sargasso Sea. It modifies the water temperature over the Newfoundland Banks fishing grounds.

snow shadow Term used in this book for a dry, sheltered region on the lee side of high mountains. The mountains block the storm clouds, and snow is deposited mostly on the windward side of the mountains. It is equivalent to a rain shadow.

Snowball Earth It is widely agreed that between 750 and 580 MYA the Earth went through several periods of intense cold It is postulated that the global temperature fell to -50°C for prolonged periods during this time, sufficiently cold to cause the global oceans to freeze over and the continents to become sheathed in ice. There is however a disagreement about the location of continents at this time: the Snowball Earth protagonists place the continents in a string around the Equator while others propose a megacontinental arrangement with its primary region over the South Pole and an arm of continental elements reaching from Antarctica to the Equator.

soft rock Term applied to all sedimentary rocks, which are composed of particles cemented together. See also *hard rock.*

source rock Any sedimentary rock, generally shale or carbonate rock, containing organic matter that with time, heat, and deep burial may produce hydrocarbons.

spear-thrower Specially designed, hand-held tool used for hurling spears. Early spear-throwers increased the distance a spear would travel, as well as its penetrating power.

species Any group of organisms that can interbreed and produce fertile offspring. A division in the hierarchy of the classification of organisms; it may be composed of one, several, or many subspecies. One or more species comprise a genus. In scientific classification, the species name is the second name used for any species; for example, in *Homo sapiens*, *Homo* is the genus name, *sapiens* the species name.

speciation The creation of one or more new species from a pre-existing species through the process of evolution.

spicules Glass-like structures of silica or calcium carbonate that form the skeletal framework for some organisms. Spicules may be simple needle-like rods, or more complex cross-shapes, or even three-dimensional shapes.

spore-bearing plant Any plant that produces spores, rather than seeds to reproduce, including true ferns, horsetails, club mosses, mosses, and liverworts. Spore-bearing plants are tied to moist environments in order to reproduce.

spreading-center Area where new crust, generally oceanic crust, is created. Older crust is displaced or moved away from both sides of the spreading center. Spreading centers form as the result of the divergence of two plates. See also *seafloor spreading.*

spritsail A square sail set from a yard slung on the underside of the bowsprit of a three-masted square-rigged ship. It was found to exert a leverage disproportionate to its size, and also had the effect of pulling the bows lower into the water, thus reducing the speed and the maneuverability of the ship, which was countered by the invention of the topsail.

square rig A ship or boat that has square sails as its main sails.

stable platform A thick, tectonically undisturbed central region of a continent covered by sediments. The sediments cover portions of a craton, which is the tectonically stable "shield" of a continent.

stadials Relatively short pulses of cooling occurring within a glacial or interglacial age that are thought to coincide with Milankovitch cycles. See also *interstadials.*

stamens In flowering plants, the male organs that produce pollen, which contains male reproductive cells (male gametes).

stegosaurs Group of dinosaurs that had triangular plates along their backbones. Defined in this book as second-regime dinosaurs.

stem-reptiles Term applied to the earliest amniotes, from which all other reptiles descended. These early reptiles, called "cotylosaurs," appeared about 300 MYA. They had anapsid skulls.

steppe A dry, grassy, and treeless plain, generally referring to the vast plains of Eastern Europe and Asia.

steppe-tundra Ecological region in which steppe and tundra are merged together without intervening taiga. Term applied to vast areas of Siberia during the Cenozoic Ice Age. See also *taiga.*

stocks (1) In geology, the term applied to chimney-shaped ore bodies which are often called "pipes"; also small, roughly circular intrusive igneous bodies. (2) In biology, an evolutionary lineage or family tree.

stratigraphy The study of the deposition and formation of rock layers, including their correlation and interrelationships, the timing and means of deposition, the succession of the fossil content, and the environment of deposition.

stratosphere The outer layer of the Earth's atmosphere where the ozone layer is located.

strike-slip fault A fault along which displacement is lateral rather than vertical. See also *transform fault*.

stromatolites Generally rounded, often columnar rock masses formed from numerous, thin layers of calcium carbonate that was either secreted or trapped by colonies of cyanobacteria (blue-green algae).

stromatoporoids A group of extinct invertebrate animals that lived about 490–125 MYA, and that built layered colonies vaguely resembling stromatolites. They were major reef-building organisms from about 480 to 440 MYA, and may have been related to sponges, but their classification is still a matter of debate.

subduction The process of one plate moving or descending beneath another plate and being remelted as it descends. The process occurs either at the boundary between two oceanic plates, such as between the Atlantic and Caribbean Plates, or at the boundary between a continental plate and an oceanic plate. In the latter case, the oceanic plate, being the denser of the two, is subducted beneath the continental plate, such as the subduction of the Pacific Plate beneath the North American Plate.

sulphides A group of minerals that contain sulphur and either a metal or semimetal. These are ore minerals from which gold, silver, lead, zinc, and other metals are extracted.

supercontinent An amalgamation of large and small continental plates into a gigantic continental mass or arrangement, such as Pangea. In this book the term is used only to describe Pangea; the term megacontinent is used to describe lesser-sized assemblies. [see also *Rodinia*— a Precambrian supercontinental clustering]

supersaturation The condition in which a solution contains a higher percentage of dissolved minerals than it would under normal conditions of temperature or pressure. As a result, the dissolved minerals rapidly precipitate out of the solution when conditions change, such as when either temperature or pressure is decreased.

suture The line or region of contact or welding between two converging continental masses.

Swanscombe Man A fossil of archaic *Homo sapiens* that lived in England around 220,000 BP. Swanscombe Man used sophisticated straight-edged hand axes.

swim bladder A small inflatable sac located just above the gut of a rayfin fish. It is filled with gas and used to adjust the fish's buoyancy as it changes water depth.

symbiotic relationship The relationship between two organisms that is beneficial to one or both of the organisms and in which neither is harmed by the relationship.

synapsids A group of reptiles that had two cranial apertures in addition to eye sockets and nostrils. The mammal-like reptiles—pelycosaurs and therapsids—that lived 323–210 MYA.

synclines Folded structures in which the oldest rocks comprise the outer rings or layers of the structure, and the youngest rocks form the inside layers or core of the structure. See also *anticlines*.

synergistic A symbiotic relationship in which both organisms benefit from the relationship.

Taconian orogeny The period of mountain building, dated about 490–440 MYA, which occurred during the early closure of the Iapetus Ocean, in the northern Appalachians in the northeastern United States and Canada. The first of a series of Pangean orogenies along the eastern coast of Laurentia.

taiga An ecological region characterized by swampy coniferous forests which generally separates steppe from tundra.

Taino-Arawak A subculture of Arawak that inhabited the islands of Cuba, Hispaniola, Jamaica, Puerto Rico, and the Antilles. They were farmers who cultivated a wide variety of plants, and produced fine-quality, decorated ceramics.

tectonic escape A process caused by the flexing of a continental platform, which reactivates ancient fault systems and causes basement rocks to be thrust upwards to form mountains in the middle of the platform. Considered to be the process that formed the Ancestral Rocky Mountains.

tectonic plates The numerous, interacting, continuously moving, large and small divisions of the lithosphere. See also *lithosphere; plates*.

tectosphere The semi-plastic upper region of the Earth consisting of a rigid outer shell, the lithosphere, and the underlying hot plastic region of the upper mantle, the asthenosphere. It is about 400 km (250 miles) thick. See also *asthenosphere; lithosphere*.

tektites Bead-like spherules of calcium silicate that are products of bolide impacts. They are a common constituent of the K/T boundary layer.

teleosts Common name applied to the "bony fish," an infraclass of rayfin fish that evolved about 220 MYA. They have expandable concertina-like gills. The group of fish that dominates modern freshwater and marine environments (numbering about 28,000 modern species).

terrane A fault-bound body of rock that has a different geological history than adjacent regions.

terrigenous Term applied to particles (sediments) produced by erosion of land surfaces.

Tertiary Period Period of geologic time, dated 65–1.8 MYA; the first period of the Cenozoic Era. It is divided into five epochs: Paleocene, Eocene, Oligocene, Miocene, and Pliocene. The name was originally proposed by Giovanni Arduino in 1760 for weakly consolidated sediments in Italy.

Tethys Ocean Name applied to the ocean that was encircled by elements of eastern Pangea. The western and eastern Tethys Ocean extended into what is now the Mediterranean area. It was largely destroyed by the collision of northern Asia with India and the other microplates that today form southern Asia, and by the collision of Africa with Europe.

tetrapod General name for the Tetrapoda, i.e., all vertebrates with four limbs—the amphibians, reptiles, birds, and mammals.

therapsids A group of mammal-like reptiles that lived 280–200 MYA. They are the descendants of the pelycosaurs that gave rise to the mammals. See also *synapsids*.

thermocline The zone of rapid temperature change that acts as a barrier to mixing of surface and deeper ocean currents. The thermocline is the zone beneath the upper, wind-driven layer of the ocean.

third-regime dinosaurs Term applied in this book to the dinosaur groups that dominated the fauna from about 90 MYA to about 65 MYA, and which were comprised largely of hadrosaurs and ceratopsians, along with ankylosaurs, pachycephalosaurs, and their attendant carnivores.

threshold level Term applied to the drop in sea level required for the exposure of the Bering Landbridge; equal to about 45 m (150 ft) below today's sea level. A drop greater than 150 feet results in even greater exposure of Beringia.

Thule culture An Arctic culture that appeared about 1,900 BP. Named after Thule in northwestern Greenland. They introduced a new technology that included large umiaks and small kayaks, bows and arrows, dog sleds, harpoons of advanced design, whale-oil lamps, and snow igloos.

tillites Sedimentary rocks formed from glacial deposits. May include sediments of all sizes from boulders to clay.

Tommotian fossils The first-known animals to have had hard parts. They evolved about 543 MYA. A diverse and widespread group of tiny invertebrates that measured a fraction of an inch in length. Some had crude mollusk-like coiled shells; others had tubes that are presumed to have contained worm-like animals; and a few had glass-like spicules.

topography The general shape of the surface of land; also the relief of the Earth's surface.

topsail A smaller sail attached to the top of the mast, which increases the headward progress of the ship by catching the more steadily blowing winds higher above the sea's surface. They countered the effect of the spritsail by lifting the bow of the ship.

Tornquist's Sea The sea that existed 470–430 MYA between Avalonia and Baltica. The line of suture formed during the destruction of that sea separates northern Germany and Scandinavia from the Low Countries. Named after A. Tornquist, the geologist who first identified the probable suture line in 1910.

toxins Substances produced by various organisms that are poisonous to other organisms.

trace element Any element or isotope of an element that is found only in trace amounts, generally considered to be less than one percent of any substance.

trachiophytes General term for all vascular plants which have a vascular system that takes the form of a bundle of tubes that transport water, carbohydrates, and trace elements throughout the plant by osmosis. They evolved about 425 MYA.

trade winds Name applied to the easterly winds that blow roughly northeast to southwest in a band north of the Equator and southeast to northwest in a band just south of the Equator. Prevailing winds in the tropics and subtropics.

transform fault A type of strike-slip fault, generally applied to the long secondary faults that are perpendicular to mid-ocean ridges, e.g., the San Andreas Fault.

trap A structure or impermeable rock layer that blocks the migration of oil or natural gas. Pools of hydrocarbons may accumulate beneath a trap. See also *cap rock; seal*.

travel time The time it takes for a seismic wave to travel from its point of origin to a subsurface rock layer or structure and to the surface where it is recorded.

Triassic Period The period of geologic time dated 248–206 MYA. The first period of the Mesozoic Era, named after the three stages of deposition that were first recognized in German rocks of this age.

trilobites Marine arthropods with an exoskeleton, a triple-lobed body, and many legs that lived 543–245 MYA. They are important index fossils throughout the Paleozoic Era.

trilobitomorphs Soft-bodied animals that resemble trilobites. Part of the Ediacaran fauna that lived off the shores of Australia 650 MYA.

triple rift A three-armed rift system that forms over a domed region beneath the Earth's crust. This system develops during the initial breakup of a continental mass as well as in ocean floors where there are unbalanced forces acting on the crust. Such phenomena are called "triple junctions." If the system continues to develop, two arms of the triple-rift continue to propagate, while the third arm, called an "aulacogen," becomes inactive. Many continental rivers flow in such failed arms of the triple-rift systems. See also *aulacogen*.

tritheledonts A group of late therapsids (cynodonts) that lived about 210–190 MYA and were probably endothermic.

Many of them were very small, with skulls only 3–6cm (1.5–2.5 inches) long, but others were as large as badgers. They are very likely candidates to have evolved into mammals.

tritylodonts A group of small, late therapsids (cynodonts) that had small pits and grooves in their snouts, implying that they had touch-sensitive whiskers and possibly fur. Their skulls and skeletons are very similar to those of mammals. They are also considered likely ancestors of mammals. They lived about 220–175 MYA.

tsunamis (Japanese for harbor waves) Very large, fast-moving waves generally caused by submarine earthquakes or volcanic eruptions. When these waves near the shallowing shore of a continent or island, they rise to extreme heights and strike far inland, causing great devastation. Often mistakenly called tidal waves.

tuff Rock formed from fine, wind-borne, or water-laid volcanic ash.

tundra A vast, treeless region that is underlain by permafrost, but its surface is often marshy in summer. It supports growths of lichens, mosses, and low-growing shrubs such as arctic willow.

turbidites Rocks or sediments formed from shallow water turbidity or from a submarine avalanche of silt, mud, and rock debris that poured down the continental slope and settled at the foot of the slope.

turbidity Term used to describe the suspended sediments in water. The greater the amount of suspended sediment, the greater the turbidity. Turbidity inhibits marine organisms in a variety of ways, including reducing the amount and depth of penetration of light; directly smothering delicate organisms such as coral polyps; and damaging delicate organs such as clam gills.

turbidity flows A submarine avalanche of mud, sand, and boulders, which often occurs in a submarine channel on the continental slope—the sediment sweeps downslope in a dense cloud, then settles out to form a fan at the base of the channel. Also called "turbidity currents."

Turgai Seaway The epicontinental seaway that covered parts of Eurasia at various times during the Mesozoic and Cenozoic Eras. The seaway separated eastern Siberia, Mongolia, and other parts of eastern Asia from the rest of Eurasia. The seaway connected to the Tethys Ocean to the south and to the Arctic Ocean in the north.

umiaks Large, open skin boats, 4.5–12 m (15–40 ft) long, and propelled by paddles, oars, or grass sails. Originally developed by the Thule.

unconformity A major break or gap in the rock record, where rocks of a given time are missing between underlying older rocks and overlying younger rocks. The break can be the result of prolonged non-deposition or of erosion of the surface above the sea.

ungulates Name applied to a diverse assemblage of herbivorous mammals that have hoofs or were once hoofed. The term is applied to even-toed ungulates such as pigs, hippopotamuses, deer, and cattle; odd-toed ungulates such as the horse and rhinoceros; the long-nosed ungulates, the elephants; cetaceans, whales and porpoises; and others including numerous extinct groups.

uplift Tectonic event that causes a region or block to be raised up relative to its former elevation and relative to adjacent regions or blocks.

upper mantle The upper layer of the Earth's mantle. It contains regions of converging convection cells that may be one of the sources of mantle plumes or hot spots. See also *asthenosphere; lithosphere; mantle; tectosphere.*

upwelling (1) In marine biology, the process by which nutrient-laden bottom-water masses rise to the ocean surface, a process that occurs along continental margins and in zones between ocean gyres, i.e., along the Equator. Upwelling areas are regions of high productivity of both plankton and the fish and other nekton that feed on the plankton. (2) In geology, a process by which magma may rise to the Earth's surface, or the process by which hot mantle material rises, forming a mantle plume or hot spot.

Variscan orogeny Period of mountain building in East Central Pangea (Western Europe) dated 345–280 MYA. It was a result of the final closing of the Rheic Ocean between Armorica-Iberia and Avalonia-Baltica. The Ancestral Pyrenees were among the mountains formed at this time. See also *Hercynian orogeny*.

varves Thin sedimentary layers that are deposited annually in lakes. Local deposits sometimes accumulate for centuries or for thousands of years. Short-term fluctuations in climate are indicated by the varying thickness and character of the individual strata. See also *rhythmites*.

vascular plants All land plants with a vascular system—bundles of tubes that transport water, carbohydrates, and trace elements throughout a plant by osmosis. Technically called "trachiophytes," they include ferns, gymnosperms, angiosperms, and many other living and extinct plants. They evolved about 420 MYA.

vertebrates All animals having an internal skeleton of bone or cartilage including a backbone. The phylum Vertebrata includes all fish, amphibians, reptiles, birds, and mammals.

vicarianism Process of evolution brought about by the separation of common lineages through the development of natural barriers. When a population becomes isolated, it tends to undergo rapid evolution. An example of this would be the isolation and subsequent evolution of mammals in South America during the Cenozoic Era. Vicarian evolution is a long-term phenomenon. See also *dispersalism; divergent evolution*.

waifs Term applied to organisms that suddenly appear in the fossil record of a region and that are unrelated to any other organisms in that region. It is thought that waifs are victims of circumstance—that they are accidentally transported to the new region from some other separate region; for example, the sudden arrival of both rodents and monkeys in South America is thought to have been the result of accidental rafting from Africa on large masses of floating vegetation.

Wallace's Line A line on a map of the Indonesia-Australian archipelago that bisects the island of Celebes. It lies between Bali and Lombok in the Indian Ocean and runs generally northeast from south of the Philippines into the Pacific Ocean. Wallace's line separated distinctly oriental (Laurasia-derived) flora-fauna from distinctly Australian (Gondwana-derived) flora-fauna.

waste chutes Channels in a fringing reef that form between masses of coral. These channels funnel detritus and other waste materials away from the coral heads, materials that otherwise would suffocate and bury the corals.

Wegener cycles Name applied in this book to cycles of continental clustering to form megacontinents and/or supercontinents such as Pangea. Complement of ocean cycles (Wilson cycles).

West Central Pangea Term used in this book to designate that portion of Central Pangea that encompassed the Florida-Bahama block and the Gulf of Mexico region.

Wilson cycles The six stages of the formation and destruction of an ocean: (1) uplift and graben formation produces a rift valley that sinks and fills with a shallow sea; (2) continued rifting and then seafloor spreading results in the formation of a mature ocean like the present Atlantic; (3) a period of decline follows, in which the failure of the spreading-center system leads to the compression of the ocean crust by the encroaching continents; (4) the ocean crust is then destroyed by subduction beneath the continents; (5) the ocean narrows to such a point that continental margins collide, leading to a prolonged period of mountain building such as that in the Mediterranean Alpine belt today; (6) the continents fuse into one contiguous landmass—like India with Asia—which results in the formation of Himalayan-type mountains. Name given to the cycles of the formation and destruction of oceans. Named after J. Tuzo Wilson, who first proposed the six stages. See also *ocean cycle*.

Wisconsin-Würm glacial age Name given to the last glacial age. It lasted from 120,000 to 10,000 BP. Wisconsin is the name used in North America; Würm (also called Weichsel) is used in Europe.

Woodland people Native American cultures that inhabited the Northeast and eastern Midwest, including many of the mound builders.

wrench faulting A type of strike-slip or lateral fault, in which the fault plane is nearly vertical. Reversed up-and-down displacement may also occur, the rock on either side pivoting like scissor blades at points along the fault.

zooplankton Term applied to small aquatic animals that have either seriously limited powers of self-propulsion or none at all, and are thus swept along by waves and wind. They include minute larvae of mollusks, such as clams and snails, and crustaceans, such as shrimp-like krill, as well as comb jellies and jellyfish. Previously included the protozoa.

zooxanthellae Protoctists (yellow algae, often dinoflagellates) that live symbiotically with other organisms. One type lives symbiotically with coral polyps, providing food for the polyps through photosynthesis and increasing the rate of calcification by the coral polyps.

Acknowledgements and Dedication

Many North Americans, Europeans, Atlantic islanders, and Caribbean people and institutions are acknowledged in alphabetical order in the lists that follow—although titles and addresses may have changed since our last meeting. Behind each entry there lies a degree of goodwill, helpfulness, and enthusiasm for their subject that is remarkable testimony to the dedicated spirit of the geoscience fraternity at large. My most sincere thanks to each and every one of these generous people.

The paleogeography reproduced in this book is the common thread around which the story is told. To provide a consistent view of an often controversial subject, it was necessary to focus on the work of one specialist. Dr. Christopher R. Scotese of the University of Texas, Arlington, was my mentor in this respect. His technical support was crucial to the fulfillment of the project and my debt to him is therefore inestimable. Dr. Scotese provided detailed paleomaps from which data have been extracted for the reproduction of many of the maps in this book (computer software and papers are detailed in the bibliography, page 341).

It has been my privilege to lead a small but extremely talented team in the completion of the research, the illustration, and the design of the book. Even so, it is impossible to progress a major project, including a three-year program of photography in widely separate locations in the Northern Hemisphere, without a secure financial base. This was largely provided by Keebler Company, Elmhurst, Illinois, who had previously underwritten several PBS/BBC television series for which I was responsible. To offer my grateful thanks to the company's management is a gross understatement of my feeling of indebtedness to Keebler.

Library research for photo-locations in the first year of the project was conducted with the help of Alison Fahley-Jones M.Sc., a graduate of the Colorado School of Mines. Research for the biological and botanical sciences, and for icon illustrations during the concluding several years of the ten-year project was undertaken by Linda G. Martin Ph.D., a doctoral graduate of CSM. In fact, her responsibilities extended well beyond library research. It included the preparation of the bibliography, writing the glossary, coordinating the flow of all work, and bringing order and control to the fulfillment of a complex project.

The artwork was completed by Gary Hincks and Mike Eaton in England. Gary Hincks, who has been my illustrator for twenty years, was responsible for the large pieces of art—the major graphics in Chapter II—and the technical illustrations throughout. Mike Eaton was responsible for the smaller pieces of art—the icons and miniature globes. But there was a degree of overlap: the artwork for Chapter XII and the Epilogue was shared. Both men are extraordinarily talented illustrators. It was a pleasure and a privilege to work with them.

The original design for the book was completed in New York, NY., by Kristen Reilly with recourse to Irwin Glusker, a well-known and distinguished New York designer, acting as consultant. Due to the author's illness and transfer to England, the process of adapting the original design to a new format was conducted by Harry Green and his colleague Bob Travis under the auspices of the Orion Publishing Group, London. These general processes required designers with special attributes: natural talent of the highest order, open and receptive minds, cool heads, and a willingness to go the extra mile to "get it right." That specification describes every one of these true professionals.

Editorial responsibility was taken over by Catherine Bradley, of Cassell & Co., London, an imprint of Orion Books. From start to finish Catherine's unflagging energy and determination to ensure the project's completion and success has been a great encouragement during a difficult transition.

And finally, I would like to pay homage to the unsung heroine of a decade-long project. Her support in the field, often acting as my "Sherpa" in remote places, or taking control of a tossing light aircraft over hazardous terrain, or showing willingness to risk life and limb without complaint, or driving thousands of fully loaded jeep-miles in "unknown parts"— indeed her general helpfulness and encouragement at all times, in all circumstances, must be exceptional by any measurement of long-standing marital relationships. So it is to my wife Joy that I unhesitatingly dedicate this book with great warmth and personal affection. I do so as a modest gesture of sincere thanks for her major contribution to the making of the book and for somehow looking after me at the same time!

Ron Redfern
Ermington, South Devon, England.
April, 2000

BAHAMAS

Glenn V. Bannister, Vice President/General Manager, Morton Thiokol, Morton Salt, Matthew Town, Inagua, Bahamas. Permissions.

Richard E. Haxby, Marine Biologist, Morton Thiokol, Morton Salt, Matthew Town, Inagua, Bahamas. Marine biology.

Basil T. Kelly, Past President, Bahamas National Trust Nature Reserve, Nassau, Bahamas. Introductions.

Jimmy Nixon, Warden, Bahamas National Trust Nature Reserve, Great Inagua, Bahamas. Lake Windsor guide.

BERMUDA

Martin S. Brewer, Ministry of Labour and Home Affairs, Hamilton, Bermuda. Geology.

Idwal Wyn Hughes, Permanent Secretary, Ministry of the Environment, Hamilton, Bermuda. Introductions.

Anthony H. Knap, Director, Bermuda Biological Station for Research, Ferry Reach, Bermuda. Marine biology.

Wolfgang E. Sterrer, Curator, Natural History Museum, Bermuda Aquarium Museum and Zoo, Smiths, Bermuda. Marine invertebrates.

Richard Winchell, Curator, Natural History Museum, Bermuda Aquarium Museum and Zoo, Hamilton, Bermuda. Sargasso Sea biology.

David Wingate, Conservation Officer, Naturalist, Nonsuch Island, Bermuda. Natural history.

CANADA

Alberta

Robert L. Christie, Geologist, Geological Survey of Canada, Institute of Sedimentary and Petroleum Geology, Calgary, Alberta. Arctic geology and tectonics, introductions.

Sylvain Dufour, Project Engineer, Geocon, Calgary, Alberta. Pingo morphology.

Martin O. Jeffries, Department of Geography, University of Calgary, Calgary, Alberta. Ice islands.

Neil J. McMillan, Chief, Geological Survey of Canada, Institute of Sedimentary and Petroleum Geology, Calgary, Alberta. Hudson Bay to Labrador Sea, ancestral rivers, tectonics.

Robert Meneley, Vice President of Exploration, Petrocan, Calgary, Alberta. Icebergs, ecology.

Franz Nantwich, Geology Department, University of Alberta, Edmonton, Alberta. Kimberlites.

W.W. Nassichuk, Geological Survey of Canada, Calgary, Alberta. Introductions.

Kirk G. Osadetz, Petroleum Geologist, Geological Survey of Canada, Institute of Sedimentary and Petroleum Geology, Calgary, Alberta. Arctic Islands, opening of the Arctic Ocean, North Slope geology.

Peter Schlederman, The Arctic Institute of North America, The University of Calgary, Calgary, Alberta. Arctic archaeology.

Ray Thorsteinsson, Paleontologist, Geological Survey of Canada, Institute of Sedimentary and Petroleum Geology, Calgary, Alberta. Placoderms, graptolites.

Hans P. Trettin, Geological Survey of Canada, Institute of Sedimentary and Petroleum Geology, Calgary, Alberta. Ellesmere Island geology.

Newfoundland

Michael A. Anderson, Paleontologist, Memorial University, St. John's, Newfoundland. Avalonian Ediacaran fossils.

Gordon Handcock, Memorial University, Department of Geography, St. John's, Newfoundland. 16th-century history, the West of England–Newfoundland.

Alan McPherson, Memorial University, Department of Geography, St. John's, Newfoundland. Past northern climates.

John Mannion, Memorial University, Department of Geography, St. John's, Newfoundland. Early Irish immigration.

Sean O'Brien, Department of Energy, Mines, and Resources, Geological Survey of Canada, St. Johns, Newfoundland. Geology and tectonics.

Harold Williams, Department of Geology, Memorial University of Newfoundland, St. John's, Newfoundland. Appalachian tectonics, ophiolites, introductions.

Northwest Territories

Rev. Laurie Dexter, Nanisivik, Northwest Territories. Inuit cultures.

Harry Finnis, Counsellor, Government of the Northwest Territories, Nanisivik, Northwest Territories. Inuit customs and history.

John D. Ostrick, Operations Manager, Inuvik Research Laboratory, Indian Affairs and Northern Development, Inuvik, NW Territories. Introductions.

W.A. Padgham, Resident Geologist, Indian and Northern Affairs, Yellowknife, Northwest Territories. Introductions.

Manley Showalter, Raecom, Yellowknife, Northwest Territories. Bush Pilot.

Bill Sutherland, Aklavik Flying Service, Inuvik, Northwest Territories. Bush Pilot.

Nova Scotia

Ken Donovan, Historian, Parks Canada, Fortess of Louisbourg, National Historic Park, Louisbourg, Nova Scotia. 17th-century history.

Charlotte Keen, Atlantic Geoscience Centre, Bedford Institute of Oceanography, Dartmouth, Nova Scotia. Seafloor geology and plate tectonics.

Michael Keen, Director, Atlantic Geoscience Centre, Bedford Institute of Oceanography, Dartmouth, Nova Scotia. Oceanography.

J. Duncan Keppie, Chief Project Geologist, Nova Scotia Department of Mines and Energy, Halifax, Nova Scotia. Tectonics, displaced terranes.

G.F.R. Lohnes, Head of Operations, Parks Canada, Fortress of Louisbourg, National Historic Park, Louisbourg, Nova Scotia. Introductions.

Bosko Loncarevic, Bedford Institute of Oceanography, Dartmouth, Nova Scotia. Seafloor geology.

K.S. Manchester, Chief, Personnel de Soutien, programmes, Geological Survey of Canada, Department of Energy, Mines and Resources, Dartmouth, Nova Scotia. Introductions.

C.E. Murray, Public Relations Manager, Ocean Information Division, Communications Branch, Institute of Oceanography at Bedford, Dartmouth, Nova Scotia. Introductions.

H. Brian Nicholls, Head, Ocean Information Division, Ocean Science and Surveys, Institute of Ocean-ography at Bedford, Dartmouth, Nova Scotia. Introductions.

William A. O'Shea, Head of Historical Resources, Parks Canada, Fortress of Louisbourg, National Historic Park, Louisbourg, Nova Scotia. 17th-century history.

David Ross, Acting Director, Bedford Institute of Oceanography, Geophysics Division, Dartmouth, Nova Scotia. Introductions.

Paul E. Schenk, Head of Geology Department, Dalhousie University, Halifax, Nova Scotia. Displaced terranes, correlationships Nova Scotia–Morocco.

Ontario

M.R. Dence, Director, Gravity, Geothermics and Geodynamics Division, Department of Energy, Mines and Resources, Ottawa, Ontario. Introductions.

Tom Frisch, Geological Survey of Canada, Ottawa, Ontario. Precambrian geology.

George D. Hobson, Director, Polar Continental Shelf Project, Department of Energy, Mines and Resources, Ottawa, Ontario. Expedition support.

Paul F. Hoffman, Geological Survey of Canada, Ottawa, Ontario. Canadian shield, Precambrian geology and tectonics, introductions.

Garth Jackson, Geological Survey of Canada, Ottawa, Ontario. Precambrian geology, expedition visit.

Pierre Lapointe, Research Scientist, Division of Seismology and Geomagnetism, Department of Energy, Mines and Resources, Ottawa, Ontario. Paleomagnetism.

John C. McGlynn, Director, Precambrian Geology Division, Geological Survey of Canada, Ottawa, Ontario. Introductions.

Digby J. McLaren, Department of Geology, University of Ottawa, Ottawa, Ontario. Geoscience, Devonian extinction, introductions.

John V. Matthews, Jr., Geological Survey of Canada, Ottawa, Ontario. Quaternary geology, Beringia, Old Crow, Blue Fish Cave.

P. Morel-à-l'Huissier, Earth Physics Branch, Department of Energy, Mines and Resources, Ottawa, Ontario. Paleomagnetism.

Richard E. Morlan, Yukon Archaeologist, National Museum of Man, Archaeological Survey of Canada, Ottawa, Ontario.

Philip Munro, Earth Physics Branch, Department of Energy, Mines and Resources, Ottawa, Ontario. Polar weather studies.

Wayne Nesbitt, Geology Department, University of Western Ontario, London, Ontario. Paleo-environments, glaciations through time.

Larry Newitt, Earth Physics Branch, Canadian Geological Survey, Ottawa, Ontario. North Magnetic Pole and Earth magnetism.

Victor Prest, Geological Survey of Canada, Ottawa, Ontario. Glaciology.

P. Blyth Robertson, Earth Physics Branch, Ottawa, Ontario. Devon Island, Haughton astrobleme, expedition visit.

Josef Svoboda, Department of Botany, University of Toronto, Erindale College, Mississauga, Ontario. Arctic botany, expedition visit, introductions.

Grant M. Young, Geology Department, University of Western Ontario, London, Ontario. Paleo-environments, glaciations through time.

Quebec

Hans Hofmann, Department of Geology, University of Montreal, Montreal, Quebec. St. Lawrence River regional geology.

Resolute Bay, Devon Island

Paddy Doyle and others, Pilots, Twin Otter STOLS.

Barry Haugh, Base Manager.

Frank Hunt, Base Manager.

Pierre Pellerin, Pilot, Helicopter Whisky, X-ray, Juliet.

DENMARK

Copenhagen

Peter R. Dawes, Geological Survey of Greenland, Copenhagen, Denmark. Geology of Greenland and Nares Strait.

GIBRALTAR

Joaquin Bensusan, Curator, Gibraltar Museum, Gibraltar. Geology of Gibraltar.

Richard J.M. Garcia, Tourist Office, Gibraltar. History of Gibraltar.

Horace J. Zammitt, Minister for Tourism, Gibraltar. Introductions.

ICELAND

Gunnar K. Bergsteinsson, Director, Icelandic Coast Guard, Iceland. Helicopter services.

Hauker Johannesson, Museum of Natural History, Reykjavik. Geology and volcanology.

Björn Rúricksson, Reykjavik. Aircraft pilot and guide.

Kaisjjan Saemundsson, Museum of Natural History, Reykjavik. Geology of Krafla region.

LESSER ANTILLES

Barbados

Sean Carrington, University of the West Indies, Barbados. Plant physiology.

Louis Clainnery, University of the West Indies, Barbados. Plant taxonomy.

Wayne Hunte, Marine Biologist, Bellairs Research Institute, McGill University, Barbados, marine biology, fish ecology.

Philip L.B. Payne, Petroleum Geologist, Barbados National Oil Company, Barbados. Accretionary lens geology.

Martinique

Yves Atlan, Directeur, Service Géologique Régional des Antilles et de la Guyane, Martinique. Lesser Antilles geology.

Jean-Pierre Viode, Directeur, Observatoire Volcanologique de la Montagne Pelée, Martinique. Volcanology.

MOROCCO

M. Ben Said, Directeur du Service Géologique, Rabat, Morocco. Introductions.

John E. Warme, Department of Geology and Geological Engineering, Colorado School of Mines, Golden, Colorado. Expedition leader, Ziz Valley.

NORWAY

Brian A. Sturt, Geologists Institutt, Bergen University, Norway. Caledonide geology.

PORTUGAL

Lisbon

Delfim de Cavalho, Director, Serviços Geológicos de Portugal, Lisbon, Portugal. Geology and tectonics, Portugal, Madeira and Azores.

Hipolito Monteiro, Marine Geologist, Serviços Geológicos de Portugal, Lisbon, Portugal. Geology and tectonics, Portugal, Madeira and Azores.

Madeira

João Gonçalves Borges, Director Regional do Turismo (Regional Tourism), Funchal, Madeira, Portugal.

G.E. Maul, Retired Director, Municipal Museum, Funchal, Madeira, Portugal. Marine biology.

Alec Zino and Francis Zino, Funchal, Madeira. Natural history, Balcões Forest.

SPAIN

Las Palmas

David Bramwell, Director, Jardin Botanico "Viera Y Clavijo," Las Palmas, Grand Canaria. Micronesia paleobotany.

Madrid

Selma Barkham, Archivist, San Sebastion. 15th–16th-century Basque history.

Dr. Manuel Díaz Curiel, Fundación Jiménez Díaz Hospital, Cercedilla, Madrid. Medical services.

Dr. Manuel L. Fernández Guerrero, Fundación Jiménez Díaz Hospital, Madrid. Medical services.

TRINDIDAD and TOBAGO

Winston M. Ali, Senior Geologist, Trinmar Limited, Trinidad. Lesser Antilles geology.

Hans E.A. Boos, Curator, Zoological Society of Tri-nidad & Tobago, Emperor Valley Zoo, Trinidad. Natural history.

Eric Patience, Manager, ASA Wright Nature Center, Trinidad. Natural history.

Krishna M. Persad, Manager Exploration and Production, Premier Consolidated Oilfields PLC, Trinidad. Lesser Antilles geology.

UNITED KINGDOM

Channel Islands

A.E. Mourant, St. Helier, Jersey. Precambrian geology.

London and Oxford

Clive Bishop, Deputy Director, British Museum of Natural History, London. Precambrian geology, Brittany and Channel Islands.

L.R.M. Cocks, Department of Paleontology, British Museum of Natural History, London. Tornquist's Sea.

John F. Dewey, Chairman, Department of Geology, Oxford University, Oxford. Tethyan tectonics, sedimentary basins.

Frederick W. Dunning, Curator, Geological Museum, London. Tectonics and geology, British Isles and Europe, introductions.

Richard A. Fortey, Department of Paleontology, British Museum of Natural History, London. Tornquist's Sea.

Ron Hedley, Director, British Museum of Natural History, London. Introductions.

Peter Ince, Deputy Curator, National Maritime Museum, Greenwich. Introductions.

Peter James, Keeper of Botany, Department of Zoology, British Museum of Natural History, London. Lichen.

John Peake, Keeper of Zoology, Department of Zoology, British Museum of Natural History, London. Fishes and rare books library.

Christopher Terrell, Curator of Navigation and History, National Maritime Museum, Greenwich. 16th-century maps and charts.

Plymouth and West Country

R.L. Atkinson, Museum Curator, Camborne School of Mines, Cornwall. Cornubian batholith geology.

Alan V. Bromley, Director, Camborne School of Mines, Cornwall. Cornubian batholith geology.

Malcolm Clarke, The Marine Biological Association U.K., Plymouth. Deep-sea biology.

UNITED STATES

Alaska

Roger C. Herrara, Alaska Operations Manager, Sohio Alaska Petroleum Company, Anchorage, Alaska. Introductions.

Gilbert Mull, Alaska Geological Survey, Anchorage, Alaska. North Slope, Brooks Range, tectonics.
Mike Ivers, Gulf Air Taxi, Yakutat, Alaska. Bush Pilot.

California
G. Brent Dalrymple, Assistant Chief Geologist, U.S. Geological Survey, Menlo Park, California. Accreted terranes.
David M. Hopkins, U.S. Geological Survey, Menlo Park, California. Bering Land Bridge, Quaternary geology.
Louis N. Marincovich, Jr., Molluscan Paleontologist, U.S. Geological Survey, Branch of Paleontology and Stratigraphy, Menlo Park, California. Arctic paleontology.
Michael J. Savage, President, Merlin Petroleum Company, San Francisco, California.

Colorado
John Andrews, Institute of Arctic/Alpine Research, University of Colorado, Denver, Colorado. Lichens.
Anthony Crone, U.S. Geological Survey, Denver, Colorado. New Madrid Fault Zone.
Beverly A. Halliwell, Marathon Oil, Denver, Colorado. Sedimentary geology, Morocco.
John D. Haun, Department of Geology and Geological Engineering, Colorado School of Mines, Golden, Colorado. Introductions.
Roy L. Jenne, Senior Scientist, National Center for Atmospheric Research, Boulder, Colorado. Physiographic paleomaps.
Harold E. Kellogg, Senior Geologist, American Petrofina Company of Texas, Denver, Colorado. Geology of Spitsbergen, Svalbard.
James D. Lowell, Consulting Geologist, Denver, Colorado. Afar triangle, tectonics.
Frank McKeown, Geologist, U.S. Geological Survey, Branch of Earthquake Hazards, Denver, Colorado. New Madrid Fault Zone.
Captain Alfred Scott McLaren, Polar Regions Consultant, United States Navy (Ret.), Boulder, Colorado. Introductions.
L. Lorraine Mintzmyer, Regional Director, National Park Service, Rocky Mountain Region, Denver, Colorado. Introductions.
A.R. Palmer, Centennial Science Program Coordinator, Geological Society of America, Boulder, Colorado. Cambrian geology and D-NAG references.
Jack Reed, GSA Coordinator, U.S. Geological Survey, Denver, Colorado. Bathymetry.
Bruce Rippeteau, Vice President for Archaeology, Colorado Historical Society, Denver, Colorado. Paleo-American archaeology.
John E. Warme, Department of Geology and Geological Engineering, Colorado School of Mines, Golden, Colorado. Geology of Morocco, Cretaceous carbonate geology.
Warren M. Washington, Director, Climate and Global Dynamics Division, National Center for Atmospheric Research, Boulder, Colorado. Climatology.
P. Webber, Institute of Arctic/Alpine Research, University of Colorado, Denver, Colorado. Lichens.
Robert J. Weimer, Department of Geology and Geological Engineering, Colorado School of Mines, Golden, Colorado. Ancestral Rockies.

Connecticut
J. Revell Carr, Director, Mystic Seaport Museum, Mystic, Connecticut. Maritime history.

Florida
Stephen W. Gard, Refuge Manager, U.S. Department of the Interior, U.S. Fish and Wildlife Service, Merritt Island National Wildlife Refuge, Titusville, Florida. Cape Hatteras.

Georgia
John Schroer, Manager, Okefenokee Swamp, National Wildlife Reserve, Folkston, Georgia.
Jimmy Spikes, Assistant Manager, Okefenokee Swamp Park, Waycross, Georgia.
Jimmy Walker, Manager, Okefenokee Swamp Park, Waycross, Georgia.

Illinois
David Lonsdale, Assistant Director, John G. Shedd Aquarium, Chicago, Illinois. Marine biology and introductions.
David M. Raup, Department Head, Geophysical Sciences Department, University of Chicago, Chicago, Illinois. Phanerozoic extinctions.
Frank Richter, Department of Earth Sciences, University of Chicago, Chicago, Illinois. Physical models of mantle convection.

Indiana
William J. Zinsmeister, Department of Geoscience, Purdue University, West Lafayette, Indiana. Geology of Spitsbergen.

Louisiana
R.J. Broussard, District Manager, Sohio Petroleum Company, Lafayette, Louisiana. Introductions.
Mitchell M. Hillman, Assistant Curator, Poverty Point, Epps, Louisiana. Archaeology.
Paul C.P. McIlhenny, Vice President, McIlhenny Company, Avery Island, Louisiana. Local archaeology.
Ann F. Ramenofsky, Assistant Professor, Department of Geography and Anthropology, Louisiana State University, Baton Rouge, Louisiana. Archaeology of Poverty Point.
William Reese, Dean, Faculty of Natural Science, University of Southwestern Louisiana, Lafayette, Louisiana. Botany, biology.

Massachusetts
Derek W. Spencer, Associate Director for Research, Woods Hole Oceanographic Institution, Woods Hole, Massachusetts. Oceanography.
Carolyn P. Winn, Research Librarian, Woods Hole Oceanographic Institution, Woods Hole, Massachusetts. Reference materials.

Missouri
Brian J. Mitchell, Department of Earth and Atmospheric Science, St. Louis University, St. Louis, Missouri. New Madrid Fault.
Otto Nuttli, Seismologist, Department of Earth and Atmospheric Science, St. Louis University, St. Louis, Missouri. New Madrid Fault.

New Jersey
Charles J. Quadri, Jr., Superintendent, Palisades Interstate Park Commission, Alpine, New Jersey. Palisade and Hudson geology.

New York
William F. Haxby, Lamont-Doherty Observatory, Palisades, New York. Seastat images.
James D. Hays, Lamont-Doherty Observatory, Palisades, New York. Paleoclimates.
Dennis V. Kent, Lamont-Doherty Observatory, Palisades, New York. Geology of Nova Scotia and Appalachia.
Malcolm R. McKenna, American Museum of Natural History, New York, New York. Eocene Arctic mammals.
Walter C. Pitman III, Lamont-Doherty Observatory, Palisades, New York. History of tectonics.
C. Barry Raleigh, Director, Lamont-Doherty Observatory, Palisades, New York. Introductions.
William B.F. Ryan, Lamont-Doherty Observatory, Palisades, New York. Geophysics, history of tectonics, and desiccation of the Mediterranean basin.
Anthony C. Watts, Lamont-Doherty Observatory, Palisades, New York. Continental margins.

North Dakota
Randolph B. Burke, Geologist, North Dakota Geological Survey, Grand Forks, North Dakota. Carbonate geology and marine biology.

Texas
Richard N. Wheatley, Senior Staff Associate, Public Affairs, Sohio Petroleum Company, Houston, Texas. Logistics for offshore drilling visit.

Washington, DC
Walter Adey, Curator, Paleobiology, Smithsonian Institute Natural History Museum, Washington, DC. Algal reefs.
Robert K. Perry, United States Naval Research Laboratory, Washington, DC. North Atlantic, bathymetry.
Peter R. Vogt, United States Naval Research Laboratory Washington, DC. North Atlantic, bathymetry.

Wisconsin
Robert H. Dott, Sedimentologist, Department of Geology and Geophysics, University of Wisconsin, Madison, Wisconsin. Wisconsin morphology.
James C. Knox, Department of Geology, University of Wisconsin, Madison, Wisconsin. Driftless zone.
John E. Kutzbach, Department of Meteorology, University of Wisconsin. Paleoclimate computer simulation, Milankovitch cycles.

Acknowledgments: Epilogue page 323
The Martin Waldseemüller globe of 1507 on page 323 was reproduced from copies of woodcut prints from the original gores now at the James, Ford, Bell Library: University of Minnesota, Minneapolis-St. Paul. The reproduction was made with the library's approval: the gores are the series of elliptical panels that form a map when pasted onto the surface of a globe. The coloring is in keeping with 16th-century cartographic style. The detail on the surface has been enhanced to correspond with the much larger (132 by 236 cm (53 inches by 93 inches) Waldseemüller woodcut map of the world of the same date, now at Schloss Wolfegg, Württemburg, Germany—just as a 16th-century cartographer would presumably have improved the original woodcut reproduction once the gores had been glued onto the surface of a plaster or wooden globe. The artist responsible for this work is Gary Hincks of Diss, a village in Norfolk, England—the principal technical illustrator of this book.

Bibliography

All sources of information and data for the text, plus details of works of reference used for the graphics of this book, are listed below. Specific permissions are noted first, followed by a list of general works of reference that were used frequently, and then by particular books and papers used for each chapter. Many of the papers were provided by and discussed with their authors, who are acknowledged individually on page 339.

Apart from Chapter I, the only sources actually quoted in the manuscript are those authors whose work may be considered controversial, or that may be in apparent contradiction to generally accepted views, or indeed where the work is so recent that it may not have received wide circulation at the time of writing.

The authors quoted in Chapter I are historical figures, mainly 19th- and 20th-century scientists who contributed to the discovery of a dynamic Earth. Some readers will be interested in the chronology of these events as the history of tectonic science unfolds in the chapter and will be as surprised as your author at the early dates of some key discoveries.

Many of the paleomaps and globes reproduced in this book are based on illustrations and figures from the undermentioned papers with the kind permission of Dr. Christopher R. Scotese.

Scotese, C.R., et al. *Atlas of Phanerozoic Plate Tectonic Reconstructions*. Washington, DC: American Geophysical Union, in press 2000.

Scotese, C.R., and W.S. McKerrow. "Paleozoic World Maps and Symposium Introduction in Paleogeography and Biogeography." In *Paleozoic Paleogeography and Biogeography*, Memoir 12, edited by W.S. McKerrow and C.R. Scotese. London: Geological Society, 1990.

Scotese, C.R. *Continental Drift, 7th edition*. Paleomap Project,Un.Texas, Arlington. 1997

Scotese, C.R., L.M. Gahagan, R.L. Larson. "Plate Tectonic Reconstructions of the Cretaceous and Cenozoic Ocean Basins." *Tectonophysics* 155 (1988): 27–48.

Scotese, C.R. "Paleomap Project" University of Texas at Arlington: website http://www.scotese.com/Info.htm April 1999 and personal communication Feb 2000

Winn, K., and C.R. Scotese. *Phanerozoic Paleogeographic Maps*, Technical Report 84. Austin: University of Texas, Institute for Geophysics, 1987.

General References

Brown, R.W. *Composition of Scientific Words*. rev. ed. Roland W. Brown: 1985 reprint.

Carroll, R.L. *Vertebrate Paleontology and Evolution*. New York: W.H. Freeman and Company, 1988 (and revised editions).

Encyclopaedia Britannica. *The New Encyclopaedia Britannica*. 15th ed. Chicago: 1987.

Gamlin, L., and G. Vines. *The Evolution of Life*. New York: Oxford University Press, 1987.

Margulis, L., and K.V. Schwartz. *Five Kingdoms: An Illustrated Guide to the Phyla of Life on Earth*. San Francisco: W.H. Freeman and Company, 3rd edition 1998.

Palmer, A.R. and Geissman.J. compilers *1999 Geologic Time Scale:* The Geological Society of America. Product code CTS004

Payne, M.M., et al. *National Geographic Atlas of the World*. 3d rev. ed. Washington, DC: National Geographic Society, 1995.

Press, F. and Siever, R: Understanding Earth: 2nd edition: W.H. Freeman and Company, 1998

Purves, K.W., Orians, G.H., Heller, H.C. and Sadava, D.: Life: *The Science of Biology*: 5th edition: W.H. Freeman and Company, 1998

The Times Atlas of the World. Comprehensive ed. New York: Times Books, 1990.

Schopf, J.W. And Klein, C, eds. *The Proterozoic Bioshpere, a multidisciplinary study.* Cambridge Univerity Press. (and UCLA) 1992 (1998).

Ziegler, P.A. *Evolution of the Arctic-North Atlantic and the Western Tethys*, Memoir 43. Tulsa, OK: American Association of Petroleum Geologists, 1988.

———. *Geological Atlas of Western and Central Europe*. Amsterdam: Elsevier Scientific Publishing Company, 1982.

I First Light

Bally, A.W., et al. *Geology of Passive Continental Margins: History, Structure and Sedimentologic Record (With Special Emphasis on the Atlantic Margin)*. AAPG Eastern Section Meeting and Atlantic Margin Energy Conference. Education Course Note Series #19. Tulsa: American Association of Petroleum Geologists, 1981.

Birkenmajer, K. "The Geology of Svalbard, the Western Part of the Barents Sea, and the Continental Margin of Scandinavia." In *The Ocean Basins and Margins: The Arctic Ocean*, vol. 5, edited by A.E.M. Nairn, M. Churkin, and F.G. Stekli. New York: Plenum Press, 1981.

Christie, R.L. "Canada—Arctic Archipelago." In *Encyclopedia of World Region Geology Part I: Western Hemisphere*, edited by R.W. Fairbridge. Stroudsburg, PA: Dowden, Hutchinson and Ross, 1975.

Dawes, P.R., and J.W. Kerr. "The Case Against Major Displacement along Nares Strait." In *Nares Strait and the Drift of Greenland: A Conflict in Plate Tectonics*. Meddelelser om Gronland, Geoscience 8, edited by P.R. Dawes and J.W. Kerr. Copenhagen: Geological Survey of Greenland, 1982.

Dewey, J.F., and W.C. Pitman III. *The Origin and Evolution of Sedimentary Basins*. Piermont, NY: Tectan, 1989.

Dott, R.H., Jr., and R.L. Batten. *Evolution of the Earth*. 3d ed. New York: McGraw-Hill Book Company, 1981.

Eldholm, O., and J. Thiede. "Cenozoic Continental Separation between Europe and Greenland." *Palaeogeography, Palaeoclimatology, Palaeoecology* 30 (1980): 243–59.

Escher, A., and W.S. Watt, eds. *Geology of Greenland*. Copenhagen: Geological Survey of Greenland, 1976.

Garner, H.F. *The Origin of Landscapes: A Synthesis of Geomorphology*. New York: Oxford University Press, 1984.

Hallam, A. *Great Geological Controversies*. New York: Oxford University Press, 1983.

Harland, W.B. "The Caledonides of Svalbard." In *Caledonian-Appalachian Orogen of the North Atlantic Region*. Paper 78–13. Ottawa: Geological Survey of Canada, 1978.

Hatcher, R.D., Jr., H. Williams, and I. Zietz, eds. *Contributions to the Tectonics and Geophysics of Mountain Chains*, Memoir 158. Boulder, CO: Geological Society of America, 1983.

Higgins, A.K., and W.E.A. Phillips. "East Greenland Caledonides—An Extension of the British Caledonides." In *The Caledonides of the British Isles*, edited by A.L. Harris, C.H. Holland, B.E. Leake. London: Geological Society of London, 1979.

Jeanloz, R. "The Earth's Core." In *The Dynamic Earth*. Readings from Scientific American. New York: W.H. Freeman and Company, 1983.

Johnson, G.L., and S.P. Srivastava. "The Case for Major Displacement along Nares Strait." In *Nares Strait and the Drift of Greenland: A Conflict in Plate Tectonics*. Meddelelser om Gronland, Geoscience 8, edited by P.R. Dawes and J.W. Kerr. Copenhagen: Geological Survey of Greenland, 1982.

Kellogg, H.E. "Tertiary Stratigraphy and Tectonism in Svalbard and Continental Drift." *The American Association of Petroleum Geologists Bulletin* 59 (March 1975): 465–85.

King, L.C. *Wandering Continents and Spreading Sea Floors on an Expanding Earth*. Chichester: John Wiley & Sons, 1983.

Monastersky, R. "Drilling Hits Birthplace of Pacific Plate." *Science News* 137 (3 February 1990): 69.

Nicholson, R. "Caledonian Correlations—British and Scandinavian." In *The Caledonides of the British Isles* edited by A.L. Harris, C.H. Holland, B.E. Leake. London: Geological Society of London, 1979.

Owen, H.G. "Constant Dimensions or an Expanding Earth?" In *The Evolving Earth*, edited by L.R.M. Cocks. Cambridge: Cambridge University Press, 1981.

Press, F., and R. Siever. *Earth*. 4th ed. New York: W.H. Freeman and Company, 1986.

Schwarzbach, M. *Alfred Wegener: The Father of Continental Drift*. Madison, WI: Science Tech, 1980.

Scientific American. *The Dynamic Earth*. Readings from Scientific American. New York: W.H. Freeman and Company, 1983.

Shea, J.H., ed. *Continental Drift*. New York: Van Nostrand Reinhold Company, 1985.

Trettin, H.P. *Reconnaissance of Lower Paleozoic Geology, Agassiz Ice Cap to Yelverton Bay, Northern Ellesmere Island*. Paper 76–1A. Ottawa: Geological Survey of Canada, 1984.

Trettin, H.P., W.D. Loveridge, and R.W. Sullivan. *U-Pb Ages on Zircon from the McClintock West Massif and the Markham Fiord Pluton, Northernmost Ellesmere Island*. Paper 82–1C. Ottawa: Geological Survey of Canada, 1984.

Wegener, A. *The Origin of Continents and Oceans*. 3d German ed. Translated by J.W. Evans. New York: E.P. Dutton and Company, 1924.

Whitten, E.H.T. "Continental-Drift Models and Correlation of Geologic Features across North Atlantic Ocean." In *North Atlantic—Geology and Continental Drift*, Memoir 12, edited by M. Kay. Tulsa: American Association of Petroleum Geologists, 1967.

Windley, B.F. *The Evolving Continents*. Chichester: John Wiley & Sons, 1987.

II New Look at an Old Planet

Bloxham, J., and D. Gubbins. "The Evolution of the Earth's Magnetic Field." *Scientific American* (December 1989): 68–75.

Brown, G.C., and A.E. Mussett. *The Inaccessible Earth*. London: George Allen & Unwin, 1981.

Carrigan, C.R. "Multiple-Scale Convection in the Earth's Mantle: A Three-Dimensional Study." *Science* 215 (19 February 1982): 965–67.

Chyba, C.F. "The Cometary Contribution to the Oceans of Primitive Earth." *Nature* 330 (17 December 1987): 632–35.

Courtillot, V., and J. Besse. "Magnetic Field Reversals, Polar Wander, and Core-Mantle Coupling." *Science* 237 (4 September 1987): 1140–45.

Crum, R. "Uncovering the Tectonic Engine." *Harvard Magazine* (March-April, 1989): 32–36.

Drake, M.J. "Siderophile Elements in Planetary Mantles and the Origin of the Moon." From the Proceedings of the Seventeenth Lunar and Planetary Science Conference, Part 2. *Journal of Geophysical Research* 92, no. B4 (30 March 1987): E377–E386.

Eberhart, J. "Halley's Whiskers: First Space Polymer Detected." *Science News* 132 (15 August 1987): 100.

Fuller, M. "Impacts That Magnetize." *Nature* 329 (22 October 1987): 674–75.

Gamlin, L., and G. Vines. *The Evolution of Life*. New York: Oxford University Press, 1987.

Hartman, W.K., and D.R. Davis. "Satellite-sized Planetesimals and Lunar Origin." *Icarus* 24 (1975): 504–15.

Hoffman, K.A. "Ancient Magnetic Reversals: Clues to the Geodynamo." *Scientific American* (May 1988): 76–83.

Holland, H.D. *The Chemical Evolution of the Atmosphere and Oceans*, Princeton University Press, 1984

Jacobs, J.A. *Reversals of the Earth's Magnetic Field*. Bristol: Adam Hilger, 1984.

Kerr, R.A. "Making the Moon from a Big Splash." *Science* 226 (November 1984): 1060–61.

———. "Where Was the Moon Eons Ago?" *Science* 221 (16 September 1983): 1166.

Mackwell, S.J. "Creep in the Mantle." *Nature* 330 (26 November 1987): 315.

Maher, K.A., and D.J. Stevenson. "Impact Frustration of the Origin of Life." *Nature* 331 (18 February 1988): 612–14.

Margulis, L. "The Origin of Plant and Animal Cells." In *Paleontology and Paleoenvironments*, edited by B.J. Skinner. Readings from American Scientist. Los Altos, CA: William Kaufmann, 1981.

———. *Symbiosis in Cell Evolution: Life and Its Evolution on the Early Earth*. San Francisco: W.H. Freeman and Company, 1981.

Margulis, L., and K.V. Schwartz. *Five Kingdoms: An Illustrated Guide to the Phyla of Life on Earth*. San Francisco: W.H. Freeman and Company, 1982.

Open University. *Evolution: Units 1–6*. Milton Keynes, UK: The Open University Press, 1981.

———. *Oceanography: Units 1–6 and Units 13–14*. Milton Keynes, UK: The Open University Press, 1984.

Penny, D. "What Was the First Living Cell?" *Nature* 331 (14 January 1988): 111–12.

Schidlowski, M. "A 3,800-Million-Year Isotopic Record of Life from Carbon in Sedimentary Rocks." *Nature* 333 (26 May 1988): 313–18.

Schopf, J.W., ed. *Earth's Earliest Biosphere: Its Origin and Evolution*. Princeton, NJ: Princeton University Press, 1983.

Schopf, J.W., and B.M. Packer. "Early Archean (3.3 Billion to 3.5-Billion-Year-Old) Microfossils from Warrawoona Group, Australia." *Science* 237 (3 July 1987): 70–72.

Scientific American. *The Biosphere*. Readings from Scientific American. San Francisco: W.H. Freeman and Company, 1980.

———. *Conditions for Life*. Readings from Scientific American. San Francisco: W.H. Freeman and Company, 1986.

———. *Continents Adrift and Continents Aground*. Readings from Scientific American. San Francisco: W.H. Freeman and Company, 1986.

———. *The Dynamic Earth*. Readings from Scientific American. New York: W.H. Freeman and Company, 1983.

———. *Earthquakes and Volcanoes*. Readings from Scientific American. San Francisco: W.H. Freeman and Company, 1980.

———. *Evolution*. Readings from Scientific American. San Francisco: W.H. Freeman and Company, 1988.

———. *Evolution and the Fossil Record*. Readings from Scientific American. San Francisco: W.H. Freeman and Company, 1988.

———. *Life: Origin and Evolution*. Readings from Scientific American. San Francisco: W.H. Freeman and Company, 1989.

———. *The Molecules of Life*. Readings from Scientific American. New York: W.H. Freeman and Company, 1985.

———. *The Physics of Everyday Phenomena*. Readings from Scientific American. San Francisco: W.H. Freeman and Company, 1989.
———. *The Planets*. Readings from Scientific American. New York: W.H. Freeman and Company, 1983.
———. *The Solar System*. Readings from Scientific American. San Francisco: W.H. Freeman and Company, 1985.
———. *The Universe of Galaxies*. Readings from Scientific American. New York: W.H. Freeman and Company, 1984.
Skinner, B.J., ed. *Paleontology and Paleoenvironments*. Readings from American Scientist. Los Altos, CA: William Kaufmann, 1981.
Stebbins, G.L. *Darwin to DNA, Molecules to Humanity*. San Francisco: W.H. Freeman and Company, 1982.
Steitz, J.A. "Snurps." *Scientific American* (June 1988): 56–63.
Stigler, S.M. "Aperiodicity of Magnetic Reversals?" *Nature* 330 (5 November 1988): 26.
Strahler, A.N., and A.H. Strahler. *Modern Physical Geography*. New York: John Wiley & Sons, 1988.
Suck, D., A. Lahm, and C. Oefner. "Structure Refined to 2Å of a Nicked DNA Octanucleotide Complex with DNase I." *Nature* 332 (31 March 1988): 464–68.

III Iapetus and Avalonia

Anderson, M.A. "Ediacaran Fauna." In *McGraw-Hill Yearbook of Science and Technology*. New York: McGraw-Hill Book Company, 1988.
Anderson, M.A., and S.C. Morris. "A Review with Descriptions of Four Unusual Forms, of the Soft-Bodied Fauna of the Conception and St. John's Groups (Late Precambrian), Avalon Peninsula, Newfoundland." Paper presented at the Third North American Paleontological Convention, Proceedings, vol.1, August 1982.
Briggs, E.G. Erwin, D.H. and Collier, F.J. *The Fossils of the Burgess Shale*, Smithsonian Institution Press, Washington D.C. 1994.
Crowley, T.J., J.G. Mengel, and D.A. Short. "Gondwanaland's Seasonal Cycle." *Nature* 329 (29 October 1988): 803–07.
Fenton, C.L., and M.A. Fenton. *The Fossil Book: A Record of Prehistoric Life*. Garden City, NY: Doubleday & Company, 1958.
Florida Escarpment Cruise Participants. "The Seeps Find at the Florida Escarpment." *Oceanus: The International Magazine of Marine Science and Policy* (Published by Woods Hole Oceanographic Institution) 27, no. 3 (Fall 1984): 32–33.
Glaessner, M.F. *The Dawn of Animal Life: A Biohistorical Study*. Cambridge: Cambridge University Press, 1984.
Grabau, A.W. *Palaeozoic Formations in the Light of the Pulsation Theory*. Peking: University Press, University of Peking, 1936.
Hofmann, H.J. "Canada's Precambrian Fossils." *GEOS* (Winter 1987): 2–5.
Hsü, K.J., ed. *Mountain Building Processes*. London: Academic Press, 1982.
King, A.F., et al. *Late Precambrian and Cambrian Sedimentary Sequences of Southeastern Newfoundland*. Geological Association of Canada and Mineralogical Association of Canada Joint Annual Meeting, Field Trip B-6 Guidebook. St. John's, Newfoundland: Geological Association of Canada and Mineralogical Association of Canada, 1984.
Margulis, L., and K.V. Schwartz. *Five Kingdoms: An Illustrated Guide to the Phyla of Life on Earth*. San Francisco: W.H. Freeman and Company, 3rd edition 1998.
Moore, R.C., C.G. Lalicker, and A.G. Fischer. *Invertebrate Fossils*. New York: McGraw-Hill Book Company, 1952.
Rast, N. "The Avalonian Plate in the Northern Appalachians and Caledonides." In *The Caledonides in the U.S.A.*, edited by D.R. Wones, International Geologic Correlation Program 1979 Conference, Proceedings. Blackburg, VA: Virginia Polytechnic Institute and State University, 1980.
Rodgers, J. "The Life History of a Mountain Range—The Appalachians." In *Mountain Building Processes*, edited by K.J. Hsü. London: Academic Press, 1983.
Rowe, F.W. *A History of Newfoundland and Labrador*. Toronto: McGraw-Hill Ryerson Limited, 1980.
Schopf, J.W., ed. *Earth's Earliest Biosphere: Its Origin and Evolution*. Princeton, NJ: Princeton University Press, 1983.
Secord, J.A. *Controversy in Victorian Geology: The Cambrian-Silurian Dispute*. Princeton, NJ: Princeton University Press, 1986.
Smith, B., and T.N. George. *British Regional Geology: North Wales*. 3d ed., completely revised by T.N. George. London: Institute of Geological Sciences, Geological Survey and Museum, 1961.
Stanley, S.M. *Earth and Life through Time*. New York: W.H. Freeman and Company, 1986.
Williams, G.E. *Megacycles: Long-Term Episodicity in Earth and Planetary History*. Benchmark Papers in Geology, no. 57. Stroudsburg, PA: Hutchinson Ross Publishing Company, 1981.
Williams, H., and A.F. King. *Trepassey Map Area, Newfoundland*, Memoir 389. Ottawa: Geological Survey of Canada, 1976.
Wilson, J.T. "Did the Atlantic Close and Then Re-open?" *Nature* 211 (1966): 676–81.

IV Tornquist's Sea

Bradley, D.C. "Tectonics of the Acadian Orogeny in New England and Adjacent Canada." *Journal of Geology* 91 (July 1983): 381–400.
Burke, C.A., and C.L. Drake, eds. *The Geology of Continental Margins*. New York: Springer-Verlag, 1974.
Carroll, R.L. *Vertebrate Paleontology and Evolution*. New York: W.H. Freeman and Company, 1988.
Chapman, C.A. *The Geology of Acadia National Park*. Old Greenwich, CT: Chatham Press, 1962.
Cocks, L.R.M., and R.A. Fortey. "Faunal Evidence for Oceanic Separations in the Palaeozoic of Britain." *Journal of the Geological Society of London* 139 (1982): 465–78.
Crowley, T.J., J.G. Mengel, and D.A. Short. "Gondwanaland's Seasonal Cycle." *Nature* 329 (29 October 1987): 803–07.
Den Tex, E. "A Geological Section across the Hesperian Massif in Western and Central Galicia." *Geologie en Mijnbouw* 60 (1981): 33–40.
Eldredge, N., and S.J. Gould. "Punctuated Equilibria: an Alternative to Phyletic Gradualism." In *Models in Paleobiology*, edited by T.J.M. Schopf. San Francisco: Freeman, Cooper and Company, 1972.
Fenton, C.L., and M.A. Fenton. *The Fossil Book: A Record of Prehistoric Life*. Garden City, NY: Doubleday & Company, 1958.
Geological Survey of Canada. *Caledonian-Appalachian Orogen of the North Atlantic Region*. Paper 78–13. Ottawa: Geological Survey of Canada, 1978.
Hatcher, R.D., Jr., H. Williams, and I. Zietz, eds. *Contributions to the Tectonics and Geophysics of Mountain Chains*, Memoir 158. Boulder, CO: Geological Society of America, 1983.
Holland, H.D., and A.F. Trendall, eds. *Patterns of Change in Earth Evolution*. Report of the Dahlem Workshop on Patterns of Change in Earth Evolution, Berlin, 1983. Berlin: Springer-Verlag, 1984.
Howell, D.G., ed. *Tectonostratigraphic Terranes of the Circum-Pacific Region*. Circum-Pacific Council for Energy and Mineral Resources, Earth Science Series, Number 1. Houston: Circum-Pacific Council for Energy and Mineral Resources, 1985.
Hsü, K.J., ed. *Mountain Building Processes*. London: Academic Press, 1982.
Huxley, A. "Address of the President Sir Andrew Huxley at the Anniversary Meeting, 30 November 1981." *Proceedings of the Royal Society of London* 379 (8 January 1982): 5–20.
Kent, D.V., and N.D. Opdyke. "The Early Carboniferous Paleomagnetic Field of North America and Its Bearing on Tectonics of the Northern Appalachians." *Earth and Planetary Science Letters* 44 (1979): 365–72.
McLaren, D.J. "Evolution Ancient and Modern." Paper presented at the Third North American Paleontological Convention, McGill University, Montreal, August 6, 1982.
———. "Bolides and Biostratigraphy." *Geological Society of America Bulletin* 94 (March 1983): 313–24.
Moseley, F., ed. *The Geology of the Lake District*. Yorkshire Geological Society, 1978.
Stanley, S.M. *Extinction*. New York: Scientific American Books, 1987.
———. *Macroevolution: Pattern and Process*. San Francisco, W.H. Freeman and Company, 1979.
Vogel, D.E. "Cabo Ortegal, Mantle Plumbe or Double Klippe?" *Geologie en Mijnbouw* 63 (1984): 131–40.
Warme, J.E., A.K. Chamberlain, and B.W. Ackman. "The Alamo Event: Devonian Cataclysmic Breccia in Southeastern Nevada." Abstract. Paper presented at Cordilleran Meeting, Geological Society of America, San Francisco, March, 1991.
Williams, H. "Paleozoic Miogeoclines and Suspect Terranes of the North Atlantic Region; Cordilleran Comparisons." In *Tectonostratigraphic Terranes of the Circum-Pacific Region*, edited by D.G. Howell, Earth Science Series, Number 1. Houston: Circum-Pacific Council for Energy and Mineral Resources, 1985.
———. "Miogeoclines and Suspect Terranes of the Caledonian-Appalachian Orogen: Tectonic Patterns in the North Atlantic Region." *Canadian Journal of Earth Science* 21 (1984): 887–901.
Williams, H., and R.D. Hatcher, Jr. "Appalachian Suspect Terranes." In *Contributions to the Tectonics and Geophysics of Mountain Chains*, Memoir 158, edited by R.D. Hatcher, Jr., H. Williams, and I. Zietz. Boulder, CO: Geological Society of America, 1983.
Williams, H., and R.K. Stevens. "The Ancient Continental Margin of Eastern North America." In *The Geology of Continental Margins*, edited by C.A. Burk and C.L. Drake. New York: Springer-Verlag, 1984.
Ziegler, P.A. *Geological Atlas of Western and Central Europe*. Amsterdam: Elsevier Scientific Publishing Company, 1982.

V The Third Age

Ager, D.V. *The Geology of Europe*. London: McGraw-Hill Book Company, 1980.
American Geological Institute. "Episodic View Now Replacing Catastrophism." *Geotimes* (November 1982): 16.
Bromley, A.V. "Ophiolitic Origin of the Lizard Complex." Paper presented at the Fourth Meeting of the Geological Societies of the British Isles, University of Sheffield, UK, September, 1979.
———. "Granites in Mobile Belts—The Tectonic Setting of the Cornubian Batholith." *Journal of the Camborne School of Mines* (Cornwall) 76 (1976): 40–47.
Bromley, A.V., and J. Holl. "Tin Mineralization in Southwest England." Camborne School of Mines, Cornwall, 1986. Photocopy.
Burg, J.P., et al. "Variscan Intracontinental Deformation: The Coimbra-Cordoba Shear Zone (SW Iberian Peninsula)." *Tectonophysics* 78 (1981): 161–77.
Carvalho, D., de. "The Metallogenetic Consequences of Plate Tectonics and the Upper Paleozoic Evolution of Southern Portugal." *Estudos, Notas e Trabalhos* 20 (1972): 297–320.
Cohen, A.D., et al. "Peat Deposits." In *The Okefenokee Swamp: Natural History, Geology and Geochemistry*, edited by A.D. Cohen, et al. Los Alamos, NM: Wetlands Surveys, 1984.
Dunning, F.W. "Caledonian-Variscan Relations in North-west Europe." *Colloque International du Centre National de la Recherche Scientifique* 243 (1977): 165–80.
Freshney, E., and R. Taylor. "The Variscides of Southwest Britain." In *United Kingdom*, edited by T.R. Owen. Paris: 26th International Geological Congress, 1980.
Gamlin, L., and G. Vines. *The Evolution of Life*. New York: Oxford University Press, 1987.
Hatcher, R.D., Jr., H. Williams, and I. Zietz, eds. *Contributions to the Tectonics and Geophysics of Mountain Chains*, Memoir 158. Boulder, CO: Geological Society of America, 1983.
Hsü, K.J., ed. *Mountain Building Processes*. London: Academic Press, 1982.
Hutton, D.H.W., and D.J. Sanderson, eds. *Variscan Tectonics of the North Atlantic Region*. Published for the Geological Society. Oxford: Blackwell Scientific Publications, 1984.
Irving, E., and D.F. Strong. "Palaeomagnetism of the Early Carboniferous Deer Lake Group, Western Newfoundland: No Evidence for Mid-Carboniferous Displacement of 'Acadia'." *Earth and Planetary Science Letters* 69 (1984) 379–90.
Julivert, M. "A Cross-Section Through the Northern Part of the Iberian Massif." *Geologie en Mijnbouw* 60 (1981): 107–28.
Keene, P. *Classic Landform of the North Devon Coast*. Classic Landform Guides, no. 6. Sheffield, UK: Geographical Association, 1970.
Keppie, J.D. *Geology and Tectonics of Nova Scotia*. Fredericton, New Brunswick: Geological Association of Canada and Mineralogical Association of Canada, 1985.
Lefort, J.P. "A New Geophysical Criterion to Correlate the Acadian and Hercynian Orogenies of Western Europe and Eastern America." In *Contributions to the Tectonics and Geophysics of Mountain Chains*, Memoir 158, edited by R.D. Hatcher, Jr., H. Williams, and I. Zietz. Boulder, CO: Geological Society of America, 1983.
Matthews, S.C., J.J. Chauvel, and M. Robardet. "Variscan Geology of Northwestern Europe." In *Géologie de l'Europe, du Précambrien aux bassins sédimentaires post-hercyniens*, Memoir 108, edited by J. Cogne, and M. Slansky. Paris: French Bureau Recherches Géologiques et Minieres, 1980.
Owen, T.R., ed. *United Kingdom*. Paris: 26th International Geological Congress, 1980.
Raup, D.M., and J.J. Sepkoski, Jr. "Mass Extinctions in the Marine Fossil Record." *Science* 215 (19 March 1982): 1501–02.
Raven, P.H., R.F. Evert, and H. Curtis. *Biology of Plants*. 3d ed. New York: Worth Publishers, 1982.
Rodgers, J. "The Life History of a Mountain Range—The Appalachians." In *Mountain Building Processes*, edited by K.J. Hsü. London: Academic Press, 1983.
Schenk, P.E. "The Meguma Terrane of Nova Scotia, Canada—An Aid in Trans-Atlantic Correlation." In *Regional Trends in the Geology of the Appalachian-Caledonian-Hercynian-Mauritanide Orogen*, edited by P.E. Schenk. Dordrecht: D. Reidel Publishing Company, 1982.
———. "The Meguma Zone of Nova Scotia—A Remnant of Western Europe, South America, or Africa?" In *Geology of the North Atlantic Borderlands*, Memoir 7, edited by J.W. Kerr and A.J. Ferguson. Calgary: Canadian Society of Petroleum Geologists, 1981.
———. "Paleogeographic Implications of the Meguma Group, Nova Scotia—A Chip of Africa?" In *The Caledonides in the U.S.A.*, edited by D.R. Wones, International Geologic Correlation Program 1979

Conference, Proceedings. Blackburg, VA: Virginia Polytechnic Institute and State University, 1980.
———. "Synthesis of the Canadian Appalachians." In *Caledonian-Appalachian Orogen of the North Atlantic Region*, Paper 78–13. Ottawa: Geological Survey of Canada, 1978.
Scotese, C.R., and W.S. McKerrow. "Paleozoic World Maps and Symposium Introduction in Paleogeography and Biogeography." In *Paleozoic Paleogeography and Biogeography*, Memoir 12, edited by W.S. McKerrow and C.R. Scotese. London: Geological Society, 1990.
Scotese, C.R., et al. *Atlas of Phanerozoic Plate Tectonic Reconstructions*. Washington, DC: American Geophysical Union, in press.
Sparks, R.S.J., and H.E. Huppert. "The Origins of Granites." *Nature* 330 (19 November 1987): 207–08.
Steitz, J.A. "Snurps." *Scientific American* (June 1988): 56–63.
Stewart, W.N. *Paleobotany and the Evolution of Plants*. London: Cambridge University Press, 1985.
Van der Voo, R. "A Plate-Tectonics Model for the Paleozoic Assembly of Pangea Based on Paleomagnetic Data." In *Contributions to the Tectonics and Geophysics of Mountain Chains*, Memoir 158, edited by R.D. Hatcher, Jr., H. Williams, and I. Zietz. Boulder, CO: Geological Society of America, 1983.
Williams, H. "Miogeoclines and Suspect Terranes of the Caledonian-Appalachian Orogen: Tectonic Patterns in the North Atlantic Region." *Canadian Journal of Earth Science* 21 (1984): 887–901.
Williams, H., and R.D. Hatcher, Jr. "Appalachian Suspect Terranes." In *Contributions to the Tectonics and Geophysics of Mountain Chains*, Memoir 158, edited by R.D. Hatcher, Jr., H. Williams, and I. Zietz. Boulder, CO: Geological Society of America, 1983.
———. "Suspect Terranes and Accretionary History of the Appalachian Orogen." *Geology* 10 (October 1982): 530–36.
Windley, B.F. *The Evolving Continents*. 2d ed. Chichester: John Wiley & Sons, 1984.
Winn, K., and C.R. Scotese. *Phanerozoic Paleogeographic Maps*, Technical Report 84. Austin: University of Texas, Institute for Geophysics, 1987.
Ziegler, P.A. *Geological Atlas of Western and Central Europe*. Amsterdam: Elsevier Scientific Publishing Company, 1982.

VI Middle-Earth

Bolt, J.R., et al. "A New Lower Carboniferous Tetrapod Locality in Iowa." *Nature* 333 (23 June 1988): 768–70.
Burton, R. *Eggs: Nature's Perfect Package*. New York: Facts on File Publications, 1987.
Colbert, E.H. "Therapsids in Pangaea and Their Contemporaries and Competitors." In *The Ecology and Biology of Mammal-like Reptiles*, edited by N. Hotton III, et al. Washington, DC: Smithsonian Institution Press, 1986.
Evans,D. and Kirschvink,J. *Neoprototerozoic paleogeogrpahy and glacial Climate: SWEAT and the Snowball Earth?* Caltech Paleomag Lanoratory. 1999.
Four Corners Geological Society. *Geology and Natural History of the Fifth Field Conference: Powell Centennial River Expedition*. Durango, CO: Four Corners Geological Society, 1969. McKee, E.D. *Paleozoic Rocks of Grans Canyon* (page 78)
Geiser, P., and T. Engelder. "The Distribution of Layer Parallel Shortening Fabrics in the Appalachian Foreland of New York and Pennsylvania: Evidence for Two Non-coaxial Phases of the Alleghanian Orogeny." In *Contributions to the Tectonics and Geophysics of Mountain Chains*, Memoir 158, edited by R.D. Hatcher, Jr., H. Williams, and I. Zietz. Boulder, CO: Geological Society of America, 1983.
Hoffman, P.F. "Did the Breakout of Laurentia Turn Gondwanaland Inside-out?" *Science* 252 (7 June 1991): 1409–12.
Hoffman, P.F., and Schrag, D.P: *A Neoprotoerozoic Snowball Earth*: Science vol 281. p.1342 (1998)
Hoffman, P.F., and Schrag, D.P: *Snowball Earth*: Scientific American Jan 2000
Hatcher, R.D., Jr., and G.W. Viele. "The Appalachian/ Ouachita Orogens: United States and Mexico." In *Perspectives in Regional Geological Synthesis*, edited by A.R. Palmer, D-NAG Special Publication 1. Boulder, CO: Geological Society of America, 1983.
Higgins, M.W., and I. Zietz. "Geologic Interpretation of Geophysical Maps of the Pre-Cretaceous 'Basement' beneath the Coastal Plain of the Southeastern United States." In *Contributions to the Tectonics and Geophysics of Mountain Chains*, Memoir 158, ed. by R.D. Hatcher, Jr., H. Williams, I. Zietz. Boulder, CO: Geological Society of America, 1983.
Hotton, N., III, et al., eds. *The Ecology and Biology of Mammal-like Reptiles*. Washington, DC: Smithsonian Institution Press, 1986.
Hunt, C.B. *Natural Regions of the United States and Canada*. San Francisco: W.H. Freeman and Co., 1974.
Kaufman, S. "Crustal Faults in North America: A Report of the COCORP Project." In *Mountain Building Processes*, edited by K.J. Hsü. London: Academic Press, 1982.
Kirschvink, J., *The Snowball Earth* in *The Protoerozoic Earth:* edited by Schopf, J.W., and Klein, C.: Cambridge University Press NY. 1992
MacLean, P.D. "Neurobehavioral Significance of the Mammal-like Reptiles (Therapsids)." In *The Ecology and Biology of Mammal-like Reptiles*, edited by N. Hotton III, et al. Washington, DC: Smithsonian Institution Press, 1986.
Ménard, G., and P. Molnar. "Collapse of a Hercynian Tibetan Plateau into a Late Palaeozoic European Basin and Range Province." *Nature* 334 (21 July 1988): 235.
Nance, D.R., T.R. Worsley, and J.B. Moody. "The Supercontinent Cycle." *Scientific American* (July 1988): 72–79.
Nelson, K.D., et al. "New COCORP Profiling in the Southeastern United States: Part 1: Late Paleozoic Suture and Mesozoic Rift Basin." *Geology* 13 (October 1985): 714–18.
Nickelsen, R.P. "Sequence of Structural Stages of the Allegheny Orogeny, at the Bear Valley Strip Mine, Shamokin, Pennsylvania." *American Journal of Science* 279 (March 1979): 225–71.
Ortega-Gutiérrez, F., and J.C. Guerrero-Garcia. "The Geologic Regions of Mexico." In *Perspectives in Regional Geological Synthesis*, edited by A.R. Palmer, D-NAG Special Publication 1. Boulder, CO: Geological Society of America, 1983.
Owen, T.R., ed. *United Kingdom*. Paris: 26th International Geological Congress, 1980.
Palmer, A.R., ed. *Perspectives in Regional Geological Synthesis*. D-NAG Special Publication 1. Boulder, CO: Geological Society of America, 1983.
Parrish, J.M., J.T. Parrish, and A.M. Ziegler. "Permian-Triassic Paleogeography and Paleoclimatology and Implications for Therapsid Distribution." In *The Ecology and Biology of Mammal-like Reptiles*, edited by N. Hotton III, et al. Washington, DC: Smithsonian Institution Press, 1986.
Petersen, M.S., J.K. Rigby, and L.F. Hintze. *Historical Geology of North America*. 2d ed. Dubuque, IA: Wm. C. Brown Company Publishers, 1980.
Rast, N. "The Alleghenian Orogeny in Eastern North America." In *Variscan Tectonics of the North Atlantic Region*, edited by D.H.W. Hutton and D.J. Sanderson. Published for the Geological Society. Oxford: Blackwell Scientific Publications, 1984.
Rodgers, J. "The Life History of a Mountain Range—The Appalachians." In *Mountain Building Processes*, edited by K.J. Hsü. London: Academic Press, 1982.
Ross, M.I., and C.R. Scotese. "Hierarchical Tectonic Analysis of the Gulf of Mexico and Caribbean Region." *Tectonophysics* 155 (1988): 130–68.
Scholle, P.A., and R.B. Halley. *Upper Paleozoic Depositional and Diagenetic Facies in a Mature Petroleum Province: A Field Guide to the Guadalupe and Sacramento Mountains*. Denver: U.S. Geological Survey, 1980.
Scotese, C.R., and W.S. McKerrow. "Paleozoic World Maps and Symposium Introduction in Paleogeography and Biogeography." In *Paleozoic Paleogeography and Biogeography*, Memoir 12, edited by W.S. McKerrow and C.R. Scotese. London: Geological Society, 1990.
Scotese, C.R., et al. *Atlas of Phanerozoic Plate Tectonic Reconstructions*. Washington, DC: American Geophysical Union, in press.
Smith, D.B. "The Permian." In *United Kingdom*, edited by T.R. Owen. Paris: 26th International Geological Congress, 1980.
Stanley, S.M. *Extinction*. New York: Scientific American Books, 1987.
Tolkien, J.R.R. *The Lord of the Rings: The Fellowship of the Ring*. 2d ed. Boston: Houghton Mifflin Co., 1982.
Wiltschko, D., and D. Eastman. "Role of Basement Warps and Faults in Localizing Thrust Fault Ramps." In *Contributions to the Tectonics and Geophysics of Mountain Chains*, Memoir 158, edited by R.D. Hatcher, Jr., H. Williams, and I. Zietz. Boulder, CO: Geological Society of America, 1983.
Winn, K., and C.R. Scotese. *Phanerozoic Paleogeographic Maps*, Technical Report 84. Austin: University of Texas, Institute for Geophysics, 1987.
Ziegler, P.A. *Geological Atlas of Western and Central Europe*. Amsterdam: Elsevier Scientific Publishing Company, 1982.

VII The New World

American Geological Institute. "Fossils of Ancient Bird Unearthed in West Texas." *Earth Science* 40 (Spring 1987): 7.
Bonatti, E. "The Rifting of Continents." *Scientific American* (March 1987): 97–103.
Briggs, J.C. *Biogeography and Plate Tectonics*. New York: Elsevier Science Publishing Company, 1987.
Burke, K.C., and J.T. Wilson. "Hot Spots on the Earth's Surface." In *Continents Adrift and Continents Aground*. Readings from Scientific American. San Francisco: W.H. Freeman and Company, 1986.
Carroll, R.L. *Vertebrate Paleontology and Evolution*. New York: W.H. Freeman and Company, 1988.
Chatterjee, S. "Skull of *Protoavis* and Early Evolution of Birds." Abstract. *Journal of Vertebrate Paleontology* 7 (16 September 1987): 14A.
Chatterjee, S. The Triassic Bird *Protoavis* . *Archaeopteryx* 13: 15-31: (1995)
Chiappe, L.M. The First 85 million years of avian evolution. *Nature*. 378: 349-355 (1995)
Cossins, A. R., and K. Bowler. *Temperature Biology of Animals*. London: Chapman and Hall, 1987.
Courtillot, V., and G.E. Vink. "How Continents Break Up." *Scientific American* (July 1983): 42–49.
Delaney, M.L. "Extinctions and Carbon Cycling." *Nature* 337 (5 January 1989): 18–19.
Dietz, R.S., and J.C. Holden. "The Breakup of Pangaea." In *Continents Adrift and Continents Aground*. Readings from Scientific American. San Francisco: W.H. Freeman and Company, 1976.
Foulger, G.R., et al. "Implosive Earthquakes at the Active Accretionary Plate Boundary in Northern Iceland." *Nature* 337 (16 February 1989): 640–42.
Friis, E.M., W.G. Chaloner, and P.R. Crane, eds. *The Origins of Angiosperms and Their Biological Consequences*. Cambridge: Cambridge University Press, 1987.
Gaffney, E.S. "Triassic and Early Jurassic Turtles." In *The Beginning of the Age of Dinosaurs: Faunal Change Across the Triassic-Jurassic Boundary*, edited by K. Padian. Cambridge: Cambridge University Press, 1986.
Gee, H. "Ruffled Feathers Calmed by Fossil Bird." *Nature* 334 (14 July 1988): 104.
Gould, S.J. *The Panda's Thumb: More Reflections in Natural History*. New York: W.W. Norton, 1980.
Gurnis, M. "Large-Scale Mantle Convection and the Aggregation and Dispersal of Supercontinents." *Nature* 332 (21 April 1988): 695–99.
Halliwell, B.A. "Deep-Water Carbonate Deposits of the Southern Margin of the Jurassic Central High Atlas Trough, Morocco." Master of Science thesis. Colorado School of Mines, Golden, CO, 1985.
Heirtzler, J.R. "The Evolution of the North Atlantic Ocean." In *Implications of Continental Drift to the Earth Science*, vol. 1, edited by D.H. Tarling, and S.K. Runcorn. London: Academic Press, 1973.
Hickey, L.J. "Summary and Implications of the Fossil Plant Record of the Potomac Group." In *Cretaceous and Tertiary Stratigraphy, Paleontology, and Structure, Southwestern Maryland and Northeastern Virginia*, edited by N.O. Frederiksen and K. Krafft, American Association of Stratigraphic Palynologists Field Trip Volume and Guidebook. Calgary: American Association of Stratigraphic Palynologists, 1984.
Hughes, N.F. *Palaeobiology of Angiosperm Origins: Problems of Mesozoic Seed-Plant Evolution*. Cambridge: Cambridge University Press, 1976.
Lamont Newsletter. "A Solution to Darwin's 'Abominable Mystery': The Age and Early History of the Angiosperms." *Lamont Newsletter* 20 (Spring 1989): 4–5.
Langston, W., Jr. "Pterosaurs." *Scientific American* (February 1981): 122–36.
Lobeck, A.K. *The Palisades*. Palisades, NY: Geographical Press (Columbia University), 1952.
Milner, A. "Late Extinctions of Amphibians." *Nature* 338 (9 March 1989): 117.
Morgan, W.J. "Hotspot Tracks and the Early Rifting of the Atlantic." *Tectonophysics* 94 (1983): 123–39.
Nance, R.D., T.R. Worsley, and J.B. Moody. "The Supercontinent Cycle." *Scientific American* (July 1988): 72–79.
Neugebauer, H.J. "Mechanical Aspects of Continental Rifting." *Tectonophysics* 94 (1983): 91–108.
Olsen, P.E. "Impact Theory: Is the Past the Key to the Future?" In *Lamont Yearbook*. Palisades, NY: Lamont Geological Observatory (Columbia University), 1985–1986.
Olsen, P.E., N.H. Shubin, and M.H. Anders. "New Early Jurassic Tetrapod Assemblages Constrain Triassic-Jurassic Tetrapod Extinction Event." *Science* (28 August 1987): 1025–29.
Ostrom, J.W. "Bird Flight: How Did It Begin?" In *Paleontology and Paleoenvironments*, edited by B.J. Skinner. Readings from American Scientist. Los Altos, CA: William Kaufmann, 1981.
Padian, K., ed. *The Beginning of the Age of Dinosaurs: Faunal Change Across the Triassic-Jurassic Boundary*. Cambridge: Cambridge University Press, 1986.
Philpotts, A.R. "Rift-Associated Igneous Activity in Eastern North America." In *Petrology and Geochemistry of Continental Rifts*, edited by E.R. Neumann and I.B. Ramberg. Dordrecht: D. Reidel Publishing Company, 1978.

Qiang, J.P., Currie, P.J., Norell, M.A. and Shu-An, J., 1998 *Two Feathered Dinosaurs from northeastern China*. Natute 393 (June 25): 753.

Robbins, E.I. "A Preliminary Account of the Newark Rift System." In *Papers Presented to the Conference on the Processes of Planetary Rifting*, Contribution no. 457. Houston: Lunar and Planetary Institute, 1981.

Ross, M.I., and C.R. Scotese. "Hierarchical Tectonic Analysis of the Gulf of Mexico and Caribbean Region." *Tectonophysics* 155 (1988): 130–68.

Russell, M.J., and D.K. Smythe. "Origin of the Oslo Graben in Relation to the Hercynian-Alleghenian Orogeny and Lithospheric Rifting in the North Atlantic." *Tectonophysics* 94 (1983): 457–72.

Ryan, W.B.F. "A Close Look at Parting Plates." *Nature* 332 (28 April 1988): 779–80.

Schlager, W., and R.N. Ginsburg. "Bahama Carbonate Platforms—The Deep and the Past." *Marine Geology* 44 (1981): 1–24.

Schreiber, B.C. "Geology of the Palisades." In *Lamont Yearbook*. Palisades, NY: Lamont Geological Observatory (Columbia University), 1977.

Scientific American. "Fossil Revisionism." *Scientific American* (October 1986): 84–85.

Taylor, D.W., and L.J. Hickey. "An Aptian Plant with Attached Leaves and Flowers: Implications for Angiosperm Origin." *Science* 247 (9 February 1990): 702–04.

Turcotte, D.L., and S.H. Emerman. "Mechanisms of Active and Passive Rifting." *Tectonophysics* 94 (1983): 39–50.

von Rad, U., et al., eds. *Geology of the Northwest African Continental Margin*. Berlin: Springer-Verlag, 1982.

Warme, J.E. "Jurassic Carbonate Facies of the Central and Eastern High Atlas Rift, Morocco." In *Studies on the Geodynamic Evolution of the Atlas System*, edited by V. Jacobshagen. New York: Springer-Verlag, 1987.

———. *Evolution of the Jurassic High Atlas Rift, Morocco: Transtension, Structural and Eustatic Controls on Carbonate Deposition, Tectonic Inversion: Field Trip 9*. 1988 AAPG Mediterranean Basins Conference and Exhibit, Nice, France. Tulsa, OK: American Association of Petroleum Geologists, Geological Survey of Morocco, and ONAREP, 1988.

Wellnhofer, P., *"Archaeopteryx."* *Scientific American*. (May 1990): 70–77.

Whitmore, T.C., ed. *Wallace's Line and Plate Tectonics*. Oxford: Clarendon Press, 1981.

Wilson, J.T. "Continental Drift." In *Continents Adrift and Continents Aground*. Readings from Scientific American. San Francisco: W.H. Freeman and Company, 1976.

Ziegler, P.A. *Geological Atlas of Western and Central Europe*. Amsterdam: Elsevier Scientific Publishing Company, 1982.

VIII Atlantic Realm

Bally, A.W., et al. *Geology of Passive Continental Margins: History, Structure and Sedimentologic Record (With Special Emphasis on the Atlantic Margin)*. AAPG Eastern Section Meeting and Atlantic Margin Energy Conference. Education Course Note Series #19. Tulsa: American Association of Petroleum Geologists, 1981.

Bathurst, R.G.C. *Carbonate Sediments and Their Diagenesis*. 2d ed. Amsterdam: Elsevier Scientific Publishing Company, 1975.

Charpal, O., de, et al. "Rifting, Crustal Attenuation and Subsidence in the Bay of Biscay." *Nature* 275 (26 October 1878): 706–11.

Cossins, A.R., and K. Bowler. *Temperature Biology of Animals*. London: Chapman and Hall, 1987.

Cracraft, J. "Early Evolution of Birds." *Nature* 331 (4 February 1988): 389–90.

Darwin, C. *The Voyage of the Beagle*. New York: Doubleday & Company, 1962.

Demaison, G.J., and G.T. Moore. "Anoxic Environments and Oil Source Bed Genesis." In *Hydrocarbon Generation and Source Rock Evaluation: (Origin of Petroleum III)*, compiled by R.M. Cluff and M.H. Barrows. Tulsa, OK: American Association of Petroleum Geologists, 1982.

Emery, K.O., and E. Uchupi. *The Geology of the Atlantic Ocean*. New York: Springer-Verlag, 1984.

Epp, D., and N.C. Smoot. "Distribution of Seamounts in the North Atlantic." *Nature* 337 (19 January 1989): 254–57.

Goreau, T.F., N.I. Goreau, and T.J. Goreau. "Corals and Coral Reefs." *Scientific American* (August 1979): 124–36.

Hedgpeth, J.W., ed. *Treatise on Marine Ecology and Paleoecology*, Memoir 67, Ecology, vol. 1. Boulder, CO: Geological Society of America, 1957.

Holland, C.H. "The Nautiloid Cephalopods: A Strange Success." *Journal of the Geological Society of London* 144 (1987): 1–15.

Holland, H.E. *The Chemical Evolution of the Atmosphere and Oceans*. Princeton, NJ: Princeton University Press, 1984.

Isaacs, J.D. "The Nature of Oceanic Life." In *The Ocean*. Readings from Scientific American. San Francisco: W.H. Freeman and Company, 1969.

Kennett, J.P. *Marine Geology*. Englewood Cliffs, NJ: Prentice-Hall, 1982.

Lefort, J.P. "Iberian-Armorican Arc and Hercynian Orogeny in Western Europe." *Geology* 7 (August 1979): 384–88.

Machado, F. "Acid Volcanoes of San Miguel, Azores." *Bulletin Volcanologique* 36 (1972): 319–27.

Margulis, L., and K.V. Schwartz. *Five Kingdoms: An Illustrated Guide to the Phyla of Life on Earth*. San Francisco: W.H. Freeman and Company, 1982.

Marshall, N.B. *Developments in Deep-Sea Biology*. Poole, UK: Blandford Press, 1979.

———. *The Life of Fishes*. London: Weidenfeld and Nicolson, 1965.

Morgan, W.J. "Hotspot Tracks and the Early Rifting of the Atlantic." *Tectonophysics* 94 (1983): 123–39.

Robertson, A.H.F., and D. Bernoulli. "Stratigraphy, Facies, and Significance of Late Mesozoic and Early Tertiary Sedimentary Rocks of Fuerteventura (Canary Islands) and Maio (Cape Verde Islands)." In *Geology of the Northwest African Continental Margin*, edited by U. von Rad, et al. Berlin: Springer-Verlag, 1982.

Ross, M.I., and C.R. Scotese. "Hierarchical Tectonic Analysis of the Gulf of Mexico and Caribbean Region." *Tectonophysics* 155 (1988): 130–68.

Rowe, G.T., ed. *Deep-Sea Biology*. The Sea, vol. 8. New York: John Wiley & Sons, 1983.

Saunders, A. "Putting Continents Asunder." *Nature* 332 (21 April 1988): 679–80.

Schlager, W., and R.N. Ginsburg. "Bahama Carbonate Platforms." *Marine Geology* 44 (1981): 1–24.

Scholle, P.A., D.G. Bebout, and C.H. Moore, eds. *Carbonate Depositional Environments*. Tulsa, OK: American Association of Petroleum Geologists, 1983.

Seibold, E., and W.H. Berger. *The Sea Floor: An Introduction to Marine Geology*. Berlin: Springer-Verlag, 1982.

Srivastava, S.P., R.K.H. Falconer, and B. MacLean. "Labrador Sea, Davis Strait, Baffin Bay: Geology and Geophysics—A Review." In *Geology of the North Atlantic Borderlands*, Memoir 7, edited by J.W. Kerr and A.J. Ferguson. Calgary: Canadian Society of Petroleum Geologists, 1981.

Stowe, K.S. *Ocean Science*. New York: John Wiley & Sons, 1979.

Vink, G.E., W.J. Morgan, and P.R. Vogt. "The Earth's Hot Spots." *Scientific American* (April 1985): 50–57.

von Rad, U., et al., eds. *Geology of the Northwest African Continental Margin*. Berlin: Springer-Verlag, 1982.

Ward, P.D. *The Natural History of Nautilus*. Boston: Allen & Unwin, 1987.

Ward, P., L. Greenwald, and O.E. Greenwald. "The Buoyancy of the Chambered Nautilus." *Scientific American* (October 1980): 190–203.

Whitmore, T.C., ed. *Wallace's Line and Plate Tectonics*. Oxford: Clarendon Press, 1981.

Ziegler, P.A. *Geological Atlas of Western and Central Europe*. Amsterdam: Elsevier Scientific Publishing Company, 1982.

IX Maritime West

Alvarez, L.W., et al. "Extraterrestrial Cause for the Cretaceous-Tertiary Extinction." *Science* 208 (6 June 1980): 1095–1108.

American Geological Institute. "Did Dinosaurs Disappear Suddenly or Gradually?" *Earth Science* (Spring 1987): 11.

Bakker, R.T. *The Dinosaur Heresies: New Theories Unlocking the Mystery of the Dinosaurs and Their Extinction*. New York: William Morrow and Company, 1986.

———. "Dinosaur Renaissance." In *Evolution and the Fossil Record*. Readings from Scientific American. San Francisco: W.H. Freeman and Company, 1978.

Berggren, W.A., and J.A. Van Couvering, eds. *Catastrophes and Earth History: The New Uniformitarianism*. Princeton, NJ: Princeton University Press, 1984.

Bohor, B.F., et al. "Dinosaurs, Spherules, and the 'Magic' Layer: A New K-T Boundary Clay Site in Wyoming." *Geology* 15: (October 1987): 896–99.

Brasier, M.D. *Microfossils*. London: George Allen and Unwin, 1980.

Briggs, J.C. *Biogeography and Plate Tectonics*. New York: Elsevier Science Publishing Company, 1987.

Brouwers, E.M., et al. "Dinosaurs on the North Slope, Alaska: High Latitude, Latest Cretaceous Environments." *Science* 237 (25 September 1987): 1608–10.

Carroll, R.L. *Vertebrate Paleontology and Evolution*. New York: W.H. Freeman and Company, 1988.

Cloud, P. *Oasis in Space: Earth History from the Beginning*. New York: W.W. Norton and Company, 1988.

Colbert, E.H. "Therapsids in Pangaea and Their Contemporaries and Competitors." In *The Ecology and Biology of Mammal-like Reptiles*, edited by N. Hotton III, et al. Washington, DC: Smithsonian Institution Press, 1986.

———. *Wandering Lands and Animals: The Story of Continental Drift and Animal Populations*. New York: Dover Publications, 1985.

———. *Dinosaurs: An Illustrated History*. Maplewood, NJ: Hammond Incorporated, 1983.

———. *Dinosaurs: Their Discovery and Their World*. New York: E.P. Dutton and Company, 1961.

Cossins, A.R., and K. Bowler. *Temperature Biology of Animals*. London: Chapman and Hall, 1987.

Crutzin, P.J. "Acid Rain at the K/T Boundary." *Nature* 330 (12 November 1987): 108–09.

Darwin, C. *The Origin of Species: By Means of Natural Selection or the Preservation of Favoured Races in the Struggle for Life*. London: Collier-Macmillan, 1962.

De Beer, G. "Darwin." In *The New Encyclopaedia Britannica*. 15th ed. Chicago: Encyclopaedia Britannica, 1987.

Dixon, D., et al. *The Macmillan Illustrated Encyclopedia of Dinosaurs and Prehistoric Animals*. New York: Macmillan Publishing Company, 1988.

Ebren, H.K., J. Hoefs, and K.H. Wedepohl. "Paleobiological and Isotopic Studies of Eggshells from a Declining Dinosaur Species." *Paleobiology* 5 (1979): 380–414.

Ebren, H.K., et. al. "Some Dinosaurs Survived the Cretaceous 'Final Event'." *Terra Cognita* 3 (Spring-Summer, 1983): 211–12, abstract.

Farlow, J.O., ed. *Paleobiology of the Dinosaurs*. Special Paper 238. Boulder, CO: Geological Society of America, 1989.

Frakes, L.A., and J.E. Francis. "A Guide to Phanerozoic Cold Polar Climates from High-Latitude Ice-Rafting in the Cretaceous." *Nature* 333 (9 June 1988): 547–49.

Friis, E.M., W.G. Chaloner, and P.R. Crane, eds. *The Origins of Angiosperms and Their Biological Consequences*." Cambridge: Cambridge University Press, 1987.

Garwin, L. "Of Impacts and Volcanoes." *Nature* 336 (22/29 December 1988): 714–15.

Gee, H. "Dinosaur Finds From China." *Nature* 340 (6 July 1989): 22.

Geological Society and Geologists' Association. "Could the Deccan Traps Have Been Responsible for the Cretaceous-Tertiary Extinctions?" *Geology Today* 4 (March/April 1988): 51.

Ginsburg, R., K. Burke, and V. Sharpton. "Scientists Debate Global Catastrophes." *Geotimes* (March 1989): 14–16.

Grieve, R.A.F. "Manson Structure Implicated." *Nature* 340 (10 August 1989): 428–29.

Hancock, J.M. "Sea-level Changes in the British Region During the Late Cretaceous." *Proceedings of the Geologists' Association of London* 100 (1989): 565–94.

Hancock, J.M., and E.G. Kauffman. "The Great Transgressions of the Late Cretaceous." *Journal of the Geological Society* 136 (March 1979): 175–86.

Hickey, L.J., K.R. Johnson., and M.R. Dawson. "The Stratigraphy, Sedimentology, and Fossils of the Haughton Formation: A Post-Impact Crater-Fill, Devon Island, N.W.T., Canada." *Meteoritics* 23 (1988): 221–31.

Hildebrand, A.R., et. al., "Chicxulub Crater: A Possible Cretaceous/Tertiary Boundary Impact Crater on the Yucatán Peninsula, Mexico." *Geology* 19 (September 1991): 867–71.

Horner, J.R. "The Nesting Behavior of Dinosaurs." *Scientific American* (April 1984): 130–37.

Hotton, N., III, et al., eds. *The Ecology and Biology of Mammal-like Reptiles*. Washington, DC: Smithsonian Institution Press, 1986.

Hughes, N.F. *Palaeobiology of Angiosperm Origins: Problems of Mesozoic Seed-Plant Evolution*." Cambridge: Cambridge University Press, 1976.

Johnson, K.R., and L.J. Hickey. "Megafloral Change Across the Cretaceous-Tertiary Boundary in the Northern Great Plains and Rocky Mountains, USA." In *Global Catastrophes in Earth History: An Interdisciplinary Conference on Impact, Vulcanism, and Mass Mortality*, Special Paper 247, edited by V.L. Sharpton and P.D. Ward. Boulder, CO: Geological Society of America, 1991.

Johnson, K.R., C.J. Orth, and D.J. Nichols. "Fossil Leaf and Palynomorph Changes Associated with an Iridium Anomaly at the Cretaceous-Tertiary Boundary in North Dakota." In *Geological Society of American Abstracts with Program*, vol. 19. Boulder, CO: Geological Society of America, 1987.

Jones, E.M., and J.W. Kodis. "Atmospheric Effects of Large Body Impacts: The First Few Minutes." In *Geological Implications of Impacts of Large Asteroids and Comets on the Earth*, Special Paper 190, edited by L.T. Silver and P.H. Schultz. Boulder, CO: Geological Society of America, 1982.

Kauffman, E.G., and W.G.E. Caldwell. "Cretaceous Transgressive and Regressive History, Sea Level Changes and Sequence Stratigraphy, Western Interior Basin, North America." In *Cretaceous Evolution of the*

Western Interior Basin of North America, Geological Association of Canada, Memoir, in press.
Kerr, R.A. "Was There a Prelude to the Dinosaurs' Demise?" *Science* 239 (12 February 1988): 729–30.
———. "Searching Land and Sea for the Dinosaur Killer." *Science* 237 (21 August 1987): 856–57.
Kunk, M.J., et al. " ^{40}Ar-^{39}Ar Dating of the Manson Impact Structure: A Cretaceous-Tertiary Boundary Crater Candidate." *Science* 244 (30 June 1989): 1565–67.
Lillegraven, J.A., Z. Kielan-Jaworowska, and W.A. Clemens, eds. *Mesozoic Mammals: The First Two-Thirds of Mammalian History*. Berkeley, CA: University of California Press, 1979.
Michener, C.D. *The Social Behavior of the Bees: A Comparative Study*. Cambridge, MA: Harvard University Press, 1984.
Monastersky, R. "Could a Cold Heart Stand a Cold Winter?" *Science News* 136 (25 November 1989): 38.
———. "New Signs of World Upheaval at K-T." *Science News* 134 (12 November 1988): 309.
———. "C-T Extinctions Without the Impact." *Science News* 133 (30 April 1988): 278.
———. "Dinosaurs in the Dark." *Science News* 133 (19 March 1988): 184–86.
———. "K-T Mass Extinctions: Abrupt or What?" *Science News* 133 (31 October 1987): 277.
Morgan, J. Warner, M. Maguire, P. et al *Chicxulub. Impacts and Extinctions: Profiling the Smoking Gun*. The British Institutions Reflection Profiling Syndicat. Website. 1996
Nitecki, M.H., ed. *Coevolution*. Chicago: University of Chicago Press, 1983.
O'Keefe, J.D., and T.J. Ahrens. "Impact Production of CO_2 by the Cretaceous/Tertiary Extinction Bolide and the Resultant Heating of the Earth." *Nature* 338 (16 March 1989): 247–48.
Parrish, J.T., and R.A. Spicer. "Late Cretaceous Terrestrial Vegetation: A Near-Polar Temperature Curve." *Geology* 16 (January 1988): 22–25.
Pessagno, E.A., Jr. "Upper Cretaceous Planktonic Foraminifera from the Western Gulf Coastal Plain." In *Palaeontographica Americana*, vol. 5. Ithaca, NY: Paleontological Research Institute, 1967.
Robertson, P.B., and J.F. Sweeney. "Haughton Impact Structure: Structural and Morphological Aspects." *Canadian Journal of Earth Sciences* 20 (1983): 1134–51.
Rocky Mountain Association of Geologists. *Atlas of the Rocky Mountain Region United States of America*. Denver, CO: Rocky Mountain Association of Geologists, 1982.
Sattler, H.R. *Dinosaurs of North America*. New York: Lothrop, Lee and Shepard Books, 1981.
Scotese, C.R., L.M. Gahagan, R.L. Larson. "Plate Tectonic Reconstructions of the Cretaceous and Cenozoic Ocean Basins." *Tectonophysics* 155 (1988) 27–48.
Scully, V., et al. *The Age of Reptiles*. New York: Harry N. Abrams, 1990.
Secord, J.A. *Controversy in Victorian Geology: The Cambrian-Silurian Dispute*. Princeton, NJ: Princeton University Press, 1986.
Silver, L.T., and P.H. Schultz, eds. *Geological Implications of Impacts of Large Asteroids and Comets on the Earth*. Special Paper 190. Boulder, CO: Geological Society of America, 1982.
Sloan, R.E. "Paleocene Dinosaur Extinction in South China." *Geological Society of America—1987 Annual Meeting—Abstracts with Programs*. Boulder, CO: Geological Society of America, 1987.
Sloan, R.E., and J.K. Rigby, Jr. "Cretaceous-Tertiary Dinosaur Extinction." *Science* 234 (5 December 1986): 1170–75.
Sloan, R.E., and L. Van Valen. "Cretaceous Mammals from Montana." *Science* 148 (9 April 1965): 220–27.
Sloan, R.E., et al. "Gradual Dinosaur Extinction and Simultaneous Ungulate Radiation in the Hell Creek Formation." *Science* 232 (2 May 1986): 629–33.
Stanley, S.M. *Extinction*. New York: Scientific American Books, 1987.
Steckler, M. "Changes in Sea Level." In *Patterns of Change in Earth Evolution*, edited by H.D. Holland and A.F. Trendall, Report of the Dahlem Workshop on Patterns of Change in Earth Evolution, Berlin, 1983. Berlin: Springer-Verlag, 1984.
Stewart, W.N. *Paleobotany and the Evolution of Plants*. Cambridge: Cambridge University Press, 1985.
Tweto, O. "Summary of Laramide Orogeny in Colorado." In *Colorado Geology*, edited by H.C. Kent and K.W. Porter. Denver, CO: Rocky Mountain Association of Geologists, 1980.
Uhl, N.W., and J. Dransfield. *Genera Palmarum*. Lawrence, KS: Allen Press, 1987.
Van Valen, L., and R.E. Sloan. "The Earliest Primates." *Science* 150 (5 November 1965): 743–45.
Weisburd, S. "Volcanoes and Extinctions: Round Two." *Science News* 131 (18 April 1987): 248–50.
Wing, S.L., and B.H. Tiffney. "Interactions of Angiosperms and Herbivorous Tetrapods Through Time." In *The Origins of Angiosperms and Their Biological Consequences*, edited by E.M. Friis, W.G. Chaloner, and P.R. Crane. Cambridge: Cambridge University Press, 1987.
Winston, M.L. *The Biology of the Honey Bee*. Cambridge, MA: Harvard University Press, 1987.
Wolbach, W.S., et al. "Global Fire at the Cretaceous-Tertiary Boundary." *Nature* 334 (25 August 1988): 665–69.
Zachos, J.C., M.A. Arthur, and W.E. Dean. "Geochemical Evidence for Suppression of Pelagic Marine Productivity at the Cretaceous/Tertiary Boundary." *Nature* 337 (5 January 1989): 61–64.
Zeuner, F.E., and F.J. Manning. "A Monograph of Fossil Bees (Hymenoptera: Apoidea)." *Bulletin British Museum of Natural History* 27: (1976) 1–268.

X Midland and Nordic Seas

Ager, D.V. *The Geology of Europe*. London: McGraw-Hill Book Company Limited, 1980.
Armi, L., et al. "The History and Decay of a Mediterranean Salt Lens." *Nature* 333 (16 June 1988): 649–51.
Asimov, I. *Asimov's Biographical Encyclopedia of Science and Technology*. Garden City, NY: Doubleday & Company, 1982.
Badgley, C.E. "Human Evolution." In *Mammals: Notes for a Short Course*. Studies in Geology 8, edited by P.D. Gingerich and C.E. Badgley. Knoxville, TN: University of Tennessee Department of Geological Sciences, 1984.
Bailey, E. "Notes on Gibraltar and the Northern Rif." In *Geological Society of London Abstracts*, paper no. 1479. London: Geological Society of London, 1951.
Biju-Duval, B., J. Dercourt, and X. Le Pichon. "From the Tethys Ocean to the Mediterranean Seas: A Plate Tectonic Model of the Evolution of the Western Alpine System." In *Structural History of the Mediterranean Basins*, edited by B. Biju-Duval and L. Montadert. Paris: Technip, 1987.
Biju-Duval, B., J. Letouzey, and L. Montadert. "Variety of Margins and Deep Basins in the Mediterranean." In *Geological and Geophysical Investigations of Continental Margins*, Memoir 19, edited by J.S. Watkins, L. Montadert, and P.W. Dickerson. Tulsa: American Association of Petroleum Geologists, 1988.
Björnsson, A., et al. "Rifting of the Plate Boundary in North Iceland 1975–1978." *Journal of Geophysical Research* 84 (10 June 1979): 3029–38.
Bonatti, E. "Rifting or Drifting in the Red Sea?" *Nature* 330 (24/31 December 1987): 692–93.
Bott, M.H.P. "The Evolution of the Atlantic North of the Faeroe Islands." In *Implications of Continental Drift to the Earth Science*, vol. 1, edited by D.H. Tarling and S.K. Runcorn. London: Academic Press, 1973.
Bourgeois, J., et al. "A Tsunami Deposit at the Cretaceous-Tertiary Boundary in Texas." *Science* 241 (29 July 1988): 567–70.
Bower, B. "Rivers in the Sand." *Science News* 136 (26 August 1989): 136–39.
Burke, K. and G.L. Wells. "Trans-African Drainage System of the Sahara: Was It the Nile?" *Geology* 17 (August 1989) 743–47.
Burke, K.C., and J.T. Wilson. "Hot Spots on the Earth's Surface." In *Continents Adrift and Continents Aground*. Readings from Scientific American. San Francisco: W.H. Freeman and Company, 1976.
Carroll, R.L. *Vertebrate Paleontology and Evolution*. New York: W.H. Freeman and Company, 1988.
Clark, D.L., "Origin, Nature, and World Climate Effect of Arctic Ocean Ice-Cover." *Nature* 300 (1982), 321–25.
Colbert, E.H. "Life on Wandering Continents." In *Rediscovery of the Earth*, edited by L. Motz. New York: Van Nostrand Reinhold Company, 1975.
Corliss, B.H., et al. "The Eocene/Oligocene Boundary Event in the Deep Sea." *Science* 226 (16 November 1984): 806–10.
Cox, K.G. "The Role of Mantle Plumes in the Development of Continental Drainage Patterns." *Nature* 342 (21/28 December 1989): 873–77.
Crough, S.T. "Hotspot Swells." *Annual Review of Earth Planetary Sciences* 11 (1983): 165–93.
Dawes, P.R., and J.W. Kerr, eds. *Nares Strait and the Drift of Greenland: A Conflict in Plate Tectonics*. Meddelelser om Grønland, Geoscience 8. Copenhagen: Geological Survey of Greenland, 1982.
Dawson, M.R., and R.M. West. "Cenozoic Vertebrates and Floras in the Arctic." *National Geographic Society Research Reports* 14 (1982): 143–48.
Day, M.H. *Guide to Fossil Man*. 4th ed. Chicago: University of Chicago Press, 1986.
Diamond, J.M. "Relationships of Humans to Chimps and Gorillas." *Nature* 334 (25 August 1988): 656.
Dietz, R.S., and M. Woodhouse. "Mediterranean Theory May Be All Wet." *Geotimes* (May 1988): 4.
Dixon, D., et al. *The Macmillan Illustrated Encyclopedia of Dinosaurs and Prehistoric Animals*. New York: Macmillan Publishing Company, 1988.
Edlund, S.A. "High Arctic Plants: New Limits Emerge." *GEOS* (Winter 1984): 10–13.
Eicher, D.L., and A.L. McAlester. *History of the Earth*. Englewood Cliffs, NJ: Prentice-Hall, 1980.
Eisenberg, J.F. *The Mammalian Radiations: An Analysis of Trends in Evolution, Adaptation, and Behavior*. Chicago: University of Chicago Press, 1981.
Epp, D., and N.C. Smoot. "Distribution of Seamounts in the North Atlantic." *Nature* 337 (19 January 1989): 254–57.
Farrelly, D. *The Book of Bamboo*. San Francisco: Sierra Club Books, 1984.
Fleagle, J.G., et al. "Age of the Earliest African Anthropoids." *Science* 234 (5 December 1986): 1247–48.
Flynn, J.J. "Ancestry of Sea Mammals." *Nature* 334 (4 August 1988): 383–84.
Fordyce, R.E. "Whales, Dolphins, Porpoises." *Earth Science* (Summer 1989): 20–23.
Foulger, G.R., et al. "Implosive Earthquakes at the Active Accretionary Plate Boundary in Northern Iceland." *Nature* 337 (16 February 1989): 640–42.
Francis, J.E., and N.J. McMillan. "Fossil Forests in the Far North." *GEOS* (Winter 1987): 6–9.
Gillispie, C.C., ed. *Dictionary of Scientific Biography*. New York: Scribner, 1970–1980.
Gingerich, P.D. "Primate Evolution." In *Mammals: Notes for a Short Course*. Studies in Geology 8, edited by P.D. Gingerich and C.E. Badgley. Knoxville, TN: University of Tennessee Department of Geological Sciences, 1984.
Gingerich, P.D., et al. "Origin of Whales in Epicontinental Remnant Seas: New Evidence from the Early Eocene of Pakistan." *Science* 220 (22 April 1983): 403–05.
Greenberg, J. "Fossils Trigger Question of Human Origins." *Science News* 121 (30 January 1982): 84.
Griffiths, M. "The Platypus." *Scientific American* (May 1988): 84–90.
Groves, C.P. *A Theory of Human and Primate Evolution*. Oxford: Clarendon Press, 1989.
Hay, R.L., and M.D. Leakey. "The Fossil Footprints of Laetoli." *Scientific American* (February 1982): 49–57.
Hoffecker,J.F., Powers,W.R., and Goebel, T. : The Colonization of Beringia and the Peopling of the New World. *Science* 259 Jan 1993
Hollister, C.D., A.R.M. Nowell, and P.A. Jumars. "The Dynamic Abyss." *Scientific American* (March 1984): 42–53.
Hopkins, D.M., ed. *The Bering Land Bridge*. Stanford, CA: Stanford University Press, 1967.
Hopkins, D.M., and L. Marincovich, Jr. "Whale Biogeography and the History of the Arctic Basin." In *Arctic Whaling: Proceedings of the International Symposium on Arctic Whaling*. The Netherlands: University of Groningen, 1984.
Hopkins, D.M., et al., eds. *Paleoecology of Beringia*. New York: Academic Press, 1982.
Houghton, J.T., ed. *The Global Climate*. Cambridge: Cambridge University Press, 1984.
Hsü, K.J. *The Mediterranean Was a Desert: The Voyage of the Glomar Challenger*. Princeton, NJ: Princeton University Press, 1983.
———. *Mountain Building Processes*. London: Academic Press, 1982.
———. "When the Mediterranean Dried Up." In *Ocean Science*. Readings from Scientific American. San Francisco: W.H. Freeman and Company, 1977.
Hsü, K.J., W.B.F. Ryan, and M.B. Cita. "Late Miocene Desiccation of the Mediterranean." *Nature* 242 (1983): 240–44.
Hurdle, B.G., ed. *The Nordic Seas*. New York: Springer-Verlag, 1986.
Illies, J.H. "An Intercontinental Belt of the World Rift System." *Tectonophysics* 8 (1969): 5–29.
Johanson, D.C., and M.A. Edey. *Lucy: The Beginnings of Humankind*. New York: Simon and Schuster, 1981.
Kennett, J.P. *Marine Geology*. Englewood Cliffs, NJ: Prentice-Hall, 1982.
Kerr, R.A. "Ancient River System Across Africa Proposed." *Science* 233 (29 August 1986): 940.
Lamont Newsletter. "Joint Lamont/Oxford/Istanbul Project for New Synthesis of the Tectonic History of the Mesozoic and Cenozoic Tethyan Region." *Lamont Newsletter* 13 (Winter 1987): 11–12.
Lamont Newsletter. "Multichannel Seismic Image Reveals the Trail of a Magma Chamber Once Beneath the Ancient Mid-Atlantic Ridge. *Lamont Newsletter* 10 (Spring 1985): 1, 6.

Laubscher, H., and D. Bernoulli. "History and Deformation of the Alps." In *Mountain Building Processes*, edited by K.J. Hsü. London: Academic Press, 1982.

Leblanc, D., and Ph. Olivier. "Role of Strike-Slip Faults in the Betic-Rifian Orogeny." *Tectonophysics* 101 (1984): 345–55.

Lewin, R. *Bones of Contention: Controversies in the Search for Human Origins.* New York: Simon and Schuster, 1987.

———. *Human Evolution: An Illustrated Introduction*. New York: W.H. Freeman and Company, 1984.

———. "Is the Orangutan a Living Fossil?" *Science* 222 (16 December 1983): 1222–23.

Lowell, J.D., and G.J. Genik. "Sea-Floor Spreading and Structural Evolution of Southern Red Sea." *American Association of Petroleum Geologists Bulletin* 56 (February 1972): 247–59.

Lowell, J.D., et al. "Petroleum and Plate Tectonics of the Southern Red Sea." In *Petroleum and Global Tectonics*, edited by A.G. Fischer and S. Judson. Princeton, NJ: Princeton University Press, 1975.

MacFadden, B.J., and R.C. Hulbert, Jr. "Explosive Speciation at the Base of the Adaptive Radiation of Miocene Grazing Horses." *Nature* 336 (1 December 1988): 466–67.

Marincovich, L., Jr., E.M. Brouwers, and D.M. Hopkins. "Paleogeographic Affinities and Endemism of Cretaceous and Paleocene Marine Faunas in the Arctic." In *U.S. Geological Survey Polar Research Symposium—Abstracts with Program*. Circular 911. Denver: U.S. Geological Survey, 1983.

McCaig, A.M. "Deep Geology of the Pyrenees." *Nature* 331 (11 February 1988): 480–81.

McCauley, J.F., et al. "Paleodrainages of the Eastern Sahara—The Radar Rivers Revisited (SIR-A/B Implications for a Mid-Tertiary Trans-African Drainage System)." *IEEE Transactions on Geoscience and Remote Sensing* GE-24 (July 1986): 624–47.

McCulloch, M.T., and P. De Deckker. "Sr Isotope Constraints on the Mediterranean Environment at the End of the Messinian Salinity Crisis." *Nature* 342 (2 November 1989): 63–65.

McKenna, M.C. "Holarctic Landmass Rearrangement, Cosmic Events, and Cenozoic Terrestrial Organisms." *Annals of the Missouri Botanical Garden* 70 (1983): 459–89.

———. "Eocene Paleolatitude, Climate, and Mammals of Ellesmere Island." *Palaeogeography, Palaeoclimatology, Palaeoecology* 30 (1980): 349–62.

Menzies, A.W. "Crustal History and Basin Development of Baffin Bay." In *Nares Strait and the Drift of Greenland: A Conflict in Plate Tectonics*, Meddelelser om Grønland, Geoscience 8, edited by P.R. Dawes and J.W. Kerr. Copenhagen: Geological Survey of Greenland, 1982.

Monahan, D., and G.L. Johnson. "Physiography of Nares Strait: Importance to the Origin of the Wegener Fault." In *Nares Strait and the Drift of Greenland: A Conflict in Plate Tectonics*, Meddelelser om Grønland, Geoscience 8, edited by P.R. Dawes and J.W. Kerr. Copenhagen: Geological Survey of Greenland, 1982.

Monastersky, R. "Clues to an Ancient Upside-Down Ocean." *Science News* 136 (29 July 1989): 71.

———. "Set Adrift by Wandering Hotspots." *Science News* 132 (17 October 1987): 250–52.

Morgan, W.J. "Hotspot Tracks and the Opening of the Atlantic and Indian Oceans." In *The Sea: The Oceanic Lithosphere*, vol. 10, edited by C. Emiliani. New York: John Wiley & Sons, 1982.

Nilsen, T.H., and S. Saxov. "The Structure and Development of the Greenland-Scotland Ridge." *Episodes* (1982): 3–8.

Nitecki, M.H., ed. *Coevolution*. Chicago: University of Chicago Press, 1983.

Olson, S.L., and D.T. Rasmussen. "Paleoenvironment of the Earliest Hominoids: New Evidence from the Oligocene Avifauna of Egypt." *Science* 233 (12 September 1986): 1202–04.

Osborn, J.W. "The Evolution of Dentitions." In *Paleontology and Paleoenvironments*, edited by B.J. Skinner. Readings from American Scientist. Los Altos, CA: William Kaufmann, 1981.

Owen, T.R., ed. *United Kingdom*. Paris: 26th International Geological Congress, 1980.

Pilbeam, D. "The Descent of Hominoids and Hominids." *Scientific American* (March 1984): 84–96.

Platt, J.P., et al. "Kinematics of the Alpine Arc and the Motion History of Adria." *Nature* 337 (12 January 1989): 158–61.

Prasad, G.V.R., and A. Sahni. "First Cretaceous Mammal from India." *Nature* 332 (14 April 1988): 638–39.

Rögl, V.F., and F.F. Steininger. "Vom Zerfal der Tethys zu Mediterran und Paratethys." *Annalen des Naturhistorischen Museums in Wien* 85 (April 1983): 135–63.

Rose, E.P.F. and Stringer, C.B.: Gibraltar Women and Neanderthal Man: *Geology Today:* 179 Sept/Oct 1997.

Ruddiman, W.F., and J.E. Kutzbach. *Forcing of Late Cenozoic Northern Hemisphere Climate by Plateau Uplift in Southeast Asia and the American Southwest*, in press.

Ruddiman, W.F., W.L. Prell, and M.E. Raymo. *History of Late Cenozoic Uplift in Southeast Asia and the American Southwest: Rationale for General Circulation Modeling Experiments*, in press.

Ryan, W.B.F. "A Close Look at Parting Plates." *Nature* 332 (28 April 1988): 779–80.

Savage, R.J.G. *Mammal Evolution: An Illustrated Guide*. New York: Facts on File Publications, 1986.

Savile, D.B.O. *Arctic Adaptations in Plants*. Monograph No. 6. Ottawa: Canada Department of Agriculture, 1972.

Schilling, J-G., P.S. Meyer, and R.H. Kingsley. "Evolution of the Iceland Hotspot." *Nature* 296 (25 March 1982): 313–20.

Schneider, S.H., and R. Londer. *The Coevolution of Climate and Life.* San Francisco: Sierra Club Books, 1984.

Schreiber, B.C., and S. Marshak. "Evaporites, Carbonates, and Oil." In *Geology of Passive Continental Margins: History, Structure and Sedimentologic Record (With Special Emphasis on the Atlantic Margin)*, edited by A.W. Bally, et al, AAPG Eastern Section Meeting and Atlantic Margin Energy Conference. Education Course Note Series #19. Tulsa: American Association of Petroleum Geologists, 1981.

Science News. "Mediterranean Salt: Isotopes and Ice." *Science News* 128 (23 November 1985): 329.

———. "Tectonics-torn Italy." *Science News* 118 (13 December 1980): 376.

Scotese, C.R., L.M. Gahagan, R.L. Larson. "Plate Tectonic Reconstructions of the Cretaceous and Cenozoic Ocean Basins." *Tectonophysics* 155 (1988) 27–48.

Scrutton, R.A. "Progress on the Margin." Review of *Early Tertiary Volcanism and the Opening of the NE Atlantic*, edited by A.C. Morton and L.M. Parson. *Nature* 340 (13 July 1989): 109.

Skinner, B.J., ed. *Climates Past and Present.* Readings from American Scientist. Los Altos, CA: William Kaufmann, 1981.

Smith, A.G., and N.H. Woodcock. "Tectonic Synthesis of the Alpine-Mediterranean Region: A Review." In *Alpine-Mediterranean Geodynamics*, edited by H. Berckhemer and K.J. Hsü. Boulder, CO: Geological Society of America, 1982.

Snow, D.W. "Coevolution of Birds and Plants." In *The Evolving Biosphere*, edited by P.L. Forey. Cambridge: Cambridge University Press, 1981.

Soderstrom, T.R., et al, eds. *Grass: Systematics and Evolution*. International Symposium held in Washington, DC, July 1986. Washington DC: Smithsonian Institution Press, 1986.

Stanley, S.M. *Earth and Life Through Time*. 2d ed. New York: W.H. Freeman and Company, 1989.

———. *Extinction*. New York: Scientific American Books, 1987.

Stebbins, G.L. *Darwin to DNA, Molecules to Humanity*. San Francisco: W.H. Freeman and Company, 1982.

Stewart, R.W. "The Atmosphere and the Ocean." In *The Ocean*. Readings from Scientific American. San Francisco: W.H. Freeman and Company, 1969.

Sweeney, J.F. "Arctic Tectonics—What We Know Today." *GEOS* (Winter 1984): 8–10.

Tankard, A.J., and H.R. Balkwill, eds. *Extensional Tectonics and Stratigraphy of the North Atlantic Margins*, Memoir 46. Tulsa, OK: American Association of Petroleum Geologists and Canadian Geological Foundation, 1989.

Tazieff, H. "The Afar Triangle." *Scientific American*. (February 1970): 32–40.

Tucholke, B.E., and V.A. Fry. "Basement Structure and Sediment Distribution in Northwest Atlantic Ocean." *The American Association of Petroleum Geologists Bulletin* 69 (December 1985): 2077–97.

Vink, G.E., W.J. Morgan, and P.R. Vogt. "The Earth's Hot Spots." *Scientific American* (April 1985): 50–57.

Vogt, P.R., and R.K. Perry. *North Atlantic Ocean: Bathymetry and Plate Tectonic Evolution*. Map and Chart Series, MC-35. Boulder, CO: Geological Society of America, 1982.

Walker, A., and M. Teaford. "The Hunt for Proconsul." *Scientific American* (January 1989): 76–82.

Weaver, K.F. "The Search for Our Ancestors." *National Geographic* (November 1985): 560–623.

Weijermars, R. "Where Does Africa Meet Europe?" *Nature* 332 (10 March 1988): 118.

Weisburd, S. "Creatures of the Dreamtime." *Science News* 133 (16 April 1988): 248–50.

West, R.M., and M.R. Dawson. "Vertebrate Paleontology and the Cenozoic History of the North Atlantic Region." *Polarforschung* 48 (1978): 103–19.

West, R.M., et al. "Upper Cretaceous and Paleogene Sedimentary Rocks, Eastern Canadian Arctic and Related North Atlantic Areas." In *Geology of the North Atlantic Borderlands*, Memoir 7, edited by J.W. Kerr and A.J. Ferguson. Calgary: Canadian Society of Petroleum Geologists, 1981.

White, R.S., and D.P. McKenzie. "Volcanism at Rifts." *Scientific American* (July 1989): 62–71.

White, R.S., et al. "Magnetism at Rifted Continental Margins." *Nature* 330 (3 December 1987): 439–44.

Wilford, J.N. Could Neanderthal Talk? Fossil Record Suggests a Resounding 'Yes": *New York Times* Science: 1998

Wilford, J.N. The Search for Early Man: *New York Times* Science: 1998

Winston, M.L. *The Biology of the Honey Bee.* Cambridge, MA: Harvard University Press, 1987.

Wolfe, J.A. "Palaeobotanical Evidence for a Marked Temperature Increase Following the Cretaceous/Tertiary Boundary." *Nature* 343 (11 January 1990): 153–56.

Wyss, A.R. "Evidence from Flipper Structure for a Single Origin of Pinnipeds." *Nature* 334 (4 August 1988): 427–28.

Ziegler, P.A. *Evolution of the Arctic-North Atlantic and the Western Tethys*, Memoir 43. Tulsa, OK: American Association of Petroleum Geologists, 1988.

———. *Geological Atlas of Western and Central Europe*. Amsterdam: Elsevier Scientific Publishing Company, 1982.

XI Fountains of Youth

Adams, J.M., and F.I. Woodward. "Patterns in Tree Species Richness as a Test of the Glacial Extinction Hypothesis." *Nature* 339 (29 June 1989): 699–701.

Appenzeller, T. "After the Deluge: Drumlins May Tell the Tale of an Ice-Age Flood." *Scientific American* (December 1989): 22–23.

Bard, E., et al. "Calibration of the ^{14}C Timescale over the Past 30,000 Years Using Mass Spectrometric U-Th Ages from Barbados Corals." *Nature* 345 (31 May 1990): 405–09.

Broecker, W.S., and G.H. Denton. "What Drives Glacial Cycles?" *Scientific American* (January 1990): 48–56.

Chappellaz, J., et al. "Ice-core Record of Atmospheric Methane Over the Past 160,000 Years." *Nature* 345 (10 May 1990): 127–131.

Climap Project Members. *Seasonal Reconstruction of the Earth's Surface at the Last Glacial Maximum.* Map and Chart Series, MC-36. Boulder, CO: Geological Society of America, 1981.

———. "The Surface of the Ice-Age Earth." *Science* 191 (19 March 1976): 1131–37.

Dansgaard, W., J.W.C. White, and S.J. Johnsen. "The Abrupt Termination of the Younger Dryas Climate Event." *Nature* 339 (15 June 1989): 532–33.

Dyke, A.S., and V.K. Prest. "Late Wisconsinan and Holocene History of the Laurentide Ice Sheet." *Géographie physique et Quaternaire* 41 (1987): 237–63.

———. *Paleogeography of Northern North America 18000–12000 Years Ago.* Map 1703A. Ottawa: Geological Survey of Canada, 1986.

Frakes, L.A. *Climates Throughout Geologic Time*. Amsterdam: Elsevier Scientific Publishing Company, 1979 (and subsequent editions).

Gribbin, J., ed. *Climatic Change.* Cambridge: Cambridge University Press, 1988.

Guiot, J., et al. "A 140,000-Year Continental Climate Reconstruction from Two European Pollen Records. *Nature* 338 (23 March 1989): 309–12.

Hallam, A. *Great Geological Controversies.* New York: Oxford University Press, 1983.

Hardy, R., et al. *The Weather Book*. Boston: Little, Brown and Company, 1982.

Hollister, C.D., A.R.M. Nowell, and P.A. Jumars. "The Dynamic Abyss." *Scientific American* (March 1984): 42–53.

Houghton, J.T., ed. *The Global Climate*. Cambridge: Cambridge University Press, 1984.

Imbrie, J., and K.P. Imbrie. *Ice Ages: Solving the Mystery*. Hillside, NJ: Enslow Publishers, 1979.

Jansen, E., and T. Veum. "Evidence for Two-Step Deglaciation and Its Impact on North Atlantic Deep-Water Circulation." *Nature* 343 (15 February 1990): 612–16.

Jones, G.A., and L.D. Keigwin. "Evidence from Fram Strait (78° N) for Early Deglaciation." *Nature* 336 (3 November 1988): 56–59.

Kasting, J., O.B. Toon, and J.B. Pollack. "How Climate Evolved on the Terrestrial Planets." *Scientific American* (February 1988): 90–97.

Kennett, J.P. *Marine Geology*. Englewood Cliffs, NJ: Prentice-Hall, 1982.

Kerr, R.A. "Linking Earth, Ocean, and Air at the AGU." *Science* 239 (15 January 1988): 259–60.

———. "Ocean Drilling Details Steps to an Icy World." *Science* 238 (22 May 1987): 912–14.

———. "Ice Cap of 30 Million Years Ago Detected." *Science* 224 (13 April 1984): 141–42.

Koerner, R.M. "Ice Core Evidence for Extensive Melting of the Greenland Ice Sheet in the Last Interglacial." *Science* 244 (26 May 1989): 964–68.

Le Roy Ladurie, E., and M. Baulant. "Grape Harvests from the Fifteenth through the Nineteenth Centuries." In *Climate and History: Studies in Interdisciplinary History*, edited by R.I. Rotberg, and T.K. Rabb. Princeton, NJ: Princeton University Press, 1981.

Lewin, R. "Impacts of Another Kind." *Science* 225 (7 September 1984): 1007.

Lockwood, J.G. *Causes of Climate*. New York: John Wiley & Sons, 1989.

Maddox, J. "Tectoclimatology Comes of Age." *Nature* 343 (8 February 1990): 507.

Monastersky, R. "Hills Point to Catastrophic Ice Age Floods." *Science News* 136 (30 September 1989): 213.

———. "Ancient Ice Reveals Sudden Climate Shift." *Science News* 135 (17 June 1989): 374.

———. "Rise of Tibet and Rockies Set Ice-Age Stage." *Science News* 135 (20 May 1989): 309.

———. "Ice Age Insights." *Science News* 134 (17 September 1988): 184–86.

Morison, S.E. *The European Discovery of America: The Southern Voyages a.d. 1492–1616*. New York: Oxford University Press, 1974.

Neale, J., and J. Flenley, eds. *The Quaternary in Britain*. Oxford: Pergamon Press, 1981.

Neftel, A., et al. "CO_2 Record in the Byrd Ice Core 50,000–5,000 Years BP." *Nature* 331 (18 February 1988): 609–11.

Nilsson, T. *The Pleistocene: Geology and Life in the Quaternary Ice Age*. Dordrecht: D. Reidel Publishing Company, 1983.

Oerlemans, J., and C.J. van der Veen. *Ice Sheets and Climate*. Dordrecht: D. Reidel Publishing Company, 1984.

Overpeck, J.T., et al. "Climate Change in the Circum-North Atlantic Region During the Last Deglaciation." *Nature* 338 (13 April 1989): 553–56.

Peel, D.A. "Ice-Age Clues for Warmer World." *Nature* 339 (15 June 1989): 508–09.

Quinn, D.B., ed. *New American World: A Documentary History of North America to 1612: America from Concept to Discovery: Early Exploration of North America*, vol. 1. New York: Arno Press, 1979.

Rampino, M.R., et al, eds. *Climate: History, Periodicity, and Predictability*. New York: Van Nostrand Reinhold Company, 1987.

Ross, M.I., and C.R. Scotese. "Hierarchical Tectonic Analysis of the Gulf of Mexico and Caribbean Region." *Tectonophysics* 155 (1988): 130–68.

Rotberg, R.I., and T.K. Rabb, eds. *Climate and History: Studies in Interdisciplinary History*. Princeton, NJ: Princeton University Press, 1981.

Science News. "Drilling Into Earth's Icy Past." *Science News* 136 (7 October 1989): 239.

———. "Pollen Provides Ancient Weather Report." *Science News* 135 (1 April 1989): 220.

———. "Wherefore the World's Wobble?" *Science News* 135 (1 April 1989): 220.

Sissons, J.B. "Ice-Dammed Lakes in Glen Roy and Vicinity: A Summary." In *The Quaternary in Britain*, edited by J. Neale and J. Flenley. Oxford: Pergamon Press, 1981.

———. "Lateglacial Marine Erosion and a Jökulhlaup Deposit in the Beauly Firth." *Scottish Journal of Geology* 17 (1981): 7–19.

Street-Perrott, F.A., and R.A. Perrott. "Abrupt Climate Fluctuations in the Tropics: The Influence of Atlantic Ocean Circulation." *Nature* 343 (15 February 1990): 607–11.

Stuiver, M. "Timescales and Telltale Corals." *Nature* 345 (31 May 1990): 387–88.

Sundquist, E.T. "Ice Core Links CO_2 to Climate." *Nature* 329 (1 October 1987): 389–90.

Vaughan, C. "As the World Wobbles." *Science News* 134 (16 July 1988): 39.

Whitehead, J.A. "Giant Ocean Cataracts." *Scientific American* (February 1989): 50–57.

Wiebe, P.H. "Rings of the Gulf Stream." *Scientific American* (March 1982): 60–70.

XII Children of the Apple Tree

Adovasio,J.M., Pedlar.D.R., Donahue, J., and Stuckenrath, R. Two Decades of debate on Meadowcroft Roickshelter: North American Archeologist: 19: 4 (1998)

Adovasio,J.M. and Hyland,D.C. *The Need to Weave*. Discovering Archeology Jan/Feb 2000

Adovasio,J.M., and Pedlar.D.R. Monte Verde and the Antiquity of humankind in the Americas: Smithsonian Institution Press: 1998

American Geological Institute. "Stone Tools of Early Americans Found." *Earth Science* (Spring 1989): 7.

Arensburg, B., et al. "A Middle Palaeolithic Human Hyoid Bone." *Nature* 338 (27 April 1989): 758–60.

Bahn, P.G. "Origins of Full-Scale Agriculture." *Nature* 339 (29 June 1989): 665.

Bahn, P.G. 50,000-year-old Americans of Pedra Furada: *Nature*: 362 114-115 March 1993

Barraclough, G., ed. *The Times Atlas of World History*. Maplewood, NJ: Hammond, 1979.

Barringer, F. "Ancient Tribes Didn't Vanish, They Just Moved." *New York Times*, 25 October 1990.

Birdsall, D., and C.M. Cipolla. *The Technology of Man*. London: Wildwood House Limited, 1980.

Borstel, C.L. "Prehistoric Site Chronology: A Preliminary Report." In *Chapters in the Archeology of Cape Cod, I: Results of the Cape Cod National Seashore Archeological Survey 1979–1981* vol. 2. Cultural Resources Management Study, no. 8, edited by F.P. McManamon. Boston: National Park Service, U.S. Department of the Interior, 1984.

Bower, B. "Talk of Ages." *Science News* 136 (8 July 1989): 24–26.

———. "Promising New Clues to Early Americans." *Science News* 133 (23 April 1988): 261.

———. "When the Human Spirit Soared." *Science News* 130 (13 December 1986): 378–79.

Chappellaz, J., et al. "Ice-core Record of Atmospheric Methane Over the Past 160,000 Years." *Nature* 345 (10 May 1990): 127–31.

Chen, C., and J.W. Olsen. "China at the Last Glacial Maximum." In *The World at 18000 BP: High Latitudes*, vol. 1, edited by O. Soffer and C. Gamble. London: Unwin Hyman, 1990.

Climap Project Members. *Seasonal Reconstruction of the Earth's Surface at the Last Glacial Maximum.* Map and Chart Series, MC-36. Boulder, CO: Geological Society of America, 1981.

Coe, M., D. Snow, and E. Benson. *Atlas of Ancient America*. New York: Facts on File Publications, 1986.

Collins.M.B., *Clovis Second: time is running out for the old paradigm* Discovering Archeology 49-51 Jan/Feb 2000

Couper, A., ed. *The Times Atlas of the Oceans*. New York: Van Nostrand Reinhold, 1983.

Creager, J.S., and D.A. McManus. "Geology of the Floor of Bering and Chukchi Seas—American Studies." In *The Bering Land Bridge*, edited by D.M. Hopkins. Stanford, CA: Stanford University Press, 1967.

Davis, D.D., and R.C. Goodwin. Island Carib Origins: Evidence and Nonevidence." *American Antiquity* 55 (1990): 37–48.

Davis, R.S. "Central Asian Hunter-Gatherers at the Last Glacial Maximum." In *The World at 18000 BP: High Latitudes*, vol. 1, edited by O. Soffer and C. Gamble. London: Unwin Hyman, 1990.

Day, M.H. *Guide to Fossil Man*. 4th ed. Chicago: University of Chicago Press, 1988.

Denver Museum of Natural History. *Ramses II: The Great Pharaoh and His Times*. Denver: 1987.

Derry, D.R. *World Atlas of Geology and Mineral Deposits*. London: Mining Journal Books, 1980.

Diamond, J.M. "The Talk of the Americas." *Nature* 344 (12 April 1990): 589–90.

———. "Who Were the First Americans?" *Nature* 329 (15 October 1987): 580–81.

Dillehay, T.D., and M.B. Collins. "Early Cultural Evidence from Monte Verde in Chile." *Nature* 332 (10 March 1988): 150–52.

Durham, J.W., and F.S. MacNeil. "Cenozoic Migrations of Marine Invertebrates through the Bering Strait Region." In *The Bering Land Bridge*, edited by D.M. Hopkins. Stanford, CA: Stanford University Press, 1967.

Einarsson, T., D.M. Hopkins, and R.R. Doell. "The Stratigraphy of Tjörnes, Northern Iceland, and the History of the Bering Land Bridge." In *The Bering Land Bridge*, edited by D.M. Hopkins. Stanford, CA: Stanford University Press, 1967.

Fagan, B.M. *Ancient North America*. New York: Thames and Hudson, 1991.

Fink, R., *Neanderthal Flute: Evidence of Natural Foundation to Diatonic Scale*: Musicological Analysis on http://www.webster.sk.ca/greenwich/FL-COMPL.HTMsEssay : 1997

Flerow, C.C. "On the Origin of the Mammalian Fauna of Canada." In *The Bering Land Bridge*, edited by D.M. Hopkins. Stanford, CA: Stanford University Press, 1967.

Frison, G.C., and D.N. Walker. "New World Palaeoecology at the Last Glacial Maximum and the Implications for New World Prehistory." In *The World at 18000 BP: High Latitudes*, vol. 1, edited by O. Soffer and C. Gamble. London: Unwin Hyman, 1990.

Frison, G.C. *Progress and Challenges: questions linger after 75 years of answers*: Discovering Archeology: 40-42 Jan/Feb 2000

Gamble, C., and O. Soffer, eds. *The World at 18000 BP: Low Latitudes*, vol. 2. London: Unwin Hyman, 1990.

———. "Pleistocene Polyphony: The Diversity of Human Adaptations at the Last Glacial Maximum." In *The World at 18000 BP: High Latitudes*, vol. 1. London: Unwin Hyman, 1990.

Garrett, W.E., ed. "The Peopling of the Earth." *National Geographic* (October 1988): 434–37.

Garrett, W.E., and S. Raymer. "Air Bridge to Siberia." *National Geographic* (October 1988): 504–09.

Gershanovich, D.E. "Late Quaternary Sediments of Bering Sea and the Gulf of Alaska." In *The Bering Land Bridge*, edited by D.M. Hopkins. Stanford, CA: Stanford University Press, 1967.

Giterman, R.E., and L.V. Golubeva. "Vegetation of Eastern Siberia During the Anthropogene Period." In *The Bering Land Bridge*, edited by D.M. Hopkins. Stanford, CA: Stanford University Press, 1967.

Giterman, R.E., A.V. Sher, and J.V. Matthews, Jr. "Comparison of the Development of Steppe-Tundra Environments in West and East Beringia: Pollen and Macrofossil Evidence from Key Sections." In *Paleoecology of Beringia*, edited by D.M. Hopkins, et al. New York: Academic Press, 1982.

Grove, J.M. *The Little Ice Age*. London: Methuen, 1988.

Groves, C.P. *A Theory of Human and Primate Evolution*. Oxford: Clarendon Press, 1989.

Hallam, A. "Continental Drift and the Fossil Record." In *Continents Adrift and Continents Aground*. Readings from Scientific American. San Francisco: W.H. Freeman and Company, 1976.

Hammond. *Past Worlds: The Times Atlas of Archaeology*. Maplewood, NJ: Hammond, 1988.

Hammond, N. *Ancient Maya Civilization*. Cambridge: Cambridge University Press, 1982.

Haynes, C.V. "Were Clovis Progenitors in Beringia?" In *Paleoecology of Beringia*, edited by D.M. Hopkins, et al. New York: Academic Press, 1982.

Holden, C. Tooling Around: Dates Show Early Siberian Settlement: *Science* 275: 1268 1997

Hollister, C.D., A.R.M. Nowell, and P.A. Jumars. "The Dynamic Abyss." *Scientific American* (March 1984): 42–53.

Hopkins, D.M. "Aspects of the Paleogeography of Beringia during the Late Pleistocene." In *Paleoecology of Beringia*, edited by D.M. Hopkins, et al. New York: Academic Press, 1982.

Hopkins, D.M., ed. *The Bering Land Bridge*. Stanford, CA: Stanford University Press, 1967.

Hopkins, D.M., et al., eds. *Paleoecology of Beringia*. New York: Academic Press, 1982.

———. "Paleoecology of Beringia—A Synthesis." In *Paleoecology of Beringia*, edited by D.M. Hopkins, et al. New York: Academic Press, 1982.

Ingstad, H. *Westward to Vinland: The Discovery of Pre-Columbian Norse House-sites in North America*. Translated from the Norwegian by Erik J. Friis. New York: St. Martin's Press, 1969.

Judge, J., and J.L. Stanfield. "The Island of Landfill." *National Geographic* (November 1986): 566–72, 578–99.

Kinoshita, June. "Maya Art for the Record." *Scientific American* (August 1990): 92–97.

Kozlowski, J.K. "Northern Central Europe *c.* 18000 BP." In *The World at 18000 BP: High Latitudes*, vol. 1, edited by O. Soffer and C. Gamble. London: Unwin Hyman, 1990.

Ladurie, E.L., and M. Baulant. "Grape Harvests from the Fifteenth through the Nineteenth Centuries." In *Climate and History: Studies in Interdisciplinary History*, edited by R.I. Rotberg and T.K. Rabb. Princeton, NJ: Princeton University Press, 1981.

Leakey, R., and A. Walker. "*Homo Erectus* Unearthed." *National Geographic* (November 1985): 625–29.

Lewin, R. *Bones of Contention: Controversies in the Search for Human Origins.* New York: Simon and Schuster, 1987.

———. "Africa: Cradle of Modern Humans." *Science* 237 (11 September 1987): 1292–95.

———. *Human Evolution: An Illustrated Introduction*. New York: W.H. Freeman and Company, 1984.

Mangelsdorf, P.C. "The Origin of Corn." *Scientific American* (August 1986): 80–86.

Marshack, A. "An Ice Age Ancestor?" *National Geographic* (October 1988): 478–81.

Marshall, E. "Clovis Counterrevolution." *Science* 249 (17 August 1990): 738–41.

Marshall, J.C. "The Descent of the Larynx?" *Nature* 338 (27 April 1989): 702–03.

Marshall, L.G., et al. "Mammalian Evolution and the Great American Interchange." *Science* 215 (12 March 1982): 1351–56.
Matthews, J.V., Jr. "East Beringia during Late Wisconsin Time: A Review of the Biotic Evidence." In *Paleoecology of Beringia*, edited by D.M. Hopkins, et al. New York: Academic Press, 1982.
McCulloch, D.S. "Quaternary Geology of the Alaskan Shore of Chukchi Sea." In *The Bering Land Bridge*, edited by D.M. Hopkins. Stanford, CA: Stanford University Press, 1967.
McGovern, T.H. "Cows, Harp Seals, and Churchbells: Adaptation and Extinction in Norse Greenland." *Human Ecology* 8 (1980): 245–75.
McManamon, F.P., ed. *Chapters in the Archeology of Cape Cod, I: Results of the Cape Cod National Seashore Archeological Survey 1979–1981*, vols. 1–2. Cultural Resources Management Study no. 8. Boston: National Park Service, U.S. Department of the Interior, 1984.
Morlan, R.E., and J. Cinq-Mars. "Ancient Beringians: Human Occupation in the Late Pleistocene of Alaska and the Yukon Territory." In *Paleoecology of Beringia*, edited by D.M. Hopkins, et al. New York: Academic Press, 1982.
Mourant, A.E. *Blood Relations: Blood Groups and Anthropology*. New York: Oxford University Press, 1983.
Müller-Beck, H. "Late Pleistocene Man in Northern Alaska and the Mammoth-Steppe Biome." In *Paleoecology of Beringia*, edited by D.M. Hopkins, et al. New York: Academic Press, 1982.
———. "On Migrations of Hunters across the Bering Land Bridge in the Upper Pleistocene." In *The Bering Land Bridge*, edited by D.M. Hopkins. Stanford, CA: Stanford University Press, 1967.
National Geographic. "The Peopling of the Earth." *National Geographic* (October 1988): 434–37.
National Maritime Museum. *The World of the Vikings: An Exhibition Mounted by the Statens Historiska Museum, Stockholm*. London: National Maritime Museum, 1972.
Nelson, D.E., et al. "New Dates on Northern Yukon Artifacts: Holocene Not Upper Pleistocene." *Science* 232 (9 May 1986): 749–51.
Nilsson, T. *The Pleistocene: Geology and Life in the Quaternary Ice Age*. Dordrecht: D. Reidel Publishing Company, 1983.
Parks Canada. *L'Anse aux Meadows National Historic Park: Newfoundland*. IAN Publication no. QS T033 000 88 A3. Canada: Minister of Supply and Services Canada, 1980.
Pfister, C. "The Little Ice Age: Thermal and Wetness Indices for Central Europe." In *Climate and History: Studies in Interdisciplinary History*, edited by R.I. Rotberg and T.K. Rabb. Princeton, NJ: Princeton University Press, 1981.
Pilbeam, D. "The Descent of Hominoids and Hominids." *Scientific American* (March 1984): 84–96.
Putman, J.J. "In Search of Modern Humans." *National Geographic* (October 1988): 439–77.
Redfern, R. *The Making of a Continent*. New York: Random House, 1983.
Rigaud, J-P. "Treasures of Lascaux Cave." *National Geographic* (October 1988): 482–99.
Ritchie, J.C., and L.C. Cwynar. "The Late Quaternary Vegetation of the North Yukon." In *Paleoecology of Beringia*, edited by D.M. Hopkins, et al. New York: Academic Press, 1982.
Rotberg, R.I., and T.K. Rabb, eds. *Climate and History: Studies in Interdisciplinary History*. Princeton, NJ: Princeton University Press, 1981.
Schele, L., and M.E. Miller. *The Blood of Kings*. New York: George Braziller, Inc., 1986.
Schledermann, P. "Ellesmere Island: Eskimo and Viking Finds in the High Arctic." *National Geographic* (May 1981): 575–601.
———. "Notes on Norse Finds from the East Cost of Ellesmere Island, N.W.T." *Arctic* 33 (September 1980): 454–63.
———. "Polynyas and Prehistoric Settlement Patterns." *Arctic* 33 (June 1980): 292–302.
———. "Preliminary Results of Archaeological Investigations in the Bache Peninsula Region, Ellesmere Island, N.W.T." *Arctic* 31 (December 1978): 459–74.
Schledermann, P., and K. McCullough. "Western Elements in the Early Thule Culture of the Eastern High Arctic." *Arctic* 33 (December 1980): 833–41.
Schmider, B. "The Last Pleniglacial in the Paris Basin (22500–17000 BP)." In *The World at 18000 BP: High Latitudes*, vol. 1, edited by O. Soffer and C. Gamble. London: Unwin Hyman, 1990.
Science News. "Stone Tips on Ancient Hunting." *Science News* 136 (1 July 1989): 13.
Scientific American. *New World Archaeology: Theoretical and Cultural Transformations*. Readings from Scientific American. San Francisco: W.H. Freeman and Company, 1974.
———. *Early Man in America*. Readings from Scientific American. San Francisco: W.H. Freeman and Company, 1973.
Sears, M., and D. Merriman. *Oceanography: The Past*. New York: Springer-Verlag, 1980.
Soffer, O. "The Russian Plain at the Last Glacial Maximum." In *The World at 18000 BP: High Latitudes*, vol. 1, edited by O. Soffer and C. Gamble. London: Unwin Hyman, 1990.
Soffer, O., and C. Gamble, eds. *The World at 18000 BP: High Latitudes*, vol.1. London: Unwin Hyman, 1990.
Soffer, O., Adovasio, J.M., and Hyland, D.C. : Dressing the "Venus": Upper Paleolithic Iconography and Perishable Technologies. Mercyhurst Archeological Institute Sept 1999
Stanley, S.M. *Earth and Life Through Time*. 2d ed. New York: W.H. Freeman and Company, 1989.
Stringer, C. "The Dates of Eden." *Nature* 331 (18 February 1988): 565–66.
Stringer, C.B., and P. Andrews. "Genetic and Fossil Evidence for the Origin of Modern Humans." *Science* 239 (11 March 1988): 1263–68.
Toth, N. "The First Technology." *Scientific American* (April 1987): 112–21
Vangengeim, E.A. "The Effect of the Bering Land Bridge on the Quaternary Mammalian Faunas of Siberia and North America." In *The Bering Land Bridge*, edited by D.M. Hopkins. Stanford, CA: Stanford University Press, 1967.
Velichko, A.A., and E.I. Kurenkova. "Environmental Conditions and Human Occupation of Northern Eurasia During the Late Valdai." In *The World at 18000 BP: High Latitudes*, vol. 1, edited by O. Soffer and C. Gamble. London: Unwin Hyman, 1990.
Vereshschagin, N.K., and G.F. Baryshnikov. "Paleoecology of the Mammoth Fauna in the Eurasian Arctic." In *Paleoecology of Beringia*, edited by D.M. Hopkins, et al. New York: Academic Press, 1982.
Weaver, K.F. "The Search for Our Ancestors." *National Geographic* (October 1985): 560–623.
Webb, C.H. *Geoscience and Man: The Poverty Point Culture*, vol. 27, 2d rev. ed. Baton Rouge, LA: Louisiana State University School of Geoscience, 1982.
Weniger, G-C. "Germany at 18000 BP." In *The World at 18000 BP: High Latitudes* vol. 1, edited by O. Soffer and C. Gamble. London: Unwin Hyman, 1990.
Wilford, J.N. "Scholars Say First Atelier Was in a Cave in France." *New York Times*, 15 May 1990.
———. "Findings Plunge Archeology of the Americas Into Turmoil." *New York Times*, 30 May 1989.
———. "Fossil Findings Fan Debate on Human Origins." *New York Times*, 14 February 1989.
Williams, R.C., et al. "GM Allotypes in Native Americans: Evidence for Three Distinct Migrations Across the Bering Land Bridge." *American Journal of Physical Anthropology*. 66 (1985): 1–19.
Wormington, H.M. *Prehistoric Indians of the Southwest*. Popular Series No. 7. Denver: The Denver Museum of Natural History, 1977.
Young, S.B. "The Vegetation of Land-Bridge Beringia." In *Paleoecology of Beringia*, edited by D.M. Hopkins, et al. New York: Academic Press, 1982.

Epilogue Between Two Waves of the Sea

Aleem, A.A. "On the History of Arab Navigation." In *Oceanography: The Past*, edited by M. Sears and D. Merriman. New York: Springer-Verlag, 1980.
Bennett, J.A. *The Divided Circle: A History of Instruments for Astronomy, Navigation and Surveying*. Oxford: Phaidon, 1987.
Boorstin, D.J. *The Discoverers*. New York: Random House, 1983.
Crone, G.R. *Maps and Their Makers: An Introduction to the History of Cartography*. 5th ed. England: Wm. Dawson & Son, 1978.
d'Ailly, P. *Ymago mundi*. Louvain, *c*.1480. In *New American World: America from Concept to Discovery*, vol. 1, edited by D.B. Quinn. New York: Arno Press, 1979.
Deagan, K.A. "Searching for Columbus's Lost Colony." *National Geographic* (November 1987): 672–75.
Ferguson, P., et. al., eds. *Imago Mundi: The Journal of the International Society for the History of Cartography*, vol. 41. London: King's College, 1989.
Goodspeed, E.J., trans. *The Apocrypha*. New York: Random House, 1959.
Guill, J.H. "Vila do Infante (Prince-Town), the First School of Oceanography in the Modern Era: An Essay." In *Oceanography: The Past*, edited by M. Sears and D. Merriman. New York: Springer-Verlag, 1980.
Ibn-Khaldûn. *The Muqaddimah*, translated by F. Rosenthal. 2d ed. Princeton, 1967. In *New American World: America from Concept to Discovery*, vol. 1, edited by D.B. Quinn. New York, Arno Press, 1979.
Jane, C., and L.A. Vigneras, trans. *The Voyages of Christopher Columbus*, by Christopher Columbus. London, 1930. In *New American World: America from Concept to Discovery*, vol. 1, edited by D.B. Quinn. New York, Arno Press, 1979.
Lyon, E. "15th-Century Manuscript Yields First Look at Niña." *National Geographic* (November 1986): 601–05.
Marden, L. "Tracking Columbus Across the Atlantic." *National Geographic* (November 1986): 572–77.
McGowan, A. *Tiller and Whipstaff: The Development of the Sailing Ship 1400–1700. The Ship*. London: National Maritime Museum, 1981.
McGrail, S. *Rafts, Boats and Ships: From Prehistoric Times to the Medieval Era. The Ship*. London: National Maritime Museum, 1981.
Meinig, D.W. *The Shaping of America: A Geographical Perspective on 500 Years of History: Atlantic America, 1492–1800*, vol. 1. New Haven, CT: Yale University Press, 1986.
Morison, S.E. *The European Discovery of America: The Southern Voyages, a.d. 1492–1616*. New York: Oxford University Press, 1974.
National Geographic. *Historical Atlas of the United States*, Centennial Edition. Washington, DC: National Geographic Society, 1988.
Nebenzahl, K. *Atlas of Columbus and the Great Discoveries*. Chicago: Rand McNally, 1990.
Quinn, D.B., ed. *New American World: A Documentary History of North American to 1612: America from Concept to Discovery: Early Exploration of North America*, vol. 1. New York: Arno Press, 1979.
———. *New American World: A Documentary History of North America to 1612: Newfoundland from Fishery to Colony: Northwest Passage Searches*, vol. 4. New York: Arno Press, 1979.
Rogers, J.G. *Origins of Sea Terms*. Mystic, CT: Mystic Seaport Museum, 1984.
Stevenson, E.L., ed. and trans. *Geography of Claudius Ptolemy*. New York: New York Public Library, 1932. In *New American World: America from Concept to Discovery*, vol. 1, edited by D.B. Quinn. New York, Arno Press, 1979.
University of Minnesota. *Antilia and America: A Description of the 1424 Nautical Chart and the Waldseemüller Globe Map of 1507 in the James Ford Bell Collection*. Minneapolis: University of Minnesota, 1955.
Wilson, D.M., ed. *The Northern World: The History and Heritage of Northern Europe, ad 400–1100*. New York: Harry N. Abrams, 1980.
Yule, H., and H. Cordier, eds. *The Book of Ser Marco Polo*, vol. 2. London, 1903–1920. In *New American World: America from Concept to Discovery*, vol. 1, edited by D.B. Quinn. New York, Arno Press, 1979.

Index

Page numbers in **boldface** indicate photo-essay references; numbers in *italics* indicate illustrations (photographs, art, and icons). All photographs are listed under the appropriate country, province or state. Individual types of plants, animals, rocks, and other such subjects are listed under their general categories.